Drug Resistance in Leishmania Parasites

Alicia Ponte-Sucre • Maritza Padrón-Nieves
Editors

Drug Resistance in Leishmania Parasites

Consequences, Molecular Mechanisms and Possible Treatments

Second Edition

Springer

Editors
Alicia Ponte-Sucre
Institute of Experimental Medicine
Universidad Central de Venezuela
Caracas, Venezuela

Maritza Padrón-Nieves
Escuela Luis Razetti Fac de Medicine
Universidad Central de Venezuela
Caracas, Venezuela

ISBN 978-3-319-74185-7 ISBN 978-3-319-74186-4 (eBook)
https://doi.org/10.1007/978-3-319-74186-4

Library of Congress Control Number: 2018942230

This Springer imprint is published by the registered company Springer Nature Switzerland AG
The registered company address is: Gewerbestrasse 11, 6330 Cham, Switzerland

To our students, their curiosity inspires us to continue in our path of searching and learning about this ancient disease.
To the patients suffering leishmaniasis, our best partners to understand and fight against this devastating disease.

Foreword

The year 2013 saw the publication of the first edition of the volume *Drug Resistance in Leishmania Parasites*, edited by Alicia Ponte-Sucre. A brand-new edition is now available in which ten of the chapters have been drastically updated, two chapters slightly revised, and two entirely new chapters added. This updated version is timely, because recent years have revealed important new information about drug resistance in *Leishmania*. Furthermore, major efforts are being made to control this problem. Knowledge has increased by studies of the mechanisms by which resistance is generated and by epidemiological and population genomic research on how drug resistance spreads. Moreover, evidence has been accumulating that the relationship between therapeutic failure and drug resistance is complicated; therapeutic failure cannot always be correlated with drug resistance. Current research is therefore being performed to understand the relationship and how both phenomena can be dissected. Furthermore, important achievements have also been made in research to develop tools for diagnosis, treatment, and control of the leishmaniases.

Alicia Ponte-Sucre and Emilia Diaz provide a general overview of the biology of *Leishmania* in the introductory chapter of this volume and highlight the challenges faced in combatting the drug-resistant forms of leishmaniases. This Introduction focuses particularly on aspects of the parasite's biology which are relevant for an understanding of topics covered by the following chapters of the book, each devoted to an important, specific aspect of drug-resistant parasites and the diseases they cause.

The different chapters are grouped into four different parts. The **first part** covers three articles that address "determinant features in leishmaniasis." Gabriele Schönian and colleagues present new insights into the evolution, taxonomy, and phylogenetic and population genetic relationships of *Leishmania*, as acquired by recent research. The power of modern approaches used in the research, such as multilocus sequence analysis, multilocus microsatellite typing, and comparative genomics for studying the inter- and intraspecies variation of *Leishmania* parasites, is discussed. In the following chapter, Lenea Campino and Carla Maia review the epidemiology, pathogenicity, and treatment of leishmaniasis in dogs and the role of dogs as reservoir hosts of *L. (L.) infantum* parasites. Additionally, the role of other mammals as potential reservoir hosts of parasites belonging to the *L. (L.) donovani* complex is addressed. These authors discuss the potential generation and spread of drug resistance by the use of the same compounds in both canine and human hosts and the measures to be taken to control human zoonotic leishmaniasis.

Jean-Claude Dujardin provides an update on the epidemiology of leishmaniasis in relation to drug-resistant and treatment-failure parasites, whose phenotypes are based on the analysis of parasites in the laboratory and on the clinical assessments of patients, respectively. Whereas in the previous edition he presented results obtained for antimonials, the work has now been extended to miltefosine and the data are compared. Risk factors for (re-)emergence and spreading of leishmaniasis are discussed by focusing on the link between drug-resistant and treatment-failure phenotypes, such as the role of asymptomatic carriers and animals, coinfection with HIV and *Leishmania* RNA viruses, human migrations, and environmental changes. Additionally, the advances made in the development of tools for epidemiological surveillance of treatment failure/drug resistance are described, ranging from clinical tools to laboratory ones. In the last chapter of this section, Lukasz Kedzierski and Krystal Evans review our current knowledge of the immune factors involved in controlling leishmaniasis and discuss the role the immune system plays in resistance to the parasitemia. The parasites have evolved a variety of strategies to evade leishmanicidal mechanisms and survive in the phagosome of macrophages. Whereas most infected individuals develop long-lasting protective immunity following primary infection, sterile immunity is rarely achieved and parasites may persist asymptomatically in the host. The authors describe the vast array of immune cells and cytokines involved in the immune response to the infection which highlights the complexity of the disease and reveals a complicated network of regulatory as well as counter-regulatory interactions that contribute to the persistence.

The **second part** of the book contains four chapters addressing the "challenges in diagnosis, treatment, and control of leishmaniasis in times of drug resistance." Combined infection by HIV and *Leishmania* is a well-known problem. Margriet den Boer and colleagues describe the epidemiology, current spread, clinical aspects, and management of this coinfection. They discuss how development of drug-resistant *Leishmania* strains complicates chemotherapy for *Leishmania*/HIV coinfection and what are the prospects for future chemotherapeutic alternatives which target *Leishmania* and HIV and tackle both infections simultaneously. Shyam Sundar and Jaya Chakravarty describe how, over the years, therapy for visceral leishmaniasis has changed because of the increased unresponsiveness for existing drugs. Whereas pentavalent antimonials have been the mainstay for treatment during most of the twentieth century, a significant subset of patients in the Indian subcontinent were apparently no longer responsive in the 1980s, even when dosage was increased substantially. Unfortunately, there is no marker yet validated for this unresponsiveness, although there are recent reports about cases where IgG1 seemed to be a good predictor of relapse when measured at the end of treatment for visceral leishmaniasis. The authors describe alternative therapeutic options that have been developed, such as conventional amphotericin-B or its lipid formulations, oral miltefosine and paromomycin, and the efficacy and recommended use of these alternative drugs.

Olga Zerpa and colleagues describe the current situation of tegumentary leishmaniasis in the American continent. It is an endemic anthropozoonosis caused by several species of both the *Leishmania* and the *Viannia* subgenera and may thus cause different pathologies. The control of this disease meets with several difficulties: the parasites have several reservoirs and use various vectors to infect humans and mammals. Current treatment involves antimonials, but the efficacy is unpredictable, probably at least in part due to drug resistance. The disease is expanding in the American continent. The authors compare some of these data with those about the disease in the Old World for a better appreciation of the unique aspects of the American tegumentary leishmaniasis. The authors argue that the precise identification of the species of the infectious agent is crucial for correct clinical diagnosis, appropriate treatment, and control of the disease, especially in relation to the challenges imposed by drug resistance. The last chapter of Part II is a new contribution by Guy Caljon and colleagues about the challenges for effective leishmaniasis treatments. The challenges are the important emergence and spread of resistance against the pentavalent antimonials in recent decades, after their successful use for over 70 years in therapy for visceral leishmaniasis, and the huge species- and strain-specific variations in drug susceptibilities that dramatically complicate effective treatment of cutaneous leishmaniasis, although this cannot be linked to development of drug resistance. Moreover, anti-leishmanial treatment failures increasingly occur with all of the currently available standard drugs. The factors probably responsible for these failures, which are related to the complex interplay between parasite, host, and drug, are discussed along with their consequences for therapy and development of new drugs.

Part III comprises three chapters about "molecular features of drug-resistant *Leishmania*." Patricia Cuervo and coworkers stress the importance of proteomic approaches in researching leishmaniasis pathogenesis and problems such as drug resistance. Genomics and transcriptomics studies are important but insufficient to reveal the full picture because posttranscriptional and posttranslational processes play a crucial role in protein expression in these parasites. The genomic data, complemented with high-throughput proteomic analysis, can shed light on resistance mechanisms and identify new drug targets against leishmaniasis. Proteomic analysis of *Leishmania* parasites has already provided information about drug resistance mechanisms. The characterization of the proteins involved has advanced, but still many fundamental questions remain to be answered. Adriano Coelho and Paulo Cotrim summarize research on ABC transporters in membranes of *Leishmania* parasites. Genome sequencing identified in different species of the genus the presence of members of all eight known different subfamilies of ABC transporters, each having specific functional characteristics. The authors discuss the work that revealed how some of these transporters are associated with drug resistance in leishmaniasis and showed their role in the pathology caused by the parasite and how the activity of these proteins affects the efficacy of the treatment. The next chapter deals with non-ABC transporters of *Leishmania* which are responsible for uptake of nutrients by the parasites but may also be exploited for mediating transport of drugs. Such transporters may become responsible for drug resistance of the parasites by mutations in their coding regions or changed expression. Scott Landfear gives an update on how the analysis of the *Leishmania* genome and recent functional studies have increased our knowledge about different classes of solute transporters involved in drug uptake and how modification of their structure or expression level confers changes in drug sensitivity and causes drug resistance.

The **fourth and last part** of the book is devoted to "tools and strategies to circumvent resistance in *Leishmania*." Shishir Gupta and Thomas Dandekar have added a new chapter in which they describe how bioinformatics is being used for querying genome, transcriptome, and proteome information to identify potential new targets for drug discovery and vaccine development in *Leishmania*. Furthermore, the authors provide information about software used in such research and give links to websites where tools can be found to examine and rank the new targets. Bruno Pradines has updated his chapter about P-glycoprotein-like transporters in *Leishmania*. Drug resistance can be due to different mechanisms that result in decreased level of the drug in the parasite. One of these mechanisms, well recognized as responsible for antimony resistance in *Leishmania,* involves an increased efflux, mediated by P-glycoprotein (Pgp)-like transporters. P-glycoproteins, well characterized in research of drug-resistant cancers, belong to the superfamily of ABC transporters. Inhibition of the drug efflux by these proteins will thus offer an attractive manner to control drug-resistant parasites in a patient. Indeed, the author describes a number of natural or synthetic compounds, some being derivatives known to modulate human Pgp, which are able to revert the resistance phenotype in parasites to a variety of drugs commonly used in both visceral and cutaneous leishmaniasis by decreasing their intracellular concentration. Concepts about the reversal mechanism of multidrug resistance by the use of chemosensitizers which alter the capacity of Pgp are discussed.

In the final chapter of this volume, Manu Vanaerschot and colleagues provide an updated chapter about "the concept of fitness and drug resistance in *Leishmania*." When pathogens develop resistance against drugs, it usually comes at the cost of making them less fit than their wild-type counterparts. This has important implications for the frequency of treatment-failure cases in endemic regions. Cases of treatment failure in patients suffering from leishmaniasis have been observed for most anti-leishmaniasis drugs. However, it is intriguing that this failure could not always be correlated with drug resistance of the infecting parasites, since cases of failure upon treatment with both pentavalent antimonials and miltefosine were accompanied with an increased fitness of the *L. L. donovani* parasites. The authors argue that these examples highlight parasite fitness as a potentially important contributor to treatment failure, at least for visceral leishmaniasis in the Indian subcontinent. They discuss available information and remaining questions about fitness for different *Leishmania* species and the different stages of their life cycle, as well as the relevance of parasite fitness for the development and spread of drug resistance and/or treatment failure in the field, and for new research toward the development of drugs for leishmaniasis and the control and elimination of the disease.

Thanks to the combined efforts of the editor and the authors of the 15 chapters, this book provides an excellent overview. It covers the current stage of our knowledge about the major problems of drug resistance of *Leishmania* parasites, as well as treatment failure in the different manifestations caused by various species of the genus. It presents the current knowledge and questions about the pathology and epidemiology of the leishmaniases in the context of *Leishmania* biology. Diagnosis, treatment, and molecular-parasitological aspects are all discussed from the perspective of drug resistance and how this could be dealt with. This book will therefore be a highly valuable source of information for both basic researchers and clinicians with interests in leishmaniasis.

Centre for Immunity, Infection and Evolution (CIIE) Paul Michels
and Centre for Translational and Chemical Biology (CTCB),
School of Biological Sciences, The University of Edinburgh
Edinburgh, Scotland

Preface

Old and New World leishmaniasis are in urgent need of reevaluation of treatment guidelines as treatment failure is an everyday growing problem. For this disease, treatment failure and drug resistance are topics that go hand in hand. Additionally, field parasites may be naturally resistant to classical drugs or might be selected as resistant by the use of current therapies. These features are (at least partially) responsible for the disappointing picture of disease persistence and death worldwide presented by leishmaniasis. A better understanding of the ailment and of drug resistance, its molecular basis, consequences, and possible treatment may help improving this depressing picture. We hope that this volume will help us to achieve this goal.

The work done by each of the authors and coauthors contributing to this volume has been awe-inspiring. Each chapter is intended to supplement well-documented texts that cover from molecular evolution to the design of compounds that may impact the drug resistance problem, as well as to the comprehension of how adaptable these parasites are. Our intention with this reedited and updated volume is to continue targeting scientists, pre- and postgraduate students, and scholars involved in the medical treatment of patients with leishmaniasis, or dedicated to the design of novel compounds and lead pharmacophores against leishmaniasis.

We acknowledge the Coordination for Research, Faculty of Medicine, and the Council for Scientific and Humanistic Research, Universidad Central de Venezuela, as well as the Missionsärztliche Institut, Würzburg, and the Alexander von Humboldt Foundation, Germany, for their support to our work for so many years. Additionally, we acknowledge the support from the Siebold-Collegium Institute for Advanced Studies, University of Würzburg, Germany, during the last steps of compilation and final organization of the content of this volume. Finally, we would like to address the excellent support of the staff at Springer, Rajeswari Balachandran, Tanja Grabner, and Claudia Panuschka for their guidance through the different steps of the publication of the book.

Caracas, Venezuela — Alicia Ponte-Sucre

Contents

About the Editors

Alicia Ponte-Sucre studied Education in Biological Sciences at the Andrés Bello Catholic University in Caracas and became Magister Scientiarum in Physiology and Biophysics in 1981 at the Venezuelan Institute of Scientific Research and Doctor of Sciences in Pharmacology in 1993 at the UCV. She spent a year (1999–2000) at the University of Würzburg with a scholarship from the Alexander von Humboldt Foundation and was a staff scientist (2003–2007) at the same university, within a multidisciplinary project from the German Research Council. Currently, she is Full Professor in Human Physiology and Coordinator for Scientific Affairs of the Faculty of Medicine, UCV. Initially, her studies were focused on the physical–chemical behavior of black lipid membranes. Later, her scientific activity was oriented toward the characterization of receptors involved in airway smooth muscle contraction. During the last 30 years, her interests have been focused on the study of parasite metabolism and membrane transporters essential for parasite survival and involved in drug resistance and the mechanisms involved in cellular differentiation and parasite–host interaction in Trypanosomatids, but especially in the *Leishmania* model. Additionally, she has characterized natural products and target-oriented designed compounds as potential therapeutic agents against diseases produced by these parasites.

Maritza Padrón-Nieves studied biology at UCV in Caracas and became Magister Scientiarum in Pharmacology in 1993 and Doctor of Sciences in Pharmacology in 2011 in the same university. From 1994 to 2000, she was head of the Department of Basic Sciences at the School of Nursery, UCV. Currently, she is Full Professor and head of the Human Pharmacology and Toxicology Chair, Faculty of Medicine, UCV. Initially, her research interests were related to the comprehension of the mechanisms involved in digoxin intoxication. Since 2006, she has dedicated her scientific interest to the identification and characterization of molecular markers of resistance in the *Leishmania* sp. infection model.

Leishmaniasis: The Biology of a Parasite 1

Emilia Díaz and Alicia Ponte-Sucre

Abstract

One of the main challenges of therapeutic tools for the treatment of parasitic diseases, including leishmaniasis, is the interwinned relationship between therapeutic failure and drug resistance. In fact, some field parasites might be naturally resistant to classical drugs and additionally, current therapies may induce drug resistance. In fact, treatment failure in leishmaniasis has multiple causes. Some are related to drugs, such as pharmacokinetic properties, toxicity, use of sub-optimal doses, or high cost of treatment. Parasite-related grounds include chemo-resistance and tolerance. Last but not least, the host plays a fundamental role in this situation since the patient's immune status and the risk of re-infection if living in an endemic region might also contribute to therapeutic failure. All these features are at least partially responsible for the disappointing persistence and re-emergence of leishmaniasis, as well as its death and disability-adjusted life year toll worldwide. A better understanding of the disease itself and of drug resistance, its molecular basis, its consequences, and the definition of possible paths for better treatments may help improve this depressing picture. In the present volume experts in the field cover current knowledge and future trends of these and many other aspects of drug resistance in *Leishmania*. This initial chapter offers a general introduction to the biology of the parasite, a piece of information fundamental for the topics included in the book and the comprehension of challenges we currently face for this disease.

E. Díaz · A. Ponte-Sucre (✉)
Laboratorio de Fisiología Molecular, Instituto de Medicina Experimental, Escuela Luis Razetti, Universidad Central de Venezuela, Caracas, Venezuela

A. Ponte-Sucre, M. Padrón-Nieves (eds.), *Drug Resistance in Leishmania Parasites*,
https://doi.org/10.1007/978-3-319-74186-4_1

1.1 Introduction

Leishmaniasis is a neglected vector-borne tropical infection still considered to be a disease of poverty [1]. It is caused by various species of an obligate intracellular parasite of the genus *Leishmania*. This parasite dwells in cells of the monocytic phagocytic system of mammals and is transmitted by female sand flies. More than 20 *Leishmania* species are pathogenic to humans and more than 30 species of sand flies function as invertebrate vectors.

The disease is endemic in several world areas, including deserts and rain forests in tropical and subtropical regions of Africa, America and Asia and sub-rural and urban areas in southern Europe [1–5]. All this makes leishmaniasis an important disease both for local inhabitants and for travellers. An estimated 350 million people worldwide are at risk of being infected; about 12 million people are infected, and annual occurrence is about 1.5–2 million cases of the cutaneous disease during the last 5 years, and 300,000 estimated cases occur annually of the visceral form of the disease with over 20,000 deaths each year [3, 6]. In anthroponotic foci, sand flies transmit parasites from human to human, and in zoonotic foci, sand flies transmit parasites between the usual local hosts and from them to humans [4, 6]. Of note, males are normally more affected than females, especially in sub-Saharan Africa.

Mortality caused by leishmaniasis is second worldwide only to malaria (among parasitic diseases) and in terms of disability-adjusted life years (DALYs), the third most common cause of morbidity after malaria and schistosomiasis. Children <15 years suffer most of the disease burden [7]. Immunosuppression secondary to HIV infection, posttransplant and chemotherapeutic agents and the recently introduced biologic therapies for chronic inflammatory conditions, has resulted in an increase in leishmaniasis burden within Europe. Additionally, global mobility has increased dramatically the number of cases of leishmaniasis in nonendemic countries [1]. All this highlights the importance that the recognition and understanding of this disease have.

1.2 Epidemiology

The disease mainly affects poor people in Africa, Asia and Latin America and is associated with malnutrition, population displacement, poor housing, weak immune system and lack of resources in 98 countries [8, 9].

Visceral leishmaniasis (VL) leads to substantial health problems or death for up to 400,000 people per year [10]. Mucocutaneous leishmaniasis (MCL) occurs mainly but not only in Central and South America, whereas most of the cases of cutaneous leishmaniasis (CL) cases occur in the Middle East and in Brazil and Peru [11, 12]. It has also been reported in Africa, Asia and Europe [1]. Disease sub-register constitutes a real problem, since only a minor proportion of the countries with endemic leishmaniasis officially report the infection [11].

According to the World Health Organization, the incidence of leishmaniasis has increased 42-fold in the last two decades. As previously mentioned, it is classified as the second worldwide cause of parasitic diseases. Rapid increases in the incidence of leishmaniasis correlates with the risk of co-infection with human immunodeficiency virus, mainly in southern Europe and in countries of Africa and Asia where antiretroviral therapy against human immunodeficiency virus is not available (or insufficient) due to its high cost to the local population [4, 13]. This theme is discussed by Alvar and colleagues in Chap. 6 of this volume. Malnutrition also increases disease severity by impairing the immune response [11, 14]. This theme is discussed by Vanaerschot and colleagues in Chap. 15 of this volume.

Environmental features such as temperature, humidity, altitude, etc. promote vector survival. Thus, climate changes affecting surrounding temperatures would positively impact the distribution of the disease and its sand fly vectors. This would mean a dramatic decrease in the geographical areas currently free from both parasites and vectors resulting in an increased world distribution of the disease [1]. Additionally, in regions where the disease is endemic (most tropical regions of the world), deforestation and urbanization, as well as increased travel (tourist and work-related reasons), migration and military operations, can also lead to an increased risk of expanding the disease to new areas [15]. This theme is discussed by Dujardin in Chap. 4 of this volume.

Clinical manifestations of disease depend, among others, on the species of infecting protozoan. Pathogenesis is based on cell destruction caused by the rupture of the "nests" of amastigotes (intracellular stage in the macrophage), accompanied by an intense inflammatory reaction. The visceral form disrupts various organs such as the liver and spleen [16]. Even self-limiting CL can leave disfiguring scars. CL can be disabling when lesions are numerous, and the most severe form, recidivans leishmaniasis, is difficult to treat, long-lasting and disfiguring. Lesions caused by MCL produce extensive destruction and distortion of oronasal and pharyngeal cavities, leading to mutilation in the face [1]. Finally, in individuals with a defective cell-mediated immune response, disseminated lesions of diffuse cutaneous leishmaniasis (DCL) resemble those of leprosy [17]. The lesions do not heal completely and may recur after some time. DCL is therefore recognized as a special public health problem, both clinically and because of its severe emotional consequences [17]. This theme is discussed by Zerpa and colleagues in Chap. 8 of this volume.

1.3 Transmission

Transmission is increasing at high rates in several world areas. This increase is a consequence of situations that boost the probability of being exposed to the sand flies such as establishment of new settlements in high-risk endemic areas or wild areas where zoonotic transmission may occur, deteriorating social and economic conditions in the poorer suburbs of some cities, and increased migration to urban areas of populations that used to live in rural areas [6, 18].

During its digenetic life cycle, *Leishmania* alternates between two extreme environments to which the parasite must adapt, i.e. the mammalian host and the insect vector (Fig. 1.1). These two environments hold physiological conditions with

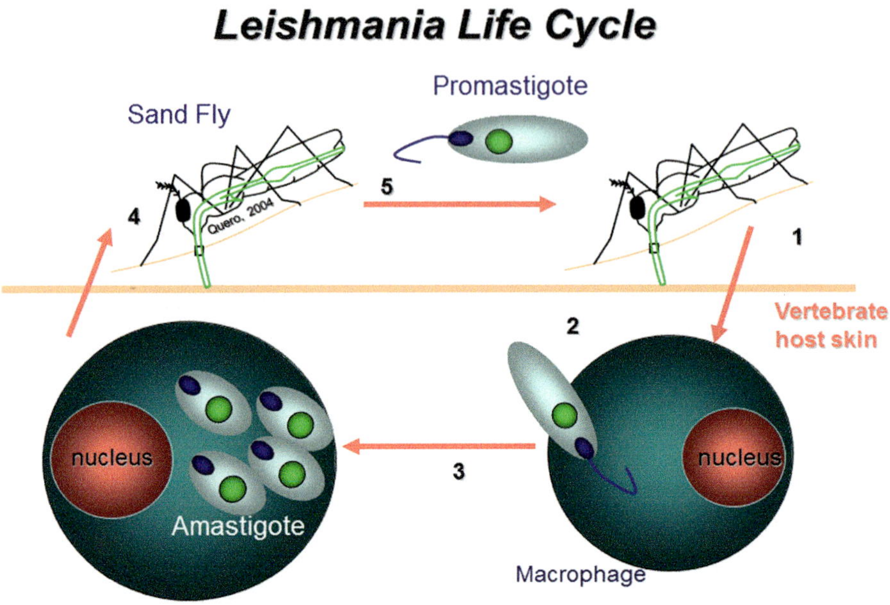

Fig. 1.1 The life cycle of *Leishmania*. The vertebrate host is infected when punctured by the infected insect vector (1). The promastigotes enter the circulating macrophages (2), infect them and reproduce there as amastigotes (3). The macrophage releases promastigotes that infect new macrophages. The insect vector ingests infected macrophages when puncturing an infected host (4). The amastigotes are released in the insect gut and reproduce as promastigotes (5)

specific—and different—characteristics. Some of these include temperature, pH, osmolarity and nutrient quantity and quality [19, 20].

The blood-sucking sand flies involved in disease transmission belong to the family Psychodidae; at least 70 known species are capable of transmitting the disease. Representatives of the genus *Lutzomyia* transmit the disease in America and of the genus *Phlebotomus* in the Old World. They have an arched chest, lanceolate wings, hairy body, long legs and delicate long and tubular abdomen. These insects measure between 1.5 and 3 mm, are yellowish and have dark eyes [21].

The habitat ranges from rain forest to very dry regions, and they are distributed from sea level to 1500 m. They live in damp and dark environments, at temperatures above 18 °C, and have crepuscular activity. They make silent, short flights with periods of rest in small jumps, although they can also cover long distances with predetermined directional flight [20, 22].

Females are hematophagous and need blood for the maturation of their offsprings. After fertilization, the females deposit between 40 and 70 eggs in damp and dark areas with abundant organic material. In this environment, larvae grow, transform to pupae and develop to the less active imago (adult insect fully grown and able to reproduce). Prophylaxis against insect-borne diseases like leishmaniasis, include, e.g. vector control by residual insecticides in areas of transmission, elimination of reservoirs and personal protection (insect nets, frequent application of insect repellents) [23]. Efficient control of the density of sand flies in endemic areas is essential for eradication of the disease. However, this measure is not always easy to implement in areas remote from cities and suburbs, and, therefore, additional methods must be implemented [24].

The study of the role of sand flies in the transmission of leishmaniasis has recently focused on the function of maxadilan, a vasodilator peptide similar in structure to the calcitonin gene-related peptide. Maxadilan is a component of insect saliva [25, 26] that inhibits oxidative metabolic processes and antigen presentation in macrophages. Additional peptides also present in the saliva favour successful inoculation of parasites to the host mammal. The insect's saliva thus facilitates transmission and increases the number of viable parasites present in the lesion [26, 27].

An accidental form of transmission is the contact with an infected vector in the laboratory. This can result from inadequate management of crop pests and exposure to samples of animals or people infected with contaminated blood or via wounds caused by contaminated needles or through pre-existing skin abrasions [28]. Despite its low frequency, it is essential to emphasize the importance of using safety codes and rules at work to ensure the protection of laboratory workers.

1.4 The Life Cycle of *Leishmania*

The life cycle of *Leishmania* alternates between two host types. *Leishmania* lives as an intracellular parasite (amastigotes) in macrophages of vertebrates such as mammals and as an extracellular parasite (promastigotes) in the gut of the insect vector [29] (Fig. 1.1). Insects ingest blood from the vertebrate host and regurgitate promastigotes when puncturing its skin. The parasites are recognized by surface receptors of macrophages and dendritic cells and are phagocytosed. Interestingly, parasites have more difficulty to establish an infection (and to survive) in mice lacking neutrophils. This result strongly suggests the relevance of a mechanism of entry into the macrophages (Trojan horse), through the use of polymorphonuclear leucocytes as the first phagocytic cells encountered in the host. The parasites promote programmed cell death in the infected neutrophiles that then are ingested by macrophages [30, 31].

Within the host cell, the parasites migrate to the phagolysosome and differentiate into amastigotes, which multiply by binary division intensely. The rupture of densely infected macrophages releases amastigotes; the liberated parasites are engulfed by naive macrophages, thus exponentially increasing the number of infected cells and spreading the disease within the host. The amastigotes ingested by insects that suck blood from an infected host are transformed into promastigotes in the digestive tract of the insect vector, where they remain 4–7 days, differentiate into infective parasites, migrate to the heart valve and obstruct the insect's proboscis [32]. Once the sand fly punctures again a host's skin, the parasites are once more liberated into the blood system of the host to close the cycle.

The clinical varieties of the disease are determined by the interplay between parasite and human host. *Leishmania* has tropism, that is, different species may prefer different tissular conditions. *Leishmania* species that cause cutaneous and mucocutaneous diseases are sensitive to temperatures over 35 °C and multiply only in exposed areas of the skin. The species that cause visceral manifestations of the disease require 37 °C for differentiation to amastigotes and migrate to the bone marrow, spleen and liver [29, 33, 34].

All trypanosomatid morphological phenotypes, except the amastigote, are motile. The parasites use a tip to base flagellum beat for swimming [35]. The single flagellum is a fundamental organelle that promotes promastigote attachment to insect host epithelia, plays a key role during cell translation, is crucial for parasite survival in the mammalian bloodstream and guides the final stages of cell division [36–39].

1.5 Biology and Plasticity of the Parasite

All *Leishmania* species are morphologically similar and display two main developmental stages through their life cycle: the amastigote that resides inside the reticuloendothelial cells of the vertebrate host and the promastigote that replicates in the gut of a phlebotomine sand fly [40]. These parasites are mostly diploid organisms with unique features that differentiate them from other eukaryotes, i.e. GPI anchoring of membrane proteins, glycosomes, kinetoplast, mitochondrial DNA editing and organization, polycistronic transcription and trans splicing (reviewed by [41]). Sexual interchange has been demonstrated in the stage that infects invertebrates; however, *Leishmania* parasites are considered to divide mainly by binary partition [42, 43].

One unique feature of these parasites is their ability to alter the copy number of individual genes or alternatively of group of genes, chromosomes and even the entire genome [41]. This "genetic plasticity" allows this flexible parasite to amplify the copy number of specific genes, an advantage that guarantees the use by *Leishmania* of these "extra copies" to increase the levels of a gene product [41, 44, 45].

In recent years, the advent of high-throughput sequencing technologies has assisted relentless progress in the genomics of human leishmaniases, with the completion of the whole genome sequences of *Leishmania* (*L.*) species. The first one was *L. (L.) major* (Ivens et al. 2005), but many others are now available (*L. (L.) infantum*, *L. (L.) mexicana*, *L. (L.) donovani* and *L. (L.) amazonenensis*) and from the *Viannia* (*V.*) species, *Leishmania* (*V.*) *braziliensis* and *L. (V.) panamensis* [46–51]. Comparison of their genetic sequences, together with the knowledge of their genetic plasticity, constitutes a fantastic tool not only to improve the comprehension of the biology of the parasite in itself but also to identify and validate targets for the rational design of the desperately needed drugs.

Moreover, genomic and transcriptomic analyses have advanced our understanding of the biology of *Leishmania* and shed new light on the complex interactions occurring within the parasite–host–vector triangle that could end up in potential new avenues to concrete tools for treatment and control programmes [51].

1.6 Communication in *Leishmania*

Chemotaxis is a communicating process fundamental to survival. The chemotactic response is a migratory activity that microorganisms use to respond to the changes that occur in the environment such as hydrostatic pressure, light, magnetic fields, osmotic pressure, temperature, etc. [52, 53].

The main ways used by *Leishmania* to acknowledge migratory behaviours are random walks, as a directional swimming interposed by tumbles which reorientate the parasite towards a high chemoattractant concentration and helical clinotaxis, a process in which a chemoattractant concentration is sampled around a curved path and modulates helical path curvature [35].

For small cells (like *Leishmania*), for which rotational diffusion speedily randomises orientation, the chemotaxis method *per excellence* is random walk; it represents a more efficient way of movement compared to deterministic chemotaxis. Furthermore, *Leishmania* swimming behaviours seem to be well adapted both for helical clinotaxis and influenced random walk chemotaxis, thus meaning that these cells lie at the threshold size in which, in their particular environments, biased random walk chemotaxis becomes favourable [35, 53–56].

Most studies analysing host–*Leishmania* interaction focus on promastigotes; these forms of the parasite are used to migrate in complex extracellular environments before being internalized by neutrophils and/or macrophages. Indeed, the initial step of infection upon the sand fly bite and injury to the structure of the skin corresponds to the inoculation of promastigotes and their deposition into the blood pool, embedded in dermal extracellular matrix components from tissue and blood [57]. Thus, the analysis of the synchronized action mediated by promastigotes and directed towards immune cells and the extracellular matrix (ECM) and basement membrane proteins conforms an important step previous to the establishment of the infection [57, 58].

In vitro studies demonstrate that promastigotes attach to and move through collagen I scaffolds. Collagen I seems to be the predominant ECM component found in the injured skin at early-stage lesions and *Leishmania*. Furthermore, some data suggest that promastigotes might need to secrete metallo- and cysteine proteinases in order to break down rigid collagen scaffolds and promote collagen remodelling and stiffness the matrix needed for migration before being internalized by a host cell [58, 59].

Leishmania promastigotes are able to evade the plasminogen/plasmin/fibrin structure (blood clot) while allowing target cells to get closer. In this interaction, anticoagulants existing in the sand fly saliva inhibit blood coagulation and neutralize promastigote procoagulant activity [60]. Additionally, plasminogen-associated vesicles trap macrophages, thus potentially allowing parasites to move further into the dermis [57].

In conclusion, the migratory response guided by chemotaxis is fundamental for *Leishmania* pathogenesis, and the comprehension of the chemical signals involved between host and parasite recognition, as well as the physiology of these events, is determinant for the fate of infection [61]. Therefore, the identification of molecules, signs and behaviours involved in these responses can be useful in the developing of alternative and successful preventive measures or treatments for this serious disease [56].

1.7 Pathogenesis

Clinical manifestations of leishmaniasis result from the interaction that exists between the infecting parasite species and the host's immune response. For example, *L. (L.) major* causes only cutaneous disease, and *L. (L.) donovani* results in visceralization, while *L. (L.)* infantum causes either a cutaneous or systemic disease. This variability may be secondary to genomic differences acquired during evolution of *Leishmania*, as has been suggested by experiments performed in murine models, making some species more adapted to target the skin and others to invade visceral organs [1]. The pathology and the outcome of these clinical manifestations is thoroughly described in Chaps. 7 and 8 of this volume and will not be referred herein.

Immunological responses to *Leishmania* infection are complex, and the infecting species can determine it. Thus, significant differences in host–parasite interactions have been described for cutaneous and visceral leishmaniasis [15]. These immunological interactions are thoroughly described in Chap. 5 of this volume and will not be referred herein.

1.8 Control and Surveillance

The primary control strategy against leishmaniasis is based on case finding and treatment; it is therefore essential to control transmission to improve diagnostic methods and to develop good treatments that can be used even in less developed countries. A parallel requirement exists for the development of cost-effective drug delivery systems, especially those that take advantage of new oral drugs [20].

Diagnosis of leishmaniasis includes clinical, epidemiological and parasitological tools among which are the case history and immunobiological, molecular and serological tests. The molecular tests allow the precise identification of suspected parasite species, and concomitant infections like, for example, the human immunodeficiency virus [13]. Genetic characterization, biochemistry and immunological analysis of *Leishmania* have contributed to the precise identification of the disease-causing parasite and have helped to dilucidate many of the processes that occur during host–parasite interactions, as well as the mechanisms of immune evasion and intracellular survival of these parasites [62–65]. This knowledge is essential in the search for effective therapeutic tools against the disease.

Even though the new serological tests such as the dipstick, the lyophilized direct agglutination test kit and the latex agglutination urine test, as well as the loop-mediated isothermal amplification (LAMP) [66], represent major progress in diagnosis of leishmaniasis, they are not yet widely used in areas of endemia. Moreover, these are indirect tests which appear promising for the prognosis of treatment outcome in VL; however, they do not provide parasitological confirmation of infection or of cure immediately following treatment [67–69].

Unfortunately, American tegumentary leishmaniasis, caused by a variety of *Leishmania* and *Viannia* species that may confuse the outcome, remains a challenge for molecular diagnosis.

Vector control is rarely carried out as a specific approach against leishmaniasis control. Domestic and peridomestic sand (*Phlebotomus*) fly vectors are more susceptible to indoor residual spraying than are other domestic vectors, such as anopheline mosquitoes or triatomine bugs; this means that transient suppression of sand fly populations is seen as an additional benefit of malaria or Chagas' disease vector control in areas where these vectors coincide [70]. This means that control of domestic and peridomestic *Leishmania*-specific sand fly vectors will probably continue as an additional benefit of programmes against other insect vectors using indoor residual spraying or insecticide-treated bed nets. In fact, insecticide-treated bed nets, which are becoming widely deployed against malaria transmission, may also become cost-effective for reducing leishmaniasis in areas of domestic transmission [71]. Finally, in areas where dogs represent reservoir hosts par excellence, the use of insecticide-treated dog collars would merit further appraisal. Such collars would reduce the likelihood of new infections in dogs and additionally could decrease the risk of transmission from dogs to humans [72].

1.9 Medicaments in Use

Since 1940, drugs such as pentavalent antimonials (Sb^V) Glucantime® and Pentostam® as well as the diamidines (e.g. pentamidine) are used against leishmaniasis. Later, antifungals like amphotericin-B [as deoxycholate (AMB-B) or in liposomal form (L-AMB)], ketoconazole, paromomycin antibiotics and dapsone were incorporated, and at the beginning of the twenty-first century, miltefosine (MIL), the first drug for oral use in leishmaniasis, was approved to be used [3]. Liver and kidney toxicity is exhibited by most of these compounds, and the cost of treatment for many of them has prompted the search for alternative drugs that meet the following conditions: oral administration, fewer side effects, lesser toxicity and lower price [73–75].

The long-term intravenous treatment with Sb^V produces severe adverse effects like pain at the injection site, cardiac arrhythmias, rashes, etc. Often side effects lead to abandonment (partial or total) of treatment by the patient, a scenario that favours the selection of drug-resistant parasites and the development of drug resistance [3]. As resistance have become widespread, this therapy has been abandoned in the Indian subcontinent (ISC). However, Sb^V are still in use Latin America and East Africa [76].

AMB-B is a polyene antibiotic that is administered as an intravenous deoxycholate and since 1997 in liposomes (AmBisome®, L-AMB). Amphotericin-B is selective against *Leishmania*. It is very toxic when injected as AMB-B and should be then used as a L-AMB formulation to decrease the toxicity risk. Additionally, its use in endemic regions is limited also by the cost of treatment (although the WHO is provided with up to 350,000 vials over the next 5 years free of charge) and the difficulties of administration. As an alternative second-line treatment, pentamidine and especially paromomycin are used since 1987, and more recently dapsone and ketoconazole have been introduced [75, 77].

Pentamidine is a diamidino compound with a relatively broad spectrum and is effective in visceral leishmaniasis and trypanosomiasis. Clinical trials of aminosidine (paromomycin) are in progress, and the use of this drug against visceral leishmaniasis may become widespread. Unfortunately, its range of targeting *Leishmania* species is restricted, and the situation regarding resistance in the field is unclear [78, 79]. Pentamidine and paromomycin are good alternatives used for oral or topical treatment in cases of cutaneous leishmaniasis [75, 77].

In 2002, MIL was approved as the first orally active drug against visceral leishmaniasis in the ISC, and in 2005 it was approved for use against cutaneous leishmaniasis in Colombia. MIL causes vomiting and diarrhoea in up to 60% of patients; it also causes reversible increases in blood transaminases and creatinine. It is contraindicated in pregnant women, and birth control measures should be taken during treatment and even 60 days thereafter [80–83]. MIL replaced antimonials in the ISC for the treatment of patients suffering VL within the Kala-azar elimination programme. However, within a decade of its introduction, it has lost efficacy, and recently some resistant clinical isolates have been described [84–87]. Finally, sitamaquine, a drug developed a decade ago [88], is now discarded due to the adverse effects it produces, as well as its lack of efficacy and the selection of sitamaquine-resistant clones of *L.* (*L.*) *donovani* [89, 90].

In summary, the pharmacopoeia against leishmaniasis nearly consists of the same medications used since the early twentieth century, except for MIL, paromomycin and novel lipid formulations of amphotericin, L-AMB [90, 91]. The drugs are highly toxic, some patients do not respond to the treatment or present relapses or therapeutic failure, and patients with diffuse cutaneous leishmaniasis are unresponsive to any form of chemotherapy [92] (Chap. 8 of this volume). Resistance against antimonials is one of the most serious problems that we face in the control of leishmaniasis, especially for the visceral leishmaniasis in areas such as North Bihar in India [3]. Important lessons must be learnt from what has happened with Sb^V in the ISC, and this is thoroughly described in Chap. 4 of this volume. These lessons might be helpful to define the use of the additional available drugs (miltefosine, paromomycin, amphotericin-B) and to monitor their efficacy to prevent the appearance of therapeutic failure as well as the emergence and selection of resistance against them, specially so since there are already described cases of resistance against MIL [85]. Such coordination is desperately needed as a guide to orient further research and prevention activities, since the increasing treatment failure rates suggest that the therapy must switch to combination therapy specially for VL [93].

1.10 Therapeutic Failure

Therapeutic failure in leishmaniasis is a common problem in endemic areas. This may be associated with multiple factors that depend both on the parasite and on the mammalian host. Regarding the mammalian host, therapeutic failure can be attributed to altered drug pharmacokinetics, reinfection or immunologic compromise. In most cases where chemotherapy fails to cure the patient, the natural susceptibility of parasites to drugs happens to be low, or, alternatively, the infecting parasite has developed chemo-resistance.

Drug resistance is defined as the decrease in efficacy of compounds in a population of cells (i.e. *Leishmania* parasites) that was previously susceptible to them. This definition assumes that the initial susceptibility of the organisms to the drugs is known, which is not always true for field strains of *Leishmania* isolated from patients [3]. There are at least three cellular mechanisms directly associated with the expression of chemo-resistance: gene amplification of drug target enzymes, structural and functional changes of drug targeted enzymes and decreased intracellular levels of the drug due to their extrusion by specific transporters [19, 93].

However, chemo-resistance is a multifactorial phenomenon. It is generally associated with increased expression of drug transporter proteins (multidrug resistance, or MDR, also known as P-glycoprotein) and multidrug resistance-associated protein (MRP) [19]. Both types of transporters modulate intracellular chemotherapeutic agent's concentrations. However, the mechanisms of chemo-resistance may be multiple and not exclusive to a single type of drug. That is, parasites (1) may use multiple mechanisms to be resistant to one drug and (2) may use one mechanism to be resistant to various drugs originating the so-called cross-reaction phenomena.

In *Leishmania* chemo-resistance may be natural, or acquired, which develops when parasites are exposed to suboptimal doses of drugs [3, 94].

Similar to what happens in cancer cells in *Leishmania*, chemo-resistance has been associated with decreased cellular accumulation of drugs, due to the extrusion by ABC-type membrane transporters [95]. Additionally, the expression of chemo-resistance is associated with changes in biochemical and physiological mechanisms that are fundamental for survival and infectivity of the parasites [19, 96]. Research into drug-induced changes in the physiology of the parasite can certainly provide clues for predicting the success of a given therapy.

What is clear is that the increase in drug resistance impacts definitively treatment outcome. Thus, it should be thoroughly dissected and comprehended. This is especially important when dealing with an intracellular parasite, *Leishmania*, a condition that in mammals impose fundamental challenges for successful control, surveillance and treatment.

1.11 Concluding Remarks and Future Trends

Until now, markers of chemo-resistance against leishmanicidal drugs have been elusive although new hopes might appear, at least for VL, by the use of whole genome sequencing of isolates might provide definitive answers to identify them [97]. Still, the reliable method for monitoring resistance of a given parasite is the isolated in vitro amastigote-macrophage model. This laborious technique requires costly and time-consuming conditions that discourage its application in the routine laboratory. Because of this and the increased incidence of cases of leishmaniasis that do not respond effectively to drugs, it has become urgent to identify cell markers for drug resistance that are easy to use in the routine laboratory and can guide leishmanicidal therapy [3, 96, 98]. These markers would improve our understanding of disease pathogenesis and would be helpful in refining the criteria for cure and in improving disease prognosis.

The present volume seeks to provide an overview of our current knowledge on chemotherapeutic failure and drug resistance in *Leishmania*. Many causes are at least partially responsible for the disappointing persistence of leishmaniasis and its death and disability-adjusted life year toll worldwide. We are convinced that a mechanistic understanding of leishmaniasis and of the drugs that are used for its handling, as well as of drug resistance, its molecular basis, consequences and possible treatments may help to improve the picture we face now days.

The organization of the book has been beautifully outlined by Paul Michels in the foreword that points out that a better understanding of the disease itself and of drug resistance, its molecular basis, its consequences and possible avenues for better treatments may help improve this depressing picture. Experts in the field cover current knowledge and future trends of these and many other aspects of drug resistance in *Leishmania* in the present volume. Each chapter comprises an exhaustive review of the particular topic and provides insights into the future of the field both from the scientific and clinical perspective.

Acknowledgments The authors are grateful for the financing support received from the Coordination for Research, Faculty of Medicine, UCV, and the Council for Scientific and Humanistic Research (CDCH), Universidad Central de Venezuela. Likewise, they are grateful for the support conferred by the Alexander von Humboldt Foundation and the University of Würzburg, Germany, to Alicia Ponte-Sucre.

References

1. Pace D. Leishmaniasis. J Infect. 2014;69(1):S10–8.
2. Davies CR, Reithinger R, Campbell-Lendrum D, Feliciangeli D, et al. The epidemiology and control of leishmaniasis in Andean countries. Cad Saude Publica. 2000;16:925–50.
3. Croft SL, Sundar S, Fairlamb AH. Drug resistance in leishmaniasis. Clin Microbiol Rev. 2006;19:111–26.
4. Rotureau B. Are new world leishmaniases becoming anthroponoses? Med Hypotheses. 2006;67:1235–41.
5. Ready PD. Leishmaniasis emergence in Europe. Euro Surveill. 2010;11:19505.
6. World Health Organization (WHO, 2016) Weekly epidemiological record. http://www.who.int/wer, 2016, 91, 285–296.
7. Karimkhani C, Wanga V, Coffeng LE, Naghavi P, et al. Global burden of cutaneous leishmaniasis: a cross-sectional analysis from the Global Burden of Disease Study 2013. Lancet Infect Dis. 2016;6:584–91. https://doi.org/10.1016/S1473-3099(16)00003-7
8. Alvar J, Vélez ID, Bern C, Herrero M, et al. Leishmaniasis worldwide and global estimates of its incidence. WHO Leishmaniasis Control Team. PLoS One. 2012;7(5):e35671.
9. World Health Organization Technical Report Series 949 (2015) Control of the leishmaniasis 2010. http://whqlibdoc.who.int/trs/WHO_TRS_949_eng.pdf.
10. Ready PD. Epidemiology of visceral leishmaniasis. Clin Epidemiol. 2014;6:147–54.
11. Desjeux P. Leishmaniasis: current situation and new perspectives. Comp Immunol Microbiol Infect Dis. 2004;27:305–18.

12. Georgiadou SP, Makaritsis KP, Dalekos GN. Leishmaniasis revisited: current aspects on epidemiology, diagnosis and treatment. J Transl Int Med. 2015;3(2):43–50.
13. Alvar J, Aparicio P, Aseffa A, Den Boer M, et al. The relationship between leishmaniasis and AIDS: the second 10 years. Clin Microbiol Rev. 2008;21:334–59.
14. Malafaia G. Protein-energy malnutrition as a risk factor for visceral leishmaniasis: a review. Parasite Immunol. 2009;31:587–96.
15. Saporito L, Giammanco G, De Grazia S, Colomba C. Visceral leishmaniasis: host–parasite interactions and clinical presentation in the immunocompetent and in the immunocompromised host. Int J Infect Dis. 2013;17:e572–6.
16. Sharma U, Singh S. Immunobiology of leishmaniasis. Indian J Exp Biol. 2009;47:412–23.
17. Ameen M. Cutaneous and mucocutaneous leishmaniasis: emerging therapies and progress in disease management. Expert Opin Pharmacother. 2010;11:557–69.
18. Romero GA, Boelaert M. Control of visceral leishmaniasis in Latin America a systematic review. PLoS Negl Trop Dis. 2010;4:e584.
19. Ponte-Sucre A. Physiological consequences of drug resistance in *Leishmania* and their relevance for chemotherapy. Kinetoplastid Biol Dis. 2003;2:14.
20. Cattand P, Desjeux P, Guzmán MJ, Jannin J, et al. Tropical diseases lacking adequate control measures: dengue, leishmaniasis, and African trypanosomiasis. In: Disease control priorities in developing countries. 2nd ed. New York: Oxford University Press; 2006. p. 451–66.
21. Feliciangeli MD, Rabinovich J. Abundance of *Lutzomyia ovallesi* but not *Lu. gomezi* (Diptera: Psychodidae) correlated with cutaneous leishmaniasis incidence in north-central Venezuela. Med Vet Entomol. 1998;12:121–31.
22. Davies CR, Reithinger R, Campbell-Lendrum D, Feliciangeli D, et al. The epidemiology and control of leishmaniasis in Andean countries. Cad Saude Publica. 2000;16(4):925–50.
23. Curtis CF. Personal protection methods against vectors of disease. Rev Med Vet. 1992;80:543–53.
24. Thakur CP. Leishmaniasis research, the challenges ahead. Indian J Med Res. 2006;123:193–4.
25. Lerner EA, Ribeiro JM, Nelson RJ, Lerner MR. Isolation of maxadilan, a potent vasodilatory peptide from the salivary glands of the sand-fly *Lutzomyia longipalpis*. J Biol Chem. 1991;261:11234–6.
26. Castro-Sousa F, Paranhos-Silva M, Sherlock I, Paixão MS, et al. Dissociation between vasodilation and *Leishmania* infection-enhancing effects of sand fly saliva and maxadilan. Mem Inst Oswaldo Cruz. 2001;96:997–9.
27. Belkaid Y, Kamhawi S, Modo G, Valenzuela J, et al. Development of a natural model of cutaneous leishmaniasis: powerful effects of vector saliva and saliva pre-exposure on the long-term outcome of *Leishmania major* infection in the mouse ear dermis. J Exp Med. 1998;188:1941–53.
28. Delgado O, Guevara P, Silva S, Belfort E, et al. Follow up of human accidental infection by *Leishmania braziliensis* using conventional immunologic techniques and polymerase chain reaction. Am J Trop Med Hyg. 1996;51:267–72.
29. Bates PA, Rogers ME. New insights into the developmental biology and transmission mechanisms of *Leishmania*. Curr Mol Med. 2004;4:601–9.
30. Peters NC, Egen JG, Secundino N, Debrabant A, et al. *In vivo* imaging reveals an essential role for neutrophils in leishmaniasis transmitted by sand flies. Science. 2008;321:970–4.
31. Ritter U, Frischknecht F, van Zandbergen G. Are neutrophils important host cells for *Leishmania* parasites? Trends Parasitol. 2009;25:505–10.
32. Killick-Kendrick R, Wallbanks KR, Molyneux DH, Lavin DR. The ultrastructure of *Leishmania major* in the foregut and proboscis of *Phlebotomus papatasi*. Parasitol Res. 1988;74 (6):586–90.
33. Ridley D. The pathogenesis of cutaneous leishmaniasis. Trans R Soc Trop Med Hyg. 1999;73:156–60.
34. Chang KP, Reed SG, McGwire BS, Soong L. *Leishmania* model for microbial virulence: the relevance of parasite multiplication and patho-antigenicity. Acta Trop. 2003;85:375–90.

35. Wheeler RJ. Use of chiral cell shape to ensure highly directional swimming in trypanosomes. PLoS Comput Biol. 2017;13(1):e1005353.
36. Gadelha C, Wickstead B, Gull K. Flagellar and ciliary beating in trypanosome motility. Cell Motil Cytoskeleton. 2007;64:629–43.
37. Rotureau B, Morales MA, Bastin P, Spath G. The flagellum-mitogen-activated protein kinase connection in Trypanosomatids: a key sensory role in parasite signaling and development? Cell Microbiol. 2009;11(5):710–8.
38. Forestier CL, Machu C, Loussert C, Pescher P, et al. Imaging host cell-*Leishmania* interaction dynamics implicates parasite motility, lysosome recruitment, and host cell wounding in the infection process. Cell Host Microbe. 2011;9:319–30.
39. Díaz E, Köhidai L, Ríos A, Vanegas O, et al. *Leishmania braziliensis*: cytotoxic, cytostatic and chemotactic effects of poly-lysine-methotrexate-conjugates. Exp Parasitol. 2013;135 (1):134–41.
40. Ponte-Sucre A. Leishmaniasis, the biology of a parasite. In: Ponte-Sucre A, Diaz E, Padrón-Nieves M, editors. Drug resistance in *Leishmania* parasites. Consequences, molecular mechanisms, and possible treatments. Wien: Springer; 2013. p. 1–12.
41. de Toledo JS, Vasconcelos EJR, Ferreira TR, Cruz AK. Using genomic information to understand *Leishmania* biology. Open Parasitol J. 2010;4:156–66.
42. Akopyants NS, Kimblin N, Secundino N, Patrick R, et al. Demonstration of genetic exchange during cyclical development of *Leishmania* in the sand fly vector. Science. 2009;324:265–8.
43. Rougeron V, De Meeûs T, Hide M, Waleckx E, et al. Extreme inbreeding in *Leishmania braziliensis*. Proc Natl Acad Sci USA. 2009;106:10224–9.
44. Sterkers Y, Crobu L, Lachaud L, Pagès M, et al. Parasexuality and mosaic aneuploidy in *Leishmania*: alternative genetics. Trends Parasitol. 2014;30(9):429–35.
45. Mannaert A, Downing T, Imamura H, Dujardin JC. Adaptive mechanisms in pathogens: universal aneuploidy in *Leishmania*. Trends Parasitol. 2012;28(9):370–6.
46. Peacock CS, Seeger K, Harris D, Murphy L, et al. Comparative genomic analysis of three *Leishmania* species that cause diverse human disease. Nat Genet. 2007;39:839–47.
47. Downing T, Imamura H, Decuypere S, Clark TG, et al. Whole genome sequencing of multiple *Leishmania donovani* clinical isolates provides insights into population structure and mechanisms of drug resistance. Genome Res. 2011;21(12):2143–56. https://doi.org/10.1101/gr.123430.111
48. Rogers MB, Hilley JD, Dickens NJ, Wilkes J, et al. Chromosome and gene copy number variation allow major structural change between species and strains of *Leishmania*. Genome Res. 2011;21(12):2129–42. https://doi.org/10.1101/gr.122945.111
49. Real F, Vidal RO, Carazzolle MF, Mondego JM, et al. The genome sequence of *Leishmania (Leishmania) amazonensis*: functional annotation and extended analysis of gene models. DNA Res. 2013;20(6):567–81. https://doi.org/10.1093/dnares/dst031
50. Llanes A, Restrepo CM, Del Vecchio G, Anguizola FJ, et al. The genome of *Leishmania panamensis*: insights into genomics of the *L. (Viannia)* subgenus. Sci Rep. 2015;5(8550). https://doi.org/10.1038/srep08550
51. Cantacessi C, Dantas-Torres F, Nolan MJ, Otranto D. The past, present, and future of *Leishmania* genomics and transcriptomics. Trends Parasitol. 2015;31(3):100–8.
52. Kohidai L. Chemotaxis as an expression of communication of *Tetrahymena*. In: Witzany G, Nowacki M, editors. Biocommunication of ciliates. Dordrecht: Springer; 2016. p. 65–82.
53. Diaz E, Zacarias AK, Pérez S, Vanegas O, et al. Effect of aliphatic, monocarboxylic, dicarboxylic, heterocyclic and sulphur-containing amino acids on *Leishmania* spp. chemotaxis. Parasitology. 2015;142(13):1621–30.
54. Bray RS. *Leishmania*: chemotaxic responses of promastigotes and macrophages *in vitro*. J Protozool. 1983;30:322–9.
55. Leslie G, Barrett M, Burchmore R. *Leishmania mexicana*: promastigotes migrate through osmotic gradients. Exp Parasitol. 2002;102:117–20.

56. Díaz E, Köhidai L, Ríos A, Vanegas O, et al. *Leishmania braziliensis*: cytotoxic and chemotactic effects of branched chain polypeptide conjugates with poly [L-Lysine] backbone. Exp Parasitol. 2013;135:134–41.
57. de Menezes JP, Koushik A, Das S, Guven C, et al. *Leishmania* infection inhibits macrophage motility by altering F-actin dynamics and the expression of adhesion complex proteins. Cell Microbiol. 2017;19(3). https://doi.org/10.1111/cmi.12668
58. Petropolis DB, Rodrigues JC, Viana NB, Pontes B, et al. *Leishmania amazonensis* promastigotes in 3D Collagen I culture: an *in vitro* physiological environment for the study of extracellular matrix and host cell interactions. Peer J. 2014;2:e317.
59. Fatoux-Ardore M, Peysselon F, Weiss A, Bastien P, et al. Large scale investigation of *Leishmania* interaction networks with host extra cellular matrix by surface plasmon resonance imaging. Infect Immun. 2014;(2):594–606.
60. Rochael NC, Lima LG, Oliveira SM, Barcinski MA, et al. *Leishmania amazonensis* exhibits phosphatidylserine-dependent procoagulant activity, a process that is counteracted by sandfly saliva. Mem Inst Oswaldo Cruz. 2013;108:679–85.
61. Pozzo LY, Fontes A, de Thomaz AA, Santos BS, et al. Studying taxis in real time using optical tweezers: applications for *Leishmania amazonensis* parasites. Micron. 2009;40(5–6):617–20.
62. Bogdan C, Gessner A, Solbach W, Röllinghoff M. Invasion, control and persistence of *Leishmania* parasites. Curr Opin Immunol. 1996;8:517–25.
63. Bañuls AL, Hide M, Tibayrenc M. Evolutionary genetics and molecular diagnosis of *Leishmania* species. Trans R Soc Trop Med Hyg. 2002;96:S9–S13.
64. Smith DF, Peacock CS, Cruz AK. Comparative genomics: from genotype to disease phenotype in the leishmaniases. Int J Parasitol. 2007;37:1173–86.
65. Schönian G, Mauricio I, Gramiccia M, Cañavate C, et al. Leishmaniases in the Mediterranean in the era of molecular epidemiology. Trends Parasitol. 2008;24:135–42.
66. Verma S, Singh R, Sharma V, Bumb RA, et al. Development of a rapid loop-mediated isothermal amplification assay for diagnosis and assessment of cure of *Leishmania* infection. BMC Infect Dis. 2017;17(1):223.
67. Tavares CA, Fernandes AP, Melo MN. Molecular diagnosis of leishmaniasis. Expert Rev Mol Diagn. 2003;3:657–67.
68. Sundar S, Agrawal S, Pai K, Chance M, et al. Detection of *Leishmania* antigen in the urine of patients with visceral leishmaniasis by a latex agglutination test. Am J Trop Med Hyg. 2005;73:269–71.
69. Salotra P, Singh R. Challenges in the diagnosis of post kala-azar dermal leishmaniasis. Indian J Med Res. 2006;123:295–310.
70. Kassi M, Kasi PM, Marri SM, Tareen I, et al. Vector control in cutaneous leishmaniasis of the old world: a review of literature. Dermatol Online J. 2008;14:1.
71. Alten B, Caglar SS, Kaynas S, Simsek FM. Evaluation of protective efficacy of K-OTAB impregnated bednets for cutaneous leishmaniasis control in Southeast Anatolia, Turkey. J Vector Ecol. 2003;28:53–64.
72. Quinnell RJ, Courtenay O. Transmission, reservoir hosts and control of zoonotic visceral leishmaniasis. Parasitology. 2009;136:1915–34.
73. Murray H. Clinical and experimental advances in treatment of visceral leishmaniasis. Antimicrob Agents Chemother. 2001;45:2185–97.
74. Melby P. Recent developments in leishmaniasis. Curr Opin Infect Dis. 2002;15:485–90.
75. Palumbo E. Current treatment for cutaneous leishmaniasis: a review. Am J Ther. 2009;16:178–82.
76. Mitropoulos P, Konidas P, Durkin-Konidas M. New world cutaneous leishmaniasis: updated review of current and future diagnosis and treatment. J Am Acad Dermatol. 2010;63(2):309–22.
77. Croft SL, Coombs GH. Leishmaniasis: current chemotherapy and recent advances in the search for novel drugs. Trends Parasitol. 2003;19:502–8.
78. Jhingran A, Chawla B, Saxena S, Barrett MP, et al. Paromomycin: uptake and resistance in *Leishmania donovani*. Mol Biochem Parasitol. 2009;164(2):111–7.

79. Bhandari V, Sundar S, Dujardin JC, Salotra P. Elucidation of cellular mechanisms involved in experimental paromomycin resistance in *Leishmania donovani*. Antimicrob Agents Chemother. 2014;58(5):2580–5.
80. Croft SL, Neal RA, Pendergast W, Chan JH. The activity of alkyl phosphorylcholines and related derivatives against *Leishmania donovani*. Biochem Pharmacol. 1987;36:2633–6.
81. Eibl H, Unger C. Hexadecylphosphocholine: a new and selective antitumor drug. Cancer Treat Rev. 1990;17:233–42.
82. Croft SL. Kinetoplastida: new therapeutic strategies. Parasite. 2008;15:522–7.
83. Soto J, Berman J. Treatment of new world cutaneous leishmaniasis with miltefosine. Trans R Soc Trop Med Hyg. 2006;100:S34–40.
84. Sundar S, Singh A, Rai M, Prajapati VK, et al. Efficacy of miltefosine in the treatment of visceral leishmaniasis in India after a decade of use. Clin Infect Dis. 2012;55(4):543–50.
85. Rijal S, Ostyn B, Uranw S, Rai K, et al. Increasing failure of miltefosine in the treatment of Kala-azar in Nepal and the potential role of parasite drug resistance, reinfection, or noncompliance. Clin Infect Dis. 2013;56(11):1530–8.
86. Mondelaers A, Sanchez-Cañete MP, Hendrickx S, Eberhardt E, et al. Genomic and molecular characterization of miltefosine resistance in *Leishmania infantum* strains with either natural or acquired resistance through experimental selection of intracellular amastigotes. PLoS One. 2016;11(4):e0154101.
87. Srivastava S, Mishra J, Gupta AK, Singh A, et al. Laboratory confirmed miltefosine resistant cases of visceral leishmaniasis from India. Parasit Vectors. 2017;10(1):49.
88. Berman J. Clinical status of agents being developed for leishmaniasis. Expert Opin Investig Drugs. 2005;14:1337–46.
89. Loiseau PM, Cojean S, Schrével J. Sitamaquine as a putative antileishmanial drug candidate: from the mechanism of action to the risk of drug resistance. Parasite. 2011;18:115–9.
90. Singh N, Kumar M, Singh RK. Leishmaniasis: current status of available drugs and new potential drug targets. Asian Pac J Trop Med. 2012;5(6):485–97.
91. Croft SL, Seifert K, Yardley V. Current scenario of drug development for leishmaniasis. Indian J Med Res. 2006;123(3):399–410.
92. Zerpa O, Ulrich M, Blanco B, Polegre M, et al. Diffuse cutaneous leishmaniasis responds to miltefosine but then relapses. Br J Dermatol. 2007;156:1328–35.
93. Kedzierski L, Sakthianandeswaren A, Curtis JM, Andrews PC, et al. Leishmaniasis: current treatment and prospects for new drugs and vaccines. Curr Med Chem. 2009;16:599–614.
94. Croft SL. PKDL – a drug related phenomenon? Indian J Med Res. 2008;128(1):10–1.
95. Higgins CF. ABC transporters: from microorganisms to man. Annu Rev Cell Biol. 1992;8:67–113.
96. Natera S, Machuca C, Padrón-Nieves M, Romero A, et al. Proficiency of drug-resistant parasites. Int J Antimicrob Agents. 2007;29:637–42.
97. Imamura H, Downing T, Van den Broeck F, Sanders MJ, et al. Evolutionary genomics of epidemic visceral leishmaniasis in the Indian subcontinent. elife. 2016;5. pii: e12613.
98. t'Kindt R, Scheltema RA, Jankevics A, Brunker K, et al. Metabolomics to unveil and understand phenotypic diversity between pathogen populations. PLoS Negl Trop Dis. 2010;4:e904.

Part I

Determinant Features in Leishmaniasis

Molecular Evolution and Phylogeny of *Leishmania*

2

Gabriele Schönian, Julius Lukeš, Olivia Stark, and James A. Cotton

Abstract

The genus *Leishmania* was first described in 1903 for the parasite *Leishmania donovani*, but many additional species have been described since then. Although recent hierarchical taxonomic schemes have increasingly used molecular or biochemical characters to assign *Leishmania* organisms into different species, they are still heirs of the first classifications based primarily on geographical distribution, vector species, and disease presentations. The current classification system, based on multilocus enzyme electrophoresis, proposes up to 53 species, although molecular phylogenies of *Leishmania* suggest that the number of species may be too large. Very recently this classification system has been revised based on multiple gene phylogenies. For many decades, there has been a controversial discussion on whether the genus *Leishmania* appeared first in the Old World or in the New World. Analyses of whole-genome data led to the supercontinent hypothesis, in which the parasites evolved from a monoxenous ancestor on Gondwana and separated into Paraleishmania and all other species around the time when Gondwana split.

Many molecular markers have demonstrated substantial intraspecies diversity and the existence of geographically and genetically isolated populations in all *Leishmania* species tested so far. In particular the idea that *Leishmania* evolve predominantly clonally with only rare sexual recombination has repeatedly been questioned by the detection of hybrids, mosaic genotypes, and gene flow between

G. Schönian (✉) · O. Stark
Charité – Universitätsmedizin Berlin, Institut für Mikrobiologie und Hygiene, Berlin, Germany
e-mail: gabriele.schoenian@t-online.de

J. Lukeš
Biology Centre, Institute of Parasitology, Czech Academy of Sciences, Ceske Budějovice (Budweis), Czech Republic

J. A. Cotton
Wellcome Trust Sanger Institute, Wellcome Trust Genome Campus, Hinxton, UK

A. Ponte-Sucre, M. Padrón-Nieves (eds.), *Drug Resistance in Leishmania Parasites*,
https://doi.org/10.1007/978-3-319-74186-4_2

populations and strong inbreeding and, finally, the detection of genetic recombination under laboratory conditions.

This chapter reviews the recent (mostly) molecular data that provide new insights into the evolution, taxonomy, phylogenetic, and population genetic relationships of *Leishmania* but also the questions raised by this knowledge. It also discusses the power of modern approaches, such as multilocus sequence analysis, multilocus microsatellite typing, and comparative genomics for studying the inter- and intraspecies variation of *Leishmania* parasites.

2.1 Introduction

Phylogenetics is the study of evolutionary relatedness among various groups of organisms (e.g., species or populations); this relatedness is revealed, for example, through morphological data and molecular data, particularly DNA sequence data. *Taxonomy*, the science of naming and classifying organisms, is enriched by phylogenetics, although both fields remain methodologically and logically distinct. They overlap, however, in the area of *phylogenetic systematics*—the science that reconstructs the pattern of evolutionary events that have led to the distribution and diversity of life.

Modern phylogenetic studies with different molecular data have transformed our knowledge of evolutionary history and, consequently, taxonomy, as phylogenies based on these data have challenged traditional classifications for many groups of organisms. This is particularly the case for the most basal groups, and a new classification system of eukaryotes has been recently proposed based on data from modern morphological approaches, biochemical pathways, and molecular phylogenetics [1].

The genus *Leishmania* has also suffered taxonomic changes. Its position within the family Trypanosomatidae has been revised, the number of species belonging to it is disputed, and geographically defined populations have been identified in many *Leishmania* species. A good definition of *Leishmania* species is crucial for correct diagnosis and prognosis of the disease as well as for making decisions regarding treatment and control measures. This is a fundamental issue since the severity and nature of the clinical manifestation in immunocompetent patients varies with the infecting organism. Different *Leishmania* species cannot be distinguished by morphological criteria and have therefore been assigned to different species primarily based on clinical, biological, geographical, and epidemiological standards and, more recently, on immunological and biochemical data. Accordingly, since the first description of the genus *Leishmania* in 1903, the number of species has increased continuously. While species based on these criteria may be clinically useful, it is unclear that they will reflect the true evolution and diversity of the genus. Although many molecular methods have been recently introduced for unraveling the phylogeny of *Leishmania* and define its taxonomy, defining a *Leishmania* species or achieving a consensus on the described species is still not easy.

2.2 Molecular Methods for *Leishmania* Phylogenetics, Identification, and Population Genetics

The selection of the molecular method or marker most suitable for its use in phylogenetic studies depends on the question needed to be addressed and the required level of resolution. While trees resulting from molecular studies of *Leishmania* are preferred, they should not be used alone, as evolution of hosts and vectors, as well as climatic and geographical features, should also be taken into account [2]. A comprehensive review has recently listed previously used markers for *Leishmania* diagnosis and strain typing [3].

2.2.1 Molecular Methods for Studying *Leishmania* Phylogeny

Currently, phylogenetic relationships at the level of Kinetoplastea, as well as at the level of the genus *Leishmania*, are mostly based on DNA sequences. Slowly evolving genomic sequences such as small subunit (SSU) rRNA genes and glycosomal glyceraldehyde phosphate dehydrogenase (gGAPDH) genes have been most widely used for establishing molecular phylogenies of these pathogenic flagellates. Numerous molecular tools have been described that distinguish species and strains of *Leishmania* parasites. Since the genus *Leishmania* is relatively homogenous, as compared to the related genus *Trypanosoma*, techniques that reveal genetic variation at a higher level of resolution are required. Multilocus enzyme electrophoresis (MLEE) has been considered for many years as the reference technique for the identification of *Leishmania* species and subspecies [4]. However, MLEE has drawbacks including the need to cultivate parasites to obtain sufficient amounts of cells for the experiments, as well as the lack of discriminatory power to differentiate the parasites below the species level [5, 6]. On the other hand, molecular approaches based on PCR or other amplification techniques have the advantage of combining high sensitivity for direct detection of the infecting parasites in various human, animal, and sand fly tissues, with the ability to distinguish *Leishmania* parasites at species and intraspecies levels [7]. The PCR-based methods include the amplification and subsequent restriction fragment length polymorphism (RFLP) or DNA sequence analysis of multicopy targets or multigene families (including coding and noncoding regions and PCR fingerprinting techniques), to the recently developed multilocus sequence analysis [8–10] and multilocus microsatellite typing (for review see [11]). These tools have been applied for the identification of the causative agent of leishmaniasis in patient isolates, for epidemiological studies in different foci endemic for the disease, as well as for taxonomic, phylogenetic, and population genetic studies in *Leishmania*.

For phylogenetic studies, differential diagnosis of species by sequencing single-copy gene targets is preferred over methods based on the evaluation of RFLP or fingerprinting patterns, although these latter methods may be useful in epidemiological studies to distinguish between a set of strains known to be circulating in a single focus. Phylogenies based on one gene are often not fully adequate to understand the phylogeny of the Trypanosomatidae or its subgroups, given some instances of recombination, or even different mutation rates between lineages. Instead, several independent genes displaying different evolutionary histories are preferable [12, 13], such as implemented in multilocus sequence analysis (MLSA).

For inferring the phylogenetic relationships and the molecular classification at the level of kinetoplastids, as well as at the *Leishmania* genus, analyses based on DNA sequence comparisons are preferred. Nucleotide sequences or predicted amino acid sequences at specific positions can be used as "characters" in a phylogenetic analysis [14]. As these characters (nucleotides) are the basic units of information encoded in the organism's genome, the potential number of informative characters is enormous. Furthermore, sequence data are highly reproducible and easy to compare between laboratories.

Multilocus sequence typing was initially developed for bacteria [15] and applied in the same manner as MLEE. In the strict bacterial context, short DNA sequences of 300–500 bp for 7–12 gene targets are generated by direct sequencing in both directions. Each sequence is scored as a haplotype, bacteria being haploid; the combination of the haplotypes for all gene targets constitutes the sequence type (ST). Gene targets must be selectively neutral, given that among the relatively small number of genes, a single gene subject to strong positive or negative selection may disrupt phylogenies. In *Leishmania*, different approaches for MLSA have been developed in which case it is the diploid sequence type that is codified, using the codes for ambiguous nucleotides. However, a publicly available database has not yet been created. The *L. (L.) donovani* complex has been studied by using ten loci for gene coding for enzymes used in MLEE [16, 17]. Five of these ten loci plus two additional conserved loci have been used for studying Chinese isolates representing different *Leishmania* species [10], and four of these loci were applied for getting new insights into the taxonomy and phylogeny of *L. (Viannia)* parasites [8]. El-Baidouri et al. have selected seven other independent loci for their MLSA approach which was applied to different Old World species of *Leishmania* [9]. All these MLSA approaches include at least partial sequencing of the selected loci and further phylogenetic analyses of the concatenated sequences. They all indicate that the same gene targets can be used through the *Leishmania* genus and will enable comparisons of genetic distances between the species but also allow to assess the degree of genetic diversity within species.

2.2.2 Molecular Methods for Population Genetic Studies in *Leishmania*

Population genetic approaches depend on highly polymorphic neutral markers that are not affected by natural selection, which must also be co-dominant to permit the detection of all three possible allele combinations in a diploid genome. Multilocus microsatellite typing (MLMT) may meet the criterion of neutrality better than MLSA. Microsatellite sequences are repeated motifs of 1–6 nucleotides that may vary in length due to the gain or loss of single-repeat units during DNA replication. This variation can easily be detected after amplification with primers annealing specifically to their flanking regions. Microsatellite markers are prone to homoplasy and the evolutionary history of a particular repeat may be uncertain. All analyses should therefore include a panel of 10–20 unlinked sequences to overcome this main obstacle in the use of microsatellite markers. Microsatellite markers have been found to be largely species-specific in *Leishmania* [18, 19] and therefore, MLMT is not suited for phylogenetic studies. In fact, comparison of DNA-based methods of strain typing shows that MLMT and PCR-RFLP of kinetoplast (k) DNA minicircles are most useful to discriminate *Leishmania* parasites at intraspecies level, with both of these methods allowing a fine-grained characterization of parasite diversity, for example, in demonstrating genetic links between remote populations of *L. (L.) infantum* and *L. (L.) donovani* [20, 21]. Given that kDNA PCR-RFLP is not co-dominant and its results are difficult to reproduce and to compare between laboratories, MLMT appears to be the current method of choice for population genetic studies in *Leishmania*.

2.2.3 Next-Generation Sequencing Used for Interspecific and Intraspecific Differentiation in *Leishmania*

New and increasingly cheaper high-throughput sequencing technologies that enable fast sequencing of large numbers of genes have opened the door for genome-wide multilocus genotyping between and within *Leishmania* species. Since the publication of the first *Leishmania* reference genome of *L. (L.) major* [22], reference genomes have been published for many other species, such as *L. (L.) infantum* and *L. (V.) braziliensis* [23], *L. (L.) donovani* [24], *L. (L.) mexicana* [25], *L. (L.) amazonensis* [26], *L. (V.) panamensis* [27], and the *Sauroleishmania L. (S.) tarentolae* [28] and *L. (S.) adleri* [29]. These data, together with unpublished assemblies for many other species—and even multiple strains for some species—are available on the kinetoplastid genome database, TriTrypDB (http://tritrypdb.org) [30]. The quality of these assemblies varies in terms of how completely they represent the true genome sequences and how contiguous they are. New sequencing technologies are now being used in *Leishmania* that allow generation of very high-quality genome assemblies more easily and from much longer sequencing reads [31], and improved genome assemblies for many species are likely to be available soon. Accurate annotation of genes and other genome features is required for making these resources useful to the research community. Although consistent, high-quality automated annotation is now possible [32], and manual inspection and improvement of annotation is still critical, particularly in ensuring that genome resources accurately reflect findings from the literature. Comparison of different leishmanial genomes revealed a remarkable conservation of gene content and synteny in orthologous chromosomes [23, 33]. Using whole-genome information for different species of *Leishmania,* MLSA could be, thus, extended to several hundreds of gene targets [34].

Next-generation sequencing allows analyses of different mutation types, such as single nucleotide polymorphisms (SNPs), insertion/deletions (indels), chromosome copy number variations (aneuploidy), and gene copy number variations (CNVs). So far, most studies in *Leishmania* have focused on analyzing SNP data which, depending on the selection criteria, can differentiate parasites at the interspecies and/or intraspecies levels (see paragraph 4.2 for more details).

Sterkers et al. [35, 36] reported that in *L. (L.) major*, chromosomal content varies not only from strain to strain but also from cell to cell creating "mosaic aneuploidy". This leads to high karyotypic plasticity and conserved intra-strain genetic heterogeneity combined with loss of heterozygosity per cell. Next-generation sequencing has confirmed the existence of remarkable chromosome copy number variations and mosaic aneuploidy for parasites belonging to the same or closely related *Leishmania* species [24, 37, 38]. Recently, Dumetz et al. [31] reported dynamic changes of aneuploidy during the parasite's life cycle. Whereas chromosome copy numbers were highly variable in a strain during in vitro cultivation, smaller yet consistent karyotype changes were noticed after a passage through a sand fly, and aneuploidy dropped significantly in a strain-specific manner in hamster amastigotes. As a consequence, all DNA-based typing methods employed earlier have the problem that they cannot decide if a cell population (or strain) consists of heterozygous cells or of homozygous cells presenting different allelic and ploidy content. Approaches to study the genomes of single cells are now available but have not yet, to our knowledge, been applied to *Leishmania* [39].

2.2.4 The Importance of Sampling for Phylogenetic and Population Genetic Studies

Sampling is crucial for phylogenetic as well as for population genetic analyses and depends on the question(s) to be addressed. None of the phylogenetic and population genetic studies published so far in *Leishmania* meet all the requirements for optimal sampling, although more recent studies increasingly try to do so. For phylogenetic inference, parasites should be ideally sampled from the whole range of geographical distribution, but most studies have analyzed only one or a few strains per species, normally reference strains that are kept in cryobanks and have been subcultured many times. Population genetic studies often suffer from the drawback that for some geographic areas, only few isolates are available. Analyzing parasites at a finer geographical scale, using sufficient numbers of isolates, has been shown to be necessary for the detection of hidden substructures within the *Leishmania* species [40]. In addition, *Leishmania* spp. pathogenic for humans are, for understandable reasons, usually overrepresented in the sample collections. It is urgently needed to include more flagellates that are collected from animal hosts or insect vectors or even asymptomatic hosts into phylogenetic and population genetic studies. The availability of parasite isolates in promastigote culture is essential for in-depth study of phenotypic differences between strains, but *Leishmania* parasites can be difficult to isolate [41], and therefore, the use of direct applications in host tissues should be preferred for molecular epidemiological and population genetic studies.

2.3 Molecular Evolution and Origin of the Genus *Leishmania*

Six basic groups of eukaryotes, similarly to the traditional "kingdoms", have been recognized in the new classification system by Adl et al. [1], and the genus *Leishmania* has been assigned to the supergroup Excavata. While groups at this highest taxonomic level share few distinguishing features, and are largely based on molecular data, excavates are ancestrally flagellated protozoa feeding on small particles via a feeding groove. *Leishmania* are kinetoplastid parasites belonging to the Trypanosomatidae (Table 2.1).

2.3.1 Molecular Phylogeny of Kinetoplastids

Kinetoplastids constitute a remarkable group of morphologically rather simple unicellular organisms that share several unusual features in their genomes. The most prominent unique structure is the kinetoplast DNA, a massive network of thousands of topologically interlocked DNA circles of two types, mini- and maxicircles, corresponding to mitochondrial DNA [44]. Other unique features include mitochondrial RNA editing of the uridine insertion/deletion type, trans-splicing of nuclear-encoded mRNA transcripts, intron poverty, presence of

Table 2.1 Taxonomic position of the genus *Leishmania* according to the classification by Adl et al. [1] and Jirku et al. [42] emended by Maslov and Lukeš [43]

Supergroup	Excavata Cavalier-Smith 2002, emend. Simpson 2003 (P?)
Phylum	Euglenozoa Cavalier-Smith 1993, emend. Simpson 1997
Class	Kinetoplastea Honigberg, 1963 emend. Vickerman 1976
Subclass	Metakinetoplastina Vickerman, 2004 (R)
Order	Trypanosomatida Kent, 1880 stat. nov. Hollande, 1952
Family	Trypanosomatidae Doflein, 1951
Subfamily	Leishmaniinae Maslov and Lukeš, 2012 emend. Shaw, Texeira and Camargo 2016
Genus	*Leishmania* Ross, 1903

(P?) Possibly paraphyletic; (R) group identified by small subunit (SSU) rRNA

Table 2.2 The morphology and hosts of Trypanosomatida modified after Stevens et al. [45]

	Genus	Morphology	Hosts	Vectors
Monoxenous	*Blastocrithidia*	Epimastigote, amastigote, cyst	Insects, ticks	
	Crithidia	Choanomastigote	Insects	
	Herpetomonas	Promastigote, opisthomastigote	Insects	
	Leptomonas	Promastigote, cyst	Insects, ciliates nematodes	
	Rhynchoidomonas	Trypomastigote—no undulating membrane	Diptera	
	Sergeia	Promastigote	Diptera	
	Blechomonas	Promastigote	Siphonaptera	
Heteroxenous	*Endotrypanum*	Amastigote, promastigote, epimastigote, trypomastigote	Sloths	Sandflies
	Leishmania	Amastigote, promastigote	Mammals, lizards	Sandflies
	Phytomonas	Promastigote	Flowering plants	Hemiptera
	Trypanosoma	Amastigote, epimastigote, trypomastigote	Vertebrates	Arthropods, leeches

hypermodified base J, and arrangement of genes in large polycistronic clusters [13]. The kinetoplastid species show a variety of life styles ranging from ubiquitous free-living organisms (some bodonids), through ecto- and endoparasites of fish (e.g., *Cryptobia*, *Trypanoplasma*, *Ichthyobodo*) to obligatory parasites of invertebrates, vertebrates, and plants (for review see [13, 43]). The species parasitizing plants (*Phytomonas*), insects (*Crithidia*, *Herpetomas*, *Leptomonas*, *Blastocrithidia*, *Rhynchoidomonas*, *Strigomonas*, *Angomonas*, *Sergeia*, *Blechomonas*, *Paratrypanosoma*), fish, amphibians, and reptiles (*Trypanosoma*) or mammals (*Trypanosoma*, *Leishmania*, *Endotrypanum*) are comprised in the Trypanosomatidae (Table 2.2) [46].

Originally, the taxonomy of kinetoplastids was based on their morphology and life cycles. With the initial molecular biological studies, it became clear that the so far existing taxonomy does not reflect the true genetic relationships of these organisms. These early molecular phylogenetic studies suffered, however, (a) from inappropriate sampling (i.e., mainly medically important trypanosomatids were included in the analysis and the diverse bodonids were ignored) and (b) from troubles with the first gene target sequence used (the SSU rRNA gene of kinetoplastids have several large fast-evolving regions which, if not removed, lead to artifacts in tree construction and, if removed, result in a faint phylogenetic signal in the alignments obtained).

A broad sampling of kinetoplastid diversity and the introduction of additional informative markers like heat-shock proteins (HSPs) and glycosomal glyceraldehyde-3-phosphate dehydrogenases (gGAPDH) revealed that the Kinetoplastea consist of Prokinetoplastina and Metakinetoplastina. The former brings together intracellular endosymbionts of fish-pathogenic amoeba, *Perkinsela* spp., and ectoparasites of fish, *Ichthyobodo* spp. [46, 47]. The latter clade is further subdivided into four subclades: the trypanosomatids (Trypanosomatida) and three clades of bodonids (Neo-, Eu-, and Parabodonida) [48]. Branching of the trypanosomatids from within the bodonids is now strongly supported [49]. As a whole, the wealth of new sequence data makes the old division of the kinetoplastids into bodonids and trypanosomatids artificial. The trypanosomatids were considered to be most closely related to the mostly free-living Eubodonida (e.g., *Bodo saltans*), a result which is congruent with an earlier study based on a partial mitochondrial DNA sequence [50]; however, the recently described *Paratrypanosoma confusum* represents the most basal branching trypanosomatid, which likely retains numerous ancestral features [49].

The current molecular phylogeny of the Trypanosomatidae is mainly based on the analyses of SSU rRNA genes [51–56] and gGAPDH genes, although to a lesser extent [53, 54, 57, 58], as well as the spliced-leader (SL) RNA gene [59]. Neither gene is, however, suitable for inferring a robust phylogeny across the entire family, and additional phylogenetic markers should be used for the trypanosomatids, such as DNA and RNA polymerase genes [60]. The current picture that has emerged from SSU rRNA and gGAPDH genes is that the genus *Trypanosoma* represents a large monophyletic clade in a sister-group relationship with the rest of the family (e.g., see Fig. 2.1). The monoxenous lineages of insect parasites currently assigned to the genera *Blastocrithidia*, *Crithidia*, *Leptomonas*, *Sergeia*, and *Wallaceina* are intermingled with dixenous lineages of parasites of mammals or reptiles (*Leishmania*) and plants (*Phytomonas*). Only *Leishmania* and *Phytomonas* form monophyletic clades, whereas all monoxenous flagellate genera have been found to be paraphyletic and widely interspersed in the phylogenetic trees.

In the SSU rRNA tree (Fig. 2.1), the root is located between the clades of trypanosomes and "non-trypanosomes". However, the recently discovered *Paratrypanosoma confusum* likely constitutes the most basal flagellate that acquired the parasitic life style [49]. While it is difficult to rigorously exclude dixenous life cycle, the available data strongly point to the fact that *P. confusum* is a monoxenous parasite of dipteran insects (Skalický et al. unpubl. data). The branching order of the main clades within Trypanosomatidae is not well supported; hence, more data is needed to confirm the basal branching of Blechomonas [64].

Monoxenous trypanosomatids of insects are not only extremely diverse but developed distinct life strategies. One clade represented by the genera *Strigomonas* and *Angomonas* invariably contains endosymbiotic bacteria in their cytoplasm [65]. It was shown that all bacteria parasitizing these globally distributed trypanosomatids are derived from a single acquisition event of a betaproteobacterium by a flagellate [66] that developed into a tight endosymbiotic relationship involving targeting of proteins from one partner into another

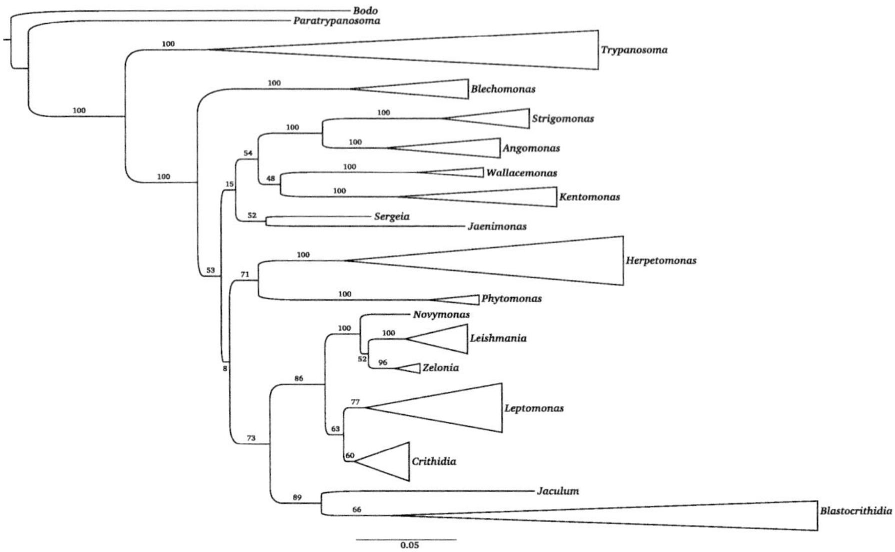

Fig. 2.1 Sequences for phylogenetic analyses were received from publicly available sources for both SSU rRNA and gGAPDH genes. The datasets for each gene were aligned by MUSCLE [61] separately and selection of relevant positions with subsequent concatenation was performed using Gblocks [62]. Phylogenetic model selection with ModelGenerator using four Γ rate categories favored GTR+Γ model and ML trees were constructed using RAxML 8.27 [63] with 1000 bootstrap replicates

[67]. Interestingly, a trypanosomatid in hemipteroid bugs captured in Ecuador hosts yet another bacterium that was acquired in an independent endosymbiotic event [68].

Another example of unique features being found in monoxenous trypanosomatids is the case of *Blastocrithidia* sp. In an unprecedented step, this flagellate repurposed all three stop codons into sense codons, and its translation machinery, therefore, has to distinguish between a multitude of in-frame stop codons and the genuine one that indeed terminates its genes [69]. These two examples demonstrate that trypanosomatids in insects constitute a group of dexterous parasites capable of altering under certain conditions their molecular and biochemical capacities.

Therefore, it is not surprising that the monoxenous parasites underwent repeated transitions to dixenous parasitism [13, 70], at least once in the *Trypanosoma* clade and once in each of the lineages leading to *Leishmania*/*Endotrypanum* and *Phytomonas*. The phylogenetic position of *Leishmania* within insect trypanosomatids as a relatively late emerging group, supports the classical "insect-first" hypothesis postulating that dixenous parasites evolved from primary insect parasites via acquisition of hematophagy [71, 72]. The discovery of two larval sand flies in Early Cretaceous Burmese amber parasitized by trypanosomatids led to the hypothesis that these protists were ingested by sand fly larvae, carried through the pupal and into the adult stage and introduced into a vertebrate during blood feeding [73]. The establishment of trypanosomatids in the vertebrate and subsequent reacquisition by sand flies finally resulted in a dixenous life cycle. Some infections of animals and humans, often immunosuppressed patients, with monoxenous trypanosomatids have been reported recently showing that acquisition of mammals as hosts by primarily insect flagellates is not a rare event of the past [74–77].

2.3.2 Molecular Phylogeny of the Genus *Leishmania*

The first phylogenetic trees of the genus *Leishmania* were based on MLEE data analyzed by phenetic and cladistic techniques [4, 78, 79]. These analyses confirmed, at the time, the monophyletic origin of the genus and its subdivision into two subgenera: *L. (Leishmania)* comprising all species from the Old World (OW), *L. (L.) mexicana* and *L. (L.) amazonensis* from the New World (NW), and *L. (Viannia)* consisting of only NW species. The lizard species were, however, excluded from these studies because the *Sauroleishmania* were then considered to be a separate genus. A concept of species complexes was proposed and later modified to group *Leishmania* species based on biological and biochemistry characteristics [72, 80]. The validity of this classification began to be questioned when the species status of some representatives of both *L. (Leishmania)* and *L. (Viannia)* subgenera as well as the concept of species complexes as a whole [81] were not supported by molecular analyses. In addition, recently discovered putative new species may belong to separate groups (for more details see paragraph 3.3).

As for the kinetoplastids, the SSU rRNA gene and mitochondrial gene sequences are most widely used for the inference of deep phylogenetic relationships within the genus *Leishmania*. The variation in the SSU rRNA gene was, however, insufficient to robustly resolve any internal branching within *Leishmania* [82], and the extensive editing of most mitochondrial genes in *Leishmania* [83] may cause problems in phylogenetic studies.

During the past 20 years, several DNA sequences have been used to investigate the phylogeny of the genus *Leishmania*. These have included single-copy genes encoding the catalytic polypeptide of DNA polymerase α (*pol*A) [60], the largest subunit of RNA polymerase II (*rpo*IILS) [60], the 7SL RNA gene [84], the noncoding multicopy ribosomal internal transcribed spacer (ITS) [85–87], the *N*-acetylglucosamine-1-phosphate transferase (NAGT) gene [88], the mitochondrial cytochrome *b* gene (*cyt*b) [89], and, more recently, sequences of the heat shock protein 70 gene (*hsp*70) subfamily [90]. Sequence analyses of these different targets have been consistent in that the subgenera *L.* (*Leishmania*) and *L.* (*Viannia*) each forms a distinct monophyletic clade and that the OW and NW species are separated within the *L.* (*Leishmania*) subgenus (Figs. 2.2 and 2.3). When *Sauroleishmania* were included, they branched off in between the *L.* (*Leishmania*) and *L.* (*Viannia*) subgenera as an independent taxon. This result suggests that lizard-hosted *Leishmania* might be derived from mammalian parasites [60] and that they should be regarded as a subgenus of *Leishmania* rather than an independent genus [91]. However, RNA and DNA polymerase genes were shown to evolve faster in the lizard *Leishmania* than in the mammalian *Leishmania* making it difficult to define the exact

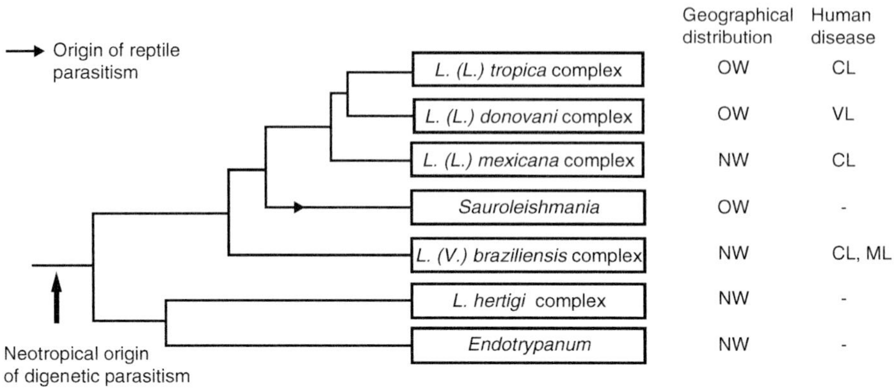

Fig. 2.2 Schematic tree showing the evolution of the *Leishmania*/*Endotrypanum* subtree of the Kinetoplastida based on POLA/ROPIILS nucleotide sequences (Croan et al. [60]). The *L. (L.) tropica* complex, as shown here, comprises sequences of *L. (L.) tropica*, *L. (L.) major*, *L. (L.) aethiopica*, and *L. (L.) arabica*; *L. (L.) donovani* complex those of *L. (L.) donovani* and *L. (L.) infantum*; *L. (L.) mexicana* complex those of *L. (L.) mexicana* and *L. (L.) amazonensis*; *L. (V.) braziliensis* complex those of *L. (V.) braziliensis* and *L. (V.) panamensis*; *L. hertigi* complex those of *L. hertigi* and *L. deanei*; and *Endotrypanum* those of *E. monterogeii* and *L. herreri*. *Sauroleishmania* were represented by the species *L. hoogstraali*, *L. tarentolae*, *L. adleri*, and *L. gymnodactyli*. For each taxon, an indication of the geographical distribution (OW, Old World; NW, New World) and typical disease pathology (CL, cutaneous; VL, visceral; MC, mucocutaneous) observed following infection is shown on the right. (Reprinted from Croan et al. [60] © 1997, with permission from Elsevier)

taxonomic position of lizard parasites [60]. In all studies, the *L.* (*Viannia*) subgenus was closest to the root, while *L.* (*Leishmania*) and *L.* (*Sauroleishmania*) formed the crown of the trees.

Based on a variety of molecular criteria, Cupolillo et al. [92] have proposed the separation of the genus *Leishmania* into two sections: Euleishmania comprising the subgenera *Leishmania*, *Sauroleishmania*, and *Viannia*, and Paraleishmania consisting of *L. hertigi*, *L. deanei*, *L. colombiensis*, *L. equatoriensis*, *L. herreri*, and strains of *Endotrypanum*. *L. hertigi*, and *L. deanei* have only been found in Neotropical porcupines and an unknown sand fly vector, and do not, or only transiently, infect humans [93]. *L. herreri* was isolated from sloths and different sand fly species in Costa Rica [94]. Comparison of DNA and RNA polymerase sequences [60] as well as PCR-RFLP of the SSU rRNA gene [93, 94] revealed that these three species are closely related to *Endotrypanum*, a parasite of Neotropical tree sloths. In the resulting trees, these species represented the most basal branches.

Several *Leishmania* isolates have been described that could not be assigned to any of the known species. Noyes et al. [95] identified a parasite, *L. martiniquensis*, isolated from human cutaneous lesions in Martinique by MLEE and sequencing of different targets, as the most divergent member of the genus *Leishmania*. Recently, a new species of *Leishmania* has been reported from a focal CL outbreak in Ghana [96] as well as from VL cases in immunocompetent and immunosupressed patients

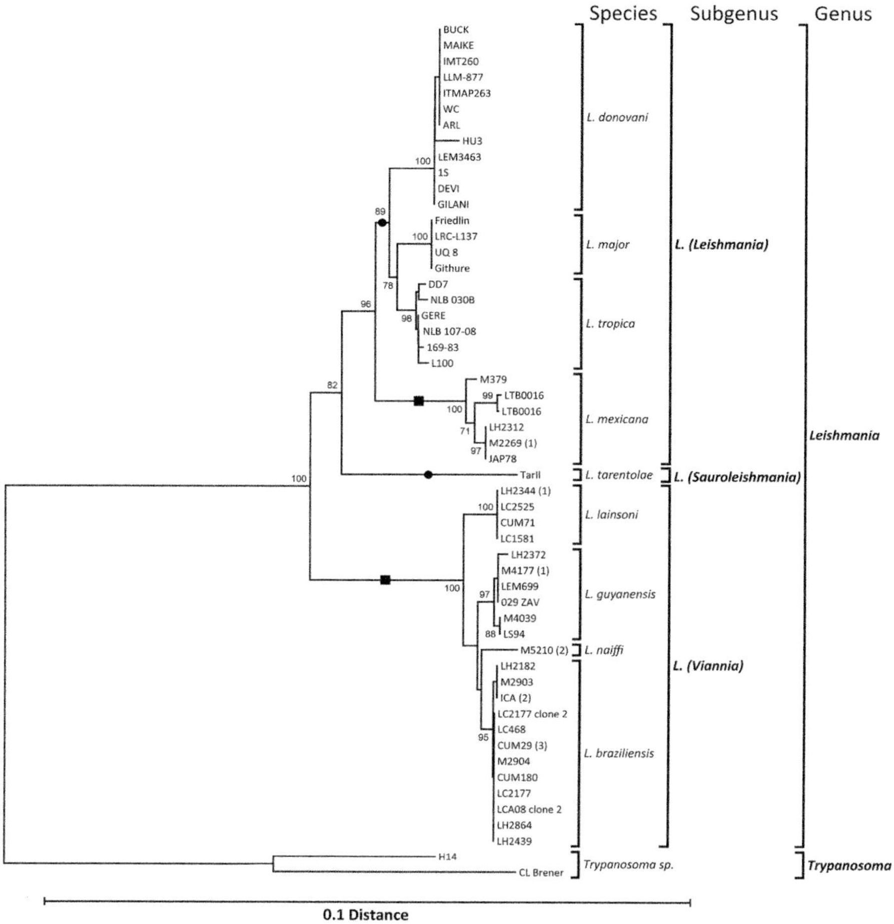

Fig. 2.3 Neighbor-joining phylogeny of hsp70 sequences of 52 strains representing 17 *Leishmania* and 2 *Trypanosoma* species, based on an alignment of 1380 nucleotides (Fraga et al. [90]). Distances were estimated using the Kimura-2 parameter model, thereby excluding all 10 sites with ambiguous nucleotides. Bootstrap support of the branches was inferred from 2000 replicates and is given in percentages at the internodes when exceeding 70%. The tree is drawn to the scale at the bottom, expressed as distance per nucleotide. Supported monophyletic species and subgenera are depicted at the right, irrespective of the species classification presented in Table 2.1 but reflecting the observations from Sect. 2.4. Old World clusters are indicated by a dot on the branch leading to the cluster, while a square is used for New World groups. The tree was rooted with the two *Trypanosoma* sequences found most related to *Leishmania* hsp70. Numbers between brackets following the strain names indicate the number of ambiguous nucleotides in the sequence. (reprinted from Fraga et al. [90] © 2010 with permission from Elsevier)

in Thailand [97], named *Leishmania* sp. *siamensis*. In those cases, parasites were, however, not isolated in culture, and the identification was based on microscopy and/or PCR using DNA extracted from clinical samples. Surprisingly, locally acquired CL was also detected in kangaroos, wallaroos, and wallabys, living in captivity in the Northern Territory of Australia, a region that was considered free of *Leishmania* parasites [98, 99]. DNA sequence analyses revealed that the parasites were genetically indistinguishable and possibly represent a novel *Leishmania* species. Autochthonous cases of CL in German and Swiss horses and in a Swiss cow have been associated by DNA sequence analyses with *Leishmania* parasites that could be classified neither as OW nor NW *Leishmania* species but were most closely related to *L.* sp. *siamensis* [100, 101]. Finally, another novel trypanosomatid has been isolated from the native Australian black fly, *Simulium* (*Morops*) *dycei* [102].

Two recent studies have analyzed different DNA sequences such as coding for RNA PolII, HSP70, gGAPDH, and V7V8 SSU rRNA and included not only parasites that fall within the *L. (Leishmania)*, *L.* (*Viannia*), and *L. (Sauroleishmania)* but also parasites earlier classified as Paraleishmania and, so far, unclassified ones. Based on their results, Espinosa et al. [103] propose a taxonomic revision of the trypanosomatids currently known as *Leishmania* and *Endotrypanum.* They adopt the principle of the subfamily Leishmaniinae within the family Trypanosomatidae [42] and define new genera and subgenera which are supported by their phylogenetic analyses (Table 2.3, see also Box 2.1). Four genera were identified within the new subfamily. The genus *Leishmania* consists of the already known subgenera *L. (Leishmania)* comprising mainly human pathogens from the Old and New Worlds, *L. (Viannia)* including exclusively NW parasites many of which are pathogenic to humans, and *L. (Sauroleishmania)* consisting of reptilian parasites occurring only in the OW. The fourth subgenus, *L.* (*Mundinia)*, was newly created for the *L. enrietti* complex. It also includes *L. (M.) martiniquensis*, the parasites isolated from a kangaroo now defined as the species *L. (M.) macropodum* [102] as well as some so far unnamed parasites. The parasites isolated from Central and South American sloths and transmitted by sand flies that eventually infect humans are assigned to the genus *Endotrypanum.* Three new genera are proposed: *Porcisia* accommodating the NW parasites isolated from porcupines, *Zelonia* comprising trypanosomatids from Neotropical hemipterans [103] and from an Australian black fly [102], and *Novymonas* harboring so far only a monoxenous trypanosomatid from the digestive tract of a hemipteran (Rhopalidae) from Ecuador [68].

The by far greatest number of taxa of the genus *Leishmania* has been included in a phylogenetic analysis based on the *hsp*70 gene [90]. Several strains per species were sequenced trying to cover the geographical distribution of different species. Species of the new subgenus *L. (Mundinia)* were, however, not analyzed. The trees were rooted using *hsp*70 sequences of the two most closely related *Trypanosoma* species (Fig. 2.3). The resulting phylogeny supported the existence of three monophyletic groups representing the subgenera *L. (Leishmania)*, *L. (Sauroleishmania)*, and *L. (Viannia)* and the basal branching of the latter. The two mammalian subgenera include 4 monophyletic clusters each corresponding to a different species or species complex.

Table 2.3 Revised taxonomy of Leishmaniinae parasites [102, 103]. The table was created based on the data taken from Espinosa et al. [103] and Barratt et al. [102]

Sub-family	Leishmaniinae							
Genus	*Leishmania*				*Porcisia*	*Endotrypanum*	*Zelonia*	*Novymonas*
Sub-genus	*L. (Leishmania)*	*L. (Viannia)*	*L. (Sauroleishm.)*	*L. (Mundinia)*				
Species	***L. (L.) donovani***[a]	***L. (V.) braziliensis***[a]	***L. (S.) tarentolae***	***L. (M.) enriettii***	***P. hertigi***	***E. schaudinni***	***Z. costaricensis***	***N. esmeraldas***
	L. (L.) infantum[a]	*L. (V.) peruviana*[a]	*L. (S.) adleri*	*L. (M.) martiniquensis*[a]	*P. deanei*	*E. monterogeii*	*Z. australiensis*	
			L. (S.) agamae	*(syn: "siamensis")*		*E. colombiensis*[a]		
	L. (L.) tropica[a]	*L. (V.) guyanensis*[a]	*L. (S.) ceramodactyli*	*L. (M.) macropodum*		*E. equatoriensis*		
	L. (L.) aethiopica[a]	*L. (V.) panamensis*[a]	*L. (S.) davidi*	*L. (M.) spp. Ghana*		*E. herreri*		
	L. (L.) major[a]		*L. (S.) gulikae*	*L. (M.) spp. Trang (TL)*				
	L. (L.) gerbilli	*L. (V.) lainsoni* [a]	*L. (S.) gymnodactyli*					
	L. (L.) arabica[a]		*L. (S.) helioscopi*					
	L. (L.) turanica	*L. (V.) shawi*[a]	*L. (S.) hemidactyli*					
			L. (S.) hoogstraali					
	L. (L.) mexicana[a]	*L. (V.) naiffi*[a]	*L. (S.) nicollei*					

(continued)

Table 2.3 (continued)

Sub-family	Leishmaniinae							
	L. (L.) amazonensis[a]		*L. (S.) phrynocephali*					
	L. (L.) aristidesi	*L. (V.) lindenbergi*[a]	*L. (S.) platycephala*					
	L. (L.) venezuelensis[a]		*L. (S.) senegalensis*					
	L. (L.) pifanoi[a]	*L. (V.) utingensis*	*L. (S.) sofieffi*					
	L. (L.) waltoni[a]		*L. (S.) smeevi*					
	L. (L.) garnhami[b]		*L. (S.) zuckermani*					
	L. (L.) forattini[b]		*L. (S.) henrici*[c]					
			L. (S.) chamaeleonensis[c]					

Type species are given in bold
[a]Species associated with human leishmaniases
[b]Species status requires conformation
[c]Most probably intestinal flagellates

2.3.3 Molecular Taxonomy of *Leishmania*

Leishmania flagellates have been assigned to different species primarily based on clinical, biological, geographical, and epidemiological criteria and, later, immunological and biochemical data, recently reviewed by Akhoundi et al. [104]. Hierarchical taxonomic schemes have been proposed using the categories of species complexes, species, subspecies, and subgenus. Compared to all other methods, MLEE has been applied to the most varied and largest number of *Leishmania* isolates in the past 25 years. The classification system resulting from the application of numerical taxonomy and cladistic techniques to electrophoretic data [4, 72, 92, 105, 106] has only very recently been replaced by new systems based on analyses of different concatenated DNA sequences [8, 9, 102, 103].

Molecular phylogenies of *Leishmania* have largely confirmed the taxonomy of the genus *Leishmania* by MLEE suggesting, however, that the number of species may be too large. The analysis of Hsp70 sequences identified only nine monophyletic groups which, according to the phylogenetic species concept, represent different species in the subgenera *L. (Leishmania)* and *L. (Viannia)* [90]. This is in good agreement with two studies that have investigated representative sample sets for the OW *L. (Leishmania)* [9] and the NW *L. (Viannia)* [8], respectively, by MLSA. For instance, two of the species earlier included in the *L. (L.) donovani* complex, namely, *L. (L.) chagasi* and *L. (L.) archibaldi*, are not supported by any molecular analyses. Strains of *L. (L.) chagasi* (NW) are undistinguishable from strains of *L. (L.) infantum* (OW) and, in fact, represent South American strains of *L. (L.) infantum* [107, 108]. *L. (L.) archibaldi* could also not be confirmed as a valid species [9, 109]; only a single nucleotide polymorphism in the glutamate oxaloacetate transaminase gene was causing the different MLEE phenotype [5]. Even more, most of the molecular phylogenies did not produce monophyletic groups for the other two species *L. (L.) donovani* and *L. (L.) infantum* [60, 86, 87, 89, 90].

Another discrepancy concerns the status of *L. (L.) killicki*, which was classified as a separate species by MLEE [4] but shown to be *L. (L.) tropica* by molecular analyses [9, 89, 110, 111]. MLSA suggests that *L. (L.) killicki* emerged from a single founder event and evolved independently from *L. (L.) tropica* but does not support a distinct taxon status for *L. (L.) killicki* [112]. According to most of the DNA-based phylogenies, *L. (L.) tropica* cannot be distinguished from *L. (L.) aethiopica* as both form a single cluster [84, 86, 89, 90]. Whether they are different subspecies of the species, *L. (L.) tropica* is debatable and needs to be investigated with a larger number of strains. The MLSA results are rather in favor of a progressive genetic isolation between the clusters representing *L. (L.) tropica* and *L. (L.) aethiopica*, suggesting that the latter might be descendants of ancestral populations that led to the *L. (L.) tropica* cluster [9].

As far as the species of the *L. (L.) mexicana* complex are concerned, most of the DNA-based phylogenies included only strains of *L. (L.) mexicana* and *L. (L.)*

amazonensis. In the hsp70 trees, which include one strain of *L. (L.) garnhami*, none of these species could be distinguished as a monophyletic clade, and *L. (L.) mexicana* was the only recognized species [90]. This is supported by some phylogenetic studies [58, 60, 87] but contrasts with others [84, 86, 89]. More strains of all species of the *L. (L.) mexicana* complex, also including *L. (L.) venezuelensis* and *L. (L.) aristidesi*, representing their whole area of distribution should be studied to evaluate the species or possible subspecies status within this complex.

Four monophyletic groups were clearly observed in an MLSA of 96 strains of the *L. (Viannia)* subgenus representing basically *L. (V.) naiffi* and *L. (V.) lainsoni*, the most divergent groups, and *L. (V.) braziliensis* and *L. (V.) guyanensis* [8]. Strains of *L. (V.) peruviana* always grouped with the strains of *L. (V.) braziliensis* and, thus, did not appear as a discrete typing unit distinct: this is in contrast to the results of MLEE, RAPD, and hsp70 phylogenies [90, 113] but is in agreement with the results of studies based on monoclonal antibodies [114] and analysis of microsatellite variation [115]. The latter study showed that strains of *L. (V.) peruviana* were intermingled with strains of *L. (V.) braziliensis* from Peru and from the Acre State, a Brazilian region bordering Peru. Strains of *L. (V.) guyanensis* and *L. (V.) panamensis* formed in different phylogenetic trees a monophyletic cluster divided into two monophyletic sub-clusters suggesting that there are two subspecies in the species *L. (V.) guyanensis*. The inclusion of *L. (V.) shawi* in the *L. (V.) guyanensis* group was demonstrated by MLEE analysis [105] and corroborated by molecular markers, as PCR-RFLP of ITS rDNA [116] and PCR-RFLP and sequence analysis of the hsp70 gene [8, 117]. Molecular phylogenies have indicated that species status is justified for *L. (V.) naiffi* and that *L. (V.) lainsoni* is a separate and the most divergent species inside the *L. (Viannia)* subgenus [8, 90, 117]. Finally, hsp70 gene analysis [117], MLMT [118] and MLSA [8], showed that the species *L. (V.) lindenbergi* [119] and *L. (V.) utingensis*, the last being represented by only one sample isolated from a *Lutzomyia tuberculata* sand fly, are quite distinct from the other *L. (Viannia)* species, although only the reference strains were analyzed.

In conclusion, the concept of species complexes for grouping *Leishmania* species which was initially proposed based on biological and biochemistry characteristics and later modified [72] is not supported by molecular phylogenies and has been abandoned in the newly defined classification scheme (Box 2.1) where the assignment to major groups across the entire genus *Leishmania* was based on gene sequences. For classification within the major groups, highly discriminatory markers such as MLST, microsatellites, or genome-wide SNPs are probably better suited.

Box 2.1 Revised classification and nomenclature of Leishmaniinae species based on molecular phylogenies of the organisms [102, 103]. Type species are underlined

The subfamily Leishmaniinae (Maslov and Lukes in [42]) consists of the genera *Leishmania, Porcisia, Endotrypanum, Zelonia* and *Novymonas.*

GENUS *LEISHMANIA* ROSS 1908

Subgenus *L.* (*Leishmania*) Safjanova, 1982 consists of the following Old World (OW) and New World (NW) species:

OW: *L. (L.) donovani* Laveran & Mesnil, 1903; *L. (L.) infantum* Nicolle 1908; *L. (L.) tropica* Wright, 1903; *L. (L.) aethiopica* Bray, Ashford & Bray, 1973; *L. (L.) major* Yakimoff & Shokhor, 1914; *L. (L.) gerbilli* Wang, Qu & Guan, 1964; *L. (L.) arabica* Peters, Elbihari & Evans, 1986; *L. (L.) turanica* Strelkova et al. 1990;

NW: *L. (L.) infantum* Nicolle 1908 (syn. *L. (L.) chagasi* Cunha & Chagas, 1937); *L. (L.) mexicana* Biagi, 1953; *L. (L.) amazonensis* Lainson & Shaw, 1972; *L. (L.) aristidesi* Lainson & Shaw, 1979; *L. (L.) venezuelensis* Bonfante-Garrido, 1980; *L. (L.) pifanoi* Medina & Romero, 1959; *L. (L.) waltoni* Shaw, Pratlong & Dedet, 2015.

Species status not yet confirmed: *L. (L.) garnhami* Scorza et al. 1979; *L. (L.) forattinii* Yoshida et al 1993.

Subgenus *L.* (*Sauroleishmania*) Ranque, 1973 consists of reptilian parasites only found in the Old World:

L. (S.) tarentolae Wenyon 1921; *L. (S.) adleri* Heisch 1954; *L. (S.) agamae* David 1929; *L. (S.) ceramodactyli* Adler & Theodor 1929; *L. (S.) davidi* Strong 1924; *L. (S.) gulikae* Ovezmuchammedov & Safjanova 1987; *L. (S.) gymnodactyli* Khodukin & Sofiev 1929; *L. (S.) helioscopi* Khodukin & Sofiev 1940; *L. (S.) hemidactyli* Mackie et al. 1923; *L. (S.) hoogstraali* McMillan 1965; *L. (S.) nicollei* Khodukin & Sofiev 1940; *L. (S.) phrynocephali* Khodukin & Sofiev 1940; *L. (S.) platycephala* Telford 2008; *L. (S.) senegalensis* Ranque 1973; *L. (S.) sofieffi* Markov et al. 1964; *L. (S.) zmeevi* Andruchko & Markov 1955; *L. (S.) zuckermani* Paperna et al. 2011;

Species status not yet confirmed: *L. (S.)* sp. I Telford 1979; *L. (S.)* sp. II Telford 1979.

(continued)

Box 2.1 (continued)

Subgenus *L. (Viannia)* Lainson & Shaw, 1987 consists of species exclusively endemic in the New World:

L. (V.) braziliensis Vianna, 1911; *L. (V.) peruviana* Velez, 1913; *L. (V.) guyanensis* Floch, 1954; *L. (V.) panamensis* Lainson & Shaw, 1972; *L. (V.) shawi* Lainson et al. 1989; *L. (V.) lainsoni* Silveira et al. 1987; *L. (V.) naiffi* Lainson & Shaw, 1989; *L. (V.) lindenbergi* Silveira et al. 2002; *L. (V.) utingensis* Braga et al. 2003.

Subgenus *L. (Mundinia)* Shaw, Camargo & Texeira 2016 consists of worldwide distributed species:

L. (M.) enrietti Muniz & Medina 1948; *L. (M.) martiniquensis* Desbois et al. 2014 (syn. *L. siamensis*); *L. (M.) macropodum* Barratt et al. 2017; *L. (M.) spp.* Ghana [MHOM/GH/2012/GH5] (LV757; *L. (M.) spp.* Trang, Thailand [MHOM/TH/2012/PVM2].

GENUS *PORCISIA* SHAW, CAMARGO & TEXEIRA 2016

Consists of parasites occurring in porcupines in the NW, previously assigned to the *L. hertigi* complex:

P. hertigi Herrer, 1971; *P. deanei* Lainson & Shaw 1977.

GENUS *ENDOTRYPANUM* MESNIL & BRIMONT 1908

Consists of NW parasites from sloths, sand flies and humans:

E. schaudinii Mesnil & Brimont 1908; *E. monterogeii* Shaw 1969; *E. colombiensis* Kreutzer et al. 1991, *E. equatorensis* Grimaldi jr. et al. 1992; *E. herreri* Zeledon, Ponce & Murillo, 1979.

GENUS *ZELONIA* SHAW, CAMARGO & TEXEIRA 2016

Consists of parasites obtained from predatory hemipterans in the NW Equatorial regions:

Z. costaricensis Yurchenko et al. 2006; *Z. costaricensis* strain G755 Noyes et al. 2002; *Z. costaricensis* strains TCC169E, 504 and 2696 Espinosa et al. 2016; *Z. australiensis* Barratt et al. 2017.

GENUS *NOVYMONAS* KOSTYGOV & YURCHENKO 2016

Contains so far only a trypanosomatid from the digestive tract of a hemipteran (Rhopalidae) from Ecuador:

N. esmeraldas Votypka, Kostygov, Maslov & Lukeš 2016.

2.4 Intraspecific Differentiation in *Leishmania*

2.4.1 Multilocus Microsatellite Typing (MLMT)

Very recently, MLMT and, to a lesser extent, MLSA provided evidence for considerable genetic structure for different *Leishmania* species at the intraspecies level. So far, microsatellite loci with high discriminatory power and suitable for characterizing closely related strains have been published for population studies in the *L. (L.) donovani* complex [120, 121], *L. (L.) major* [18, 122], *L. (L.) tropica* [110], and for species of the subgenus *L. (Viannia)* [115, 123, 124]. Furthermore, a searchable database of microsatellite loci within the genome has been established at http://www.genomics.liv.ac.uk/tryps/Microsatellites.V1.html, which allows the development of additional microsatellite markers for the *L. (L.) donovani complex*, *L. (L.) major*, and *L. (V.) braziliensis* [125]. Indeed, novel microsatellite loci can be successfully identified from even very incomplete draft genome assemblies (e.g., [126, 127] and microsatellite polymorphisms at those loci inferred from sequence data [128]) before further testing.

Most of the MLMT studies published so far have addressed epidemiological and population genetic questions related to the *L. (L.) donovani* complex. When strains of *L. (L.) donovani* and *L. (L.) infantum* isolated from the main regions endemic for VL were investigated, seven main genetically distinct populations were identified: three populations of *L. (L.) infantum* from the Mediterranean area, South America, and Asia comprising the strains representing the zymodeme (MLEE type) MON-1 (v, vi, and vii) and one (iv) of other zymodemes (taken together as non-MON-1), as well as two populations of *L. (L.) donovani* from East Africa (ii and iii) and one of *L. (L.) donovani* MON-2 from India (i) (Fig. 2.4). The highest microsatellite diversity was observed for *L. (L.) infantum* from the Mediterranean Basin. MLMT distinguished strains of *L. (L.) infantum* belonging to the predominating isoenzyme type, MON-1, and revealed the existence of genetically different populations, often with geographical associations, on different hierarchical levels [6, 129–134]. MON-1 strains from the Western Mediterranean differed from those of the Eastern Mediterranean and North Africa. Different genetic groups within MON-1 strains were also prevalent in the Spanish mainland and the Balearic Islands, respectively, as well as in Israeli and Palestinian foci. MLMT detected gene flow between different populations of *L. (L.) infantum* and hybrids between populations representing different zymodemes. New World strains of *L. (L.) infantum* (syn. *L. (L.) chagasi*) were less diverse and most closely related to the strains from southwest Europe [108].

Microsatellite diversity was also remarkable for East African strains of *L. (L.) donovani*. Two main populations have been identified, one comprising strains from Sudan and northern Ethiopia and the other strains from southern Ethiopia and Kenya, which were both further divided into two subpopulations. The presence of two geographically and genetically isolated populations of *L. (L.) donovani* in Ethiopia is supported by differences in clinical behavior and biology of the strains from the two foci [135]. Four putative hybrids detected in this study were retyped

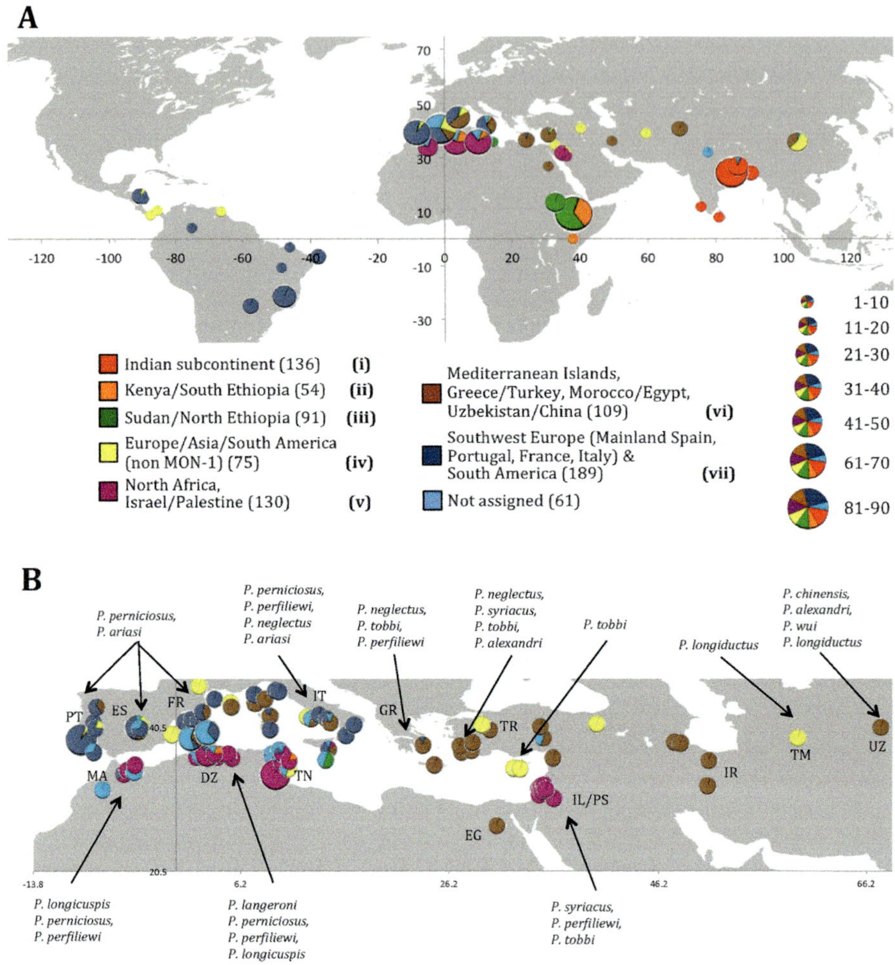

Fig. 2.4 (**a**) Geographical distribution of the identified population clusters inferred by MLMT for 845 individual isolates, of which 784 were clearly assigned to clusters (i) to (vi) and 61 were "not assigned" (Stark, Schönian et al. unpublished data). Colors refer to the population assignments obtained by BCA. Pie chart sizes are classified to illustrate the relative frequency of samples in the respective focus. (**b**) Focus on the Mediterranean Basin. Arrows indicate the phlebotomine sand fly species present in these regions. Country abbreviations are *PT* Portugal, *ES* Spain, *FR* France, *IT* Italy, *GR* Greece, *TR* Turkey, *IR* Iran, *TM* Turkmenistan, *UZ* Uzbekistan, *IL* Israel, *PS* Palestine, *EG* Egypt, *TN* Tunisia, *DZ* Algeria, *MA* Morocco

using a combination of MLMT and MLSA and shown to be true genetic hybrids. Each of them possessed heterozygous markers consistent with inheritance of divergent alleles from genetically different Ethiopian *L. (L.) donovani* lineages [136]. MLMT of 124 Sudanese strains of *L. (L.) donovani* revealed significant genetic diversity, minor structuring between years, and highlighted the role of dogs as important local reservoirs of visceral leishmaniasis [137]. In contrast, strains of *L. (L.) donovani* from the Indian subcontinent were surprisingly homogenous with over 80% of the strains tested sharing an identical MLMT profile regardless of their geographical origin, clinical manifestation, and whether they presented in vitro or in vivo susceptibility to antimonial drugs (Sb^V) [138, 139].

MLMT exposed three main populations of *L. (L.) major*, in Central Asia, the Middle East, and Africa [122]. Studies in Iran and Pakistan describe three and two clusters of *L. (L.) major*, respectively, most of which differ from the three main populations [140, 141]. This might be related to the existence of different transmission cycles involving different vector and/or reservoir host species. In Tunisia, MLMT of *L. (L.) major* revealed genetically differentiated populations of the parasites which spread according to a geographical gradient most probably resulting from human activities [142].

The existence of genetically different populations with geographical associations was also shown for *L. (L.) tropica* [110]. The population structure of *L. (L.) tropica* was found to be more complex, with genetically isolated sympatric populations in rather small territories, e.g., in Israel and the Palestinian Authority, and in Morocco, and the emergence of new variants and foci. In Israel and the Palestinian Authority, increased prevalence of human CL could be linked by MLMT to the recent emergence of genetically similar strains of *L. (L.) tropica* [143]. The expansion of these strains seems to result from the reemergence of a previously existing genotype. A second cluster of strains in this study differed from all other *L. (L.) tropica* in their serological, biochemical, and molecular parameters and by the involvement of a new vector species and was closely related to African strains of *L. (L.) tropica*. More strains from this area fell into different genetic entities mostly related to Asian strains of *L. (L.) tropica*. These many locally encountered genetic variants in the Israeli-Palestinian region have been, most likely, imported during numerous migrations of humans and, eventually, infected animal reservoirs from the past until now. Moroccan strains of *L. (L.) tropica* were separated into two phylogenetic clusters independent from their geographical origin [144]. MLMT has, thus, confirmed the intrafocal distribution of genetic variants of *L. (L.) tropica* observed earlier in MLEE studies [145]. Indian strains of *L. (L.) tropica* regardless whether they were isolated from human cases of CL or VL grouped always together and with strains from other Asian foci [146]. The dermatotropic and viscerotropic strains were, however, not genetically identical. Whether this reflects their different pathogenicities remains to be established. A recent MLMT study of *L. (L.) killicki* (syn. *L. (L.) tropica*) supports its assignment to the *L. (L.) tropica* complex and reveals strong structuring in the parasites between Tunisia and Algeria and within different Tunisian regions, suggesting low dispersion of these parasites [147]. MLMT of *L. (L.) aethiopica* confirmed their close relationship to *L. (L.) tropica* but was unable to answer the question whether these parasites represent two separate species or rather different variants of the same *Leishmania* species [148].

MLMT analyses have been applied to different sample sets of the *L. (Viannia)* subgenus. Variation in 15 microsatellite markers has been studied in 120 strains of *L. (Viannia)* from different Brazilian foci of CL [118]. The strains of *L. (V.)*

braziliensis isolated along the Atlantic coast and those of *L. (V.) guyanensis*, mainly from the Amazonas region, formed two clearly separated populations both exhibiting significant levels of recombination. MLMT identified an epidemic clone inside the Atlantic coast population consisting of 13 strains from a CL outbreak in Minas Gerais. Strains of *L. (V.) braziliensis* from northern Brazil did not group with those from the Atlantic coast but were found to be very polymorphic. They seemed to be more closely related to strains of other subgenus *L. (Viannia)* species, such as *L. (V.) shawi*, *L. (V.) naiffi*, and *L. (V.) lainsoni*, also isolated in northern Brazil CL foci. Bias due to inadequate sampling strategies cannot be excluded for the analyses of strains derived from northern Brazil. Using the same MLMT approach, high genetic diversity, with multilocus genotypes strongly differentiated from each other, were observed for 24 strains of *L. (V.) braziliensis* from Peru [149]. The sample set consisted of strains for which the in vitro susceptibility toward antimonial drug or the clinical treatment outcome was known. No correlation could, however, be found between genotypes and resistance phenotypes.

The MLMT of Bolivian and Peruvian *L. (V.) braziliensis* revealed a strong population structure at a microgeographical scale as the populations within the different countries were genetically heterogenous [40]. The substantial heterozygote deficiency and extreme inbreeding found in this study is not consistent with strict clonal reproduction as previously proposed [150] but rather point to frequent sexual crosses of genetically related parasites or even of individuals from the same strain (endogamy). A high level of sexual recombination and substantial endogamy together with strong Wahlund effects (sampling strains from different subpopulations) was reported in a study of 153 strains of *L. (V.) guyanensis* from French Guyana investigating variation in 12 microsatellite loci [151]. The significant isolation observed suggests an important role for natural hosts and/or vectors in the dispersion of parasites across the country.

2.4.2 Whole-Genome Sequencing

So far, only a few studies have investigated whole-genome SNPs, chromosome, and gene copy number variations for a significant number of strains of the same *Leishmania* species that would allow conclusions on intraspecific diversity. Imamura et al. [37] have recently investigated the history of VL on the Indian subcontinent (ISC) by analyzing whole-genome sequences of 204 *L. (L.) donovani* isolated from VL cases in Nepal, India, and Bangladesh. They reported that most of these parasites first appeared in the nineteenth century which is matching the first historical records of VL epidemics in the area. As shown earlier, the parasite genomes are indeed genetically similar, but whole-genome SNP analyses identified three divergent genetic lineages circulating on the ISC: a core group of 191 closely related parasites occurring in the lowlands of all three countries, a small group of 12 strains from Nepalese highlands, and a single divergent isolate from Nepal (Fig. 2.5). The core population could be clustered into six discrete monophyletic groups which first appeared in the 1960s. Thus, whole-genome analyses confirm

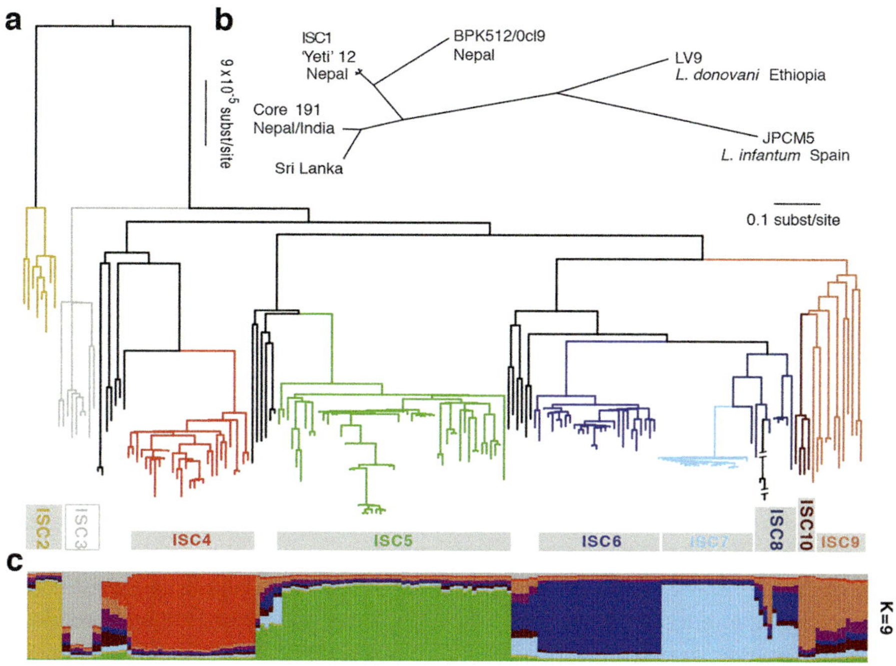

Fig. 2.5 Genealogical history of *L. (L.) donovani* from the ISC (Imamura et al. [37]). (**a**) Maximum-likelihood tree based on SNPs called for 191 strains from the core population in the Indian subcontinent. Samples are colored by population assignment, with putative hybrid strains not clustered in the main groups in black. Further analysis confirms the hybrid ancestry of some of these isolates. (**b**) Unrooted phylogenetic network of the *L. (L.) donovani* complex based on split decomposition of maximum-likelihood distances between isolates described here, reference genome isolates, and two published Sri Lankan isolates (Zhang et al. 2014). (**c**) Model-based clustering of 191 isolates from the core population reveals six discrete monophyletic groups and some groups and other samples of less certain ancestry. Colored bars show the fraction of ancestry per strain assigned to a given cluster, with colors assigned to the population most closely related to each cluster. Reproduced under CC BY license (http://creativecommons.org/licenses/by/4.0/)

earlier hypotheses of sustained and ancient reproductive isolation from other *L. (L.) donovani* lineages due to a recent bottleneck event on the ISC related to the insecticide spraying under the Malaria Control Program in the 1960s. Parasites in one of genetically distinct groups were found to be frequently resistant to antimonial treatment. High plasticity was observed for these *L. (L.) donovani* genomes, gene copy number variants cover ~11% of the genome, most of the isolates were aneuploid, and almost all chromosomes show some variations.

Using double-drug resistance markers, genetic recombination among *Leishmania* parasites was unequivocally demonstrated to occur in the sand fly vector under laboratory conditions [152]. The detection of natural hybrids and mosaic genotypes [16, 37, 129, 134, 136, 153–157], gene flow between populations [129, 134], and strong inbreeding [40, 151] have repeatedly posed questions about the role and extent of sexual recombination in natural populations of *Leishmania*. Rogers et al.

[158] have applied whole-genome sequencing to 12 *L. (L.) infantum* isolated in a CL focus in the Cukurova province of southeast Turkey, mainly from sand fly vectors, to investigate the frequency of sexual reproduction in these parasites. They observed a genome-wide pattern of patchy heterozygosity both within individual strains and across the whole group symptomatic of hybrid ancestry. Comparisons with other *L. (L.) donovani* and *L. (L.) infantum* genomes led to the assumption that the Cukurova isolates derived from a single relatively recent cross of two diverse strains with subsequent recombination within the population. After the original hybridization event, the population reproduced primarily clonally, but some recombination also occurred. The frequency of mating has been estimated as ca. 1.3×10^{-5} meioses per mitosis suggesting that sexual crosses might be rare in natural populations of *Leishmania*.

2.5 Origin of *Leishmania* Parasites

Whether the genus *Leishmania* appeared first in the Old World or in the New World has been controversially discussed during the last decades. Tuon et al. [159] have pointed out that regardless of its origin, the spread of *Leishmania* most likely followed the migration of vectors and hosts together, although *Leishmania* are quite capable of jumping hosts. The earliest fossil sand flies (ca. 120 Mya) were reported in Lebanon [160], which formed part of Gondwana, and reptiles or primitive mammals may have been the hosts of primitive *Leishmania*. The different vector-parasite-host theories of dissemination are summarized in Table 2.4.

The Palaearctic origins hypothesis suggests that the first association of the parasites with vertebrates occurred in the Old World with Cretaceous reptiles. Infections of Old World rodents then appeared in the Palaeocene and were carried by vertebrate hosts and sand fly vectors across Beringia to the Neoarctic in the Eocene. During the Pliocene, infected sigmodontine rodents brought the parasites to the Neotropics via the Panamanian land bridge. There, endemic vectors introduced the parasites to caviomorph rodents, sloths, armadillos, and anteaters [2, 161].

Alternatively, it has been proposed that *Leishmania* originated in the Neotropics during the Palaeocene with sloths as the first vertebrate hosts. After adaptation to rodents in the Eocene, infected porcupines would have carried the parasites across the Panamanian land bridge to the Neoarctic. From there the parasites were transported by other mammals across Beringia during the Miocene [162]. This hypothesis is supported by host-based area cladograms which use patterns of origination and dispersal of hosts and vectors to infer the phylogeny of the parasites. However, *Leishmania* are often not host- or even vector-specific. Recently, the first apparent fossil member of the genus *Leishmania*, the ~100 my old *Paleoleishmania proterus*, was detected in reptilian blood which was inside the body of the extinct sand fly, *Palaeomyia burmitis*, in Early Cretaceous Burmese amber [73, 163, 164]. Thus, protozoan-vector associations seem to have been established by the Early Cretaceous (100–110 my), reptiles were early hosts of *Leishmania*-like parasites, and the adaptation to mammals occurred later when reptiles declined

Table 2.4 Geological timescale and the theory of the evolution of *Leishmania* adapted from Tuon et al. [159] and modified

Eon	Era		Period		Epoch	Duration	Evolution
Phanerozoic	Cenozoic		Neogene		Holocene	0–11.5 ty	*Leishmania* (*L*) spread to Neotropic through primitive rodents from Neoartic
					Pleistocene	11.5–1.806 ty	
					Pliocene	1.8–5.3 my	Isthmus of Panama formation and physical unification between Neoartic and Neotropic allowing further spread of *Leishmania (L.)* to South America
					Miocene	5.3–23.0 my	
			Paleogene		Oligocene	23.0–33.9 my	
					Eocene	33.9–55.8 my	*Leishmania* (*L*) into Neoartic after Bering Strait formation by primitive rodents from Paleoartic, appearance of genus *Phlebotomus*, confirmed vector of *Leishmania*
					Paleocene	55.8–65.5 my	Placental mammals, ancestral hosts of *Leishmania*
	Mesozoic		Cretaceous		Upper	65.5–100 my	First *Leishmania* descendant in a reptile host, first hematophagous winged insect, separation of Africa and South America, split between Paraleishmania and the other *Leishmania* species (~90-100 my)
					Lower	100–145 my	
				Jurassic		145–199 my	The first digenetic protozoan, a possible ancestor of *Leishmania*
				Triassic		199–251 my	
	Paleozoic			Permian		251–299 my	Division of Trypanosomatidae following the evolution of Hemiptera and Diptera (vector of the *Leishmania*)
				Carboniferous		299–360 my	First winged insect

(continued)

Table 2.4 (continued)

Eon	Era		Period		Epoch	Duration	Evolution
				Devonian		360–416 my	Formation of first digenetic protozoan, ancestor of other *Trypanosoma*, not *Leishmania*. Parasite of a primitive fish
				Silurian		416–444 my	
				Ordovician		444–488 my	
				Cambrian		488–542 my	
Proterozoic		Neoproterozoic				542–1000 my	
		Mesoproterozoic				1–1.6 by	Possible origin of the Phylum Protista
		Paleoproterozoic				1.6–2.5 by	
Archean						3.5–2.5 by	

by, billions of years; my, millions of years; ty, thousands of years during the Eocene to Oligocene transition [161, 165]. This hypothesis implies that *Sauroleishmania* form a sister clade to all other leishmanial species [166, 167]. In contrast, rooted sequence-based phylogenetic trees of currently known *Leishmania* parasites favor a neotropical origin, showing the New World species branching off close to the base of the trees and the Old World species being at the crown of the subgenus *L. (Leishmania)* [45, 58, 60, 84, 89, 90, 102]. In these phylogenies, NW species emerged 46–34 mya and are ancestral to the OW species [109, 167]. The parasites were then dispersed by their hosts to the Nearctic via the Panamanian land bridge and further to the Palaearctic via the Bering land bridge. This view is further supported by the higher diversity found in the New World species of *Leishmania* [72], as well as by latest analysis of trypanosomatids from Australia [102]. The Neotropical origins hypothesis is, however, in discordance with the position of Old World *L. (Sauroleishmania)* closer to *L. (Leishmania)* than to *L. (Viannia)* but branching off within the New World taxa. It further suggests that reptilian species are derived from mammalian parasites which is in contrast to the Palaearctic hypothesis, and assumes two intercontinental migrations, first of the ancestral *Leishmania/Sauroleishmania* to the Palaearctic and then of a member of *L. (Leishmania)* subgenus back to the Neotropics [167].

The multiple origins hypothesis considers the great genetic difference between the parasites assigned to Euleishmania, comprising the parasites of the genus *L. (Leishmania)*, and Paraleishmania [92], according to the new taxonomy (see Box 2.1 and Table 2.3), and favors an ancient divergence between these two groups. It has been speculated that the two sections of the genus *Leishmania* became separated before the split of Gondwana [166]. The same authors concluded that, with the separation of Gondwana in the Mesozoic, the Euleishmania evolved into *L.* (*Leishmania*) and *L. (Sauroleishmania)* in the OW and *L.* (*Viannia*) in the NW. This conclusion is supported by the great genetic distance between the *L.* (*Leishmania*) and *L.* (*Viannia*) subgenera and the high genetic diversity within *L.* (*Viannia*) [116]. This theory, however, does not explain why the American branches of the subgenus *L.* (*Leishmania*) appear more ancient than the OW branches.

The supercontinent hypothesis is a variation of the multiple origins theory discussed earlier by Yurchenko et al. [58] but received phylogenetic support more recently in the study published by Harkins et al. [167]. These authors applied a phylogenomic approach analyzing more than 200,000 variable sites and 49 genes from across the genome for 24 leishmanial species. In their scenario, Leishmaniinae evolved from monoxenous ancestor on Gondwana, and the split between the Paraleishmania and all other species occurred ~90–100 mya, around the time when Gondwana split. This is in agreement with earlier speculations, that parasites adapted to mammals during the radiation of the latter around 90 mya [168]. Genetic diversification between the OW and NW parasites reflects the vicariance after the separation of South America and Africa [167, 169]. Only the migration of the NW lineage in the *L. (Leishmania)* subgenus is needed by this hypothesis, which took place 30 mya during the mid-Miocene when temperatures were warm enough for sand fly survival. The results of Harkins et al. are consistent with the early Cretaceous fossils of *Paleoleishmania proterus* found in sand flies trapped in Burmese amber ~100 mya [73] and with the finding that parasites isolated in different geographical regions, such as South America, Australia, Africa, and Asia, are members of the newly defined subgenus *L. (Mundinia)* [103]. Finally, a new Australian species, *Zelonia australiensis*, was found to be related to a parasite isolated in Costa Rica, *Zelonia costaricensis* (earlier *Leptomonas costaricensis*), suggesting a divergence time between the two of ~40 mya when Australia and South America became completely separated. Using this vicariance event for calibrations, it was confirmed that the common ancestor of the Leishmaniinae emerged around 90 mya on Gondwana [102].

The New World species of *L.* (*Leishmania*) most likely have originated in the Old World. *L. (L.) mexicana* has many similarities to Asian *L. (L.) major* and has been proposed to have dispersed to the Neoarctic together with its rodent reservoirs during the Eocene via the Bering land bridge [78, 161] and could then have entered the Neotropics during the Pliocene either via island hopping or after the Panamanian land bridge had been formed. There, further speciation could have taken place leading to the occurrence of the currently known species related to *L. (L.) mexicana*, namely, *L. (L.) amazonensis*, *L. (L.) aristidesi*, *L. (L.) venezuelensis*, and *L. (L.) forattinii* [166].

The etiological agent of New World visceral leishmaniasis, named *L. (L.) chagasi*, has been introduced relatively recently in the American continent, by the European conquistadores, along with multiple, and perhaps ongoing, introductions [107, 108, 170]. Numerous molecular studies have revealed a very restricted diversity within strains of *L. (L.) chagasi* and could not distinguish them from *L. (L.) infantum* indicating a very recent geographical separation. Studies on microsatellite variation have finally proven that strains of *L. (L.) chagasi*, or better of South American *L. (L.) infantum*, were most similar with populations of *L. (L.) infantum* from southwest Europe and arrived in the New World about 500 years ago [108, 171].

In conclusion, a revised classification scheme and nomenclature of *Leishmaniinae* species has been proposed based on molecular phylogenies of the organisms [102, 103]. It represents a useful simplification of the parasites' taxonomy, particularly for the clinician, without losing the detailed knowledge built up over the last 20 years, which is particularly relevant for epidemiological studies. In the future, assignment to and within major groups across the entire genus should be based on whole-genome analyses which are congruent and uncontroversial and explore the significance of variable aneuploidy for the biology and evolution of the parasites. In *Leishmania*, changes in aneuploidy are likely adaptive and depending on the life stage [31]. Clinical samples with only minimal in vitro passaging or, preferentially, without passaging at all should be, therefore, used in future studies linking genomic adaptations to treatment failure, drug resistance, immune, and other environmental pressures.

Acknowledgments We thank the many authors that have contributed to what we know today about the molecular evolution and phylogeny of *Leishmania* and whose work is not cited herein, in particular those who have worked with us, and all of those with whom we have had discussions about this subject. We acknowledge the financial support of the European Union (grants QLK2-CT-2001-01810, INCO-CT2005-015407, FP7-222895); of Deutsche Forschungsgemeinschaft (SCHO 448/6 and 448/8); of Wellcome Trust (078742/Z05/Z and core support of the WT Sanger Institute WT098051 and WT206194); and of the Czech Grant Agency (16-18699S) and the European Research Council (CZ LL1601).

References

1. Adl SM, Simpson AG, Farmer MA, Andersen RA, et al. The new higher level classification of eukaryotes with emphasis on the taxonomy of protists. J Eukaryot Microbiol. 2005;52(5):399–451.
2. Kerr SF. Molecular trees of trypanosomes incongruent with fossil records of hosts. Mem Inst Oswaldo Cruz. 2006;101(1):25–30.
3. Akhoundi M, Downing T, Votypka J, Kuhls K, et al. *Leishmania* infections: molecular targets and diagnosis. Mol Asp Med. 2017;57:1–29.
4. Rioux JA, Lanotte G, Serres E, Pratlong F, et al. Taxonomy of *Leishmania*. Use of isoenzymes. Suggestions for a new classification. Ann Parasitol Hum Comp. 1990;65(3):111–25.
5. Jamjoom MB, Ashford RW, Bates PA, Chance ML, et al. *Leishmania donovani* is the only cause of visceral leishmaniasis in East Africa; previous descriptions of *L. infantum* and "*L. archibaldi*" from this region are a consequence of convergent evolution in the isoenzyme data. Parasitology. 2004;129(Pt 4):399–409.
6. Kuhls K, Keilonat L, Ochsenreither S, Schaar M, et al. Multilocus microsatellite typing (MLMT) reveals genetically isolated populations between and within the main endemic regions of visceral leishmaniasis. Microbes Infect. 2007;9(3):334–43.
7. Van der Auwera G, Bart A, Chicharro C, Cortes S, et al. Comparison of *Leishmania* typing results obtained from 16 European clinical laboratories in 2014. Euro Surveill. 2016;21(49):30418.
8. Boite MC, Mauricio IL, Miles MA, Cupolillo E, et al. New insights on taxonomy, phylogeny and population genetics of *Leishmania* (*Viannia*) parasites based on multilocus sequence analysis. PLoS Negl Trop Dis. 2012;6(11):e1888.
9. El Baiduri F, Diancourt L, Berry V, Chevenet F, et al. Genetic structure and evolution of the *Leishmania* genus in Africa and Eurasia: what does MLSA tell us. PLoS Negl Trop Dis. 2013;7(6):e2255.
10. Zhang CY, Lu XJ, Du XQ, Jian J, et al. Phylogenetic and evolutionary analysis of Chinese *Leishmania* isolates based on multilocus sequence typing. PLoS One. 2013;8(4):e63124.
11. Schonian G, Kuhls K, Mauricio IL. Molecular approaches for a better understanding of the epidemiology and population genetics of *Leishmania*. Parasitology. 2011;138(4):405–25.
12. Philippe H. Molecular phylogeny of kinetoplastids. In: Coombs GH, et al., editors. Evolutionary relationships among Protozoa. Dordrecht, Boston, London: Kluwer Academic Publishers; 1998. p. 195–212.
13. Simpson AG, Stevens JR, Lukes J. The evolution and diversity of kinetoplastid flagellates. Trends Parasitol. 2006;22(4):168–74.
14. Hillis DM, Moritz C, editors. Molecular systematics. Sunderland, MA: Sinnauer Associates; 1990.
15. Maiden MC, Bygraves JA, Feil E, Morelli G, et al. Multilocus sequence typing: a portable approach to the identification of clones within populations of pathogenic microorganisms. Proc Natl Acad Sci U S A. 1998;95(6):3140–5.
16. Mauricio IL, Yeo M, Baghaei M, Doto D, et al. Towards multilocus sequence typing of the *Leishmania donovani* complex: resolving genotypes and haplotypes for five polymorphic metabolic enzymes (ASAT, GPI, NH1, NH2, PGD). Int J Parasitol. 2006;36(7):757–69.

17. Zemanova E, Jirku M, Mauricio IL, Horak A, et al. The *Leishmania donovani* complex: genotypes of five metabolic enzymes (ICD, ME, MPI, G6PDH, and FH), new targets for multilocus sequence typing. Int J Parasitol. 2007;37(2):149–60.
18. Jamjoom MB, Ashford RW, Bates PA, Kemp SJ, et al. Polymorphic microsatellite repeats are not conserved between *Leishmania donovani* and *Leishmania major*. Mol Ecol Notes. 2002;2:104–6.
19. Schwenkenbecher JM, Frohlich C, Gehre F, Schnur LF, et al. Evolution and conservation of microsatellite markers for *Leishmania tropica*. Infect Genet Evol. 2004;4(2):99–105.
20. Bhattarai NR, Dujardin JC, Rijal S, De Doncker S, et al. Development and evaluation of different PCR-based typing methods for discrimination of *Leishmania donovani* isolates from Nepal. Parasitology. 2010;137(6):947–57.
21. Botilde Y, Laurent T, Quispe Tintaya W, Chicharro C, et al. Comparison of molecular markers for strain typing of *Leishmania infantum*. Infect Genet Evol. 2006;6(6):440–6.
22. Ivens AC, Peacock CS, Worthey EA, Murphy L, et al. The genome of the kinetoplastid parasite, *Leishmania major*. Science. 2005;309(5733):436–42.
23. Peacock CS, Seeger K, Harris D, Murphy L, et al. Comparative genomic analysis of three *Leishmania* species that cause diverse human disease. Nat Genet. 2007;39(7):839–47.
24. Downing T, Imamura H, Decuypere S, Clark TG, et al. Whole genome sequencing of multiple *Leishmania donovani* clinical isolates provides insights into population structure and mechanisms of drug resistance. Genome Res. 2011;21(12):2143–56.
25. Rogers MB, Hilley JD, Dickens NJ, Wilkes J, et al. Chromosome and gene copy number variation allow major structural change between species and strains of *Leishmania*. Genome Res. 2011;21(12):2129–42.
26. Real F, Vidal RO, Carazzolle MF, Mondego JM, et al. The genome sequence of *Leishmania (Leishmania) amazonensis*: functional annotation and extended analysis of gene models. DNA Res. 2013;20(6):567–81.
27. Llanes A, Restrepo CM, Del Vecchio G, Anguizola FJ, et al. The genome of *Leishmania panamensis*: insights into genomics of the L. (Viannia) subgenus. Sci Rep. 2015;5:8550.
28. Raymond F, Boisvert S, Roy G, Ritt JF, et al. Genome sequencing of the lizard parasite *Leishmania tarentolae* reveals loss of genes associated to the intracellular stage of human pathogenic species. Nucleic Acids Res. 2012;40(3):1131–47.
29. Coughlan S, Mulhair P, Sanders M, Schönian G, et al. The genome of *Leishmania adleri* from a mammalian host highlights chromosome fission in Sauroleishmania. Sci Rep. 2017;7:43747.
30. Aslett M, Aurrecoechea C, Berriman M, Brestelli J, et al. TriTrypDB: a functional genomic resource for the Trypanosomatidae. Nucleic Acids Res. 2010;38(Database issue):D457–62.
31. Dumetz F, Imamura H, Sanders M, Seblova V, et al. Modulation of aneuploidy in *Leishmania donovani* during adaptation to different in vitro and in vivo environments and its impact on gene expression. MBio. 2017;8(3):e00599–17.
32. Steinbiss S, Siva-Franco F, Brunk B, Foth B, et al. Companion: a web server for annotation and analysis of parasite genomes. Nucleic Acids Res. 2016;44(W1):W29–34.
33. de Toledo JS, Vasconselos EJR, Ferreira TR, Cruz AK. Using genomic information to understand *Leishmania* biology. Open Parasitol J. 2010;4:156–66.
34. Valdivia HO, Almeida LV, Roatt BM, Reis-Cunha JL, et al. Comparative genomics of canine-isolated *Leishmania (Leishmania) amazonensis* from an endemic focus of visceral leishmaniasis in Governador Valadares, southeastern Brazil. Sci Rep. 2017;7:40804.
35. Sterkers Y, Lachaud L, Crobu L, Bastien P, et al. FISH analysis reveals aneuploidy and continual generation of chromosomal mosaicism in *Leishmania major*. Cell Microbiol. 2011;13(2):274–83.
36. Sterkers Y, Lachaud L, Bourgeois N, Crobu L, et al. Novel insights into genome plasticity in Eukaryotes: mosaic aneuploidy in Leishmania. Mol Microbiol. 2012;86(1):15–23.
37. Imamura H, Downing T, Van den Broeck F, Sanders MJ, et al. Evolutionary genomics of epidemic visceral leishmaniasis in the Indian subcontinent. elife. 2016;5:e12613.

38. Valdivia HO, Reis-Cunha JL, Rodrigues-Luiz GF, Baptista RP, et al. Comparative genomic analysis of *Leishmania (Viannia) peruviana* and *Leishmania (Viannia) braziliensis*. BMC Genomics. 2015;16:715.
39. Dujardin JC, Mannaert A, Durrant C, Cotton JA. Mosaic aneuploidy in *Leishmania*: the perspective of whole genome sequencing. Trends Parasitol. 2014;30(12):554–5.
40. Rougeron V, De Meeus T, Hide M, Waleckx E, et al. Extreme inbreeding in *Leishmania braziliensis*. Proc Natl Acad Sci U S A. 2009;106(25):10224–9.
41. El Tai NO, El Fari M, Mauricio I, Miles MA, et al. *Leishmania donovani*: intraspecific polymorphisms of Sudanese isolates revealed by PCR-based analyses and DNA sequencing. Exp Parasitol. 2001;97(1):35–44.
42. Jirku M, Yurchenko VY, Lukes J, Maslov DA. New species of insect trypanosomatids from Costa Rica and the proposal for a new subfamily within the Trypanosomatidae. J Eukaryot Microbiol. 2012;59(6):537–47.
43. Maslov DA, Podlipaev SA, Lukes J. Phylogeny of the kinetoplastida: taxonomic problems and insights into the evolution of parasitism. Mem Inst Oswaldo Cruz. 2001;96(3):397–402.
44. Shapiro TA, Englund PT. The structure and replication of kinetoplast DNA. Annu Rev Microbiol. 1995;49:117–43.
45. Stevens JR, Noyes HA, Schofield CJ, Gibson W, et al. The molecular evolution of Trypanosomatidae. Adv Parasitol. 2001;48:1–56.
46. Lukes J, Skalicky T, Tyc J, Votypka J, et al. Evolution of parasitism in kinetoplastid flagellates. Mol Biochem Parasitol. 2014;195(2):115–22.
47. Tanifuji G, Archibald JM. Actin gene family dynamics in cryptomonads and red algae. J Mol Evol. 2010;71(3):169–79.
48. Moreira D, Lopez-Garcia P, Vickerman K. An updated view of kinetoplastid phylogeny using environmental sequences and a closer outgroup: proposal for a new classification of the class Kinetoplastea. Int J Syst Evol Microbiol. 2004;54(Pt 5):1861–75.
49. Flegontov P, Votypka J, Skalicky T, Logacheva MD, et al. Paratrypanosoma is a novel early-branching trypanosomatid. Curr Biol. 2013;23(18):1787–93.
50. Blom D, de Haan A, van den Berg M, Sloof P, et al. RNA editing in the free-living bodonid Bodo saltans. Nucleic Acids Res. 1998;26(5):1205–13.
51. Hollar L, Lukes J, Maslov DA. Monophyly of endosymbiont containing trypanosomatids: phylogeny versus taxonomy. J Eukaryot Microbiol. 1998;45(3):293–7.
52. Lukes J, Jirku M, Dolezel D, Kral'ova I, et al. Analysis of ribosomal RNA genes suggests that trypanosomes are monophyletic. J Mol Evol. 1997;44(5):521–7.
53. Hamilton PB, Stevens JR, Gaunt MW, Gidley J, et al. Trypanosomes are monophyletic: evidence from genes for glyceraldehyde phosphate dehydrogenase and small subunit ribosomal RNA. Int J Parasitol. 2004;34(12):1393–404.
54. Maslov DA, Yurchenko VY, Jirku M, Lukes J, et al. Two new species of trypanosomatid parasites isolated from Heteroptera in Costa Rica. J Eukaryot Microbiol. 2010;57(2):177–88.
55. Merzlyak E, Yurchenko V, Kolesnikov AA, Alexandrov K, et al. Diversity and phylogeny of insect trypanosomatids based on small subunit rRNA genes: polyphyly of *Leptomonas* and *Blastocrithidia*. J Eukaryot Microbiol. 2001;48(2):161–9.
56. Stevens JR, Noyes HA, Dover GA, Gibson WC, et al. The ancient and divergent origins of the human pathogenic trypanosomes, *Trypanosoma brucei* and *T. cruzi*. Parasitology. 1999;118 (Pt 1):107–16.
57. Svobodova M, Zidkova L, Cepicka I, Obornik M, et al. *Sergeia podlipaevi* gen. nov., sp. nov. (Trypanosomatidae, Kinetoplastida), a parasite of biting midges (Ceratopogonidae, Diptera). Int J Syst Evol Microbiol. 2007;57(Pt 2):423–32.
58. Yurchenko VY, Lukes J, Jirku M, Zeledon R, et al. *Leptomonas costaricensis* sp. n. (Kinetoplastea: Trypanosomatidae), a member of the novel phylogenetic group of insect trypanosomatids closely related to the genus *Leishmania*. Parasitology. 2006;133(Pt 5):537–46.

59. Votypka J, Maslov DA, Yurchenko V, Jirku M, et al. Probing into the diversity of trypanosomatid flagellates parasitizing insect hosts in South-West China reveals both endemism and global dispersal. Mol Phylogenet Evol. 2010;54(1):243–53.
60. Croan DG, Morrison DA, Ellis JT. Evolution of the genus *Leishmania* revealed by comparison of DNA and RNA polymerase gene sequences. Mol Biochem Parasitol. 1997;89(2):149–59.
61. Edgar RC. MUSCLE: a multiple sequence alignment method with reduced time and space complexity. BMC Bioinformatics. 2004;5:113.
62. Talavera G, Castresana J. Improvement of phylogenies after removing divergent and ambiguously aligned blocks from protein sequence alignments. Syst Biol. 2007;56(4):564–77.
63. Stamatakis A. RAxML-VI-HPC: maximum likelihood-based phylogenetic analyses with thousands of taxa and mixed models. Bioinformatics. 2006;22(21):2688–90.
64. Votypka J, et al. Diversity of trypanosomatids (Kinetoplastea: Trypanosomatidae) parasitizing fleas (Insecta: Siphonaptera) and description of a new genus *Blechomonas* gen. n. Protist. 2013;164(6):763–81.
65. de Souza W, Motta MC. Endosymbiosis in protozoa of the Trypanosomatidae family. FEMS Microbiol Lett. 1999;173(1):1–8.
66. Alves JM, Klein CC, da Silva FM, Costa-Martins AG, et al. Endosymbiosis in trypanosomatids: the genomic cooperation between bacterium and host in the synthesis of essential amino acids is heavily influenced by multiple horizontal gene transfers. BMC Evol Biol. 2013;13:190.
67. Morales J, Kokkori S, Weidauer D, Chapman J, et al. Development of a toolbox to dissect host-endosymbiont interactions and protein trafficking in the trypanosomatid *Angomonas deanei*. BMC Evol Biol. 2016;16(1):247.
68. Kostygov AY, Dobakova E, Grybchuk-Ieremenko A, Vahala D, et al. Novel trypanosomatid-bacterium association: evolution of endosymbiosis in action. MBio. 2016;7(2):e01985.
69. Zahonova K, Kostygov AY, Sevcikova T, Yurchenko V, et al. An unprecedented non-canonical nuclear genetic code with all three termination codons reassigned as sense codons. Curr Biol. 2016;26(17):2364–9.
70. Maslov DA, Votypka J, Yurchenko V, Lukes J, et al. Diversity and phylogeny of insect trypanosomatids: all that is hidden shall be revealed. Trends Parasitol. 2013;29(1):43–52.
71. Baker JR. Speculations on the evolution of the family Trypanosomatidae Doflein, 1901. Exp Parasitol. 1963;13:219–33.
72. Lainson R, Shaw JJ. Evolution, classification and geographical distribution. In: Peters W, Killick-Kendrick R, editors. The leishmaniases in biology and medicine. New York: Academic Press; 1987. p. 1–120.
73. Poinar G Jr. Early Cretaceous trypanosomatids associated with fossil sand fly larvae in Burmese amber. Mem Inst Oswaldo Cruz. 2007;102(5):635–7.
74. Chicharro C, Alvar J. Lower trypanosomatids in HIV/AIDS patients. Ann Trop Med Parasitol. 2003;97(Suppl 1):75–8.
75. Jimenez MI, Lopez-Velez R, Molina R, Canavate C, et al. HIV co-infection with a currently non-pathogenic flagellate. Lancet. 1996;347(8996):264–5.
76. Dedet JP, Pratlong F. *Leishmania*, *Trypanosoma* and monoxenous trypanosomatids as emerging opportunistic agents. J Eukaryot Microbiol. 2000;47(1):37–9.
77. Kraeva N, Butenko A, Hlavacova J, Kostygov A, et al. *Leptomonas seymouri*: adaptations to the dixenous life cycle analyzed by genome sequencing, transcriptome profiling and co-infection with *Leishmania donovani*. PLoS Pathog. 2015;11(8):e1005127.
78. Thomaz-Soccol V, Lanotte G, Rioux JA, Pratlong F, et al. Monophyletic origin of the genus *Leishmania* Ross, 1903. Ann Parasitol Hum Comp. 1993;68(2):107–8.
79. Thomaz-Soccol V, Lanotte G, Rioux JA, Pratlong F, et al. Phylogenetic taxonomy of New World Leishmania. Ann Parasitol Hum Comp. 1993;68(2):104–6.
80. Vickerman K. The diversity of the kinetoplastid flagellates. In: Lumsden WHR, Evans DA, editors. Biology of the Kinetoplastida. London: Academic Press; 1976. p. 1–34.

81. Dedet JP. Current status of epidemiology of leishmaniases. In: Farrell JP, editor. *Leishmania* series: world class parasites. Londin: Kluwer Academis; 2002. p. 1–10.
82. van Eys GJ, Schoone GJ, Kroon NC, Ebeling SB, et al. Sequence analysis of small subunit ribosomal RNA genes and its use for detection and identification of *Leishmania* parasites. Mol Biochem Parasitol. 1992;51(1):133–42.
83. Maslov DA, Avila HA, Lake JA, Simpson L, et al. Evolution of RNA editing in kinetoplastid protozoa. Nature. 1994;368(6469):345–8.
84. Zelazny AM, Zhivotovsky L, Weir BS, et al. Evaluation of 7SL RNA gene sequences for the identification of *Leishmania* spp. Am J Trop Med Hyg. 2005;72(4):415–20.
85. Davila AM, Momen H. Internal-transcribed-spacer (ITS) sequences used to explore phylogenetic relationships within *Leishmania*. Ann Trop Med Parasitol. 2000;94(6):651–4.
86. Berzunza-Cruz M, Cabrera N, Crippa-Rossi M, Sosa Cabrera T, et al. Polymorphism analysis of the internal transcribed spacer and small subunit of ribosomal RNA genes of *Leishmania mexicana*. Parasitol Res. 2002;88(10):918–25.
87. Spanakos G, Piperaki ET, Menounos PG, Tegos N, et al. Detection and species identification of Old World *Leishmania* in clinical samples using a PCR-based method. Trans R Soc Trop Med Hyg. 2008;102(1):46–53.
88. Waki K, Dutta S, Ray D, Kolli BK, et al. Transmembrane molecules for phylogenetic analyses of pathogenic protists: *Leishmania*-specific informative sites in hydrophilic loops of trans-endoplasmic reticulum N-acetylglucosamine-1-phosphate transferase. Eukaryot Cell. 2007;6 (2):198–210.
89. Asato Y, Oshiro M, Myint CK, Yamamoto Y, et al. Phylogenic analysis of the genus *Leishmania* by cytochrome b gene sequencing. Exp Parasitol. 2009;121(4):352–61.
90. Fraga J, Montalvo AM, De Doncker S, Dujardin JC, et al. Phylogeny of *Leishmania* species based on the heat-shock protein 70 gene. Infect Genet Evol. 2010;10(2):238–45.
91. Noyes HA, Chance ML, Croan DG, Ellis JT, et al. *Leishmania (Sauroleishmania)*: a comment on classification. Parasitol Today. 1998;14(4):167.
92. Cupolillo E, Medina-Acosta E, Noyes H, Momen H, et al. A revised classification for *Leishmania* and *Endotrypanum*. Parasitol Today. 2000;16(4):142–4.
93. Noyes HA, Arana BA, Chance ML, Maingon R, et al. The *Leishmania hertigi* (Kinetoplastida; Trypanosomatidae) complex and the lizard Leishmania: their classification and evidence for a neotropical origin of the *Leishmania-Endotrypanum* clade. J Eukaryot Microbiol. 1997;44 (5):511–7.
94. Noyes HA, Camps AP, Chance ML. *Leishmania herreri* (Kinetoplastida; Trypanosomatidae) is more closely related to Endotrypanum (Kinetoplastida; Trypanosomatidae) than to *Leishmania*. Mol Biochem Parasitol. 1996;80(1):119–23.
95. Noyes H, Pratlong F, Chance M, Ellis J, et al. A previously unclassified trypanosomatid responsible for human cutaneous lesions in Martinique (French West Indies) is the most divergent member of the genus *Leishmania* ss. Parasitology. 2002;124(Pt 1):17–24.
96. Villinsky JT, Klena JD, Abbassy M, Hoel DF, et al. Evidence for a new species of *Leishmania* associated with a focal disease outbreak in Ghana. Diagn Microbiol Infect Dis. 2008;60 (3):323–7.
97. Sukmee T, Siripattanapipong S, Mungthin M, Worapong J, et al. A suspected new species of *Leishmania*, the causative agent of visceral leishmaniasis in a Thai patient. Int J Parasitol. 2008;38(6):617–22.
98. Rose K, Curtis J, Baldwin T, Mathis A, et al. Cutaneous leishmaniasis in red kangaroos: isolation and characterisation of the causative organisms. Int J Parasitol. 2004;34(6):655–64.
99. Dougall A, Shilton C, Low Choy J, Alexander B, et al. New reports of Australian cutaneous leishmaniasis in Northern Australian macropods. Epidemiol Infect. 2009;137(10):1516–20.
100. Lobsiger L, Muller N, Schweizer T, Frey CF, et al. An autochthonous case of cutaneous bovine leishmaniasis in Switzerland. Vet Parasitol. 2010;169(3–4):408–14.
101. Muller N, Welle M, Lobsiger L, Stoffel MH, et al. Occurrence of *Leishmania* sp. in cutaneous lesions of horses in Central Europe. Vet Parasitol. 2009;166(3–4):346–51.

102. Barratt J, Kaufer A, Peters B, Craig D, et al. Isolation of novel Trypanosomatid, *Zelonia australiensis* sp. nov. (Kinetoplastida: Trypanosomatidae) provides support for a Gondwanan origin of dixenous parasitism in the Leishmaniinae. PLoS Negl Trop Dis. 2017;11(1): e0005215.
103. Espinosa OA, Serrano MG, Camargo EP, Texeira MM, et al. An appraisal of the taxonomy and nomenclature of trypanosomatids presently classified as *Leishmania* and *Endotrypanum*. Parasitology. 2016:1–13.
104. Akhoundi M, Kuhls K, Cannet A, Votypka J, et al. A historical overview of the classification, evolution, and dispersion of *Leishmania* parasites and sandflies. PLoS Negl Trop Dis. 2016;10 (3):e0004349.
105. Cupolillo E, Grimaldi G Jr, Momen H. A general classification of New World *Leishmania* using numerical zymotaxonomy. Am J Trop Med Hyg. 1994;50(3):296–311.
106. Thomaz-Soccol V, Velez ID, Pratlong F, Agudelos S, et al. Enzymatic polymorphism and phylogenetic relationships in *Leishmania* Ross, 1903 (Sarcomastigophora: Kinetoplastida): a case study in Colombia. Syst Parasitol. 2000;46(1):59–68.
107. Mauricio IL, Stothard JR, Miles MA. The strange case of *Leishmania chagasi*. Parasitol Today. 2000;16(5):188–9.
108. Kuhls K, Alam MZ, Cupolillo E, Ferreira GE, et al. Comparative microsatellite typing of New World *Leishmania infantum* reveals low heterogeneity among populations and its recent old world origin. PLoS Negl Trop Dis. 2011;5(6):e1155.
109. Lukes J, Mauricio IL, Schonian G, Dujardin JC, et al. Evolutionary and geographical history of the *Leishmania donovani* complex with a revision of current taxonomy. Proc Natl Acad Sci U S A. 2007;104(22):9375–80.
110. Schwenkenbecher JM, Wirth T, Schnur LF, Jaffe CL, et al. Microsatellite analysis reveals genetic structure of *Leishmania tropica*. Int J Parasitol. 2006;36(2):237–46.
111. Schönian G, Schnur L, el Fari M, Oskam L, et al. Genetic heterogeneity in the species *Leishmania tropica* revealed by different PCR-based methods. Trans R Soc Trop Med Hyg. 2001;95(2):217–24.
112. Chaara D, Banuls AL, Haouas N, Talignani L, et al. Evolutionary history of *Leishmania killicki* (synonymous *Leishmania tropica*) and taxonomic implications. Parasit Vectors. 2015a;8:198.
113. Banuls AL, Jonquieres R, Guerrini F, Le Pont F, et al. Genetic analysis of *Leishmania* parasites in Ecuador: are *Leishmania (Viannia) panamensis* and *Leishmania (V.) guyanensis* distinct taxa? Am J Trop Med Hyg. 1999;61(5):838–45.
114. Grimaldi G Jr, Tesh RB. Leishmaniases of the New World: current concepts and implications for future research. Clin Microbiol Rev. 1993;6(3):230–50.
115. Oddone R, Schweynoch C, Schonian G, de Sousa Cdos S, et al. Development of a multilocus microsatellite typing approach for discriminating strains of *Leishmania (Viannia)* species. J Clin Microbiol. 2009;47(9):2818–25.
116. Cupolillo E, Grimald j G, Momen H, Beverley SM, et al. Intergenic region typing (IRT): a rapid molecular approach to the characterization and evolution of *Leishmania*. Mol Biochem Parasitol. 1995;73(1–2):145–55.
117. da Silva LA, de Sousa Cdos S, da Graca GC, Porrozzi R, et al. Sequence analysis and PCR-RFLP profiling of the hsp70 gene as a valuable tool for identifying *Leishmania* species associated with human leishmaniasis in Brazil. Infect Genet Evol. 2010;10(1):77–83.
118. Kuhls K, Cupolillo E, Silva SO, Schweynoch C, et al. Population structure and evidence for both clonality and recombination among Brazilian strains of the subgenus *Leishmania (Viannia)*. PLoS Negl Trop Dis. 2013;7(10):e2490.
119. Silveira FT, Ishikawa EA, De Souza AA, Lainson R, et al. An outbreak of cutaneous leishmaniasis among soldiers in Belem, Para State, Brazil, caused by *Leishmania (Viannia) lindenbergi* n. sp. A new leishmanial parasite of man in the Amazon region. Parasite. 2002;9 (1):43–50.

120. Jamjoom MB, Ashord RW, Bates PA, Kemp SJ, et al. Towards a standard battery of microsatellite markers for the analysis of the *Leishmania donovani* complex. Ann Trop Med Parasitol. 2002;96(3):265–70.
121. Ochsenreither S, Kuhls K, Schaar M, Presber W, et al. Multilocus microsatellite typing as a new tool for discrimination of *Leishmania infantum* MON-1 strains. J Clin Microbiol. 2006;44 (2):495–503.
122. Al-Jawabreh A, Diezmann S, Mueller M, Wirth T, et al. Identification of geographically distributed sub-populations of *Leishmania (Leishmania) major* by microsatellite analysis. BMC Evol Biol. 2008;8(1):183.
123. Russell R, Iribar MP, Lambson B, Brewster S, et al. Intra and inter-specific microsatellite variation in the *Leishmania* subgenus *Viannia*. Mol Biochem Parasitol. 1999;103(1):71–7.
124. Rougeron V, Waleckx E, Hide M, De Meeus T, et al. A set of 12 microsatellite loci for genetic studies of *Leishmania braziliensis*. Mol Ecol Resour. 2008;8:351–3.
125. Fakhar M, Motazedian MH, Daly D, Lowe CD, et al. An integrated pipeline for the development of novel panels of mapped microsatellite markers for *Leishmania donovani* complex, *Leishmania braziliensis* and *Leishmania major*. Parasitology. 2008;135(5):567–74.
126. Restrepo CM, Llanes A, De La Guardia C, Lleonart R, et al. Genome-wide discovery and development of polymorphic microsatellites from *Leishmania panamensis* parasites circulating in central Panama. Parasit Vectors. 2015;8:527.
127. Pajuelo MJ, Eguiluz M, Dahlstrom E, Rquena D, et al. Identification and characterization of microsatellite markers derived from the whole genome analysis of *Taenia solium*. PLoS Negl Trop Dis. 2015;9(12):e0004316.
128. Zhan L, Paterson IG, Fraser BA, Watson B, et al. megasat: automated inference of microsatellite genotypes from sequence data. Mol Ecol Resour. 2017;17(2):247–56.
129. Chargui N, Amro A, Haouas N, Schonian G, et al. Population structure of Tunisian *Leishmania infantum* and evidence for the existence of hybrids and gene flow between genetically different populations. Int J Parasitol. 2009;39(7):801–11.
130. Seridi N, Amro A, Kuhls K, Belkaid M, et al. Genetic polymorphism of Algerian *Leishmania infantum* strains revealed by multilocus microsatellite analysis. Microbes Infect. 2008;10(12–13):1309–15.
131. Amro A, Schonian G, Al-Sharabati MB, Azmi K, et al. Population genetics of *Leishmania infantum* in Israel and the Palestinian Authority through microsatellite analysis. Microbes Infect. 2009;11(4):484–92.
132. Alam MZ, Nakao R, Sakurai T, Kato H, et al. Genetic diversity of *Leishmania donovani/infantum* complex in China through microsatellite analysis. Infect Genet Evol. 2014;22:112–9.
133. Gouzelou E, Haralambous C, Antoniou M, Christodoulou V, et al. Genetic diversity and structure in *Leishmania infantum* populations from southeastern Europe revealed by microsatellite analysis. Parasit Vectors. 2013;6:342.
134. Kuhls K, Chicharro C, Canavate C, Cortes S, et al. Differentiation and gene flow among European populations of *Leishmania infantum* MON-1. PLoS Negl Trop Dis. 2008;2(7):e261.
135. Gelanew T, Kuhls K, Hurissa Z, Weldegebreal T, et al. Inference of population structure of *Leishmania donovani* strains isolated from different Ethiopian visceral leishmaniasis endemic areas. PLoS Negl Trop Dis. 2010;4(11):e889.
136. Gelanew T, Hailu A, Schonian G, Lewis MD, et al. Multilocus sequence and microsatellite identification of intra-specific hybrids and ancestor-like donors among natural Ethiopian isolates of *Leishmania donovani*. Int J Parasitol. 2014;44(10):751–7.
137. Baleela R, Llewellyn MS, Fitzpatrick S, Kuhls K, et al. *Leishmania donovani* populations in Eastern Sudan: temporal structuring and a link between human and canine transmission. Parasit Vectors. 2014;7:496.
138. Alam MZ, Kuhls K, Schweynoch C, Sundar S, et al. Multilocus microsatellite typing (MLMT) reveals genetic homogeneity of *Leishmania donovani* strains in the Indian subcontinent. Infect Genet Evol. 2009;9(1):24–31.

139. Downing T, Stark O, Vanaerschrot M, Imamura H, et al. Genome-wide SNP and microsatellite variation illuminate population-level epidemiology in the *Leishmania donovani* species complex. Infect Genet Evol. 2012;12(1):149–59.
140. Mahnaz T, Al-Jawabreh A, Kuhls K, Schonian G, et al. Multilocus microsatellite typing shows three different genetic clusters of *Leishmania major* in Iran. Microbes Infect. 2011;13 (11):937–42.
141. Alam MZ, Bhutto AM, Soomro FR, Baloch JH, et al. Population genetics of *Leishmania (Leishmania) major* DNA isolated from cutaneous leishmaniasis patients in Pakistan based on multilocus microsatellite typing. Parasit Vectors. 2014;7:332.
142. Harrabi M, Bettaieb J, Ghawar W, Toumi A, et al. Spatio-temporal genetic structuring of *Leishmania major* in Tunisia by microsatellite analysis. PLoS Negl Trop Dis. 2015;9(8): e0004017.
143. Azmi K, Krayter L, Nasereddin A, Ereqat S, et al. Increased prevalence of human cutaneous leishmaniasis in Israel and the Palestinian Authority caused by the recent emergence of a population of genetically similar strains of Leishmania tropica. In: Infect Genet Evol; 2016.
144. Krayter L, Alam MZ, Rhajaoui K, Schnur LF, et al. Multilocus Microsatellite Typing reveals intra-focal genetic diversity among strains of *Leishmania tropica* in Chichaoua Province, Morocco. Infect Genet Evol. 2014;28:233–9.
145. Pratlong F, Rioux JA, Dereure J, Mahjour J, et al. *Leishmania tropica* in Morocco. IV – Intrafocal enzyme diversity. Ann Parasitol Hum Comp. 1991;66(3):100–4.
146. Krayter L, et al. Multilocus microsatellite typing reveals a genetic relationship but, also, genetic differences between Indian strains of *Leishmania tropica* causing cutaneous leishmaniasis and those causing visceral leishmaniasis. Parasit Vectors. 2014;7:123.
147. Chaara D, Banuls AL, Haouas N, Talignani L, et al. Comparison of *Leishmania killicki* (syn. *L. tropica*) and *Leishmania tropica* population structure in Maghreb by microsatellite typing. PLoS Negl Trop Dis. 2015b;9(12):e0004204.
148. Krayter L, Schnur LF, Schonian G. The genetic relationship between *Leishmania aethiopica* and *Leishmania tropica* revealed by comparing microsatellite profiles. PLoS One. 2015;10(7): e0131227.
149. Adaui V, Maes I, Huyse T, Van den Broeck F, et al. Multilocus genotyping reveals a polyphyletic pattern among naturally antimony-resistant *Leishmania braziliensis* isolates from Peru. Infect Genet Evol. 2011;11(8):1873–80.
150. Tibayrenc M, Kjellberg F, Ayala FJ. A clonal theory of parasitic protozoa: the population structures of *Entamoeba, Giardia, Leishmania, Naegleria, Plasmodium, Trichomonas*, and *Trypanosoma* and their medical and taxonomical consequences. Proc Natl Acad Sci U S A. 1990;87(7):2414–8.
151. Rougeron V, Banuls AL, Carme B, Simon S, et al. Reproductive strategies and population structure in Leishmania: substantial amount of sex in *Leishmania Viannia guyanensis*. Mol Ecol. 2011;20(15):3116–27.
152. Akopyants NS, Kimblin N, Secundino N, Patrick R, et al. Demonstration of genetic exchange during cyclical development of *Leishmania* in the sand fly vector. Science. 2009;324 (5924):265–8.
153. Bonfante-Garrido R, Melendez E, Barroeta S, de Alejos MA, et al. Cutaneous leishmaniasis in western Venezuela caused by infection with *Leishmania venezuelensis* and *L. braziliensis* variants. Trans R Soc Trop Med Hyg. 1992;86(2):141–8.
154. Belli AA, Miles MA, Kelly JM. A putative *Leishmania panamensis/Leishmania braziliensis* hybrid is a causative agent of human cutaneous leishmaniasis in Nicaragua. Parasitology. 1994;109(Pt 4):435–42.
155. Ravel C, Cortes S, Pratlong F, Morio F, et al. First report of genetic hybrids between two very divergent *Leishmania* species: *Leishmania infantum* and *Leishmania major*. Int J Parasitol. 2006;36(13):1383–8.

156. Nolder D, Roncal N, Davies CR, Llanos-Cuentas A, et al. Multiple hybrid genotypes of *Leishmania (Viannia)* in a focus of mucocutaneous leishmaniasis. Am J Trop Med Hyg. 2007;76(3):573–8.
157. Darce M, Moran J, Palacios X, Belli A, et al. Etiology of human cutaneous leishmaniasis in Nicaragua. Trans R Soc Trop Med Hyg. 1991;85(1):58–9.
158. Rogers MB, Downing T, Smith BA, Imamura H, et al. Genomic confirmation of hybridisation and recent inbreeding in a vector-isolated *Leishmania* population. PLoS Genet. 2014;10(1): e1004092.
159. Tuon FF, Neto VA, Amato VS. *Leishmania*: origin, evolution and future since the Precambrian. FEMS Immunol Med Microbiol. 2008;54(2):158–66.
160. Lewis D. A taxonomic review of the genus *Phlebotomus* (Diptera: Psychodidae). Bull Br Mus Nat Hist (Ent). 1982;45:121–209.
161. Kerr SF. Palaearctic origin of *Leishmania*. Mem Inst Oswaldo Cruz. 2000;95(1):75–80.
162. Noyes HA, Morrison DA, Chance ML, Ellis JT, et al. Evidence for a neotropical origin of *Leishmania*. Mem Inst Oswaldo Cruz. 2000;95(4):575–8.
163. Poinar G Jr, Poinar R. Evidence of vector-borne disease of Early Cretaceous reptiles. Vector Borne Zoonotic Dis. 2004a;4(4):281–4.
164. Poinar G Jr, Poinar R. *Paleoleishmania proterus* n. gen., n. sp., (Trypanosomatidae: Kinetoplastida) from Cretaceous Burmese amber. Protist. 2004b;155(3):305–10.
165. Goldner A, Herold N, Huber M. Antarctic glaciation caused ocean circulation changes at the Eocene-Oligocene transition. Nature. 2014;511(7511):574–7.
166. Momen H, Cupolillo E. Speculations on the origin and evolution of the genus *Leishmania*. Mem Inst Oswaldo Cruz. 2000;95(4):583–8.
167. Harkins KM, Schwartz RS, Cartwright RA, Stone AC, et al. Phylogenomic reconstruction supports supercontinent origins for *Leishmania*. Infect Genet Evol. 2016;38:101–9.
168. Shaw JJ. Ecological and evolutionary pressures on leishmanial parasites. Braz J Genet. 1997;20:123–8.
169. Fernandes AP, Nelson K, Beverley SM. Evolution of nuclear ribosomal RNAs in kinetoplastid protozoa: perspectives on the age and origins of parasitism. Proc Natl Acad Sci U S A. 1993;90 (24):11608–12.
170. Momen H, Pacheco RS, Cupolillo E, Grimaldi G Jr, et al. Molecular evidence for the importation of Old World *Leishmania* into the Americas. Biol Res. 1993;26(1-2):249–55.
171. Leblois R, Kuhls K, Francois O, Schonian G, et al. Guns, germs and dogs: on the origin of *Leishmania chagasi*. Infect Genet Evol. 2011;11(5):1091–5.

The Role of Reservoirs: Canine Leishmaniasis

3

Lenea Campino and Carla Maia

Abstract

Canine leishmaniasis caused by *Leishmania* (*Leishmania*) *infantum* species (syn. *L.* (*L.*) *chagasi* species in Latin America), which is transmitted by the bite of phlebotomine sand flies, is endemic and affects millions of dogs in Asia, Europe, North Africa, and South America and is considered as an emergent disease in North America. Domestic dogs (*Canis familiaris*) are the major hosts for these parasites and the main reservoir host for human infection.

Recent years have seen important advances on the epidemiology, pathology, and canine genetic factors linked with animal resistance or susceptibility to leishmaniasis. Despite the lack of pathognomonic manifestations, infection by *Leishmania* can be suspected if a combination of clinical signs is present, namely, lymphadenomegaly, cutaneous alterations, loss of body weight, ocular disturbs, epistaxis, onychogryphosis, and lameness. However, the definitive diagnosis of canine leishmaniasis is complex since not all infected animals develop signs of disease. This fact cannot be ignored since asymptomatic (without clinical signs) dogs are infectious to phlebotomine vectors, although at a lower risk than symptomatic (with clinical signs) dogs. The fact that dogs never achieve parasitological cure together with the widespread use of the available anti-*Leishmania* drugs for both canine and human treatment certainly contributes to the spread of drug-resistant parasites with the natural consequences for the clinical outcome of the disease.

Early detection of infection and close surveillance or treatment of infected animals together with the development of effective molecules for therapy (ideally different from the ones used for humans) and, more importantly, for immunoprophylaxis are essential to control the dissemination of the disease

L. Campino (✉) · C. Maia
Global Health and Tropical Medicine, GHTM, Instituto de Higiene e Medicina Tropical, IHMT, Universidade Nova de Lisboa, UNL, Lisboa, Portugal
e-mail: Campino@ihmt.unl.pt

A. Ponte-Sucre, M. Padrón-Nieves (eds.), *Drug Resistance in Leishmania Parasites*,
https://doi.org/10.1007/978-3-319-74186-4_3

among other dogs, being also a crucial element for the control of human zoonotic leishmaniasis.

This chapter reviews the role of dogs as reservoir hosts of *L.* (*L.*) *infantum* and as accidental hosts of other *Leishmania* species, as well as the role of other mammals as potential reservoir hosts of parasites belonging to the *L.* (*L.*) *donovani* complex. The potential generation and spread of drug resistance by the use of the same compounds in both canine and human hosts are also discussed.

3.1 Introduction

Leishmaniases are parasitic diseases caused by protozoa belonging to the genus *Leishmania* Ross, 1903, transmitted by the bite of an insect vector, the phlebotomine sand fly (Diptera; Psychodidae), from the genus *Phlebotomus* (Old World) or *Lutzomyia* (New World).

By definition, a reservoir host is an animal in which an infectious agent survives persistently in a way that the animal may serve as a source of parasites to the vectors. A good reservoir host for leishmaniasis should be in close contact with man via the phlebotomine sand fly, should be susceptible to the pathogenic agent, and should make the parasite available to the vector in sufficient load to cause infection. A good reservoir host should be the main meal source for the phlebotomine sand fly and both (reservoir and vector) should rest and breed in the same habitat. Disease should present a chronic evolution allowing the animal to survive at least until the next transmission season [1–3].

Visceral leishmaniasis (VL) caused by parasites of *Leishmania (L.) donovani* complex is a severe human disease which often leads to death if left untreated [3]. Domestic dogs (*Canis familiaris*) are considered the major host of *Leishmania (L.) infantum* Nicolle, 1908, one of the species from the *L. (L.) donovani* complex (please see Chap. 2 for a detailed description) and the main reservoir host for zoonotic human infection. Canine leishmaniasis (CanL) is endemic and affects millions of dogs in Latin America, the Mediterranean Basin, and Asia. It is an emergent disease in North America [4, 5].

3.2 Epidemiology of Canine Leishmaniasis

3.2.1 Canine Leishmaniasis in Endemic Regions

L. (L.) infantum, the etiological agent for zoonotic leishmaniasis in the Old World, is synonymous with *L. (L.) chagasi* in Central and South America. In fact, it is thought that *L. (L.) infantum* was introduced in America by infected dogs carrying these parasites by the European conquistadores [6] (please see Chap. 2 for a detailed description). CanL is endemic in approximately 50 countries among the 98 where human leishmaniases are present, affecting two major geographic regions, the Mediterranean Basin and Brazil, with distinct rates of prevalence, which vary significantly within a small area, depending on ecological and climatic conditions that determine the abundance of vectors [7]. When favorable conditions for transmission (such as high phlebotomine sand fly vectors and canine-host densities) exist, leishmaniasis spreads quickly and extensively among the dog population [8]. However, the percentage of infected dogs that develop patent disease is low even in areas of high endemicity. On the other hand, it is difficult to compare prevalence rates since different epidemiological studies have used various and different methods to detect infection. Early epidemiological studies were based in direct parasitological tests, which were later replaced by more sensitive serological techniques followed by molecular methods. In addition, infection rates obtained by means of passive detection cannot be compared with those determined from house-to-house surveys [9]. Moreover, prevalence rates in the same geographical regions can fluctuate over time. Large variations in prevalence may be the result of several factors, such as the number of animals included in the studies and the decision to eliminate or treat infected dogs, as well as the natural waves in vector populations.

Although zoonotic VL was considered a rural disease, it is becoming more and more frequent in urban areas. Urbanization of leishmaniasis is associated with an increased number of detached houses with gardens in the peripheries of cities. Dogs are commonly kept in these gardens, which provide natural habitats for phlebotomine sand flies. Moreover, high numbers of stray dogs, present in urban/peri-urban settlements, may contribute to the spread of infection since these animals are an easy target for phlebotomine sand fly biting and are more susceptible for infection due to their precarious physical conditions and outdoor living habits [10, 11].

Although some studies [12, 13] suggest a positive relationship between prevalence of leishmaniasis in canine and human populations, CanL is more prevalent and more widely distributed than VL, and does not strongly correlate with prevalence in humans. For example, in Southern European countries, CanL is highly prevalent, while human leishmaniasis is hypoendemic.

3.2.2 Canine Leishmaniasis in Non-endemic Regions

CanL occurs in non-endemic regions of the world. For instance, in recent years, about 700 imported CanL cases have been reported from traditionally non-endemic European countries of leishmaniasis [14]. Most infected dogs from those regions had been living or traveled to endemic countries of the Mediterranean Basin. In Great Britain (UK), 257 cases of CanL were diagnosed between 2005 and 2007. About 15% of the dogs had been rescued from re-homing centers in the country of origin, and about 15% entered the UK with confirmed leishmaniasis [15]. Despite phlebotomine sand fly transmission of *Leishmania* parasites in Northern European countries has not yet been confirmed, autochthonous cases of leishmaniasis have been reported in dogs from Czech Republic [16], Finland [17], Germany [18], Hungary [19], Romania [20], the Netherlands [21], and the UK [15]. Nevertheless, the route of transmission, i.e., vectorial, transplacental, or even by direct contact, of most of the cases remains unknown [22]. Autochthonous cases of CanL in North America had rarely been reported before 2000, when it was diagnosed in 41% (46/112) of foxhounds from a kennel in New York State, USA [23]. A retrospective study performed from April 2000 to December 2003 by the Centers for Disease Control and Prevention revealed that canine infection was present in 18/35 US states and in 2/4 Canadian provinces [4].

Changes in the seasonal dynamics of phlebotomine sand fly populations together with the presence of infected dogs may lead to the establishment of new foci in areas previously defined as non-endemic. Major issues of CanL in non-endemic regions include its diagnosis, decision to treat, the therapeutic regimen, and disease monitoring [24]. An expansion of *L. (L.) infantum* transmission toward northern latitudes includes the cases in the foothills of the Alps in Northern Italy [25]; of French Pyrenees, in southern France [26]; of Catalonia, in northeastern Spain [27]; and of Galicia, in northern Spain [28]. In the last decade, CanL has also spread from northern Tunisia to previously non-endemic southern areas [29].

3.2.3 Transmission of Leishmaniasis

Phlebotomine sand flies are the only proven vectors of *Leishmania* parasites. However, at least a dozen phlebotomine sand fly species of the subgenus *Larroussius* have been incriminated as vectors of *L. (L.) infantum* in the Old World, and in Latin America, the most important phlebotomine vector is *Lutzomyia longipalpis*. In the USA, vector-borne transmission has not been identified despite the high prevalence of leishmaniasis in foxhound dogs. A potential phlebotomine sand fly vector, *Lutzomyia shannoni*, is present in southern, midwestern, and southeastern areas of the country [4]. Nonvector-based mechanisms postulated for transmission of CanL in the USA include vertical and mechanical transmission or by direct dog-to-dog contact through bites or wounds [5, 30]. Rosypal et al. [31] in the USA and da Silva et al. [32] in Brazil demonstrated vertical transmission in puppies born from experimentally and naturally infected female dogs, respectively, and Gibson-Corley et al. [33] described systemic *L. (L.) infantum* infection in two sibling American foxhounds from the USA, probably caused by vertical transmission. The presence of parasites in the semen of infected dogs and venereal transmission has also been reported in Spain and Brazil [34, 35]. Transmission of infection from infected to uninfected dogs through blood transfusion has been documented in endemic regions, alerting that canine blood donors should be monitored for the presence of *L. (L.) infantum* [36, 37].

The capacity of infected ticks and fleas to act as potentially additional vectors of CanL has been evaluated. For example, hamsters inoculated with macerate of fleas (*Ctenocephalides felis*) and ticks (*Rhipicephalus sanguineus*) blood-fed on infected dogs were *Leishmania* PCR positive. *L. (L.) infantum* transovarial transmission by infected ticks has also been observed by Ferreira et al. [38] and Dantas-Torres et al. [39].

The epidemiological significance of these potential modes of *Leishmania* transmission among dogs remains uncertain since they probably cannot sustain transmission in the absence of phlebotomine sand flies. As with dogs, most of the occasional VL transmission routes described for humans (sexual, congenital, blood transfusion) are not significant in the maintenance of the *Leishmania* life cycle with the exception of the artificial anthroponotic cycle (through the share of syringes contaminated with *Leishmania* among intravenous drug users) [40].

3.2.4 Infectiousness of Symptomatic Versus Asymptomatic Dogs

Infectivity of dogs to phlebotomine sand flies has been determined by xenodiagnoses using colonized vector species from the Old and New Worlds. The infection rate of reared *P. perniciosus* fed on infected dogs was as high as 92% [41], while the infection rate of reared *L. longipalpis* was up to 51.9% [42]. This discrepancy might be due to the fact that the minimum number of parasites necessary to infect *P. perniciosus* is lower than that necessary to infect *L. longipalpis* [42]. A xenodiagnostic meta-analysis made on dogs varying in clinical disease severity in both Europe and South America confirmed that infectiousness increases with clinical severity, i.e., symptomatic dogs are more able to transmit parasites to the vectors than those animals infected but without clinical signs [43]. In fact, the detection of parasites in skin biopsies has been found to be significantly more frequent in dogs presenting cutaneous lesions, whether they have been naturally or experimentally infected [44]. Nevertheless, even the low infectiousness of asymptomatic dogs seems to be relevant from an epidemiological point of view, so that control needs to be directed at both dogs, i.e., with and without clinical signs [41, 43].

3.2.5 Genetic Factors Related with Susceptibility/Resistance to Disease

Epidemiological studies in canine populations suggest a role for genetics in the resistance to disease [8]. However, so far, only two genes have been implicated in susceptibility/resistance to *Leishmania* infection outcome.

The natural resistance-associated macrophage protein 1 (*NRAMP1*) gene encodes an ion transporter protein involved in the control of intraphagosomal replication of parasites and in macrophage activation, including increased expression of chemokine and cytokine genes, such as tumor necrosis factor (TNF), interleukin-1-alpha (IL-1α), and inducible nitric oxide synthase (i-NOS) [45]. The association between the *NRAMP1* gene and disease susceptibility was demonstrated using both resistant and susceptible naturally and experimentally infected dogs [46, 47]. The haplotype of T antigen epitope TAG-8-141 has been associated with Boxer breed predisposition to CanL [47], although Bueno et al. [48] did not find differences in the expression of this gene between phenotypically resistant and susceptible dogs. Additionally, Turchetti et al. [49] have not found significant differences in basal transcription of genes associated with innate immunity (i.e., *NRAMP1*, *nucleotide oligomerization domain* (*NOD*)*1*, *NOD2*, *toll-like receptor* (*TLR*)*1*, *TLR2*, *TLR3*, *TLR4*, *TLR5*, *TLR6*, *TLR7*, and *TLR9*) between primary canine monocyte-derived macrophages and *Leishmania*-free dogs with higher or lower resistance to intracellular survival of the parasites.

The canine genetic variation in major histocompatibility complex class II termed "the dog leukocyte antigen (DLA) system" suggests a significant association between the presence of the beta chain allele *DLA–DRB1**01502 and susceptibility to CanL in a group of mongrel dogs [50].

These findings associating genetics with susceptibility/resistance to CanL are pieces of a complex multigene puzzle that determines the individual dog's natural predisposition to disease. Nevertheless, the outcome of infection is also influenced by nongenetic factors such as nutritional status, concomitant infections, ectoparasitism, parasite virulence, and previous exposure to *Leishmania* parasites.

3.2.6 Risks Factors: Breed, Age, Gender, and Lifestyle of Dogs

As just mentioned, susceptibility or resistance to disease is partially genetically controlled; however, it is not known how age, breed, concomitant infections, gender, nutrition, habits, and immunological status affect CanL outcome. Large epidemiological surveys and retrospective studies have revealed that some breeds such as German shepherds, Boxers, Dobermans, and Rottweilers are more susceptible to disease than other breeds like the Poodle and the Yorkshire terrier [9, 51–54]. All dog breeds are potentially susceptible to *Leishmania* infection, although it is accepted that autochthonous breeds and mongrels from endemic areas develop variable levels of resistance, such as reported by Solano-Gallego et al. [52] in the Ibizan hound dogs, which develop primarily a cellular immune response and thus rarely develop clinical disease.

França-Silva et al. [53] and Cortes et al. [55] found that short-/medium-fur dogs were at a higher risk of acquiring CanL than the dogs with long hair. In fact, phlebotomine sand flies feed preferentially on the margin of the muzzle and the nose of the dogs [56]. On the other hand, Gálvez et al. [57] and Miró et al. [28] described an association between larger breed size and weight with the increase of CanL seroprevalence due to a greater body surface area susceptible to phlebotomine sand fly bites.

The prevalence of *L. (L.) infantum* infection has also been associated with the animal's age, maybe because older animals have been exposed for longer times to phlebotomines [58]. The age distribution of infected animals has a bimodal distribution, with a first peak biased toward 2–4-year-old animals and a second peak found among 78-year-old dogs [57, 59]. The lower prevalence of infection in adult dogs between 4 and 7 years old might be related with the immune equilibrium between inoculation of parasites and an efficacious host response, while the high prevalence of anti-*Leishmania* antibodies found in the older animals may be the result of a depression of their immune system, or the reactivation of an old latent infection, or the presence of concomitant infections or geriatric diseases.

Gender predisposition to the infection has been a field of discussion, as some studies have reported that there is no link between sex and infection [28, 53, 55], while others have reported a higher prevalence in male dogs [51, 54, 59]. According to Fisa et al. [60], the gender-related differences could be due to a shorter time of exposure to infection of female dogs, due to their increased mortality during pregnancy and nursing. For Queiroz et al. [61], the main reason is the preference of owners for male dogs used as guard or hunting dogs. In fact, living or spending the night outdoors was pointed out as the main risk factor for *L. (L.) infantum* infection; thus, the rate of infection for outdoor working breeds was found to be higher than for pet dogs [54, 57]. However, Zivicnjak et al. [59] did not find differences in the prevalence of infection between hunting dogs (20.5%; 35/171), guard dogs (11.8%; 9/76), and pets (15.4%; 2/13). In the same line of reason, as stray dogs live outdoors and additionally are more likely to experience deficient health and nutritional conditions, they are an easier target for phlebotomine sand fly biting and for infection. However, in a study made in Lisbon, Portugal, the prevalence of infection in domestic dogs (18.4%; 51/277) and in stray dogs (21.6%; 21/97) was not significantly different [10].

3.2.7 Other Potential Reservoir Hosts of *Leishmania (L.) donovani* Complex

Despite dogs being the main domestic reservoir host for *L. (L.) infantum*, in some regions, mammals such as foxes, jackals, and wolves have also been incriminated, particularly as wild reservoirs. Red foxes (*Vulpes vulpes*) seem to have a role in the maintenance of an autonomous or semiautonomous sylvatic cycle in the Mediterranean Basin [62, 63], while in Brazil, the prevalence of infection in crab-eating foxes (*Cerdocyon thous*) did not demonstrate a transmission cycle independently of domestic dogs [43].

The evidence that wild lagomorphs (hares and, to a lesser extent, rabbits) can play a role as reservoir hosts of *L. (L.) infantum* has recently been proposed in a new focus of visceral and cutaneous leishmaniasis in Fuenlabrada, Spain [64, 65]. Different factors related to human-induced environmental changes (land cover and land use) have been involved in this leishmaniasis outbreak, which have favored the creation of an urban periphery where both lagomorphs and phlebotomine sand fly vectors (*P. perniciosus*) had the optimal conditions to increase in numbers. Such conditions resulted in a high detection of *L. (L.) infantum* infection in hares as parasite was detected by nested PCR in the spleen or in skin samples of 43 of the 148 animals studied (29%). *Leishmania* DNA was also detected in one of the 66 spleen samples from rabbits (1.5%). Canine seroprevalence for *Leishmania*, studied during the same period in the same area in 2070 dogs, was found to be 1.64% (reviewed by [14]). Xenodiagnoses carried out on hares and rabbits collected from the outbreak area revealed the ability of both lagomorphs to transmit *L. (L.) infantum* to reared *P. perniciosus* [64, 65]. In addition, studies of blood meal preferences revealed that *P. perniciosus* caught in the area prefer to feed on these lagomorphs [65, 66]. Altogether, data support the idea of the urbanization of a sylvatic transmission cycle and that infected dogs are not essential to maintaining the transmission cycle of *L. (L.) infantum*.

Among reports on mammal hosts infected with *L. (L.) infantum*, the ones regarding domestic cats (*Felis catus domesticus*) deserve special public health attention. *L. (L.) infantum* infection and feline leishmaniasis have been reported in cats from several endemic geographic areas in Europe, the Middle East, and Brazil [67–77].

As previously defined, a reservoir host is regarded as the living system in which the parasite population is maintained for long periods of time. Cats can act as a reservoir host for *L. (L.) infantum*, rather than being an accidental host, since they (1) can be infected by *Leishmania*, (2) present parasites in peripheral blood, (3) are a blood source and can transmit parasites to competent vectors, (4) cohabit with humans, and (5) only in a few cases develop patent disease [67, 69, 78–82]. These attributes may allow their classification at least as a "good" reservoir [1]. Nevertheless, the epidemiological importance of cats in zoonotic leishmaniasis is still poorly understood [75, 83]. Thus, it would be very important to determine the proportion of transmission in endemic areas attributable to cats to clarify if they are able to sustain and spread *Leishmania* infection [83].

Wild rats [84, 85] and domestic horses [86–88] have been found infected with *L. (L.) infantum* in the Old and New Worlds. These animals seem to display resistance to disease, as they are able to control infection without development of clinical signs and, even when cutaneous lesions are present, they self-heal. However, this dogma was recently challenged as concomitant cutaneous and visceral infection in three horses from Belo Horizonte, Brazil [89], has been reported.

In contrast to *L. (L.) infantum*, which is a zoonosis, *L. (L.) donovani* infection has generally been considered an anthroponosis. However, in an epidemiological study in Nepal, Bhattarai et al. [90] found *L. (L.) donovani* DNA in blood of goats (16%), cows (5%), and buffaloes (4%). Similar results were obtained in northwestern Ethiopia as antibodies to and/or DNA of *L. (L.) donovani* complex have been detected in the blood of several domestic animals (i.e., goats, sheep, cows, dogs, and donkeys) [91]. Despite these findings, further investigation is required to confirm their possible role in leishmaniasis transmission.

3.2.8 Dog as a Reservoir or Accidental Host for *Leishmania* spp. (Other than *L. (L.) infantum*)

Although dogs have been found naturally infected by several species of *Leishmania*, their role in the transmission of species other than *L. (L.) infantum* is not known, and it is probably negligible (Table 3.1).

Table 3.1 *Leishmania* species, other than *L. (L.) infantum*, that have been isolated from dogs

	Leishmania species	References
Old World	*L. (L.) arabica*	Peters et al. [92], Elbihari et al. [93]
	L. (L.) donovani	Dereure et al. [94, 95]
	L. (L.) major	Elbihari et al. [93], Peters et al. [96], Morsy et al. [97]
	L. (L.) tropica	Dereure et al. [98], Guessous-Idrissi et al. [99], Lemrani et al. [100], Ntais et al. [101]
New World	*L. (L.) amazonensis*	Tolezano et al. [102], Ramirez et al. [103]
	L. (V.) braziliensis	Aguilar et al. [104], Vélez et al. [105]
	L. colombiensis	Delgado et al. [106]
	L. (L.) mexicana	Hashiguchi et al. [107]
	L. (V.) panamensis	Ramirez et al. [103], Vélez et al. [105], Dereure et al. [108]
	L. (V.) peruviana	Llanos-Cuentas et al. [109]

A canine survey performed in Morocco at the end of the 1980s demonstrated a few cases of leishmaniasis due to *L. (L.) tropica* [98]. Although the zymodemes isolated from skin lesions were the same as those found in humans in the same focus, the small number of animals infected (7 out of 313), together with the apparent short duration of the lesions, suggested that infection by this species in dogs was probably accidental.

Dereure et al. [94, 95] have also found, in an endemic focus of anthroponotic VL in eastern Sudan, a CanL seroprevalence between 42.9% and 74.3%. The same zymodemes of the *L. (L.) donovani* complex were present in both humans and dogs. However, in a study performed later in the same region, a low number of dogs were found to have specific antibodies against *Leishmania* or to harbor parasites [110]. Nevertheless, the results obtained by these authors concerning host attractiveness of *Phlebotomus orientalis*, the only proven vector of *Leishmania* parasites responsible for VL in that area, suggested that dogs could play a role in the transmission dynamics of infection. All these data reinforce the need of more extensive studies to clarify the epidemiological roles of humans, dogs, and potential sylvatic hosts in eastern Sudan, specifically, whether:

1. Canine infection is largely accidental and a consequence of high infection rate in humans; the highest seroprevalence found in dogs was found during an outbreak of human VL [95].
2. Dogs are the local reservoir host of human VL.
3. Dogs act as links between the possible sylvatic cycle and humans.

Concerning human American cutaneous leishmaniasis (ACL) caused by *L. (Viannia) braziliensis* complex, the postulate about domestic dogs acting as reservoir hosts is based on the detection of a relatively high prevalence of cutaneous lesions in dogs surveyed in numerous and widespread endemic sites and on the identification of a large number of indistinguishable *Leishmania* strains isolated from both human and dogs. In spite of a positive correlation observed between the risk of human ACL and CanL prevalence in Huanuco, Peru, and in Salta, Argentina, the scarcity of parasites in cutaneous lesions (responsible for decreased infectiousness to the phlebotomine sand fly vector) together with the high serorecovery rates suggests that dogs are able to control infection and thus may not be the main reservoir host of the parasite [111, 112].

3.3 Canine Disease

3.3.1 Clinical Signs and Pathological Parameters

Dogs get in contact with the infective forms of *Leishmania* through the bite of a phlebotomine sand fly. However, not all exposed dogs develop clinical manifestations, and nowadays it is recognized that asymptomatic infections (without clinical signs) are much more frequent than symptomatic ones (with clinical signs). After infection, the time until appearance of the disease signs varies extensively (from 3 months up to 7 years) and ranges from the total absence of signs to severe systemic disease. In the early stage of disease, there is no precise symptomatology, but once the disease becomes patent, it rapidly progresses to death within weeks or months or, more frequently, to a chronic phase lasting several years. Despite the lack of pathognomonic manifestations, clinical diagnosis is achieved through the combination of clinical signs particularly lymphadenomegaly, cutaneous alterations, loss of body weight, ocular disturbs, epistaxis, onychogryphosis, and lameness.

The typical histopathological finding is a granulomatous inflammatory infiltration with macrophages, lymphocytes, and plasma cells in cell-rich organs of the mononuclear phagocytic system such as the liver, spleen, lymph nodes, bone marrow, gastrointestinal tract, and skin.

Among cutaneous alterations, localized, multifocal, or diffuse exfoliative dermatitis with a symmetrical distribution is usually seen in dogs with CanL. Cutaneous ulcers are observed over the margins of ears, pressure points (i.e., parts of the body that sustain the animal's weight when it is resting, e.g., elbows), legs, and mucocutaneous junctions and have been attributed to local trauma and vascular damage [51, 58]. Blepharitis and conjunctivitis have been described as frequent signs [51, 58], although anterior uveitis has been described as the most frequent one [113].

Lymphadenopathy is common and facilitates palpation of the superficial lymph nodes, such as popliteal, prescapular, and submandibular. This sign is caused by the increased number and size of lymphoid follicles and the hypertrophy and hyperplasia of medullary macrophages in the cords and sinuses [114]. Splenomegaly may be mild and difficult to detect upon abdominal palpation. It is caused by the disorganization of normal lymphoid tissue, loss of normal spleen leukocyte diversity via replacement of leukocytes by plasma cells, and the associated hyperplasia of white and red pulp, as by changes in the microvascular structure [115]. Hepatomegaly has also been documented. Histological liver changes (inflammation of the hepatic capsule, portal inflammation, formation of granulomas, hypertrophy, and hyperplasia of the Kupffer cells) and increased biochemical alterations (plasmatic globulin) have been associated with the progression of the disease [116]. Protein electrophoresis reveals a significant decrease in albumin levels and a combined amplification in beta- and gamma globulin concentrations, both characteristic but nonspecific of CanL [9].

Renal disease might be the only apparent abnormality present in virtually all dogs infected with CanL. The kidney lesions can progress from asymptomatic proteinuria to nephrotic syndrome or chronic renal failure with glomerulonephritis, with tubulointerstitial nephritis, and more rarely with amyloidosis [58]. Chronic renal failure is a severe manifestation of the disease, and it is the principal cause of animal death in CanL [7]. Glomerular lesions are frequently associated with glomerular deposition of immune complexes [117]. Nevertheless, treatment with pentavalent antimonials (Sb^V) has sometimes been pinpointed as responsible for the deterioration of renal conditions of already affected kidneys in leishmaniotic dogs. Mild to severe anemia, although generally normocytic and normochromic, is the most common hematological abnormality and may be caused by blood loss, hemolysis, decreased erythropoiesis, and chronic renal failure [51, 58]. CanL-associated nasal bleed (epistaxis) is less common than other signs. It appears to be the result of multiple and variable pathogenic factors such as thrombocytopathy, hyperglobulinemia-induced serum hyperviscosity, and nasal mucosa ulceration [118].

Atypical forms of CanL include mucosal lesions, erosive and nonerosive polyarthritis, osteolytic and osteo-proliferative bone lesions, chronic colitis and disorders of the cardiovascular, respiratory, and neurologic and musculoskeletal systems [119]. Since *Leishmania* infections share many clinical and pathological features with other canine diseases, laboratorial confirmation of infection is necessary.

3.3.2 Diagnosis

CanL diagnosis is still a challenge in spite of advances made in the development of parasitological, serological, and molecular techniques. Reasons for attempting laboratory diagnosis are the confirmation of (1) disease, (2) presence of *Leishmania* infection (in epidemiological studies, to prevent blood transfusion from infected donors and importation of dogs to non-endemic countries), or (3) therapeutic control, so requirements may vary with regard to the selection of the adopted laboratory test and the biological material. An ideal technique should have high sensitivity and specificity, must be reproducible and easy to perform, and adaptable for use in local laboratories without sophisticated equipment, and it should detect all *Leishmania*-infected dogs in an initial stage, preferentially using noninvasive procedures to obtain the samples. In dogs with clinical signs compatible with CanL, analytical diagnosis should be performed to confirm the presence or absence of the infection (Fig. 3.1).

Serological diagnosis is widely and frequently used as specific humoral response in CanL and is, in general, very intense with high levels of specific immunoglobulins. The presence of anti-*Leishmania* antibodies alone is not

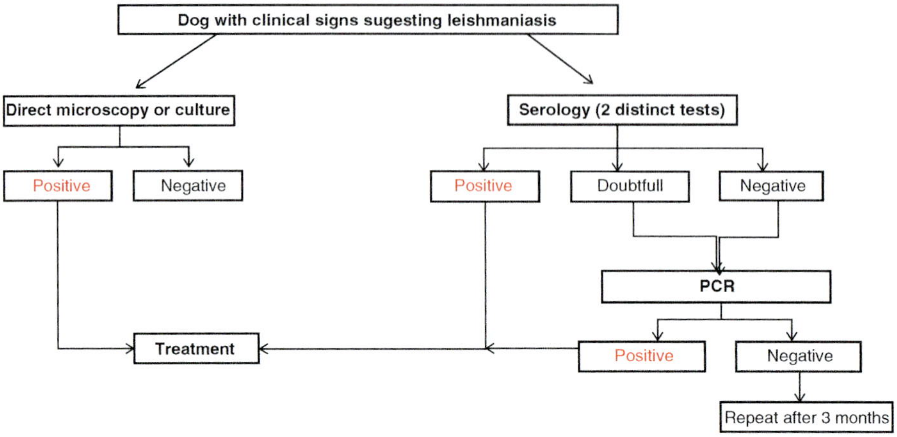

Fig. 3.1 Example of diagnosis methodologies in dogs with clinical signs compatible with leishmaniasis

conclusive of *Leishmania* infection, as it may simply reflect exposure to the parasite. In addition, serological assays have several intrinsic drawbacks including the persistence of specific antibodies after recovery or cross-reactions with antibodies against other pathogens such as *Trypanosoma cruzi* or other *Leishmania* species (in South and Central America) and *Ehrlichia canis* [120]. High levels of sensitivity and specificity are necessary to avoid false-negative results, which underestimate *Leishmania* infection rate in dog populations in endemic areas, and to minimize false-positive reactions, which can lead to unnecessary euthanasia of noninfected dogs. It is, thus, advisable to perform more than one serological test to gain more certainty in the diagnosis of CanL [8]. Several quantitative (e.g., enzyme-linked immunosorbent assay (ELISA), immunofluorescent antibody test (IFAT)) and qualitative (e.g., immunochromatographic tests) techniques are available for CanL diagnosis [120, 121]. ELISA using immuno-dominant recombinant proteins as antigen is very sensitive and specific. Nevertheless, IFAT is still considered the "gold standard" based on the high sensitivity and specificity. The rapid immunochromatographic kits are very attractive because of their single-test format, ease of use, and very quick response times allowing immediate intervention by the veterinarian.

Definitive diagnosis can be achieved using microscopic examination of stained smears, by culture or by detecting *Leishmania* DNA from infected organs/tissues. However, heterogeneous distribution of parasites in the organs together with low parasitism can lead to false-negative results. In vitro culture of tissue biopsies with replicate inoculation in several tubes is more sensitive than direct microscopy, particularly for low parasite loads, and it is 100% specific. Nowadays, in vitro culture is seldom used for diagnosis due to several drawbacks, including cost, time for diagnosis, low sensitivity, and risk of contamination. However, it is still the method of choice to obtain sufficient number of parasites for (1) isoenzymatic or even molecular identification, (2) in vitro drug susceptibility/resistance studies, (3) experimental infections, and (4) to get antigen for serological diagnostics [120].

Polymerase chain reaction (PCR) methods (e.g., conventional PCR, nested PCR) are more reliable in detecting the presence and the characterization of *Leishmania* parasites, not only in active cases but also for monitoring parasitological cure after chemotherapy. PCR can be carried out on a broad range of clinical specimens. Maia et al. [122] considered that popliteal lymph node PCR is useful as a first-line primary diagnosis or for therapeutic follow-up. Bone marrow PCR is recommended if the lymph node is too small to allow a safe biopsy. Quantitative PCR (qPCR), compared to other PCR techniques, enables reduction in assay time, lowers the risk of contamination, and improves sensitivity. Quantitative PCR can be very useful for the diagnosis of CanL since it facilitates the monitoring of parasite load during and after treatment in different samples allowing the prediction of recurrences associated with tissue loads of residual parasites after treatment [120, 121, 123].

Less invasive sampling (e.g., blood) would be desirable to facilitate diagnosis. Unfortunately, the duration, consistency, and intensity of parasitemia in CanL are still largely unknown, and false-negative results, especially in asymptomatic dogs, are frequent. On the other hand, during the transmission season, false-positive results may appear due to transient infections. A sensitive, noninvasive, painless, and fast sampling method, e.g., applying conjunctival oral, ear, and nasal swabs or hair and cerumen samples, coupled with a sensitive and specific PCR-based methods, has shown promising results for diagnosis, for treatment follow-up, and/or for assessing *Leishmania* exposure in dogs [124–127].

3.3.3 Treatment and Drug Resistance

The same drugs are used for treatment of CanL and human leishmaniasis; however, the method of administration and dosage differ. The most commonly used drugs for the treatment of CanL are allopurinol, the pentavalent meglumine antimoniate, and miltefosine (MIL, Table 3.2). Whereas the health of infected dogs improves and dogs are apparently cured, available treatments do not seem to eliminate the parasites.

An evaluation of 47 clinical trials assessing 14 different protocols with single or multiple molecules [128] concluded that the use of meglumine antimoniate, at a

Table 3.2 Drugs most commonly used for the treatment of canine leishmaniasis

Drug	Dose and duration	Side effects
Allopurinol	10–30 mg/kg/BID, at least 6–12 months; PO	Xantine urolithiasis
Meglumine antimoniate (alone or with allopurinol)	75–100 mg/kg/SID, 4–8 weeks; SC	Injection site reaction, nephrotoxicity, vomiting
Miltefosine (alone or with allopurinol)	2 mg/kg/SID, 4 weeks; PO	Diarrhea, teratogenic, vomiting

BID twice a day, *PO per os*, *SC* subcutaneous, *SID* once a day minimum dosage of 100 mg/kg daily for at least 3–4 weeks, combined with a long-term use of allopurinol had a good clinical efficacy and a reduced relapse rate. MIL, alone or in combination with allopurinol, has been suggested as an alternative therapy for CanL treatment [129]. Moreover, MIL seems to be safer for liver and kidney function than meglumine antimoniate [130, 131]. Despite the lack of evidence for recommending the use of allopurinol alone [128], its use as monotherapy has been advocated in dogs with mild clinical signs and on those with nephrotic syndrome or at end-stage kidney disease [7]. Conventional amphotericin-B deoxycholate (AMB-B) has also a good efficacy, but it is nephrotoxic and might endanger dogs with CanL that already have a renal pathology [132]). Regardless, liposomal amphotericin-B (L-AMB), which is used as first-line drug in many endemic regions to human VL, is highly efficient against the disease with minimum toxicity for the host. Despite its high cost, the liposomal formulation has been used to treat dogs with leishmaniasis in Europe. Aminosidine has also severe side effects (nephrotoxicity and ototoxicity) and therefore is not recommended as first-line therapy of CanL. Other drugs reported to have some efficacy against CanL include ketoconazole, pentamidine, marbofloxacin, metronidazole with spiramycin, and metronidazole with enrofloxacin [7, 128]. The use of immunomodulators, such as prednisone and prednisolone, associated with specific treatment prevents the severe immunological changes that occur during CanL. They activate both cellular and humoral immunity and are recommended only when there are lesions as a consequence of immunocomplex deposition. Immunostimulants, such as levamisole and domperidone, are also used as they activate macrophages and the cellular immunity to control/prevent *Leishmania* multiplication [162, 133].

In vivo and in vitro parasitic resistance to pentavalent antimonials, amphotericin-B, aminosidine, and miltefosine has been reported in human infection [134]. However, scarce data is available regarding the occurrence of drug resistance in CanL. Decreased in vitro sensitivity to (Sb^V) in *L. (L.) infantum* parasites isolated from dogs during and after several treatment courses has been reported [135–138]. Additionally, strains of *L. (L.) infantum* isolated from dogs that have undergone several courses of (Sb^V) therapy were able to grow in laboratory-reared *P. perniciosus* sand flies used for xenodiagnosis in these dogs [139]. Furthermore, antimony-resistant *L. (L.) infantum* strains have been isolated from dogs living in areas of low antimony drug pressure [140], and the resistant phenotype was maintained after the passage through *P. perniciosus* and *L. longipalpis* experimentally infected [141]. Low in vitro susceptibility of a *L. (L.) infantum* strain isolated from a dog under treatment with allopurinol [138] and allopurinol resistance of *L. (L.) infantum* strains isolated from dogs with disease relapse have also been reported [142]. All these data question the use of human drugs to treat dogs and highlight the need of combined therapy in CanL in order to avoid or reduce not only relapses but also the potential development of parasite resistance to antileishmanial drugs.

The lack of parasitological cure in dogs and the widespread use of the few available anti-*Leishmania* drugs in both canine and human treatment may lead to the emergence and circulation of resistant parasites. Thus, the World Health Organization strongly discourages the use in veterinary practice of L-AMB-B and paromomycin in order to avoid drug resistance to the first-line drugs used for treatment of human leishmaniasis in Europe [143]. Nevertheless, drugs belonging to different classes from those used in human leishmaniasis treatment should be design to optimize the treatment and achieve clinical cure as well as clearance of parasites in dogs with CanL.

3.3.4 Prevention and Control Measures

Control of CanL must be multidisciplinary and should address vertebrate hosts, vectors, and parasites. It should target individual prevention of new infections, through the use of insecticides with anti-vector effect, immunoprophylaxis, and early treatment of leishmaniasis cases.

Although the WHO still recommends culling dogs infected with *L. (L.) infantum*, which from our point of view is incomprehensible, this measure is difficult to implement in countries where dogs are considered part of the family. In Brazil, seropositive dogs are eliminated as part of a control program although its effectiveness in the control of infection is not clear-cut [144, 145]. Failure may occur due to the (1) high incidence of infection, (2) high infectiousness of dogs to phlebotomine sand flies, (3) poor sensitivity of diagnostic methods, (4) inability to reach and test the entire canine population, (5) delay between diagnosis and culling, and (6) rapid replacement of culled dogs by new susceptible animals.

Laboratory and field evaluations have demonstrated that the interruption of *Leishmania* life cycle can be achieved through the use of impregnated dog collars and topical application of insecticide with repellent and insecticidal effects against phlebotomine sand flies [146, 147]. The manufacturer-recommended residual activity time is typically 3–4 weeks for pour on and 5–6 months for collars, although the collars are often damaged or lost at a very high rate. Furthermore, a significant decrease in the incidence of zoonotic leishmaniasis in children [148] and dogs has been observed in areas where most of them were treated with pyrethroids, such as deltamethrin collars or spot-on solution of permethrin [149]. Additional measures to control phlebotomine sand flies include reducing microhabitats favorable to them in the vicinity of the houses, indoor insecticide spraying, and housing dogs at dusk [150].

The best strategy to control leishmaniasis would be a canine effective vaccine, as the effective immunization of dogs in endemic areas should significantly reduce CanL and potentially the incidence in humans. An effective vaccine should control both infection progression and the parasite transmissibility via the vector [151]. In the last few years, the efficacy of several vaccine candidates in protecting dogs against *Leishmania* infection has been tested [151] (Table 3.3). Two canine vaccines (Leishmune® and Leish-tec®) have been commercialized in Brazil. The first vaccine launched in 2004, Leishmune®, was shown to induce a significant, long-lasting, and strong protective effect against CanL in phase III of clinical trials [152, 153]. This vaccine was also proposed to be used as immune therapy of infected dogs and as a transmission-blocking vaccine [159, 160]. However, in 2014, the Brazilian Ministry

Table 3.3 Clinical trials (phase III) made with vaccine candidates against canine leishmaniasis that shows more than 50% of efficacy against clinical disease

Vaccine candidate	Efficacy against clinical disease (%)	Reference
FML	92	Silva et al. [152]
FML+saponina QuilA	80	Borja-Cabrera et al. [153]
Alum-ALM+ aluminum hydroxide	69.3	Mohebali et al. [154]
*Li*ESAp-MDP	92	Lemesre et al. [155]
Gentamicin-attenuated *L. (L.) infantum*	92	Daneshvar et al. [156]
*Li*ESAp-QA-21	63	Oliva et al. [157]
L. (L.) donovani A2+ saponin	71.4	Regina-Silva et al. [158]

FML Fucose mannose ligand antigen purified from *L. (L.) donovani*, *QuilA Quillaja saponaria saponin*, *Alum-ALM* Aluminum hydroxide (alum) precipitated *L. (L.) major*, *LiESAp* Purified excreted/secreted antigens promastigotes *L. (L.) infantum*, *MDP* Muramil dipeptide, *L. Leishmania*, *QA Quillaja saponaria*, *A2* recombinant protein A2 of *L. (L.) donovani*

of Agriculture, Livestock and Food Supply suspended its commercialization due to noncompliance with all the requirements for phase III studies (http://www.agricultura.gov.br/assuntos/politica-agricola/arquivos/nota-tecnica-dfip-38-14-leishmune.pdf/view). The second vaccine launched in Brazil in 2008, Leish-tec®, conferred a significant reduction in the number of cases of CanL with a 71.4% efficacy (only 7 of the 195 vaccinated dogs in comparison with 24 of the 192 dogs from the placebo group were considered infected according to parasitological exams), and the infectiousness to reared *L. longipalpis* of vaccinated dogs was 46.6% lower in comparison with non-vaccinated animals [158]. In Europe, the first commercialized vaccine for CanL consisting of purified excreted-secreted proteins of *L. (L.) infantum* and with QA-21 saponin as adjuvant (CaniLeish®) was launched in 2011. In a field study with 90 naïve beagles (46 vaccinated and 44 controls) naturally exposed to the parasite, CaniLeish® has provided a significant reduction in the risk of progressing to active infection or overt disease, with a clinical efficacy of 68% [157]. Additionally, the infectiousness to reared *P. perniciosus* of vaccinated dogs was significantly lower when compared to matched controls [161]. A second vaccine (Letifend®) consisting of a recombinant Protein Q from *L. (L.) infantum* MON-1 has recently (in 2017) been commercialized in Europe. According to the product information available at the European Medicines Agency (https://ec.europa.eu/health/documents/community-register/2016/20160420134483/anx_134483_en.pdf), a vaccinated dog has five times less risk to develop clinical disease than a non-vaccinated dog.

3.4 Conclusions

In recent years, important advances have been made concerning leishmaniasis, including its epidemiology, pathology, diagnosis, clinical management, and genetic factors related with resistance or susceptibility to *Leishmania* infection. Taking into account that CanL diagnosis is still often problematic and that dogs never reach parasitological cure, early detection of infection and close surveillance and treatment of these animals are very important measures to decrease infection incidence and spreading. Development of effective molecules (ideally different from those used for humans) for therapy and immunoprophylaxis is needed to control the spread of the infection among other dogs and as an essential component of the control of human zoonotic leishmaniasis.

3.5 Future Trends

Considerable progress has already been made in the diagnosis, treatment, and clinical management of infected dogs. However, sensitive methods are necessary for *Leishmania* detection at an early stage of infection and in asymptomatic animals. Furthermore, new drugs different from those used against human leishmaniasis are needed, and combined therapies should be designed to achieve parasitological cure and to block transmission. Nevertheless, control strategies should be based on prevention of infection, ideally through the development of an efficacious vaccine against CanL in association with insecticides with prolonged residual action against *Leishmania* vectors. Finally, an interdisciplinary network between veterinarians, researchers, physicians, public health entities, dogs' owners, and the general public would generate knowledge, tools, and education packages, which would contribute toward a significant reduction of the burden of canine and human leishmaniasis.

To sum up, it is important to keep in mind that one of the main control methods for leishmaniasis includes prolonged treatment of the major domestic reservoir host. Current knowledge about the epidemiology and transmission of zoonotic leishmaniasis suggests that selection of resistant parasites is still not an emerging problem. However, the chances of emergence and spread of resistant parasites in the canine and human populations should not be neglected, especially when dogs and humans are treated with the same drugs. Several methodologies, such as in vitro and ex vivo systems, which are already being used or developed for monitoring drug resistance in human leishmaniasis, can easily be adapted to CanL. Surveillance systems using these techniques would allow early control measures to diminish the impact of the introduction of resistant *L. (L.) infantum* strains in endemic regions.

Acknowledgments The authors wish to thank A. Pereira for his work on the references. C. Maia, DVM, PhD, holds a FCT Investigator Starting Grant (IF/01302/2015) from Fundação para a Ciência e a Tecnologia, Ministério da Ciência, Tecnologia e Ensino Superior, Portugal.

References

1. Bray R. The zoonotic potential of reservoirs of leishmaniasis in the Old World. Ecol Dis. 1982;1:257–67.
2. Ashford R. Leishmaniasis reservoirs and their significance in control. Clin Dermatol. 1996;14:523–32.

3. WHO. Working to Overcome the Global Impact of Neglected Tropical Diseases; 2010. http://www.who.int/neglected_diseases/2010report/NTD_2010report_web.pdf
4. Duprey ZH, Steurer FJ, Rooney JA, Kirchhoff LV, et al. Canine visceral leishmaniasis, United States and Canada, 2000–2003. Emerg Infect Dis. 2006;12:440–6.
5. Petersen CA. Leishmaniasis, an emerging disease found in companion animals in the United States. Top Companion Anim Med. 2009;24:182–8.
6. Lukes J, Mauricio IL, Schönian G, Dujardin JC, et al. Evolutionary and geographical history of the *Leishmania donovani* complex with a revision of current taxonomy. Proc Natl Acad Sci USA. 2007;104:9375–80.
7. Solano-Gallego L, Koutinas A, Miró G, Cardoso L, et al. Directions for the diagnosis, clinical staging, treatment and prevention of canine leishmaniosis. Vet Parasitol. 2009;165:1–18.
8. Baneth G, Koutinas AF, Solano-Gallego L, Bourdeau P, et al. Canine leishmaniosis – new concepts and insights on an expanding zoonosis: part one. Trends Parasitol. 2008;24:324–30.
9. Campino L. Canine reservoirs and leishmaniasis: epidemiology and disease. In: Farrel JP, editor. World Class Parasites *Leishmania*, vol. 4. Boston, Dordrecht, London: Kluwer Academic; 2002. p. 45–57.
10. Cortes S, Afonso MO, Alves-Pires C, Campino L. Stray dogs and leishmaniasis in urban areas, Portugal. Emerg Infect Dis. 2007;13:1431–2.
11. Miró G, Cardoso L, Pennisi MG, Oliva G, et al. Canine leishmaniosis-new concepts and insights on an expanding zoonosis: part two. Trends Parasitol. 2008;24:371–7.
12. Marty P, Le Fichoux Y, Giordana D, Brugnetti A. Leishmanin reaction in the human population of a highly endemic focus of canine leishmaniasis in Alpes-Maritimes, France. Trans R Soc Trop Med Hyg. 1992;86:249–50.
13. Cunha S, Freire M, Eulalio C, Cristóvão J, et al. Visceral leishmaniasis in a new ecological niche near a major metropolitan area of Brazil. Trans R Soc Trop Med Hyg. 1995;89:155–8.
14. Antoniou M, Gramiccia M, Molina R, Dvorak V, et al. The role of indigenous phlebotomine sandflies and mammals in the spreading of leishmaniasis agents in the Mediterranean region. Euro Surveill. 2013;18:20540.
15. Shaw S, Langton D, Hillman T. Canine leishmaniosis in the United Kingdom: a zoonotic disease waiting for a vector? Vet Parasitol. 2009;163:281–5.
16. Svobodova V, Svoboda M, Friedlaenderova L, Drahotsky P, et al. Canine leishmaniosis in three consecutive generations of dogs in Czech Republic. Vet Parasitol. 2017;237:122–4.
17. Karkamo V, Kaistinen A, Näreaho A, Dillard K, et al. The first report of autochthonous non-vector-borne transmission of canine leishmaniosis in the Nordic countries. Acta Vet Scand. 2014;56:84.
18. Naucke T, Lorentz S. First report of venereal and vertical transmission of canine leishmaniosis from naturally infected dogs in Germany. Parasit Vectors. 2012;5:67.
19. Tánczos B, Balogh N, Király L, Biksi I, et al. First record of autochthonous canine leishmaniasis in Hungary. Vector Borne Zoonotic Dis. 2012;12:588–94.
20. Mircean V, Dumitrache MO, Mircean M, Bolfa P. Autochthonous canine leishmaniasis in Romania: neglected or (re)emerging? Parasit Vectors. 2014;7:135.
21. Teske E, van Knapen F, Beijer EGM, Slappendel RJ. Risk of infection with *Leishmania* spp. in the canine population in the Netherlands. Acta Vet Scand. 2002;43:195–201.
22. Maia C, Cardoso L. Spread of *Leishmania infantum* in Europe with dog travelling. Vet Parasitol. 2015;213:2–11.
23. Gaskin AA, Schantz P, Jackson J, Birkenheuer A, et al. Visceral leishmaniasis in a New York foxhound kennel. J Vet Intern Med. 2002;16:34–44.
24. Best MP, Ash A, Bergfeld J, Barrett J. The diagnosis and management of a case of leishmaniosis in a dog imported to Australia. Vet Parasitol. 2014;202:292–5.
25. Maroli M, Rossi L, Baldelli R, Capelli G, et al. The northward spread of leishmaniasis in Italy: evidence from retrospective and ongoing studies on the canine reservoir and phlebotomine vectors. Tropical Med Int Health. 2008;13:256–64.

26. Dereure J, Vanwambeke SO, Malé P, Martinez S, et al. The potential effects of global warming on changes in canine leishmaniasis in a focus outside the classical area of the disease in southern France. Vector Borne Zoonotic Dis. 2009;9:687–94.
27. Ballart C, Alcover MM, Picado A, Nieto J, et al. First survey on canine leishmaniosis in a non-classical area of the disease in Spain (Lleida, Catalonia) based on a veterinary questionnaire and a cross-sectional study. Prev Vet Med. 2013;109:116–27.
28. Miró G, Checa R, Montoya A, Hernández L, et al. Current situation of *Leishmania infantum* infection in shelter dogs in northern Spain. Parasit Vectors. 2012;5:60.
29. Chargui N, Haouas N, Gorcii M, Akrout Messaidi F, et al. Increase of canine leishmaniasis in a previously low-endemicity area in Tunisia. Parasite. 2007;14:247–51.
30. Schantz P, Steurer FJ, Duprey ZH, Kurpel KP, et al. Autochthonous visceral leishmaniasis in dogs in North America. J Am Vet Med Assoc. 2005;226:1316–22.
31. Rosypal AC, Troy GC, Zajac AM, Frank G, et al. Transplacental transmission of a North American isolate of *Leishmania infantum* in an experimentally infected beagle. J Parasitol. 2005;91:970–2.
32. Da Silva S, Ribeiro VM, Ribeiro RR, Tafuri WL, et al. First report of vertical transmission of *Leishmania (Leishmania) infantum* in a naturally infected bitch from Brazil. Vet Parasitol. 2009;166:159–62.
33. Gibson-Corley KN, Hostetter JM, Hostetter SJ, Mullin K, et al. Disseminated *Leishmania infantum* infection in two siblings' foxhounds due to possible vertical transmission. Can Vet J. 2008;49:1005–8.
34. Riera C, Valladares J. Viable *Leishmania infantum* in urine and semen in experimentally infected dogs. Parasitol Today. 1996;12:412.
35. Diniz SA, Melo MS, Borges AM, Bueno R, et al. Genital lesions associated with visceral leishmaniasis and shedding of *Leishmania* sp. in the semen of naturally infected dogs. Vet Pathol. 2005;42:650–8.
36. De Freitas E, Melo MN, Pimenta Da Costa-Val A, Marques Michalick MS. Transmission of *Leishmania infantum* via blood transfusion in dogs: potential for infection and importance of clinical factors. Vet Parasitol. 2006;137:159–67.
37. Tabar MD, Roura X, Francino O, Altet L, et al. Detection of *Leishmania infantum* by real-time PCR in a canine blood bank. J Small Anim Pract. 2008;49:325–8.
38. Ferreira MG, Reinaldo Fattori K, Souza F, Marçal V, et al. Potential role for dog fleas in the cycle of *Leishmania* spp. Vet Parasitol. 2009;165:150–4.
39. Dantas-Torres F, Martins TF, Paiva-Cavalcanti M, Figueredo LA, et al. Transovarial passage of *Leishmania infantum* kDNA in artificially infected *Rhipicephalus sanguineus*. Exp Parasitol. 2010;125:184–5.
40. Campino L, Santos-Gomes G, Pratlong F, Dedet JP, et al. The isolation of *Leishmania donovani* MON-18, from an AIDS patient in Portugal: possible needle transmission. Parasite. 1994;1:391–2.
41. Molina R, Amela C, Nieto J, San-Andrés M, et al. Infectivity of dogs naturally infected with *Leishmania infantum* to colonized *Phlebotomus perniciosus*. Trans R Soc Trop Med Hyg. 1994;88:491–3.
42. Travi BL, Tabares CJ, Cadena H, Ferro C, et al. Canine visceral leishmaniasis in Colombia: relationship between clinical and parasitologic status and infectivity for sand flies. Am J Trop Med Hyg. 2001;64:119–24.
43. Quinnell R, Courtenay O. Transmission, reservoir hosts and control of zoonotic visceral leishmaniasis. Parasitology. 2009;136:1915–34.
44. Campino L. Leishmanioses em Portugal. Características emergentes da epidemiologiae do diagnóstico. Universidade Nova de Lisboa, Instituto de Higiene e Medicina Tropical, 1998; p. 192.
45. Blackwell J, Goswami T, Evans CAW, Sibthorpe D, et al. *SLC11A1* (formerly *NRAMP1*) and disease resistance. Cell Microbiol. 2001;3:773–84.

46. Altet L, Francino O, Solano-Gallego L, Renier C, et al. Mapping and sequencing of the canine *NRAMP1* gene and identification of mutations in leishmaniasis susceptible dogs. Infect Immun. 2002;70:2763–71.
47. Sanchez-Robert E, Altet L, Sanchez A, Francino A. Polymorphism of *SLC11A1* (*NRAMP1*) gene and canine leishmaniasis in a case-control study. J Hered. 2005;96:755–8.
48. Bueno R, Carvalho Neta AV, Xavier MN, Oliveira RG, et al. cDNA sequencing and expression of *NRAMP1* (*SLC11A1*) in dogs phenotypically resistant or susceptible to visceral leishmaniasis. Vet Immunol Immunopathol. 2009;127:332–9.
49. Turchetti A, da Costa LF, Romão Ede L, Fujiwara RT, et al. Transcription of innate immunity genes and cytokine secretion by canine macrophages resistant or susceptible to intracellular survival of *Leishmania infantum*. Vet Immunol Immunopathol. 2015;163:67–76.
50. Quinnell R, Kennedy LJ, Barnes A, Courtenay O, et al. Susceptibility to visceral leishmaniasis in the domestic dog is associated with MHC class II polymorphism. Immunogenetics. 2003;55:23–8.
51. Ciaramella P, Oliva G, De Luna R, Gradoni L, et al. A retrospective clinical study of canine leishmaniasis in 150 dogs naturally infected by *Leishmania infantum*. Vet Rec. 1997;141:539–43.
52. Solano-Gallego L, Llull J, Ramos G, Riera C, et al. The Ibizian hound presents a predominantly cellular immune response against natural *Leishmania* infection. Vet Parasitol. 2000;90:37–45.
53. França-Silva JC, da Costa RT, Siqueira AM, Machado-Coelho GL, et al. Epidemiology of canine visceral leishmaniosis in the endemic area of Montes Claros Municipality, Minas Gerais State, Brazil. Vet Parasitol. 2003;111:161–73.
54. Miranda S, Roura X, Picado A, Ferrer L, et al. Characterization of sex, age, and breed for a population of canine leishmaniosis diseased dogs. Res Vet Sci. 2008;85:35–8.
55. Cortes S, Vaz Y, Neves R, Maia C, et al. Risk factors for canine leishmaniasis in an endemic Mediterranean region. Vet Parasitol. 2012;189:189–96.
56. Killick-Kendrick R. The biology and control of phlebotomine sand flies. Clin Dermatol. 1999;17:279–89.
57. Gálvez R, Miró G, Descalzo MA, Nieto J, et al. Emerging trends in the seroprevalence of canine leishmaniasis in the Madrid region (central Spain). Vet Parasitol. 2010;169:327–34.
58. Koutinas AF, Polizopoulou ZS, Saridomichelakis MN, Argyriadis D, et al. Clinical considerations on canine visceral leishmaniasis in Greece: a retrospective study of 158 cases (1989–1996). J Am Anim Hosp Assoc. 1999;35:376–83.
59. Zivicnjak T, Martinković F, Marinculić A, Mrljak V, et al. A seroepidemiologic survey of canine visceral leishmaniosis among apparently healthy dogs in Croatia. Vet Parasitol. 2005;131:35–43.
60. Fisa R, Gállego M, Castillejo S, Aisa MJ, et al. Epidemiology of canine leishmaniosis in Catalonia (Spain) the example of the Priorat focus. Vet Parasitol. 1999;83:87–97.
61. Queiroz PVS, Monteiro GRG, Macedo VPS, Rocha MAC, et al. Canine visceral leishmaniasis in urban and rural areas of Northeast Brazil. Res Vet Sci. 2009;86:267–73.
62. Abranches P, Conceição-Silva F, Silva-Pereira M. Kala-azar in Portugal. V. The sylvatic cycle in the enzootic endemic focus of Arrábida. J Trop Med Hyg. 1984;87:197–200.
63. Mancianti F, Mignone W, Galastri F. Serologic survey for leishmaniasis in free-living red foxes (*Vulpes vulpes*) in Italy. J Wildl Dis. 1994;30:454–6.
64. Molina R, Fernández BI, Santos Sanz S, Sierra Moros MJ, et al. The hare (*Lepus granatensis*) as potential sylvatic reservoir of *Leishmania infantum* in Spain. Vet Parasitol. 2012;190:268–71.
65. Jiménez M, González E, Martín-Martín I, Hernández S, et al. Could wild rabbits (*Oryctolagus cuniculus*) be reservoirs for *Leishmania infantum* in the focus of Madrid, Spain? Vet Parasitol. 2014;202:296–300.

66. Jimenez M, González E, Iriso A, Marco E, et al. Detection of *Leishmania infantum* and identification of blood meals in *Phlebotomus perniciosus* from a focus of human leishmaniasis in Madrid, Spain. Parasitol Res. 2013;112:2453–9.
67. Ozon C, Marty P, Pratlong F, Breton C, et al. Disseminated feline leishmaniosis due to *Leishmania infantum* in Southern France. Vet Parasitol. 1998;75:273–7.
68. Martín-Sánchez J, Acedo C, Muñoz-Pérez M, Pesson B, et al. Infection by *Leishmania infantum* in cats: epidemiological study in Spain. Vet Parasitol. 2007;145:267–73.
69. Nasereddin A, Salant H, Abdeen Z. Feline leishmaniasis in Jerusalem: serological investigation. Vet Parasitol. 2008;158:364–9.
70. Hatam GR, Hosseini SMH, Ardehali S, Fallah E, et al. First report of natural infection in cats with *Leishmania infantum* in Iran. Vector Borne Zoonotic Dis. 2010;10:313–6.
71. Vides J, Schwardt TF, Sobrinho LS, Marinho M, et al. *Leishmania chagasi* infection in cats with dermatologic lesions from an endemic area of visceral leishmaniosis in Brazil. Vet Parasitol. 2011;178:22–8.
72. Pennisi MG, Lupo T, Malara D, Masucci M, et al. Serological and molecular prevalence of *Leishmania infantum* infection in cats from Southern Italy. J Feline Med Surg. 2012;14:656–7.
73. Chatzis M, Andreadou M, Leontides L, Kasabalis D, et al. Cytological and molecular detection of *Leishmania infantum* in different tissues of clinically normal and sick cats. Vet Parasitol. 2014;202:217–25.
74. Maia C, Ramos C, Coimbra M, Bastos F, et al. Bacterial and protozoal agents of feline vector-borne diseases in domestic and stray cats from southern Portugal. Parasit Vectors. 2014;7:115.
75. Maia C, Sousa C, Ramos C, Cristóvão JV, et al. First case of feline leishmaniosis caused by *Leishmania infantum* genotype E in a cat with concurrent nasal squamous cell carcinoma. J Feline Med Surg. 2015;1:2055116915593969.
76. Pennisi MG, Cardoso L, Baneth G, Bordeau P, et al. Leish Vet update and recommendations on feline leishmaniosis. Parasit Vectors. 2015;8:302.
77. Can H, Döşkaya M, Gökhan Özdemir H, Atalay Şahar E, et al. Seroprevalence of *Leishmania* infection and molecular detection of *Leishmania tropica* and *Leishmania infantum* in stray cats of İzmir, Turkey. Exp Parasitol. 2016;167:109–14.
78. Poli A, Abramo F, Barsotti P, Leva S, et al. Feline leishmaniosis due to *Leishmania infantum* in Italy. Vet Parasitol. 2002;106:181–91.
79. Savani E, de Oliveira Camargo MC, de Carvalho MR, Zamperi RA, et al. The first record in the Americas of an autochthonous case of *Leishmania (Leishmania) infantum chagasi* in a domestic cat (*Felix catus*) from Cotia County, São Paulo State, Brazil. Vet Parasitol. 2004;120:229–33.
80. Maroli M, Pennisi MG, Di Muccio T, Khoury C, et al. Infection of sandflies by a cat naturally infected with *Leishmania infantum*. Vet Parasitol. 2007;145:357–60.
81. Maia C, Nunes M, Campino L. Importance of cats in zoonotic leishmaniasis in Portugal. Vector Borne Zoonotic Dis. 2008;8:555–9.
82. Maia C, Gomes J, Cristóvão J, Nunes M, et al. Feline *Leishmania* infection in a canine leishmaniasis endemic region, Portugal. Vet Parasitol. 2010;174:336–40.
83. Maia C, Campino L. Can domestic cats be considered reservoir hosts of zoonotic leishmaniasis? Trends Parasitol. 2011;27:341–4.
84. Bettini S, Pozio E, Gradoni L. Leishmaniasis in Tuscany (Italy): (II) *Leishmania* form wild Bodentia and Carnivora in a human and canine leishmaniasis focus. Trans R Soc Trop Med Hyg. 1980;74:77–83.
85. Papadogiannakis E, Spanakos G, Kontos V, Menounos PG, et al. Molecular detection of *Leishmania infantum* in wild rodents (*Rattus norvegicus*) in Greece. Zoonoses Public Health. 2009;57:e23–5.
86. Koehler K, Stechele M, Hetzel U, Domingo M, et al. Cutaneous leishmaniosis in a horse in Southern Germany caused by *Leishmania infantum*. Vet Parasitol. 2002;109:9–17.
87. Solano-Gallego L, Fernández-Bellon H, Serra R, Gállego M, et al. Cutaneous leishmaniosis in three horses in Spain. Equine Vet J. 2003;35:320–3.

88. Rolão N, Martins MJ, João A, Campino L. Equine infection with *Leishmania* in Portugal. Parasite. 2005;12:183–6.
89. Soares I, Silva SO, Moreira FM, Prado LG, et al. First evidence of autochthonous cases of *Leishmania (Leishmania) infantum* in horse (*Equus caballus*) in the Americas and mixed infection of *Leishmania infantum* and *Leishmania (Viannia) braziliensis*. Vet Parasitol. 2013;197:665–9.
90. Bhattarai NR, Van der Auwera G, Rijal S, Picado A, et al. Domestic animals and epidemiology of visceral leishmaniasis, Nepal. Emerg Infect Dis. 2010;16:231–7.
91. Rohousova I, Talmi-Frank D, Kostalova T, Polanska T, et al. Exposure to *Leishmania* spp. and sand flies in domestic animals in northwestern Ethiopia. Parasit Vectors. 2015;8:360.
92. Peters W, Elbihari S, Evans D. *Leishmania* infecting man and wild animals in Saudi Arabia. 2. *Leishmania arabica* n. sp. Trans R Soc Trop Med Hyg. 1986;80:497–502.
93. Elbihari S, Cheema A, el-Hassan A. *Leishmania* infecting man and wild animals in Saudi Arabia. 4. Canine cutaneous leishmaniasis in the Eastern Province. Trans R Soc Trop Med Hyg. 1987;81:925–7.
94. Dereure J, Boni M, Pratlong F, El Hadi Osman M, et al. Visceral leishmaniasis in Sudan: first identifications of *Leishmania* from dogs. Trans R Soc Trop Med Hyg. 2000;94:154–5.
95. Dereure J, El-Safi SH, Bucheton B, Boni M, et al. Visceral leishmaniasis in Eastern Sudan: parasite identification in humans and dogs; host-parasite relationships. Microbes Infect. 2003;5:1103–8.
96. Peters W, Elbihari S, Liu C, Le Blancq SM, et al. *Leishmania* infecting man and wild animals in Saudi Arabia. 1. General survey. Trans R Soc Trop Med Hyg. 1985;79:831–9.
97. Morsy T, Schnur LF, Feinsod FM, Salem AM, et al. Natural infections of *Leishmania major* in domestic dogs from Alexandria, Egypt. Am J Trop Med Hyg. 1987;37:49–52.
98. Dereure J, Rioux JA, Gallego M, Perières J, et al. *Leishmania tropica* in Morocco: infection in dogs. Trans R Soc Trop Med Hyg. 1991;85:595.
99. Guessous-Idrissi N, Berrag B, Riyad M, Sahibi H, et al. Short report: *Leishmania tropica*: etiologic agent of a case of canine visceral leishmaniasis in northern Morocco. Am J Trop Med Hyg. 1997;57:172–3.
100. Lemrani M, Nejjar R, Pratlong F. A new *Leishmania tropica* zymodeme causative agent of canine visceral leishmaniasis in northern Morocco. Ann Trop Med Parasitol. 2002;96:637–8.
101. Ntais P, Christodoulou V, Tsirigotakis N, Dokianakis E, et al. Will the introduction of *Leishmania tropica* MON-58, in the island of Crete, lead to the settlement and spread of this rare zymodeme? Acta Trop. 2014;132:125–30.
102. Tolezano JE, Uliana SR, Taniguchi HH, Araújo MF, et al. The first records of *Leishmania (Leishmania) amazonensis* in dogs (*Canis familiaris*) diagnosed clinically as having canine visceral leishmaniasis from Araçatuba County, São Paulo State, Brazil. Vet Parasitol. 2007;149:280–4.
103. Ramírez J, Hernández C, León CM, Ayala MS, et al. Taxonomy, diversity, temporal and geographical distribution of cutaneous leishmaniasis in Colombia: a retrospective study. Sci Rep. 2016;6:28266.
104. Aguilar C, Fernandez E, Fernandez R, Deane L. Study of an outbreak of cutaneous leishmaniasis in Venezuela. The role of domestic animals. Mem Inst Oswaldo Cruz. 1984;79:181–95.
105. Vélez I, Carrillo LM, López L, Rodríguez E, et al. An epidemic outbreak of canine cutaneous leishmaniasis in Colombia caused by *Leishmania braziliensis* and *Leishmania panamensis*. Am J Trop Med Hyg. 2012;86:807–11.
106. Delgado O, Castes M, Clinton White A Jr, Kreutzer RD. *Leishmania colombiensis* in Venezuela. Am J Trop Med Hyg. 1993;48:145–7.
107. Hashiguchi Y, Gomez EA, de Coronel VV, Mimori T, et al. Andean leishmaniasis in Ecuador caused by infection with *Leishmania mexicana* and *L. major*-like parasites. Am J Trop Med Hyg. 1991;44:205–17.
108. Dereure J, Espinel I, Barrera C, Guerrini F, et al. Leishmaniasis in Ecuador. 4. Natural infestation of the dog by *Leishmania panamensis*. Ann Soc Belg Med Trop. 1994;74:29–33.

109. Llanos-Cuentas EA, Roncal N, Villaseca P, Paz L, et al. Natural infections of *Leishmania peruviana* in animals in the Peruvian Andes. Trans R Soc Trop Med Hyg. 1999;93:15–20.
110. Hassan M, Osman FO, El-Raba'a F, Schallig H, et al. Role of the domestic dog as a reservoir host of *Leishmania donovani* in Eastern Sudan. Parasit Vectors. 2009;2:26.
111. Padilla A, Marco JD, Diosque P, Segura MA, et al. Canine infection and the possible role of dogs in the transmission of American tegumentary leishmaniosis in Salta, Argentina. Vet Parasitol. 2002;110:1–10.
112. Reithinger R, Espinoza J, Davies C. The transmission dynamics of canine American cutaneous leishmaniasis in Huánuco, Peru. Am J Trop Med Hyg. 2003;69:473–80.
113. Peña M, Roura X, Davidson M. Ocular and periocular manifestations of leishmaniasis in dogs: 105 cases (1993–1998). Vet Ophtalmol. 2000;1:35–41.
114. Giunchetti RC, Martins-Filho OA, Carneiro CM, Mayrink W, et al. Histopathology, parasite density and cell phenotypes of the popliteal lymph node in canine visceral leishmaniasis. Vet Immunol Immunopathol. 2008a;121:23–33.
115. Santana C, Vassallo J, De Freitas LA, Oliveira G, et al. Inflammation and structural changes of splenic lymphoid tissue in visceral leishmaniasis: a study on naturally infected dogs. Parasite Immunol. 2008;30:515–24.
116. Giunchetti RC, Mayrink W, Carneiro CM, Corrêa-Oliveira R, et al. Histopathological and immunohistochemical investigations of the hepatic compartment associated with parasitism and serum biochemical changes in canine visceral leishmaniasis. Res Vet Sci. 2008b;84:269–77.
117. Nieto C, Navarrete I, Habela MA, Serrano F, et al. Pathological changes in kidneys of dogs with natural *Leishmania* infection. Vet Parasitol. 1992;45:33–47.
118. Petanides T, Koutinas AF, Mylonakis ME, Day MJ, et al. Factors associated with the occurrence of epistaxis in natural canine leishmaniasis (*Leishmania infantum*). J Vet Intern Med. 2008;22:866–72.
119. Blavier A, Keroack S, Denerolle P, Goy-Thollot I, et al. Atypical forms of canine leishmaniosis. Vet J. 2001;162:108–20.
120. Maia C, Campino L. Methods for diagnosis of canine leishmaniasis and immune response to infection. Vet Parasitol. 2008;158:274–87.
121. Paltrinieri S, Gradoni L, Roura X, Zatelli A, et al. Laboratory tests for diagnosing and monitoring canine leishmaniasis. Vet Clin Pathol. 2016;45:552–78.
122. Maia C, Ramada J, Cristóvão JM, Gonçalves L, et al. Diagnosis of canine leishmaniasis: conventional and molecular techniques using different tissues. Vet J. 2009;179:142–4.
123. Francino O, Altet L, Sánchez-Robert E, Rodriguez A, et al. Advantages of real-time PCR assay for diagnosis and monitoring of canine leishmaniosis. Vet Parasitol. 2006;137:214–21.
124. Gramiccia M, Di Muccio T, Fiorentino E, Scalone A, et al. Longitudinal study on the detection of canine *Leishmania* infections by conjunctival swab analysis and correlation with entomological parameters. Vet Parasitol. 2010;171:223–8.
125. Belinchón-Lorenzo S, Iniesta V, Parejo JC, Fernández-Cotrina J, et al. Detection of *Leishmania infantum* kinetoplast minicircle DNA by real time PCR in hair of dogs with leishmaniosis. Vet Parasitol. 2013;192:43–50.
126. Belinchón-Lorenzo S, Parejo JC, Iniesta V, Fernández-Cotrina J, et al. First detection of *Leishmania* kDNA in canine cerumen samples by qPCR. Vet Parasitol. 2016;228:65–8.
127. Ferreira A, Almeida GG, de Oliveira Silva S, Vogas GP, et al. Nasal, oral and ear swabs for canine visceral leishmaniasis diagnosis: new practical approaches for detection of *Leishmania infantum* DNA. PLoS Negl Trop Dis. 2013;7:e2150.
128. Noli C, Auxilia S. Treatment of canine Old World visceral leishmaniasis: a systematic review. Vet Dermatol. 2005;16:213–32.
129. Miró G, Oliva G, Cruz I, Canavate C, et al. Multicentric, controlled clinical study to evaluate effectiveness and safety of miltefosine and allopurinol for canine leishmaniosis. Vet Dermatol. 2009;20:397–404.

130. Bianciardi P, Brovida C, Valente M, Aresu L, et al. Administration of miltefosine and meglumine antimoniate in healthy dogs: clinicopathological evaluation of the impact on the kidneys. Toxicol Pathol. 2009;37:770–5.
131. Mateo M, Maynard L, Vischer C, Bianciardi P, et al. Comparative study on the short-term efficacy and adverse effects of miltefosine and meglumine antimoniate in dogs with natural leishmaniosis. Parasitol Res. 2009;105:155–62.
132. Oliva G, Gradoni L, Ciaramella P, De Luna R, et al. Activity of liposomal amphotericin B (AmBisome) in dogs naturally infected with *Leishmania infantum*. J Antimicrob Chemother. 1995;36:1013–9.
133. Sabaté D, Llinás J, Homedes J, Sustc M, et al. A single-centre, open-label, controlled, randomized clinical trial to assess the preventive efficacy of a domperidone-based treatment programme against clinical canine leishmaniasis in a high prevalence area. Prev Vet Med. 2014;115:56–63.
134. Croft S, Sundar S, Fairlamb A. Drug resistance in leishmaniasis. Clin Microbiol Rev. 2006;19:111–26.
135. Gramiccia M, Gradoni L, Orsini S. Decreased sensitivity to meglumine antimoniate (Glucantime) of *Leishmania infantum* isolated from dogs after several courses of drug treatment. Ann Trop Med Parasitol. 1992;86:613–20.
136. Carrió J, Portús M. *In vitro* susceptibility to pentavalent antimony in *Leishmania infantum* strains is not modified during *in vitro* or *in vivo* passages but is modified after host treatment with meglumine antimoniate. BMC Pharmacol. 2002;2:11.
137. Luz R, Melo M, Dujardin J, Maes L. Post-treatment Sb-susceptibility of *Leishmania infantum* clinical isolates of dogs. In: Fourth World Congress on Leishmaniasis. Lucknow: 4th Worldleish; 2009. p. 273.
138. Maia C, Nunes M, Marques M, Henriques S, et al. *In vitro* drug susceptibility of *Leishmania infantum* isolated from humans and dogs. Exp Parasitol. 2013;135:36–41.
139. Gradoni L, Maroli M, Gramiccia M, Mancianti F. *Leishmania infantum* infection rates in *Phlebotomus perniciosus* fed on naturally infected dogs under antimonial treatment. Med Vet Entomol. 1987;4:339–42.
140. Aït-Oudhia K, Gazanion E, Sereno D, Oury B, et al. *In vitro* susceptibility to antimonials and amphotericin B of *Leishmania infantum* strains isolated from dogs in a region lacking drug selection pressure. Vet Parasitol. 2012;187:386–93.
141. Seblova V, Oury B, Eddaikra N, Aït-Oudhia K, et al. Transmission potential of antimony-resistant *Leishmania* field isolates. Antimicrob Agents Chemother. 2014;58:6273–6.
142. Yasur-Landau D, Jaffe CL, David L, Baneth G. Allopurinol resistance in *Leishmania infantum* from dogs with disease relapse. PLoS Negl Trop Dis. 2016;10:e0004341.
143. WHO. Report of a WHO informal consultation on liposomal amphotericin Bin the treatment of visceral leishmaniasis Rome, Italy; 2005. WHO/CDS/NTD/IDM/2007.4
144. Courtenay O, Quinnell RJ, Garcez LM, Shaw JJ, et al. Infectiousness in a cohort of Brazilian dogs: why culling fails to control visceral leishmaniasis in areas of high transmission. J Infect Dis. 2002;186:1314–20.
145. Nunes CM, Pires MM, da Silva KM, Assis FD, et al. Relationship between dog culling and incidence of human visceral leishmaniasis in an endemic area. Vet Parasitol. 2010;170:131–3.
146. Killick-Kendrick R, Killick-Kendrick M, Focheux C, Dereure J, et al. Protection of dogs from bites of phlebotomine sandflies by deltamethrin collars for control of canine leishmaniasis. Med Vet Entomol. 1997;11:105–11.
147. Courtenay O, Kovacic V, Gomes PA, Garcez LM, et al. A long-lasting topical deltamethrin treatment to protect dogs against visceral leishmaniasis. Med Vet Entomol. 2009;23:245–56.
148. Gavgani A, Hodjati MH, Mohite H, Davies CR. Effect of insecticide-impregnated dog collars on incidence of zoonotic visceral leishmaniasis in Iranian children: a matched-cluster randomised trial. Lancet. 2002;360:374–9.

149. Manzillo VF, Oliva G, Pagano A, Manna L, et al. Deltamethrin-impregnated collars for the control of canine leishmaniasis: evaluation of the protective effect and influence on the clinical outcome of *Leishmania* infection in kenneled stray dogs. Vet Parasitol. 2006;142:142–5.
150. Maroli M, Gradoni L, Oliva G, Castagnaro M, et al. Guidelines for prevention of leishmaniasis in dogs. J Am Vet Med Assoc. 2010;236:1200–6.
151. Gradoni L. Canine *Leishmania* vaccines: still a long way to go. Vet Parasitol. 2015;208:94–100.
152. Silva V, Borja-Cabrera GP, Correia Pontes NN, de Souza EP, et al. A phase III trial of efficacy of the FML-vaccine against canine kala-azar in an endemic area of Brazil (São Gonçalo do Amaranto, RN). Vaccine. 2000;19:1082–92.
153. Borja-Cabrera G, Correia Pontes NN, da Silva VO, Paraguai de Souza E, et al. Long lasting protection against canine kala-azar using the FML-Quil A saponin vaccine in an endemic area of Brazil (São Gonçalo do Amarante, RN). Vaccine. 2002;20:3277–84.
154. Mohebali M, Khamesipour A, Mobedi I, Zarei Z, et al. Double-blind randomized efficacy field trial of alum precipitated autoclaved *Leishmania major* vaccine mixed with BCG against canine visceral leishmaniasis in Meshkin-Shahr district, I.R. Iran. Vaccine. 2004;22:4097–100.
155. Lemesre JL, Holzmuller P, Gonçalves RB, Bourdoiseau G, et al. Long-lasting protection against canine visceral leishmaniasis using the LiESAp-MDP vaccine in endemic areas of France: double-blind randomised efficacy field trial. Vaccine. 2007;25(21):4223–34.
156. Daneshvar H, Namazi MJ, Kamiabi H, Burchmore R, et al. Gentamicin-attenuated *Leishmania infantum* vaccine: protection of dogs against canine visceral leishmaniosis in endemic area of southeast of Iran. PLoS Negl Trop Dis. 2014;8(4):e2757.
157. Oliva G, Nieto J, Foglia Manzillo V, Cappiello S, et al. A randomised, double-blind, controlled efficacy trial of the LiESP/QA-21 vaccine in naïve dogs exposed to two *Leishmania infantum* transmission seasons. PLoS Negl Trop Dis. 2014;8:e3213.
158. Regina-Silva S, Feres AM, França-Silva JC, Dias ES, et al. Field randomized trial to evaluate the efficacy of the Leish-Tec® vaccine against canine visceral leishmaniasis in an endemic area of Brazil. Vaccine. 2016;34:2233–9.
159. Borja-Cabrera G, Cruz Mendes A, Paraguai de Souza E, Hashimoto Okada LY, et al. Effective immunotherapy against canine visceral leishmaniasis with the FML-vaccine. Vaccine. 2004;22:2234–43.
160. Saraiva E, de Figueiredo Barbosa A, Santos FN, Borja-Cabrera GP, et al. The FML-vaccine (Leishmune) against canine visceral leishmaniasis: a transmission blocking vaccine. Vaccine. 2006;24:2423–31.
161. Bongiorno G, Paparcone R, Foglia Manzillo V, Oliva G, et al. Vaccination with LiESP/QA-21 (CaniLeish®) reduces the intensity of infection in *Phlebotomus perniciosus* fed on *Leishmania infantum* infected dogs – a preliminary xenodiagnosis study. Vet Parasitol. 2013;197:691–5.
162. Alvar J, Cañavate C, Molina R, Moreno J, et al. Canine leishmaniasis. Adv Parasitol. 2004;57:1–88.

Epidemiology of Leishmaniasis in the Time of Drug Resistance (the Miltefosine Era)

4

Jean-Claude Dujardin

Abstract

In the first edition of this chapter (Dujardin J-C, Decuypere S, Epidemiology of leishmaniasis in the time of drug resistance. In: Ponte-Sucre A, Padron-Nieves M, Diaz E (eds) Drug resistance in *Leishmania* parasites: consequences, molecular mechanism and possible treatments. Spinger, pp 65–83), we updated various aspects of leishmaniasis epidemiology, with a particular emphasis on their relation with parasite drug resistance (DR), with a focus on antimonials (SSG). We made a clear distinction between DR, a parasite phenotype measured in the laboratory and treatment failure (TF), a clinical phenotype assessed in the patient. In this second edition, the objective is to update knowledge (whatever the drug) in this domain and to focus on miltefosine (MIL), contrasting wherever relevant with SSG. In the first part, we present data on the current efficacy of MIL, highlighting the increase in TF, and only a few cases of DR. Then, we update information on the risk factors for (re)emergence and spreading of leishmaniasis, focusing on the link between DR and TF: among others, we discuss the role of asymptomatics and animals, the importance of co-infection (considering the usual suspect HIV but also newcomers as *Leishmania* RNA viruses (LRV)), and the risk related with massive human migrations and environmental changes. Finally, we review the advances made about tools for epidemiological surveillance of TF/DR, ranging from clinical ones to laboratory ones. Recommendations and perspectives for further research will be discussed at the end.

J.-C. Dujardin (✉)
Molecular Parasitology Unit, Institute of Tropical Medicine, Antwerp, Belgium
e-mail: jcdujardin@itg.be

A. Ponte-Sucre, M. Padrón-Nieves (eds.), *Drug Resistance in Leishmania Parasites*,
https://doi.org/10.1007/978-3-319-74186-4_4

4.1 Introduction

In the first edition of this chapter [1], various aspects of leishmaniasis epidemiology were reviewed in the particular context of drug resistance (DR). Three years after, the best reference on the incidence of the disease worldwide remains the paper of Alvar et al. ([2], published while our first chapter was in the publication pipeline). The study from Alvar, based on different sources and taking into account among others underreporting of cases, provided an incidence estimate of 0.2–0.4 cases and 0.7–1.2 million cases of visceral (VL) and cutaneous leishmaniasis (CL), respectively, with an estimated mortality of 20,000–40,000 cases each year. However, the figures in the Indian subcontinent (ISC) changed as a consequence of the Kala-azar Elimination Program (KAEP) that aimed to decrease the incidence of VL below 1/10,000 at district or subdistrict level by 2017 [3]. A recent update of the efficacy of this regional program showed that Nepal has met the target, Bangladesh 90% of it, and India two thirds of it [3, 4]. This together with the remaining challenges encountered by the program will be further detailed in the chapter on visceral leishmaniasis in the present volume (Chap. 7).

Our intention in this present chapter is not to repeat the review of the main risk factors for (re)emergence and spreading of leishmaniasis but well to highlight new epidemiological findings, concepts, and tools particularly relevant in the context of surveillance of treatment failure (TF) and DR. After a focus on antimonials (SSG) in the first edition, here we will focus wherever possible on miltefosine (MIL), the drug that replaced SSG in the ISC and for which we now have more data. We will describe that despite a fear for rapid emergence of DR against that drug, this still occurs rarely in the ISC where it was used massively in the frame of the KAEP but that, in contrast, TF rate increased significantly, highlighting again the major conceptual difference between these two concepts, TF and DR.

With respect to transmission, we will highlight the growing attention given to asymptomatics and animals and the new insights provided by mathematical modeling. Human-made and environmental changes remain a critical issue that will be discussed in the setting of the migrants' crisis and (peri-)urban development.

The role of co-infections and their impact on immune status and emergence of DR and/or TF remain critical: this will be described in the context of HIV co-infections, as the first case of clinical MIL resistance was found in a leishmanial/HIV-co-infected patient; however, we will also discuss the recent findings about the role of the endosymbiotic *Leishmania* RNA virus (LRV) and other pathogens in treatment outcome. A major attention will be given to the new diagnostic and epidemiological tools developed in the last years, for applications in both clinical and laboratory settings.

4.2 Epidemiology of Miltefosine Resistance and Treatment Outcome

We previously insisted on the need to clearly distinguish (1) DR, a parasite phenotype characterized by a decreased susceptibility to drug, following successful molecular adaptation under drug pressure, and (2) TF, the clinical phenotype of a patient not responding to a given treatment or presenting a relapse [1]. In the case of SSG, it was clear that DR was indeed playing a major (but not unique) role in TF, at least in the ISC. In the case of MIL, the situation is (still) rather different.

Analysis of the most extensive trials made on MIL efficacy since 2002 (Table 4.1) reveals highly variable outcomes (from 0% to 66% of failure), depending among others on clinical forms, *Leishmania* species, dosage, or clinical protocols. While all these reports may be interesting from a clinical point of view, a few of them only provide useful epidemiological information, and we will focus on three of them. **Firstly**, studies of Sundar et al. [8] and Rijal et al. [9] in the ISC demonstrated—for a similar dosage—an increase in TF rate of VL cases due to *L. (L.) donovani*, after a decade of MIL use. Noteworthy, increasing the duration of follow-up from 6 to 12 months showed a doubling of TF rate in Nepal (from 10.8% to 20% after 6 months and 12 months, respectively), suggesting potential underreporting in studies where follow-up was limited to 6 months. **Secondly**, in Ethiopia, Ritmeijer et al. [13] reported a significant increase in TF rate of VL cases depending on the immune status of the patients (0% and 35.5% TF in HIV-negative and HIV-positive patients, respectively), highlighting the intertwining between the epidemiology of HIV-*Leishmania* co-infection and leishmaniasis TF, a feature that was already described with other drugs [23]. **Thirdly**, within the same study, Soto et al. [16] analyzed the efficacy of MIL in Neotropical CL and found differences in TF rate depending on species (35% and 66% in subjects infected with *L. (L.) mexicana* and *L. (V.) braziliensis*, respectively, both in Guatemala); this species effect is well known and was already observed in the case of SSG, where *L. (V.) braziliensis* was also the species associated with the highest TF rate [24]. Interestingly, another report showed that patients infected with *L. (V.) braziliensis* in Bolivia responded much better to MIL than in Guatemala (same dosage, 12% TF [17]), suggesting the occurrence of confounding factors, like different variants of that parasite species, ethnic differences, or clinical features (like LRV infection; see below).

Reports on the MIL susceptibility of clinical isolates are scantier (Table 4.2), and a series of considerations are required for their interpretation. On one hand, a cutoff needs to be defined to identify miltefosine-resistant (MIL-R) strains (if any) within a given species. In our own practice, we use the ED_{50} value of a laboratory strain of the same species, for which MIL resistance had been experimentally selected. For *L. (L.) donovani*, we used 32 and 74 μM for assays based on amastigotes or promastigote, respectively [32]. For *L. (V.) braziliensis* and *L. (V.) panamensis*, 32 μM was used as the cutoff value (amastigote-based assays [30]). For *L. (L.) infantum*, 20 μM was the selected cutoff [31]. We propose to consider as resistant isolates only those that display an ED_{50} positioned within the same range as that of the experimental resistant strains; if ED_{50} is higher in clinical isolates than the chosen baseline but

Table 4.1 Selection of reports on the outcome of MIL treatment in different clinical forms and regions; papers organized according to the period of recruitment to follow the evolution of treatment outcome

Report	Clinical form (species), country	No of patients[a]	Treatment regimen	Period of recruitment	Treatment failure rate[b]
ISC	VL				
[5]	VL, India	291 (adults)	28 days of 50 mg/day (≤25 kg) or 100 mg/day (>25 kg)	1999–2000	3% (6 months)
[6]	VL, India	79 (children)	28 days of 2.5 mg/kg/day	2001–2002	5% (6 months)
[7]	VL, India	71	28 days, 2.5 mg/kg body weight/day	2007–2009	7%
[8]	VL, India	551 adults and children	28 days of 2.5 mg/kg/day (<12 years), 50 mg/day (≤25 kg) or 100 mg/day (>25 kg) MLF	2009–2010	6.9% (6 months)
[9]	VL, *L. (L.) donovani*, Nepal	115	Individuals aged ≥12 years weighing >25 kg: 100 mg daily; those weighing <25 kg: 50 mg daily; children aged 2–11 years: 2.5 mg/kg daily, for 28 days	2008–2011	10.8% (6 months); 20% (12 months)
ISC	PKDL				
[10]	PKDL, India	18	12 weeks 50-mg capsules: two capsules for patients >25 kg or one capsule for <25 kg	2007–2009	5.5% (12 months)
[10]	PKDL, India	17	2.5 mg/kg/day for 8 weeks	2007–2009	17.6% (12 months)
[11]	PKDL, India	57	50 mg twice daily for 90 days	2008–2012	10.5% (18 months)
[11]	PKDL, India	17	50 mg thrice for 60 days	2008–2012	31% (18 months)
[12]	PKDL, India	29	50 mg, if <25 kg or 100 mg in divided doses if ≥ 25 kg; 2.5 mg/kg for children	2009–2010	3.4% (12 months)
East Africa	VL				
[13]	VL, non-HIV, Ethiopia	106	100 mg/day for 28 days	1997–	0% (6 months)
[13]	VL, HIV, Ethiopia	45	100 mg/day for 28 days	1997–	35.5% (6 months)

[14]	VL, HIV-neg, Sudan & Kenya	49	2.5 mg/kg/day for 28 days	2010–2012	28% (6 months)
Middle East	ZCL				
[15]	ZCL, Iran	28	Unavailable	Unavailable	7.1% (6 months)
New World	CL				
[16]	CL, *L. (V.) panamensis*, Colombia	44		2000–2002	9% (6 months)
[16]	CL, Guatemala	38		2000–2002	47.4% (6 months)
[16]	*L. (V.) braziliensis*	15		2000–2002	66% (6 months)
[16]	*L. (L.) mexicana*	14		2000–2002	35% (6 months)
[17]	CL, *L. (V.) braziliensis*, Bolivia	41	2.5 mg/kg/day for 28 days	2005–2007	12% (6 months)
[18]	CL, *L. (V.) braziliensis or L. (V.) panamensis*, Colombia	122	50 mg, 3 times per day for 28 days	2006–2008	30.2% (6 months)
[19]	CL, *L. (V.) braziliensis* or *L. (V.) panamensis*, Colombia	122	3 × 50 mg/day/28 days	2006–2008	26% (6 months)
	L. (V.) panamensis	30			40%
	L. (V.) braziliensis	51			50.9%
[20]	CL, *L. (V.) guyanensis*, Brazil	56	2.5 mg/kg of body weight (maximum daily dose of 150 mg) for 28 consecutive days	2007–2008	25% (6 months)
[21]	CL, *L. (V.) braziliensis*, Brazil	58	2.5 mg/kg of body weight (maximum daily dose of 150 mg) for 28 consecutive days	2007–2008	25.8% (6 months)
New World	MCL				
[22]	ML, *L. (V.) braziliensis* or *L. (L.) amazonensis*), Argentina	8	2.5–3.3 mg/kg/day (maximal dose is 150 mg/day), during 28–35 days	2010–2012	12.5%

VL visceral, *CL* cutaneous, *ML* (mucosal) leishmaniasis, *PKDL* post-Kala-azar dermal leishmaniasis
[a]Patients who got complete treatment, excluding dead ones and those lost to follow-up
[b]Calculated on the number of patients

Table 4.2 MIL susceptibility of *Leishmania* clinical isolates

Report	Species	Country	Strains	Stage	Susceptibility	Molecular characterization
		ISC				
[25]	*L. (L.) donovani*	Nepal	24, VL, 2002–2004	Ama	6.3[a]	nd
[26]	*L. (L.) donovani*	India	– 13, VL, cured, 2009–2010	Ama	– 2.4[a]	No SNPs
			– 3 relapses, VL, 2009–2010		– 4.7[a]	
			– 5 pre-tx, PKDL, 1998–2011		– 8.6[a]	
			– 3 post-tx, PKDL, 1998–2011)		– 16.1[a]	
			– 22 VL (2009–2010)		– 2.6[a]	
			– 8 PKDL (1998–2011)		– 11.4[a]	
[27]	*L. (L.) donovani*	India	– 24, VL, pre-tx, 2009–2010	Pro	– 3.7[a]	nd
			– 26, VL, post-tx, 2009–2010		– 6.1[a]	
[9]	*L. (L.) donovani*	Nepal	– 17, VL, cure, 2008–2011	Pro	– 4.3[a]	nd
			– 26, VL, relapses, 2008–2011		– 2.6[a]	
[28]	*L. (L.) donovani*	India	2, VL, failure, 2011	Pro	– >100	2 non synonymous SNPs in LdMT
		LAT				
[25]	*L. (V.) braziliensis*	Peru	– 7, CL, 2002–2004	Ama	– 67.4[a]	nd
	L. (V.) guyanensis		– 2, CL, 2002–2004		– 47.2[a]	
	L. (V.) lainsoni		– 4, CL, 2002–2004		– 6.6[a]	
	L. (L.) mexicana		– 1, CL, 2002–2004		– >73.8[a]	
[29]	*L. (V.) braziliensis*	Colombia	– 63	Ama	– 68%[b]	nd
	L. (V.) panamensis		– 107, 1980–2010		– 20%[b]	
	L. (V.) guyanensis		– 34		– 3%[b]	

[30]	*L. (V.) panamensis*	Colombia	– 1 pair, both pre- and post-tx	Ama	– From 10 to >32[a]	In post-tx: no SNPs, lower expression of LbMT
	L. (V.) braziliensis		– 1 pair, both pre- and post-tx		– From 23 to >32[a]	
		EUR				
[31]	*L. (L.) infantum*	France	– 1 HIV/VL, 2005	Ama	> 20[a]	SNPs in LiMT and LiRos3

ISC Indian subcontinent, *LAT* Latin America; tested strains: number, clinical form (*VL* visceral leishmaniasis, *CL* cutaneous leishmaniasis), MIL-treatment outcome of the corresponding patient (when available), moment of isolation (pre- or post-tx, respectively, before the onset of treatment or after failure), period of sampling; stage: Ama, intracellular amastigotes; Pro, axenic promastigotes; susceptibility: [a]ED_{50} in μM (average), [b]% of strains considered to be resistant to MIL (i.e., resistance criterion being a reduction of parasite burden <44% at 16 μM MIL); molecular characterization: nd, not done; SNPs (Single Nucleotide Polymorphisms), after sequencing of LxMT (*Leishmania* miltefosine transporter, x standing for the species: d, donovani, I, infantum, b, braziliensis) or LxRos3 (β subunit of LxMT); expression, mRNA level of LxMT or LxRos3 lower than the defined cutoff, we propose to consider them as MIL-tolerant. On the other hand, as in the case of other drugs, innate insensitivity to MIL should be distinguished from MIL resistance. Indeed, Yardley et al. [25] report that *L. (V.) braziliensis*, *L. (V.) guyanensis*, and *L. (L.) mexicana* isolates all collected in Peru in 2002–2004 expressed already a high ED_{50} (>32 μM), while MIL was not yet in use in Peru at that time. This highlights the importance of baseline susceptibility studies before the implementation of any drug.

Keeping this in mind, the following epidemiological information can be extracted from Table 4.2. **Firstly**, the baseline susceptibility of *L. (L.) donovani* in the ISC before implementation of MIL is rather high (low ED_{50} values); later measurements made during the MIL era on VL isolates from relapsing cases did not express significant changes in ED_{50}. Accordingly, MIL resistance is not (yet) a major issue in the ISC, and causes of the increasing MIL-TF were so far essentially independent of drug susceptibility: parasite virulence [33], pharmacokinetics [34], and age/gender of the patients [35]. Noteworthy, although isolates from post-Kala-azar dermal leishmaniasis (PKDL)-TF showed higher ED_{50}, these should still be considered as MIL-tolerant. It is not surprising that the less susceptible strains are isolated from PKDL, as patients suffering of PKDL require longer therapeutic schemes, with a consequent higher drug pressure. Current findings are anyway rather concerning: considering the potential role of PKDL patients as reservoir for the parasite, PKDL patients might be included as potential contributors to the emergence and spreading of MIL resistance in the future. Unfortunately, this future seems to be already here, as at the moment of finalizing this chapter, already a first report has been published, describing two cases of MIL resistance in India [28]: both clinical isolates showed high IC_{50} values together with non-synonymous mutations in the gene encoding the *L. (L.) donovani* miltefosine transporter (LdMT). **Secondly**, the situation is rather complex in Latin America where innate insensitivity and resistance are possibly intertwined. Noteworthy, Fernandez et al. [29] reported a breakpoint in the distribution of the MIL susceptibility of *L. (V.) panamensis*, with less susceptible strains being described after 2005, year of implementation of MIL as second-line therapy against leishmaniasis in Colombia. Furthermore, Obonaga et al. [30] compared the susceptibility of paired samples (before treatment and at the time of failure) and found a reduced susceptibility at failure. It is not clear if this reflects (1) a progressive adaptation of the strains under MIL pressure and the emergence of acquired MIL resistance or (2) the selection of parasites with an innate lower susceptibility. The report of Obonaga et al. [30] could support acquired resistance, if it could be shown that the same strain was present at the onset of treatment (low ED_{50}). **Thirdly**, immunosuppression could accelerate the emergence of MIL resistance, a hypothesis supported by the detection of the first case of clinical resistance in a European HIV-co-infected patient [31].

4.3 Transmission Patterns

Accurate description of transmission patterns is essential for the design of efficient control programs. These are based on a series of assumptions—if not dogmas—often originating from studies undertaken decades ago, and it might be relevant to revisit these with modern tools that are now available. The best illustration of this need is anthroponotic VL in the ISC. The main assumptions underlying the KAEP can be summarized as follows: (1) humans constitute the reservoir of *L. (L.) donovani*, VL patients playing a major role here as suggested by the higher risk of infection among household contacts of patients [36], (2) with PKDL cases constituting the main reservoir during inter-epidemic periods [37], and (3) *Phlebotomus argentipes* is the vector and infects people within their houses. Accordingly, main pillars of the KAEP are diagnosis and treatment of patients together with insecticide spraying of the houses. Since the launching of the KAEP in 2005, new information was gathered, which questions each of these assumptions and could justify some fine-tuning of control programs, if they were confirmed.

Firstly, two studies based on polymerase chain reaction (PCR) [38] and serology [39], respectively, were performed in Nepal and demonstrated that goats living in VL endemic villages were infected with *L. (L.) donovani* (16% of PCR positivity in blood and 21.6% of direct agglutination test positivity). Classification tree analysis demonstrated that among several risk factors for asymptomatic infection among humans living in the neighborhood, the proximity of infected goats ranked first [38]. Similarly, in India, *L. (L.) donovani* DNA was encountered in the blood of goats [40], and ownership of domestic animals was shown to constitute a risk factor for human infection in high-transmission foci [41]. Last but not the least, a report suggested the possible epidemiological role of dogs in domestic foci of VL in Bangladesh [42]. However, new dogmas should not be created from these reports. Indeed, these studies did not demonstrate that animals could constitute a reservoir of *L. (L.) donovani* in the ISC, as they could equally constitute a "sink," i.e., a dead-end host infected from a human reservoir. To answer that question, studies on the infectiousness of *L. (L.) donovani*-infected domestic animals would be required, ideally by xenodiagnosis. **Secondly**, a series of mathematical modeling studies provided more detailed information on the human reservoir itself. A basic transmission model for anthroponotic VL was elaborated, and simulation results suggested that transmission of *L. (L.) donovani* is predominantly driven by asymptomatically infected humans [43]. This makes sense as these are indeed more numerous than VL patients (infection: disease ratio of 8.9 to 1, estimation based on serology [44], but like for animals, their infectiousness should be demonstrated by xenodiagnosis [45]). In the absence of diagnostics and treatment of asymptomatics (toxicity and cost of current drugs), these data strengthen the importance of vector control in the frame of the KAEP, as it may interfere with the transmission to/from asymptomatics, which was verified by mathematical modeling [43]. Asymptomatic infections might thus constitute a threat for the elimination program [46], and more than ever, research is required on these areas of knowledge. **Thirdly**, while there is no doubt about the role of *P. argentipes* in the transmission of *L. (L.) donovani* (prevalence of infection ranging between 4.9% and 17.4% in Bihar [47], around 0.5% in Nepal [48]), some reports may raise questions about the intra-domiciliary behavior of this vector. Indeed, after a long trial on the effectiveness of long-lasting insecticidal bed nets (supposed to protect at night inside the houses) in the ISC, the authors concluded that these were not providing an additional protection against VL in comparison with control measures in place [49]: this could reflect a lack of power in the study but could also be explained by an unexpected behavior of the sand flies, like biting outside the houses. Here again, further studies on the biology and ecology of sand flies are required.

The basic VL model described above was also used directly in the context of DR. More specifically, authors tried to explain the observed increase in the SSG-TF rate in the ISC from about 5% in 1980 to about 64% in 1997 [50]. The model suggests that such a quick rise in TF could not be reproduced even if first-line treatment would fail in 100% of cases infected with the resistant strain. Thus, the authors concluded that additional assumptions were required, for instance, that SSG-resistant parasites were transmitted more effectively than nonresistant parasites. This prediction was verified experimentally by several studies linking SSG resistance with higher metacyclogenesis and greater capacity to cause in vivo infections [51–54]. This higher fitness was also shown by strains resistant to drug combination [55]. In the case of MIL, we mentioned previously that clinical isolates from MIL relapses (well all MIL-sensitive) were more virulent than those from treatment success: the fact that many of the isolates from MIL relapse cases were also SSG-R might suggest that the increased fitness of MIL relapse isolates could be a heritage of the SSG period [33]. These data are further discussed in Chap. 15 on this volume, related to parasite fitness.

4.4 Human-Made and Environmental Changes

Since the first edition of this chapter, new studies have further documented the tight link existing between human-made and environmental changes and the epidemiology of leishmaniasis: we will mention here a few striking examples related to (1) human migration, (2) urbanization of foci as a consequence of environmental changes, and (3) new geographical presentations of foci.

In the past, human migration has been extensively documented as a main risk factor for emergence and spreading of leishmaniasis. A classic example is the migration (probably through infected dogs) of *L. (L.) infantum* from Portugal to the New World during the post-Conquista era. Together with the colonization of permissive sand flies, like *Lutzomyia longipalpis*, this has created the dramatic combination of conditions for the importation and installation of VL in the New World [56]. Nowadays, as a consequence of social, political, and economic instability, migrants are massively coming from/to leishmaniasis-endemic regions. In northwestern Ethiopia, a shift from sporadic cases of VL to a real epidemic occurred in 2005 and was associated with the return of migrant workers from a region bordering Sudan [57]. A recent review found a strong correlation between VL epidemics and civil unrest in South Sudan, related to immunologically naïve migrants entering in contact with the parasites in endemic regions or to infected migrants establishing new foci in leishmaniasis-free region [58]. Europe, after being the source of migration and exportation of *L. (L.) infantum*, has become in turn a target of migration. From the 283,532 migrants listed by Frontex in 2014 [59], many came from leishmaniasis-endemic countries like Syria and Afghanistan (historical foci of anthroponotic CL due to *L. (L.) tropica*) or Somalia (anthroponotic VL due to *L. (L.) donovani*) [60]. We lack information on the number of these who are ill at the time of arrival in Europe, but VL belongs to the most common etiologies in severely infected migrants [61], and recommendations for general practitioners are being formulated, like in Germany [62]. Besides ensuring a correct management of the clinical cases, a reflection is also required about the risk of installation of new leishmaniasis foci in Southern Europe, where sand flies are endemic, including permissive ones like *P. perniciosus* [63] or *P. tobbi* [64]. While the risk of colonization by new *Leishmania* species is probably very low, it is not null, and surveillance might be recommended: in this context, the availability of standardized species typing tools all over Europe is highly recommended (see section on tools).

Besides traveling and migrating, humans are also known to modify the environment, and this in turn might create the conditions for the emergence of new leishmaniasis epidemics. An excellent illustration of it is the recent CL and VL epidemics in the suburbs of Madrid, in Spain, known as the Fuenlabrada outbreak. Between July 2009 and December 2012, 446 cases were reported in four cities, but most of them clustered in Fuenlabrada, at the border of a park area [65]. Epidemiological investigations rapidly identified *L. (L.) infantum* as the etiological agent in that focus and *P. perniciosus* as the vector, but interestingly, the human outbreak was not accompanied by an increase in canine leishmaniasis, the usual suspect, which led the epidemiologists to look for alternative reservoir, which they rapidly found by PCR and xenodiagnosis: rabbits and hares [66]. Both animal species were abundant in the fields before their transformation into parks, and additional animals could have invaded them as a consequence of constructions in the area [65]. Burrows create ideal habitats for sand flies and logically a *L. (L.) infantum* transmission cycle could have developed near the houses. On the basis of these epidemiological findings, environmental control measures were taken, including destructions of burrows, treatment with insecticides, and capture of hares and rabbits. The outbreak of Fuenlabrada is a perfect illustration of how leishmaniasis' epidemiology should be analyzed and highlights several lessons: it is dynamic, the disease can be encountered in unexpected areas, transmission of the parasite may involve unexpected reservoir, and it can be controlled by measures integrating environment control. As such, scientists working on leishmaniasis must remain open and not stick only to what is written in textbooks, including the present one.

Environmental factors may also concern treatment outcome and DR. This is highlighted by a series of studies that were made around the role of environmental contamination with arsenic (As) and its possible role in the emergence/spreading of SSG resistance in the ISC. Considering (1) the important As contamination in Asia since 1970, as a consequence of the installation of tube wells, (2) the high endemicity of SSG-R in the ISC, and (3) the structural and chemical similarities between As and SSG, Perry et al. proposed the hypothesis that As contamination could have played a role in the emergence/spreading of SSG resistance [67]. The hypothesis was demonstrated experimentally, by chronic exposure of laboratory mice to As and serial passage of *L. (L.) donovani* in these, resulting in the acquisition of SSG resistance [68]. An epidemiological validation of the hypothesis was attempted in India by analyzing treatment outcome in patients treated with SSG between 2006 and 2010 and by measuring As level at proximity of patient's houses. This showed that patients using well water with high concentrations of As had a higher risk of treatment failure than patients using wells with low concentrations [69]: however, the study was underpowered and retrospective [69]; hence, further work would be required to assess the contribution of As in the collapse of SSG.

A last example highlighting how dynamic the epidemiology of leishmaniasis is concerns the recent description of VL foci in hilly districts of Nepal, which was so far not considered as endemic for *L. (L.) donovani* transmission [70]. The authors concluded that there was local transmission of the parasite because of (1) the occurrence of VL in habitants who never traveled, (2) a large number of asymptomatic residents, and (3) the detection of *Leishmania* sp. in *P. argentipes* collected in the area. In the ISC, *L. (L.) donovani* is generally endemic in the lowlands, and therefore, the KAEP is essentially focusing on these regions. The Nepalese report is preoccupying as it reminds us that leishmaniasis can occur outside the "classical" areas, where health staff does not necessarily have the needed training and where there are no prevention campaigns [70]. This type of foci could jeopardize the KAEP and constitute sources of reinvasion after the elimination phase. In addition, these new foci may be associated with new variants of the parasite, characterized by distinct phenotypes. This is illustrated (1) in the hilly districts of Nepal, where we found the so-called "Yeti" variants (genotype ISC1 [71, 72]), which are genomically very different from the lowland variants of *L. (L.) donovani* [71], and (2) in Sri Lanka where new foci were also described, associated with another genetically distant variant of *L. (L.) donovani*, which is essentially dermotropic [73].

4.5 Epidemiology and Immune Status

After DR and human-made and environmental changes, immune status of the host represents the third major risk factor for (re)emergence and spreading of leishmaniasis. The usual suspect is obviously the immunosuppression associated with HIV, which was already covered extensively in the first edition of this chapter. In the context of MIL resistance, it is of utmost relevance, as the first few cases of clinical MIL resistance were indeed detected in HIV-*Leishmania*-co-infected patients (see above). However, in the present chapter, we would like to address two additional aspects of co-infection particularly relevant in the context of treatment outcome, i.e., the infection of *Leishmania* by special viruses and the co-infection between *Leishmania* and other parasites.

Leishmania RNA viruses (LRV) are endosymbionts that were reported so far essentially in Latin America, in 74% of the *Leishmania* sp. isolates in French Guiana [74]; in 11% and 45% of *L. (V.) braziliensis* isolates in Bolivia and Peru, respectively [75]; in *L. (V.) braziliensis* (44%), *L. (V.) guyanensis* (27%), *L. (V.) lainsoni* (33%), and *L. (L.) amazonensis* (100%) in Brazil [76], albeit in some regions of that country it is was very rare [77]. It is still difficult to conclude about the non-endemicity of LRV in other regions of the world, as it was not yet systematically searched. However, case reports pop up here and there like in Iran [78] or Ethiopia [79]. The interest for these viruses is growing since the early discovery of the subverting capacity of LRV: indeed, the nucleic acids of the virus were shown to be potent immunogens and to play a role in the pathogenicity, in particular, the capacity for metastasis in experimental models [80]. The link between the presence of LRV and mucocutaneous (MCL) leishmaniasis was supported in a clinical context in Brazil [76] but not in Peru and Bolivia [75]. However, two sister and converging studies highlighted a link between LRV positivity and treatment outcome. On one hand, in Guiana, 27% of the patients infected with LRV-positive *L. (V.) guyanensis* showed a TF with pentamidine, while those infected with LRV-negative parasites were all cured [81]. On the other hand, in Peru, the presence of LRV1 in *L. (V.) braziliensis* was associated with a fourfold increased risk of SSG-TF [75]. The latter report completed a series of studies done on the same material and provided a new light on the epidemiology of treatment failure. A first study reported a high prevalence of treatment outcome and SSG resistance in Peru but a low correlation between the two parameters, thereby questioning the validity of the laboratory tools or the biological role of resistance in treatment outcome [82]. A second study analyzed different risk factors for TF and identified *Leishmania* species as a major risk factor: in particular, *L. (V.) braziliensis* was associated with most failures in Peru [24]. The third study was the one showing the high prevalence of LRV1 in *L. (V.) braziliensis* from Peru [75], thereby demonstrating the importance of confounding factors in the epidemiology of TF. Accordingly, in Peru, results may lead to the hypothesis that TF is not related with DR, well most likely to a virus more abundant in *L. (V.) braziliensis* and interfering with the immune response of the host known to be essential for the treatment efficacy.

Besides LRV which constitutes a particular case of "co-infection" and the well-documented HIV-*Leishmania* co-infection, there are very few systematic studies on co-infection with other pathogens and its effect on treatment outcome. A systematic review on that topic is in preparation [83], and it highlights the restricted knowledge we have on that topic and the need for further research. Two parasites might deserve further attention, *Trypanosoma cruzi* and helminths, respectively, co-infecting 41% of the leishmaniasis patients in Argentina [84] and 14–88% of leishmaniasis patients in Brazil [85]. A recent report showed a different immune response in *T. cruzi*-infected CL and MCL patients as compared with *Leishmania* single infections, which could have an effect on disease duration [86]. Co-infection with intestinal helminths was associated with a poor response to therapy in Brazil [87], which is probably also associated with disturbance of the immune response by the helminths, known, for instance, to be associated with a strong Th2 response. Altogether, the different examples of co-infection addressed in this section highlight the need for broadening the scope of epidemiological surveillance of leishmaniasis and integrating other pathogens or conditions in the monitoring.

4.6 Tools for Epidemiological Surveillance in the Context of Treatment Failure and Drug Resistance

Considering the fact that in many cases, including in the context of MIL, there is no correlation so far between TF and DR, it is important to distinguish surveillance tools according to their application for monitoring treatment outcome or DR. Our aim in this section is not to review all the available tools but well to illustrate major concepts that could underlie surveillance activities and guide further research.

In the context of treatment outcome, two types of tools—clinical and laboratory—deserve a particular attention; these could be integrated to feed data in a national or regional system of epidemiological surveillance, as developed for malaria. On the clinical side, a pilot project successfully evaluated the relevance of retrospective quarterly cohort monitoring for following clinical outcome of VL at the level of Primary Health Centers [88]. The tool was evaluated in the context of MIL therapy, provided an early evaluation of the effectiveness of treatment strategies—not per se the efficacy of the drug itself—and could be applied in any VL treatment program [88]. Another example of clinical tools concerns algorithms, which can guide screening, clinical handling, and follow-up of treatment efficacy. In Sri Lanka, for instance, ten clinical markers predicted CL with more than 90% of accuracy, but without reaching 100% of sensitivity/specificity, highlighting the need of complementary laboratory tests [89]. In Peru, a series of parameters were analyzed for their capacity to predict SSG-TF in CL context: a new risk factor was identified, and the presence of concomitant distant lesions together with other parameters allowed to build a prognostic score for SSG-TF with a sensitivity of 77.78% and a specificity of 95.52% [90]. On the laboratory side, a recent report showed that the detection of IgG1 at the end of treatment of VL cases was a good predictor of relapse: this result was validated in India and Sudan, and a lateral flow rapid diagnostic test was developed to detect anti-*Leishmania* IgG1 [91]. More sophisticated are the molecular tools, like real-time quantitative PCR, which appear to be promising not only to measure drug efficacy but also to assess cure, as shown in the context of Indian VL [92]. Loop-mediated isothermal amplification (LAMP) constitutes a first step toward simplification of molecular tools and showed potential for application as test of cure [93].

As shown above, *Leishmania* species was shown to constitute a risk factor for TF; hence, species identification tools should also be considered in this present conceptual inventory. A large variety of tests exist for this application, each laboratory generally having its own target or procedure [94]. However, for quality diagnosis as well as for surveillance, it is more than time to agree on one method, well validated, standardized, and subjected to quality control. While isoenzyme electrophoresis, a method only applied in a few labs in the world, is still considered as a golden standard for *Leishmania* typing, molecular tools targeting the heat-shock protein

70 (Hsp70) emerge as the method of choice for robust species identification. Different studies are illustrative of the pipeline that should be followed for the development, validation, and implementation of typing tools. Firstly, the target was taxonomically validated by a phylogenetic analysis considering 17 species [95]. Secondly, Hsp70 PCR-RFLP was developed and experimentally validated as a universal tool for species identification in the New and Old World [96]. It was then further improved in terms of sensitivity [97] and evaluated in a clinical context in Latin America [98], in the Old World [99] and in the context of imported pathologies in Europe ([100]; in this case PCR sequencing approach was used). Later on, PCR sequencing of Hsp70 genes was compared to three other single-locus markers, revealing that Hsp70 PCR was one of the methods giving the best resolution [101]. Finally, it was included, together with other methods, in a large study comparing the accuracy of species identification in 16 European clinical laboratories in 2014: results confirmed the robustness of Hsp70 PCR, while they demonstrated errors in some laboratories using other methods [102], thereby highlighting the need for a single method with standard operating procedures. Considering the role of co-infecting pathogens in TF, laboratory tools should also allow simultaneous detection of other pathogens, possibly in a multiplex format. While this approach is common for viral or bacterial diseases, in leishmaniasis, this concept is generally used for differential diagnosis [103] or the identification of different *Leishmania* species [104]. Last but not the least, in the New World, the systematic detection of LRV, now facilitated by the use of a monoclonal antibody [105], should be integrated with the diagnosis of leishmaniasis, given the link between this specific case of co-infection, pathogenicity, and TF (see above).

In the context of DR, we will distinguish in vitro susceptibility assays from molecular tests detecting markers of resistance. As discussed in the previous edition of this chapter, for several drugs including SSG, in vitro susceptibility should be measured on intracellular amastigotes in macrophages. These assays are time-consuming and reductionist (because they do not integrate the immunological context), and they are difficult to apply on clinical isolates which are often not very infectious in these in vitro assays. In this context, MIL might constitute an exception; indeed, the MIL susceptibility of promastigotes of Indian *L. (L.) donovani* strains was shown to correlate with the susceptibility measured on intracellular amastigotes [106]. Accordingly, the authors proposed a simple resazurin-based promastigote assay for the routine monitoring of MIL susceptibility in clinical isolates. A similar assay based on promastigote susceptibility recently demonstrated the two first cases of MIL resistance in India [28]. Molecular assays could rapidly complement and replace the in vitro susceptibility assays, once a clear idea is available on the mechanisms of resistance active in a clinical context. Indeed, so far, most molecular studies done on molecular mechanisms leading to MIL resistance were done in the context of experimentally selected resistance. These reports highlight the role of MIL transporters and more specifically the complex LdMT and LdROS (the β subunit of LdMT) in the development of resistance [107]. Experimentally selected MIL-R strains show mutations in one or both genes [31, 32]. The studies made on the first few MIL-R clinical isolates indicate that the same two molecules can be affected in the field [28, 31], hereby supporting the use of assays targeting these genes. However, as highlighted in a recent opinion paper [108], the parasite may follow different strategies to inactivate these transporters, a concept called "the many roads to drug resistance": different single nucleotide polymorphisms (SNPs) affecting the functionality of the gene, complete deletions, or decrease in the somy of the chromosomes bearing the respective genes [108]. Consequently, even if the mechanism and the target are known, it is currently impossible to use a PCR assay susceptible to detect all kinds of molecular alterations. In this context, whole genome sequencing (WGS) is probably the most promising avenue. The study by Shaw et al. [32] illustrates the polyvalence of this tool: following the genomic adaptations of two *L. (L.) donovani* strains upon experimental selection of MIL resistance, the authors first detected decrease in the somy of chromosome 13 (bearing LdMT gene), before the installation of more structural alterations, like deletions or single-base mutations. WGS can be applied for high-throughput analyses, as recently shown by the publication of a study based on the sequencing of 204 clinical isolates [71]. However, two recent studies showed that the somy and heterozygosity of parasites could differ between life stages [109, 110], highlighting the need—in the future—to sequence genomes directly in clinical samples. This is now possible thanks to genome capture methods, and we recently demonstrated the proof of principle of this new technology, by obtaining a high-quality genomic sequence of amastigotes from 21/24 bone marrow samples (Malgorzata et al. in preparation).

Besides a direct application for the detection of multiple genomic changes in genes specifically involved in DR, WGS also demonstrated its power for molecular epidemiology, especially in the context of microevolution of young populations. This was highlighted by the study of 204 clinical isolates of *L. (L.) donovani* from the ISC [71]. Indeed, in the past, it was impossible to study the molecular epidemiology of parasites from this region, and even multi-locus microsatellite typing (MLMT) could not discriminate the strains of this genetically very homogeneous population [111]. As expected, WGS was much more discriminatory: (1) all strains could be distinguished, (2) different subpopulations were identified (namely, ISC1 to ISC10, among which there were six congruent monophyletic groups and other groups showing signs of recombination), and (3) the history of the whole population could be tracked back until its probable emergence around 1850 during the first VL epidemics. Interestingly, the study showed that parasites from the lowlands (ISC2–10) did all contain an intrachromosomal amplicon of the H-locus, containing among others the MRPA gene, known to play a major role in sequestration of Sb^{III}: work in progress showed that this amplification probably provided a preadaptation to SSG for all lowland parasites [112]. A second layer of adaptations was encountered in one subpopulation of parasites associated with SSG-R and SSG-TF (ISC5): all of them contained a "fatal" 2nt-indel in the gene encoding AQP1, a known transporter involved in the uptake of Sb^{III}. This large-scale WGS study did not show any evidence of alterations of LdMT or LdROS in the sequenced parasites, even in parasites coming from MIL-TF patients. Besides offering a new insight on the natural history of the parasite and on the understanding of DR, this study also provided a reference frame for epidemiological surveillance of *L. (L.) donovani* in the ISC, particularly relevant in the context of KAEP and its post-elimination phase. Indeed, if new epidemics occur in the future, we will be able to associate them with one of the described genotypes or to detect new ones; both pieces of information are essential for the long-term monitoring of the program. We made a first attempt in simplification of the molecular tools, by developing a single-locus genotyping tool to track the main *L. (L.) donovani* groups of the ISC [72]: this method consisted in the PCR amplification of regions containing group-specific SNPs, followed by sequencing of the amplicon. The method was evaluated in Nepal and allowed the correct classification of 58% of the samples, the unclassified samples representing genotypes undetected in the WGS study or populations for which no ISC-SLG test was developed or used [72]. This limit is inherent to "closed"/targeted genotyping tools as SLG and shows the importance to complement them with "open"/untargeted genotyping tools, like WGS.

4.7 Conclusions and General Recommendations

More than ever, the epidemiology of leishmaniasis appears to be extremely dynamic as a consequence of three major risk factors, human-made and environmental changes, immune status of the host, and TF accompanied or not by DR. New foci appear in unexpected areas, involving unexpected hosts, highlighting the need to revisit some assumptions and dogmas, in order to further guide control programs. More than ever, the confrontation of clinical and laboratory studies shows that TF is not at all a synonymous of DR. It is known that TF can be due to the quality of the drug and the quality of the health systems or immune-genetic features of the host. However, we demonstrated here the importance of pathogen factors other than DR, like the virulence of *Leishmania* or its infection by LRV. In the case of MIL, several alarming reports show a decrease in the efficacy, which is particularly preoccupying in the ISC, where this drug has been used in monotherapy for a decade, in the context of the KAEP. Despite the early warnings of scientists on the risk of rapid emergence of DR against MIL, the first and rare cases of DR only start to emerge and monitoring of the further evolution of the distribution of these MIL-R strains is more than ever required. A battery of tools, clinical and laboratory, are available for this endeavor, and major efforts should be undertaken to validate them adequately, standardize them, and disseminate them. In the context of a small arsenal of available drugs and with only few compounds in the final stage of the pipeline, it is more than ever required to protect the existing drugs, and surveillance of the disease, treatment outcome, and drug resistance may contribute to it.

Acknowledgments The research in the Molecular Parasitology Unit here quoted received funding from the European Union's Horizon 2020 research and innovation program under the Marie Sklodowska-Curie grant agreement No 642609, the Flemish Fund for Scientific Research (12Q8115N and G.0.B81.12), the Kaladrug-R consortium from the European Union Framework Program (FP7-222895), the Belgian Science Policy Office (TRIT, P7/41), the Department of

Economy, Science and Innovation in Flanders (ITM-SOFIB) and the Belgian Cooperation Agency (DGD; institutional collaboration with Instituto de Medicina Tropical A.von Humboldt in Lima and with Instituto Pedro Kouri in La Havana).

References

1. Dujardin J-C, Decuypere S. Epidemiology of leishmaniasis in the time of drug resistance. In: Ponte-Sucre A, Padron-Nieves M, Diaz E, editors. Drug resistance in *Leishmania* parasites. Consequences, molecular mechanism and possible treatments. Wienen: Springer; 2013. p. 65–83.
2. Alvar J, Vélez ID, Bern C, Herrero M, et al. Leishmaniasis worldwide and global estimates of its incidence. PLoS One. 2002;7:e35671.
3. SEARO/WHO. Health Ministers commit to eliminating kalar azar [media advisory]. New Delhi: WHO Regional Office for South-East Asia; 2014. http://www.searo.who.int/mediacentre/releases/2014/pr1581/en/
4. WHO-TDR. Sustainable visceral leishmaniasis elimination requires research on infection reservoirs. TDR news item; 2016. http://www.who.int/tdr/news/2016/visceral-leishmaniasis-research/en/
5. Sundar S, Jha TK, Thakur CP, Bhattacharya SK, et al. Oral miltefosine for Indian visceral leishmaniasis. N Engl J Med. 2002;347:1739–46.
6. Bhattacharya SK, Jha TK, Sundar S, Thakur CP, et al. Efficacy and tolerability of miltefosine for childhood visceral leishmaniasis in India. Clin Infect Dis. 2004;38:217–21.
7. Patra P, Guha SK, Maji AK, Saha P, et al. Efficacy of oral miltefosine in visceral leishmaniasis in rural West Bengal, India. Indian J Pharm. 2012;44:500–3.
8. Sundar S, Singh A, Rai M, Prajapati VK, et al. Efficacy of miltefosine in the treatment of visceral leishmaniasis after a decade of use in India. Clin Infect Dis. 2012;55:543–50.
9. Rijal S, Ostyn B, Uranw S, Rai K, et al. Increasing failure of miltefosine in the treatment of Kala-azar in Nepal and the potential role of parasite drug resistance, reinfection, or noncompliance. Clin Infect Dis. 2013;56:1530–8.
10. Sundar S, Sinha P, Jha TK, Chakravarty J, et al. Oral miltefosine for Indian post-kala-azar dermal leishmaniasis: a randomised trial. Tropical Med Int Health. 2013;18:96–100.
11. Ramesh V, Singh R, Avishek K, Verma A, et al. Decline in clinical efficacy of oral miltefosine in treatment of Post Kala-azar Dermal Leishmaniasis (PKDL) in India. PLoS Negl Trop Dis. 2015;9:e0004093.
12. Sundar S, Singh A, Chakravarty J, Rai M. Efficacy and safety of miltefosine in treatment of post-kala-azar dermal leishmaniasis. Sci World J. 2015;2015:414378.
13. Ritmeijer K, Dejenie A, Assefa Y, Hundie TB, et al. A comparison of miltefosine and sodium stibogluconate for treatment of visceral leishmaniasis in an Ethiopian population with high prevalence of HIV infection. Clin Infect Dis. 2006;43:357–64.
14. Wasunna M, Njenga S, Balasegaram M, Alexander N, et al. Efficacy and safety of AmBisome in combination with sodium stibogluconate or miltefosine and miltefosine monotherapy for African visceral leishmaniasis: phase II randomized trial. PLoS Negl Trop Dis. 2016;10: e0004880.
15. Mohebali M, Fotouhi A, Hooshmand B, Zarei Z, et al. Comparison of miltefosine and meglumine antimoniate for the treatment of zoonotic cutaneous leishmaniasis (ZCL) by a randomized clinical trial in Iran. Acta Trop. 2007;103:33–40.
16. Soto J, Arana BA, Toledo J, Rizzo N, et al. Miltefosine for new world cutaneous leishmaniasis. Clin Infect Dis. 2004;38:1266–72.
17. Soto J, Rea J, Balderrama M, Toledo J, et al. Efficacy of miltefosine for Bolivian cutaneous leishmaniasis. Am J Trop Med Hyg. 2008;78:210–1.
18. Vélez I, López L, Sánchez X, Mestra L, et al. Efficacy of miltefosine for the treatment of American cutaneous leishmaniasis. Am J Trop Med Hyg. 2010;83:351–6.

19. López L, Cruz C, Godoy G, Robledo SM, et al. Thermotherapy effective and safer than miltefosine in the treatment of cutaneous leishmaniasis in Colombia. Rev Inst Med Trop Sao Paulo. 2013;55:S0036-46652013000300197.
20. Chrusciak-Talhari A, Dietze R, Chrusciak Talhari C, da Silva RM, et al. Randomized controlled clinical trial to access efficacy and safety of miltefosine in the treatment of cutaneous leishmaniasis Caused by *Leishmania (Viannia) guyanensis* in Manaus, Brazil. Am J Trop Med Hyg. 2011;84:255–60.
21. Machado PR, Ampuero J, Guimarães LH, Villasboas L, et al. Miltefosine in the treatment of cutaneous leishmaniasis caused by *Leishmania* braziliensis in Brazil: a randomized and controlled trial. PLoS Negl Trop Dis. 2010;4:e912.
22. Garcia Bustos MF, Barrio A, Parodi C, Beckar J. Miltefosine versus meglumine antimoniate in the treatment of mucosal leishmaniasis. Medicina (B Aires). 2014;74:371–7.
23. Cota GF, de Sousa MR, de Mendonça AL, Patrocinio A, et al. *Leishmania*-HIV co-infection: clinical presentation and outcomes in an urban area in Brazil. PLoS Negl Trop Dis. 2014;8: e2816.
24. Arevalo J, Ramirez L, Adaui V, Zimic M, et al. Influence of *Leishmania (Viannia)* species on the response to antimonial treatment in patients with American tegumentary leishmaniasis. J Infect Dis. 2007;195:1846–51.
25. Yardley V, Croft SL, De Doncker S, Dujardin JC, et al. The sensitivity of clinical isolates of *Leishmania* from Peru and Nepal to miltefosine. Am J Trop Med Hyg. 2005;73:272–5.
26. Bhandari V, Kulshrestha A, Deep DK, Stark O, et al. Drug susceptibility in *Leishmania* isolates following miltefosine treatment in cases of visceral leishmaniasis and post kala-azar dermal leishmaniasis. PLoS Negl Trop Dis. 2012;6:e1657.
27. Prajapati VK, Sharma S, Rai M, Ostyn B, et al. *In vitro* susceptibility of *Leishmania donovani* to miltefosine in Indian visceral leishmaniasis. Am J Trop Med Hyg. 2013;89:750–4.
28. Srivastava S, Mishra J, Gupta AK, Singh A, et al. Laboratory confirmed miltefosine resistant cases of visceral leishmaniasis from India. Parasit Vectors. 2017;10:49.
29. Fernández OL, Diaz-Toro Y, Ovalle C, Valderrama L, et al. Miltefosine and antimonial drug susceptibility of *Leishmania Viannia* species and populations in regions of high transmission in Colombia. PLoS Negl Trop Dis. 2014;8:e2871.
30. Obonaga R, Fernández OL, Valderrama L, Rubiano LC, et al. Treatment failure and miltefosine susceptibility in dermal leishmaniasis caused by *Leishmania* subgenus *Viannia* species. Antimicrob Agents Chemother. 2014;58:144–52.
31. Mondelaers A, Sanchez-Cañete MP, Hendrickx S, Eberhardt E, et al. Genomic and molecular characterization of miltefosine resistance in *Leishmania infantum* strains with either natural or acquired resistance through experimental selection of intracellular amastigotes. PLoS One. 2016;11:e0154101.
32. Shaw CD, Lonchamp J, Downing T, Imamura H, et al. *In vitro* selection of miltefosine resistance in promastigotes of *Leishmania donovani* from Nepal: genomic and metabolomic characterization. Mol Microbiol. 2016;99:1134–48.
33. Rai K, Cuypers B, Bhattarai NR, Uranw S, et al. Relapse after treatment with miltefosine for visceral leishmaniasis is associated with increased infectivity of the infecting *Leishmania donovani* strain. MBio. 2013;4:e00611–3.
34. Dorlo TP, Rijal S, Ostyn B, de Vries PJ, et al. Failure of miltefosine in visceral leishmaniasis is associated with low drug exposure. J Infect Dis. 2014;210:146–53.
35. Ostyn B, Hasker E, Dorlo TP, Rijal S, et al. Failure of miltefosine treatment for visceral leishmaniasis in children and men in South-East Asia. PLoS One. 2014;9:e100220.
36. Hirve S, Boelaert M, Matlashewski G, Mondal D, et al. Transmission dynamics of visceral Leishmaniasis in the Indian subcontinent – a systematic literature review. PLoS Negl Trop Dis. 2016;10:e0004896.
37. Nandy AM. Ten years of kala-azar in west Bengal, Part I. Did post-kala-azar dermal leishmaniasis initiate the outbreak in Parganas? Bull World Health Organ. 1992;70:341–6.

38. Bhattarai NR, Van der Auwera G, Rijal S, Picado A, et al. Domestic animals and epidemiology of visceral leishmaniasis, Nepal. Emerg Infect Dis. 2010;16:231–7.
39. Khanal B, Picado A, Bhattarai NR, Van Der Auwera G, et al. Spatial analysis of *Leishmania donovani* exposure in humans and domestic animals in a recent kala azar focus in Nepal. Parasitology. 2010;137:1597–603.
40. Singh N, Mishra J, Singh R, Singh S. Animal reservoirs of visceral leishmaniasis in India. J Parasitol. 2013;99:64–7.
41. Singh SP, Picado A, Boelaert M, Gidwani K, et al. The epidemiology of *Leishmania donovani* infection in high transmission foci in India. Tropical Med Int Health. 2010;15(Suppl 2):12–20.
42. Alam MZ, Yasin G, Kato H, Sakurai T, et al. PCR-based detection of *Leishmania donovani* DNA in a stray dog from a visceral Leishmaniasis endemic focus in Bangladesh. J Vet Med Sci. 2013;75:75–8.
43. Stauch A, Sarkar RR, Picado A, Ostyn B, et al. Visceral leishmaniasis in the Indian subcontinent: modelling epidemiology and control. PLoS Negl Trop Dis. 2011;5:e1405.
44. Ostyn B, Gidwani K, Khanal B, Picado A, et al. Incidence of symptomatic and asymptomatic *Leishmania donovani* infections in high-endemic foci in India and Nepal: a prospective study. PLoS Negl Trop Dis. 2011;5:e1284.
45. Chapman LA, Dyson L, Courtenay O, Chowdhury R, et al. Quantification of the natural history of visceral leishmaniasis and consequences for control. Parasit Vectors. 2015;8:521.
46. Das S, Matlashewski G, Bhunia GS, Kesari S, et al. Asymptomatic *Leishmania* infections in northern India: a threat for the elimination programme? Trans R Soc Trop Med Hyg. 2014; 108:679–84.
47. Tiwary P, Kumar D, Singh RP, Rai M, et al. Prevalence of sand flies and *Leishmania donovani* infection in a natural population of female *Phlebotomus argentipes* in Bihar State, India. Vector Borne Zoonotic Dis. 2012;12:467–72.
48. Bhattarai NR, Das ML, Rijal S, van der Auwera G, et al. Natural infection of *Phlebotomus argentipes* with *Leishmania* and other trypanosomatids in a visceral leishmaniasis endemic region of Nepal. Trans R Soc Trop Med Hyg. 2009;103:1087–92.
49. Picado A, Singh SP, Rijal S, Sundar S, et al. Longlasting insecticidal nets for prevention of *Leishmania donovani* infection in India and Nepal: paired cluster randomised trial. BMJ. 2010; 341:c6760.
50. Stauch A, Duerr HP, Dujardin JC, Vanaerschot M, et al. Treatment of visceral leishmaniasis: model-based analyses on the spread of antimony-resistant *L. donovani* in Bihar, India. PLoS Negl Trop Dis. 2012;6:e1973.
51. Vanaerschot M, Maes I, Ouakad M, Adaui V, et al. Linking *in vitro* and *in vivo* survival of clinical *Leishmania donovani* strains. PLoS One. 2010;5:e12211.
52. Vanaerschot M, Decuypere S, Berg M, Roy S, et al. Drug-resistant microorganisms with a higher fitness – can medicines boost pathogens? Crit Rev Microbiol. 2013;39:384–94.
53. Vanaerschot M, Huijben S, Van den Broeck F, Dujardin JC. Drug resistance in vectorborne parasites: multiple actors and scenarios for an evolutionary arms race. FEMS Microbiol Rev. 2014;38:41–55.
54. Ouakad M, Vanaerschot M, Rijal S, Sundar S, et al. Increased metacyclogenesis of antimony-resistant *Leishmania donovani* clinical lines. Parasitology. 2011;138:1392–9.
55. García-Hernández R, Gómez-Pérez V, Castanys S, Gamarro F. Fitness of *Leishmania donovani* parasites resistant to drug combinations. PLoS Negl Trop Dis. 2015;9:e0003704.
56. Leblois R, Kuhls K, François O, Schönian G, et al. Guns, germs and dogs: On the origin of *Leishmania chagasi*. Infect Genet Evol. 2011;11:1091–5.
57. Alvar J, Bashaye S, Argaw D, Cruz I, et al. Kala-azar outbreak in Libo Kemkem, Ethiopia: epidemiologic and parasitologic assessment. Am J Trop Med Hyg. 2007;77:275–82.
58. Al-Salem W, Herricks JR, Hotez PJ. A review of visceral leishmaniasis during the conflict in South Sudan and the consequences for East African countries. Parasit Vectors. 2016;9:460.
59. "Annual Risk Analysis 2015" Frontex. 27 April 2015. p. 59.

60. Postigo JA. Leishmaniasis in the World Health Organization Eastern Mediterranean Region. Int J Antimicrob Agents. 2010;36(Suppl 1):S62–5.
61. Alp E, Erdem H, Rello J. Management of septic shock and severe infections in migrants and returning travelers requiring critical care. Eur J Clin Microbiol Infect Dis. 2016;35:527–33.
62. Papan C, Hübner J, von Both U. Infectious diseases in refugees and their minors arriving in Germany – what the GP needs to know [Article in German]. MMW Fortschr Med. 2017;158:58–62.
63. Svárovská A, Ant TH, Seblová V, Jecná L, et al. *Leishmania major* glycosylation mutants require phosphoglycans (lpg2-) but not lipophosphoglycan (lpg1-) for survival in permissive sand fly vectors. PLoS Negl Trop Dis. 2010;4:e580.
64. Seblova V, Myskova J, Hlavacova J, Votypka J, et al. Natural hybrid of *Leishmania infantum/ L. donovani*: development in *Phlebotomus tobbi, P. perniciosus* and *Lutzomyia longipalpis* and comparison with non-hybrid strains differing in tissue tropism. Parasit Vectors. 2015;8:605.
65. Arce A, Estirado A, Ordobas M, Sevilla S, et al. Reemergence of leishmaniasis in Spain: community outbreak in Madrid, Spain, 2009 to 2012. Euro Surveill. 2013;18:20546.
66. Molina R, Jiménez MI, Cruz I, Iriso A, et al. The hare (*Lepus granatensis*) as potential sylvatic reservoir of *Leishmania infantum* in Spain. J Vet Parasitol. 2012;190:268–71.
67. Perry MR, Wyllie S, Prajapati VK, Feldmann J, et al. Visceral leishmaniasis and arsenic: an ancient poison contributing to antimonial treatment failure in the Indian subcontinent? PLoS Negl Trop Dis. 2011;5:e1227.
68. Perry MR, Wyllie S, Raab A, Feldmann J, et al. Chronic exposure to arsenic in drinking water can lead to resistance to antimonial drugs in a mouse model of visceral leishmaniasis. Proc Natl Acad Sci USA. 2013;110:19932–7.
69. Perry M, Wyllie S, Prajapati V, Menten J, et al. Arsenic, antimony, and *Leishmania*: has arsenic contamination of drinking water in India led to treatment-resistant kala-azar? Lancet. 2015;385(Suppl 1):S80.
70. Ostyn B, Uranw S, Bhattarai NR, Das ML, et al. Transmission of *Leishmania donovani* in the hills of Eastern Nepal, an outbreak investigation in Okhaldhunga and Bhojpur districts. PLoS Negl Trop Dis. 2015;9:e0003966.
71. Imamura H, Downing T, Van den Broeck F, Sanders MJ, et al. Evolutionary genomics of epidemic visceral leishmaniasis in the Indian subcontinent. elife. 2016;5:e12613.
72. Rai K, Bhattarai NR, Vanaerschot M, Imamura H, et al. Single locus genotyping to track *Leishmania donovani* in the Indian subcontinent: application in Nepal. PLoS Negl Trop Dis. 2017;11:e0005420.
73. Siriwardana HV, Noyes HA, Beeching NJ, Chance ML, et al. *Leishmania donovani* and cutaneous leishmaniasis, Sri Lanka. Emerg Infect Dis. 2007;13:476–8.
74. Ginouvès M, Simon S, Bourreau E, Lacoste V, et al. Prevalence and distribution of *Leishmania* RNA virus 1 in *Leishmania* parasites from French Guiana. Am J Trop Med Hyg. 2016; 94:102–6.
75. Adaui V, Lye LF, Akopyants NS, Zimic M, et al. Association of the endobiont double-stranded RNA virus LRV1 with treatment failure for human leishmaniasis caused by *Leishmania braziliensis* in Peru and Bolivia. J Infect Dis. 2016;213:112–21.
76. Cantanhêde LM, da Silva Júnior CF, Ito MM, Felipin KP, et al. Further evidence of an association between the presence of *Leishmania* RNA virus 1 and the mucosal manifestations in tegumentary leishmaniasis patients. PLoS Negl Trop Dis. 2015;9:e0004079.
77. Macedo DH, Menezes-Neto A, Rugani JM, Rocha AC, et al. Low frequency of LRV1 in *Leishmania braziliensis* strains isolated from typical and atypical lesions in the State of Minas Gerais, Brazil. Mol Biochem Parasitol. 2016;210:50–4.
78. Hajjaran H, Mahdi M, Mohebali M, Samimi-Rad K, et al. Detection and molecular identification of *Leishmania* RNA virus (LRV) in Iranian *Leishmania* species. Arch Virol. 2016;161: 3385–90.

79. Zangger H, Hailu A, Desponds C, Lye LF, et al. *Leishmania aethiopica* field isolates bearing an endosymbiotic dsRNA virus induce pro-inflammatory cytokine response. PLoS Negl Trop Dis. 2014;8:e2836.
80. Hartley MA, Ronet C, Zangger H, Beverley SM, et al. *Leishmania* RNA virus: when the host pays the toll. Front Cell Infect Microbiol. 2012;2:99.
81. Bourreau E, Ginouves M, Prévot G, Hartley MA, et al. Presence of *Leishmania* RNA virus 1 in *Leishmania guyanensis* increases the risk of first-line treatment failure and symptomatic relapse. J Infect Dis. 2016;213:105–11.
82. Yardley V, Ortuno N, Llanos-Cuentas A, Chappuis F, et al. American tegumentary leishmaniasis: is antimonial treatment outcome related to parasite drug susceptibility? J Infect Dis. 2006;194:1168–75.
83. Martínez DY, Verdonck K, Kaye PM, Adaui V, et al. Tegumentary leishmaniasis and coinfections other than HIV. PLOS Neglect Trop Dis. 2018;12(3):e0006125. https://doi.org/10.1371/journal.pntd.0006125
84. Chiaramonte MG, Zwirner NW, Caropresi SL, Taranto NJ, et al. *Trypanosoma cruzi* and *Leishmania* spp. human mixed infection. Am J Trop Med Hyg. 1996;54:271–3.
85. O'Neal SE, Guimarães LH, Machado PR, Alcântara L, et al. Influence of helminth infections on the clinical course of and immune response to *Leishmania braziliensis* cutaneous leishmaniasis. J Infect Dis. 2007;195:142–8.
86. Parodi C, García Bustos MF, Barrio A, Ramos F, et al. American tegumentary leishmaniasis: T-cell differentiation profile of cutaneous and mucosal forms-co-infection with *Trypanosoma cruzi*. Med Microbiol Immunol. 2016;205:353–69.
87. Azeredo-Coutinho RB, Pimentel MI, Zanini GM, Madeira MF, et al. Intestinal helminth co-infection is associated with mucosal lesions and poor response to therapy in American tegumentary leishmaniasis. Acta Trop. 2016;154:42–9.
88. Ostyn B, Malaviya P, Hasker E, Uranw S, et al. Retrospective quarterly cohort monitoring for patients with visceral leishmaniasis in the Indian subcontinent: outcomes of a pilot project. Tropical Med Int Health. 2013;18:725–33.
89. Siriwardana HV, Senarath U, Chandrawansa PH, Karunaweera ND. Use of a clinical tool for screening and diagnosis of cutaneous leishmaniasis in Sri Lanka. Pathog Glob Health. 2015; 109:174–83.
90. Valencia C, Arévalo J, Dujardin JC, Llanos-Cuentas A, et al. Prediction score for antimony treatment failure in patients with ulcerative leishmaniasis lesions. PLoS Negl Trop Dis. 2012; 6:e1656.
91. Bhattacharyya T, Ayandeh A, Falconar AK, Sundar S, et al. IgG1 as a potential biomarker of post-chemotherapeutic relapse in visceral leishmaniasis, and adaptation to a rapid diagnostic test. PLoS Negl Trop Dis. 2014;8:e3273.
92. Sudarshan M, Weirather JL, Wilson ME, Sundar S. Study of parasite kinetics with antileishmanial drugs using real-time quantitative PCR in Indian visceral leishmaniasis. J Antimicrob Chemother. 2011;66:1751–5.
93. Verma S, Singh R, Sharma V, Bumb RA, et al. Development of a rapid loop-mediated isothermal amplification assay for diagnosis and assessment of cure of *Leishmania* infection. BMC Infect Dis. 2017;17:223.
94. Van der Auwera G, Dujardin JC. Species typing in dermal leishmaniasis. Clin Microbiol Rev. 2015;28:265–94.
95. Fraga J, Montalvo AM, De Doncker S, Dujardin JC, et al. Phylogeny of *Leishmania* species based on the heat-shock protein 70 gene. Infect Genet Evol. 2010;10:238–45.
96. Montalvo AM, Fraga J, Monzote L, Montano I, et al. Heat-shock protein 70 PCR-RFLP: a universal simple tool for *Leishmania* species discrimination in the New and Old World. Parasitology. 2010;137:1159–68.
97. Montalvo AM, Fraga J, Maes I, Dujardin JC, et al. Three new sensitive and specific heat-shock protein 70 PCRs for global *Leishmania* species identification. Eur J Clin Microbiol Infect Dis. 2012;31:1453–61.

98. Fraga J, Veland N, Montalvo AM, Praet N, et al. Accurate and rapid species typing from cutaneous and mucocutaneous leishmaniasis lesions of the New World. Diagn Microbiol Infect Dis. 2012;74:142–50.
99. Montalvo AM, Fraga J, El Safi S, Gramiccia M, et al. Direct *Leishmania* species typing in Old World clinical samples: evaluation of 3 sensitive methods based on the heat-shock protein 70 gene. Diagn Microbiol Infect Dis. 2014;80:35–9.
100. Van der Auwera G, Maes I, De Doncker S, Ravel C, et al. Heat-shock protein 70 gene sequencing for *Leishmania* species typing in European tropical infectious disease clinics. Euro Surveill. 2013;18:20543.
101. Van der Auwera G, Ravel C, Verweij JJ, Bart A, et al. Evaluation of four single-locus markers for *Leishmania* species discrimination by sequencing. J Clin Microbiol. 2014;52:1098–104.
102. Van der Auwera G, Bart A, Chicharro C, Cortes S, et al. Comparison of *Leishmania* typing results obtained from 16 European clinical laboratories in 2014. Euro Surveill. 2016;21:30418.
103. Rodríguez-Brito S, Camacho E, Mendoza M, Niño-Vega GA. Differential identification of *Sporothrix* spp. and *Leishmania* spp. by conventional PCR and qPCR in multiplex format. Med Mycol. 2015;53:22–7.
104. Karani M, Sotiriadou I, Plutzer J, Karanis P. Bench-scale experiments for the development of a unified loop-mediated isothermal amplification (LAMP) assay for the *in vitro* diagnosis of *Leishmania* species' promastigotes. Epidemiol Infect. 2014;142:1671–7.
105. Zangger H, Ronet C, Desponds C, Kuhlmann FM, et al. Detection of *Leishmania* RNA virus in *Leishmania* parasites. PLoS Negl Trop Dis. 2013;7:e2006.
106. Kulshrestha A, Bhandari V, Mukhopadhyay R, Ramesh V, et al. Validation of a simple resazurin-based promastigote assay for the routine monitoring of miltefosine susceptibility in clinical isolates of *Leishmania donovani*. Parasitol Res. 2013;112:825–8.
107. Pérez-Victoria FJ, Sánchez-Cañete MP, Seifert K, Croft SL, et al. Mechanisms of experimental resistance of *Leishmania* to miltefosine: implications for clinical use. Drug Resist Updat. 2006;9:26–39.
108. Hefnawy A, Berg M, Dujardin JC, De Muylder G. Exploiting knowledge on *Leishmania* drug resistance to support the quest for new drugs. Trends Parasitol. 2017;33:162–74.
109. Dumetz F, Imamura H, Sanders M, Seblova V, et al. Modulation of aneuploidy in *Leishmania donovani* during adaptation to different *in vitro* and *in vivo* environments, and its impact on gene expression. MBio. 2017;8(3):pii e00599–17.
110. Prieto Barja P, Pescher P, Bussotti G, Dumetz F, et al. Asexual maintenance of genetic diversity in the protozoan pathogen *Leishmania donovani*. Nat Ecol Evol. 2017;1(12):1961–9.
111. Alam MZ, Kuhls K, Schweynoch C, Sundar S, et al. Multilocus microsatellite typing (MLMT) reveals genetic homogeneity of *Leishmania donovani* strains in the Indian subcontinent. Infect Genet Evol. 2009;9:24–31.
112. Dumetz F, Cuypers B, Imamura H, Zander D, et al. (2018) Molecular pre-adaptation to antimony resistance in *Leishmania donovani* of the Indian subcontinent. mSphere (in press)

The Role of the Immune System in Resistance to Infection

5

Lukasz Kedzierski and Krystal J. Evans

Abstract

Leishmaniasis is a spectrum of diseases with clinical symptoms ranging in severity from skin lesions to serious disfigurement and fatal systemic infection. The outcome of infection depends on the parasite species as well as host genetic factors and immune competence. In order to develop a successful infection, *Leishmania* must evade both the innate and adaptive immune responses. Whilst protective immunity has been driven by Th1-type T cell responses, the role of Th2-type cytokines is not entirely clear, although it has been implicated in susceptibility to leishmaniasis. A successful treatment of all the forms of leishmaniasis depends on efficient elimination of parasites by activated macrophages. Paradoxically, *Leishmania* species have evolved a variety of strategies to evade leishmanicidal mechanisms and survive in macrophages in the phagosome. Interestingly, most infected individuals develop long-lasting protective immunity following primary infection; however, sterile immunity is hardly ever achieved, and parasites are believed to persist asymptomatically in the host. The vast array of immune cells and cytokines involved in the immune response to *Leishmania* clearly highlights the complexity of the disease and reveals a complicated net of regulatory and counter-regulatory interactions. This chapter outlines our current knowledge of the immune factors implicated in the disease and discusses the role the immune system plays in resistance to infection.

L. Kedzierski (✉)
Faculty of Veterinary and Agricultural Sciences, Department of Microbiology and Immunology, The Peter Doherty Institute for Infection and Immunity, The University of Melbourne, Melbourne, Australia
e-mail: lukaszk@unimelb.edu.au

K. J. Evans
BioMelbourne Network, Melbourne, Australia

A. Ponte-Sucre, M. Padrón-Nieves (eds.), *Drug Resistance in Leishmania Parasites*,
https://doi.org/10.1007/978-3-319-74186-4_5

5.1 Introduction

The control of *Leishmania* infection is mediated by cellular immune responses leading to macrophage activation and parasite killing. Although humoral response is also present during the infection, antibodies play no role in protection and are associated with the non-healing disease. Antileishmanial immunity is mediated via both innate (macrophages, neutrophils, dendritic cells (DC)) and adaptive (T cells) immunities, but the $CD4^+$ T cell subset is crucial for resistance. Experimental studies using leishmaniasis mouse model of disease gave rise to the Th1/Th2 paradigm of resistance and susceptibility associated with intracellular infection. This clear-cut dichotomy is mostly associated with the cutaneous leishmaniasis (CL) but is not so well defined in the visceral leishmaniasis (VL). Nevertheless, it is universally accepted that the nature of the T cell response is a crucial factor in resistance to the disease, despite evident differences in the responses observed between mouse experimental infection and human leishmaniasis.

5.2 Cutaneous Leishmaniasis

5.2.1 Innate Immune Responses

5.2.1.1 Macrophages

Macrophages play a pivotal role in *Leishmania* infection. Macrophages are professional phagocytes, and *Leishmania* utilises their phagocytic function as a strategy for internalisation and subsequent replication within the macrophage phagolysosomes. Thus, macrophages act as both the host cells for *Leishmania* replication and effector cells that kill the parasites. Internalisation of *Leishmania* by macrophages triggers the production of reactive oxygen species [1] and leads to generation of nitric oxide (NO) [2] and *N*-hydroxy-L-arginine (LOHA) [3] as mediators of parasite killing. Nevertheless, there appear to be different requirements for effective killing of leishmanial species causing CL. Whilst NO and LOHA are sufficient for elimination of *L. (L.) major* [4], a successful anti-*L. (L.) amazonensis* response also requires superoxide production [5]. Additionally, infection of macrophages leads to the production of pro-inflammatory cytokines that are implicated in parasite killing. Interleukin (IL)-12 is necessary for the leishmanicidal activity of macrophages, as it leads to upregulation of interferon gamma (IFN-γ) by T cells and NK cells, generation of Th1-type responses and T cell-dependent and -independent macrophage activation leading to an increase of inducible nitric oxide synthase (iNOS) and NO production and subsequent parasite elimination [6]. A subversive activity of *Leishmania* parasites in this process is the inhibition of IL-12 production, which downregulates the immune response to infection [7]. Production of pro-inflammatory cytokines by macrophages results in the recruitment of pro-inflammatory cells to the site of infection, involved in granuloma formation aimed at isolating the microbial growth foci.

5.2.1.2 Neutrophils

Neutrophils are among the first cells recruited to the site of infection and are thought to participate in the containment of *Leishmania* parasites [8]. Published data on the involvement of neutrophils in *Leishmania* infection are contradictory, indicating either their role in resistance to leishmaniasis or disease exacerbatory activities [9]. However, it has been shown that in the context of infection initiated by the bite of a sand fly, neutrophils are recruited to the site of infection and phagocytose parasites, a process that is vital for disease progression [10]. Indeed, subsequent studies demonstrated that the uptake of parasitised neutrophils by dermal DCs leads to inhibition of activation of parasite-specific $CD4^+$ T cells [11]. Moreover, capture of infected, apoptotic neutrophils by DCs completely inhibited their ability to cross-present leishmanial antigen to $CD8^+$, thus indicating that a cross-talk between neutrophils and DCs is central to the early immune evasion strategies [12]. These findings suggest that neutrophils play a role in promoting disease progression, rather than resistance. However, other studies have found that neutrophils contribute to parasite killing through the release of neutrophil extracellular traps (NETs) [13]. Therefore, the role of neutrophils in *Leishmania* infection is multifaceted with both protective and permissive roles during the acute phase and immunomodulatory role in chronic phase affecting lesion progression and inflammatory milieu [14, 15]. Subsequent production of mast cell-derived mediators, IgG-mediated mechanisms and cytokine/chemokine released by macrophages and neutrophils results in the recruitment of DCs, an important component linking the innate with the adaptive immune response against *Leishmania* [16].

5.2.1.3 Dendritic Cells

The main function of DCs is the recognition and processing of foreign antigens, and subsequent presentation to T cells [17], and as such, they are considered to be gatekeepers in the defence against invading pathogens. Skin DCs, Langerhans cells and dermal DCs are the most efficient antigen-presenting cells (APC) in the body [18]. In case of *Leishmania* infection, dermal DCs appear to present antigen directly to T cells [19]. Small numbers of parasites are taken up directly by dermal DCs shortly after infection [20], but majority of the DCs become infected through contact with parasitised neutrophils [9]. Several weeks postinfection, the number of DCs ($CD11c^+$ cells) in the lesion increases due to their recruitment [8], and infected DCs are able to prime naïve $CD4^+$ and $CD8^+$ T cells [21]. Activated DCs migrate to draining lymph nodes where apart from T cells, they also activate resting NK cells and trigger IFN-γ production [22]. However, *Leishmania* parasites evolved complex mechanisms to avoid DC functions, which lead to downregulation of DC activation. This is further compounded by dichotomic role of DCs in promoting either susceptibility or resistance in CL [23].

Amastigote infection of DCs results in reduced phosphorylation and degradation of vital molecules in Janus kinase/signal transducer and activator of transcription (JAK/STAT), nuclear factor (NF)-κB and interferon regulatory factor (IRF) pathways [24], which in turn cause inadequate DC activation, T cell priming, impaired NK cells activation and suppression of IL-12 and IFN-γ production.

These might be *Leishmania*-related general phenomena; however, there are species- and stage-specific differences in modulation of DC functions. Whilst infection with *L. (L.) major* or *L. (L.) donovani* promastigotes led to production of IL-12 by murine DCs [25, 26], infection with *L. (L.) mexicana* amastigotes did not lead to DC activation or IL-12 and other pro-inflammatory cytokines production [27]. Similarly, infection with *L. (L.) amazonensis* amastigotes leads to downregulation of signalling events and impaired DC function [28], and in humans *L. (L.) amazonensis* has been shown to use Langerhans cells to skew $CD4^+$ T cell function towards regulatory T cells (T reg) and to suppress protective responses [29].

Amastigote uptake by DCs at the site of infection results in the upregulation of IL-12 [25], which is essential for parasite elimination within DCs [26] and for the effector functions of macrophages [30]. Uptake of amastigotes by DCs also leads to surface upregulation of MHC class I, MHC II and co-stimulatory molecules. The ability of DCs to present antigens through the MHC I and II pathways leads to stimulation of *Leishmania*-specific $CD4^+$ and $CD8^+$ T cell responses [26, 30, 31]. A recent report demonstrated that $CD103^+$ Batf3-dependent DCs are the major source of IL-12 and are crucial in immunity against *L. (L.) major* [32]. This process appears to be regulated by IRF-4 transcription factor that has the ability to inhibit IL-12 production in DCs [33]. Although other cell subsets, including macrophages and B cells, are able to present leishmanial antigens, antigen presentation by DCs is essential for acquired resistance to *Leishmania*.

5.2.2 Adaptive Immune Response

5.2.2.1 $CD4^+$ T Cells

T cells play an essential role in generation of effector and memory responses to intracellular pathogens. In case of leishmaniasis, protective immunity is associated with a cell-mediated immune response, whereas nonprotective responses have a strong humoral component in the absence of cell-mediated immunity. The protection against CL is intimately linked to development of Th1-type immunity and IFN-γ production. Experimental studies established a clear-cut dichotomy between Th1-mediated protection and Th2-mediated susceptibility. In resistant C57BL/6 mice, resolution of the disease is mediated as a consequence of IFN-γ release by Th1 cells and upregulation of NO in macrophages that harbour parasites [34]. Conversely, persistence of lesions in susceptible BALB/C mice is due to $CD4^+$ T cell differentiation to Th2-type effector cells and the production of IL-4, which in turn promotes antibody responses and suppresses macrophage activation, resulting in parasite survival and replication [35]. The Th1 response is linked to IFN-γ production; however, it is functionally heterogeneous. It has been shown that a high frequency of $CD4^+$ T cells producing IFN-γ only is not sufficient for resistance to infection. The quality and magnitude of the response are crucial factors influencing protective outcome and are controlled by the type of antigen-presenting cells (APCs), amount of antigen and duration it is being presented to the immune system as well as cytokine milieu [36–38]. The Th1 response mounted by $CD4^+$ T cells which are single-positive, i.e. producing IFN-γ only or tumour necrosis factor (TNF), has limited aptitude to develop into memory cells compared to IL-2 producing cells. Hence, their capacity to provide long-term durable protection is rather limited. On the other hand, IFN-γ and TNF are known to synergise in order to more efficiently kill parasites [39]; therefore, multifunctional $CD4^+$ T cells that simultaneously produce multiple cytokines are more likely to be involved in resistance to infection. Indeed, the frequency of multifunctional $CD4^+$ T cells (IFN-γ^+ TNF^+ IL-2^+) correlates with the degree of protection following vaccination [40]. These data indicate that functional heterogeneity of Th1 response to *Leishmania* plays a significant role in resistance to infection.

The Th1/Th2 dichotomy has been questioned in recent times since there is accumulating evidence that early IL-4 responses might not be required to promote susceptibility, and there is considerable complexity in the mechanisms responsible for acquired immunity [41]. Resistant C57BL/6 mice produce IL-4 early at the onset of the infection. This increase in IL-4 did not impact the mounting of an unimpaired Th1 response and disease resolution [42]. In several cases, resistance to infection in BALB/C mice following immunisation has not been linked to strong Th1 response [43, 44], or high pre-challenge IFN-γ levels did not correlate with protection [45]. Keratinocytes and epidermal cytokine expression have been implicated as decisive factors in generation of Th1 immunity [46]. The critical events that influence Th1/Th2 differentiation were thought to occur in the lymph nodes early during infection; however, it was also acknowledged that the skin, as primary site of infection, could influence the immune response [47]. During the first few hours of infection, *Leishmania* induces several cytokines in keratinocytes, and the gene expression profile of cells differs in susceptible and resistant mice. In particular, production of IL-4 by epidermal cells can explain the somewhat controversial role this cytokine plays in induction of Th1/Th2 responses. Whilst IL-4 is associated with a Th2 response and susceptibility to leishmaniasis [48], it is also able to induce the production of IL-12 by DCs but only when present early during the infection [49]. Therefore, an early, transient IL-4 production by keratinocytes is essential for induction of Th1 response against *L. (L.) major*, by acting in a paracrine fashion on DCs, which then produce IL-12 upon migration to the lymph node [46]. It has been also shown that IL-6, a major inflammatory cytokine, plays an important role in *Leishmania* protection. High levels of IL-6 of keratinocyte origin have been detected in resistant strains, and mice with IL-6 deficiency in the non-haematopoietic compartment display Th2 skewing and non-healing phenotype [46].

5.2.2.2 T Regulatory Cells

Susceptibility and resistance to infection are also influenced by T reg cells ($CD4^+$ $CD25^+$), which reside in the skin where they suppress harmful immune responses to infectious agents, counteract inflammatory responses and limit tissue damage [50]. During *L. (L.) major* infection, T reg cells accumulate in the dermis where they suppress the ability of the effector T cells to eliminate parasites. This process has been linked to the production of IL-10 [51], a cytokine that is also implicated in the maintenance of parasite persistence [52]. High levels of IL-10 produced by antigen-driven T reg cells lead to lack of vaccine efficacy despite the presence of strong Th1 responses [45]. In humans, T reg cells have been found in lesions of CL patients [53] and have been implicated in immunopathogenesis of the cutaneous infection [54, 55]. It has been demonstrated that $CD4^{+}CD25^{+}Foxp3^{+}$ T regulatory cells are involved in a rapid loss of resistance to infection in immune animals following inoculation with a killed parasite vaccine [56]. These data clearly point to the important regulatory role that T reg cells play in resistance and susceptibility to cutaneous leishmaniasis.

5.2.2.3 $CD8^{+}$ T Cells

Cytotoxic activity and cytokine production are two major effector functions of $CD8^{+}$ T cells that contribute to the disease outcome in *Leishmania* infections. Majority of data do not indicate a protective role for $CD8^{+}$ T cells in controlling primary infection [57]. However, they clearly play a role in resistance to infection by inducing Th1 response via cytokine production (IFN-γ) or in recall responses to secondary infection [58]. IFN-γ producing $CD8^{+}$ T cells are fundamental for the development of a Th1 response and thus contribute to healing in C57BL/6 mice [59, 60]. Besides cytokine production, $CD8^{+}$ T cells are thought to participate in controlling the infection through cytotoxic mechanisms, such as granzyme and perforin production and Fas/FasL pathways [61]. However, activation of $CD8^{+}$ T cell cytolytic responses is harmful to the host and drives the development of metastatic lesions in CL [62], mainly due to inflammasome activation and IL-1β release [63]. In human CL, recruitment of $CD8^{+}$ T cells producing granzyme A to the site of infection is associated with tissue damage, albeit the fact that this is a consequence of antiparasitic action [64]. Also in *L. (V.) braziliensis*-caused CL, $CD8^{+}$ T cells were shown to play harmful role contributing to disease immunopathology via their cytotoxic activity leading to tissue destruction [65]; however, in patients with subclinical *L. (V.) braziliensis* infection, $CD8^{+}$ T cells have been shown to be the major source of IFN-γ and were suggested to help to control the infection [66]. Nevertheless, in the context of *L. (L.) braziliensis* infection, $CD8^{+}$ T cell appears to be extremely detrimental. It is still not known what is the exact route of $CD8^{+}$ T cell activation in leishmaniasis, since the parasites reside in a parasitophorous vacuole inside the host macrophages, and it is not clear how these cells present antigen through MHC I [67]. The most likely mechanism is cross-presentation, which has been well documented for macrophages and DCs [68, 69], a process suggested to occur during *Leishmania* infection [67] and one which the parasite is also able to block in order to evade immunity [70]. The evidence and the importance of cross-presentation in leishmaniasis have been demonstrated in a recent study, where depletion of cross-presenting, Batf3-dependent DCs increased susceptibility to *L. (L.) major* infection [71].

5.2.2.4 Humoral Immune Response

Development of humoral immune responses is often linked to susceptibility to *Leishmania* infection, and in general antibodies are not considered to be a major factor in resistance to disease. B cell depletion using anti-IgM antibodies enhanced resistance to *Leishmania* in BALB/C mice [72]. Administration of IL-7, a B cell stimulant, to BALB/C mice increases B cell numbers and enhances disease severity [73]. Furthermore, B cell-deficient (μMT) mice lacking B cells through the targeted disruption of the immunoglobulin M locus are more resistant to infection than their wild-type counterparts [74]. In addition, the adoptive transfer studies of B cells and serum into BALB/C μMT mice have shown that it is the antigen presentation of specific B cells rather than Ig effector functions that is involved in the susceptible phenotype of BALB/C [75]. B cells were shown to be required for susceptibility and Th2 cell development in BALB/C mice infected with *L. (L.) major* [75]. The ability of B cells to skew the immune response towards a Th2 phenotype was linked to their capacity to present antigen to T cells. In addition, it has been shown that IL-10 produced by B cells plays a role in the development of susceptibility to cutaneous infection by inhibiting (in vitro) IL-12 production by DCs [76]. Although the involvement of B cell-mediated responses in *Leishmania* infection is controversial, and evidence points towards promoting susceptibility at least in the mouse model of CL, some data indicate that B cell can present antigen and activate $CD4^+$ T cells, thus enhancing resistance to infection [77].

5.2.3 Role of Cytokines in Resistance to *Leishmania* Infection

As described above a whole range of cytokines and chemokines are involved in the immune response to *Leishmania*, including but not limited to IL-4, IL-10, IL-12, IL-13, TNF and IFN-γ. The profile and timing of cytokine production correlate with the clinical outcome of *Leishmania* infection. A variety of immune cells express cytokines, mostly $CD4^+$ T cells (Th1 and Th2), but also $CD8^+$ T cells, $CD4^-CD8^-$ double negative T cells [78], NK cells, DCs and macrophages [79], mast cells (Maurer et al. 2006), regulatory B cells [76] and eosinophils [80].

5.2.3.1 Th2-Type and Anti-inflammatory Cytokines

The exemplary Th2 cytokine in leishmaniasis is interleukin-4. IL-4 drives Th2 response and promotes susceptibility through inhibition of macrophage activation and abrogation of IL-12 expression. The role of IL-4 in susceptibility to *Leishmania* has been illustrated in studies using transgenic or knockout mice. C57BL/6 IL-4 transgenic mice are more susceptible to infection than wild-type mice. Targeted disruption of the IL-4 gene or depletion of IL-4 in susceptible BALB/C mice renders them more resistant to infection with *L. (L.) major* [81]. Additionally, disruption of the IL-4 receptor on $CD4^+$ T cells promotes resistance in BALB/C mice [82]. However, some studies indicated that BALB/C IL-4-deficient mice remained susceptible to disease in the absence of this cytokine [83], whereas other studies showed that the same mice were resistant to *Leishmania* infection [81, 84]. These data question whether cytokines other than IL-4 might affect Th1 development during the infection. Recently, IL-4 has been identified as a negative regulator of chemokine production involved in Th1-type cell recruitment to the site of infection [85]. Short-term blocking of IL-4 led to changes in Th1-associated chemokine gene expression and correlated with increased accumulation of IFN-γ producers. It has been also shown that IL-4 signalling via IL-4Rα plays a crucial role in resistance to infection and is required for promoting Th1 responses [86].

IL-13 shares a number of characteristics with IL-4, and both share a common signalling pathway through IL-4Rα [87]. IL-13 has been demonstrated to have disease-promoting properties and to act independently of IL-4 [84, 88], indicating that IL-13 and IL-4 effects might be additive. High levels of IL-13 might prevent the onset of Th1 response by inhibiting IL-12 production by macrophages and skewing the response towards deleterious Th2-type. In *L. (L.) mexicana*-induced disease, studies with IL-13 knockout mice implicated this cytokine in preventing disease resolution by inhibiting IL-12R expression [84]. IL-13 also has the ability to render specific $CD4^+$ T cells unresponsive to IL-12, hence promoting parasite resistance [89].

IL-10 is a major immunosuppressive cytokine in leishmaniasis and, as already discussed, is essential for parasite persistence [52] and can exacerbate infection [51, 90]. It is a potent suppressor of macrophage activation, inhibits DC maturation [91] and is produced by a plethora of cells of the immune system [92]; however, a major source of IL-10 in CL (due to *L. (V.) braziliensis*) in humans has been identified as T regulatory type 1 (Tr-1)-like cells [93]. The ability of vaccinated mice to downregulate IL-10 secretion has been linked to protection following inoculation with SIR-2-deficient *L. (L.) infantum* parasites [94] and a phosphomannomutase (PMM) knockout line of *L. (L.) major* [44]. IL-10 knockout mice are highly resistant to *L. (L.) major*, whereas IL-10 transgenic mice on the resistant background become susceptible [95, 96]. IL-10's crucial role in suppression of the immune response has been demonstrated in *L. (L.) mexicana* and *L. (L.) amazonensis* infections, although effective resolution of infection with these New World species requires neutralisation of both IL-4 and IL-10 [97]. It has been shown that IL-10 differentially influences the quality, magnitude and protective efficacy of Th1 cells depending on the vaccine platform [98]. Interestingly, co-expression of IL-10 and IFN-γ by Th1 $CD4^+$ cells prevents pathogen clearance and contributes to chronic infection [99]. IL-10 secreted by T cells has been shown to affect immune activation early in infection, and a lack of T cell-specific IL-10 leads to enhanced protection following vaccination [100].

IL-22 is a member of IL-10 family of cytokines and in a mouse model of CL has been shown to exert a protective effect by limiting tissue damage [101] and skewing response towards protective Th1 when administered during the course of infection [102].

IL-9 has been shown to play a role in disease susceptibility. It is mainly produced by Th2 clones [103], and its induction can be either IL-4 dependent or independent [104, 105]. During *L. (L.) major* infection, IL-9 synthesis was observed from 4 weeks onward only in susceptible BALB/C but not in resistant C57BL/6 mice [106]. IL-9 neutralisation in BALB/C mice resulted in a diminished Th2 response and a shift towards protective Th1 responses leading to enhanced effector functions (increased NO production by macrophages) implicating IL-9 as a susceptibility factor in leishmaniasis [107].

Transforming growth factor-β (TGF-β) is a regulatory cytokine that controls initiation and resolution of inflammatory responses [108]. Different *Leishmania* species have been shown to induce TGF-β production from macrophages and release the active form of TGF-β from the latent complex [109]. This cytokine is important for determining susceptibility to experimental leishmaniasis [110], and anti-TGF-β treatment promotes resolution of *L. (L.) major* infection in mice by augmenting NO production [111]. Overall, mouse studies have indicated that TGF-β inhibits Th1 responses and leads to increased susceptibility to leishmaniasis. This is achieved by suppression of NO production and inhibition of TNF and IFN-γ.

5.2.3.2 Th1-Type and Pro-inflammatory Cytokines

Cytokines with the ability to influence Th1 development, such as IL-12 or IFN-γ, play a protective role in leishmaniasis. IL-12 promotes resistance through macrophage activation and NO production and is necessary for the priming of naïve T cells towards the Th1 pathway. Resistant mice depleted of IL-12 through the use of anti-IL-12 antibodies become more susceptible to infection, and administration of IL-12 to susceptible BALB/C mice promotes resistance to infection [112]. In addition, genetic disruption of IL-12 gene leads to upregulation of deleterious IL-4 response and establishment of progressive disease [113]. It has been suggested that IL-12 might be required for optimal proliferation and IFN-γ production by Th1 cells, both of which are significantly enhanced in the presence of IL-12 or can promote Th1 cell survival [114]. The memory $CD4^+$ T cells generated during *L. (L.) major* infection requires IL-12 for IFN-γ production and differentiation into Th1-type, whereas in the absence of IL-12, these cells became IL-4 producers [115]. The majority of IL-12 is produced by antigen-presenting cells such as macrophages, DCs and neutrophils [116]; however, *L. (L.) major* has the ability to selectively block its production in macrophages [117]. Thus, DCs appear to be the major source of IL-12 in leishmaniasis acting in combination with DC-derived IL-1α/β to influence Th1 development and promote resistance to cutaneous infection [118].

Similar to IL-12 deficiency, in case of IFN-γ deficiency, the immune response will default to Th2-type and lead to susceptibility to *L. (L.) major* [119]. NK cells are the primary early source of IFN-γ [120], which plays a role in rapid development of Th1 response. Nevertheless, these cells are not essential for resistance to the cutaneous infection, since efficient IL-12-dependent IFN-γ production by $CD4^+$ T cells has been reported in the absence of NK cells [121]. IFN-γ is a key cytokine triggering the antileishmanial functions of macrophages via induction of NO production and can activate macrophages alone or in synergy with TNF or IL-7 [122, 123]. Resistant mice display elevated levels of IFN-γ compared to susceptible mice, whilst targeted disruption of the IFN-γ gene [119] or the ligand binding chain of the IFN-γ receptor [124] in C57BL/6 mice results in increased susceptibility to *Leishmania* infection. However, contradictory data exist on the role of IFN-γ in *Leishmania* infection as some studies show that administration of IFN-γ to BALB/C mice at the time of infection does not affect susceptibility of BALB/C mice to leishmaniasis [125]. Additionally, non-healing lesions in C57BL/6 mice are observed despite a strong Th1 response characterised by high IFN-γ, NO expression and low IL-4 production [126].

TNF is a pro-inflammatory cytokine produced primarily by activated macrophages but is also produced by fibroblasts and T and B cells. It mediates resistance by controlling intracellular pathogen replication as well as limiting the duration of the inflammatory response [127]. Synergising with IFN-γ, TNF activates macrophages to exert iNOS-dependent leishmanicidal activity [2]. Mice resistant to *Leishmania* produce high levels of TNF in the draining lymph nodes, whereas susceptible mice produce none or minimal TNF [128].

IL-17 is a strong pro-inflammatory cytokine, and increased levels of IL-17 have been detected in patients with CL [129]. The most prominent role of IL-17 is the induction of pro-inflammatory responses via production of cytokines such as IL-6, TGF-β or TNF. In the absence of IL-10, *L. (L.) major*-infected mice display increased levels of IL-17 and neutrophil infiltration. It has been postulated that IL-17 exacerbates pathology, and its production is upregulated by IFN-γ and controlled by IL-10 [130]. However, a recent analysis of cytokine profile in healing and non-healing lesions due to *L. (L.) major* indicated that patients with healing lesions had higher levels of IL-17 suggesting its role in resistance to infection [131]. Thus, it appears that IL-17 might have some protective role; its high levels are involved in augmented immunopathology.

IL-27 is a cytokine produced upon exposure to inflammatory stimuli and is functionally and structurally related to IL-12 [132]. It has been implicated in the regulation of T cell functions and IFN-γ production and, as a consequence, in promoting Th1 responses [133]. Resistant mice lacking WSX-1 (a component of the IL-27 receptor) produce increased levels of IL-4 following *L. (L.) major* infection and a delayed Th1 response [134]. However, the requirement for IL-27 appears to be transient and important only in early infection since WSX-1 knockout mice are able to control lesion development and resolve infection [135]. IL-23 is a pro-inflammatory cytokine that also shows homology to IL-12 [136]. IL-23-deficient mice showed increased susceptibility to bacterial and parasitic infections [137], and IL-23 is involved in the regulation of IFN-γ production [138]. In leishmaniasis, IL-27 and IL-23 might play a complementary protective role with other Th1 cytokines. Human patients with *L. (L.) major* infection and a healing CL lesions display elevated levels of IL-27 and IL-23 compared to patients with non-healing lesions [139].

Type I interferons (IFN-α/β) are pro-inflammatory cytokines that are involved early in *L. (L.) major* infection as regulators of the innate response, NO production and IFN-γ expression [140]. Administration of recombinant IFN-β is sufficient to promote resistance in otherwise susceptible host [141]. Due to their role in the resistance to infection, *Leishmania* developed strategies to overcome their effects. Parasite proteases are responsible for degradation of STAT2 [28] and mechanistic target of rapamycin (mTOR) [142], both essential for Type I IFNs signalling.

IL-1β is an important mediator of an inflammatory response and has been identified as a major player controlling resistance and susceptibility to leishmaniasis. Increased IL-1β production has been linked to disease severity in *L. (L.) mexicana*-infected patients [143]. However, NLRP3 inflammasome-driven production of IL-1β enhanced host resistance to infection in C57BL6 model [144], an observation contradicted by another report indicating that inflammasome-deficient BALB/C mice with defective IL-1β (and IL-18) production were resistant to CL infection [145]. The inflammasome-dependent IL-1β and persistent recruitment of neutrophils have been identified as essential components of the non-healing response [146], and IL-1β signalling and NLRP3 activation were linked to $CD8^+$ T cell induced pathology [63]. Thus, it appears that IL-1β is detrimental to the host resistance to infection.

Taken together, the vast array of immune cells and cytokines as well as co-inhibitory molecules [147] and chemokines [148] involved in the immune response to *Leishmania* clearly highlights the complexity of the disease. To compound the matter even more, several of the immunomodulators and cell types described here appear to have a dual role in promoting both the susceptibility and resistance to infection. The murine model of cutaneous leishmaniasis, which mimics many aspects of the human disease, has been used to dissect the role of cytokines and T helper responses. In human cutaneous leishmaniasis, a clear dichotomy in T cell responses has not been reported; instead the patients revealed mixed Th1 and Th2 immunity [149]. Similarly, in human visceral leishmaniasis, there is no strong association between Th1 responses and resistance to disease; instead patients showed co-existing Th1- and Th2-type responses [150]. It appears that in humans, the outcome of disease is influenced by the balance between the two T cell populations and is further affected by the host genetic factors, inoculum size, parasite strain and cytokine milieu.

5.3 Visceral Leishmaniasis

VL results from infection with the *Leishmania* species *L. (L.) donovani* and *L. (L.) infantum* (*chagasi*). Parasites disseminate from the site of infection in the skin to reside and multiply within macrophages of the liver, spleen and bone marrow [151]. The majority of people infected with visceralising *Leishmania* species experience asymptomatic infection, and only a small proportion of infections lead to clinically severe disease. However, when left untreated, clinical VL manifests as systemic chronic, unresolving infection, which is usually fatal. Patients who recover from VL display immunity to reinfection, which suggests that the development of vaccines that provide clinical protection is a feasible goal. Immunocompromised individuals are susceptible to infection, and VL species are significant opportunistic pathogens during HIV infection [152]. Together this indicates an important role for the host immune response during infection. The underlying factors that influence disease susceptibility are not entirely understood, but host genetic factors clearly play a role in determining the outcome of infection. The presence of the *Slc11a1* gene is associated with protection against *Leishmania* infection, as well as other intracellular pathogens [153]. Slc11a1 is integral for regulating many cellular functions in macrophages, including cytokine production and antigen processing [154] and may also play a role in MHC class II expression in DCs [155]. *Sca11a1* mutant mice are susceptible to *Leishmania* infection, and experimental VL infection of these mice leads to high parasite burdens in the visceral organs. Interestingly, parasite infection resolves in the liver in a manner determined by MHC haplotype [151], indicating a role for acquired immune responses in the control of parasite burden. In contrast to mice, polymorphisms in humans are confined to the promoter region of the *Sca11a* gene [156]. Genetic linkage analysis has demonstrated an association between VL patients and polymorphisms in 5′ (CA) repeat in the *Slc11a1* promoter [157].

5.3.1 Experimental Murine Models

Clinical studies examining the immune response to VL infection are limited by the difficulty in directly accessing infected tissues in patients. Many studies have investigated the systemic response to VL infection by examining circulating peripheral blood mononuclear cells (PBMCs) and serum cytokine levels. This has limited the utility of in vitro approaches; however, the recent development of whole blood assays to detect cytokine production from infected patient samples [158] has enabled the study of immune correlates of disease status. To explore the immune response to VL, experimental murine models of infection have been developed, and rodents are competent hosts for both *L. (L.) donovani* and *L. (L.) infantum*. Establishment of rodent infections via i.v. infection with amastigotes has provided much insight into the organ-specific immune responses generated in the bone marrow, liver and spleen during VL. Low-dose dermal infection models using the infective metacyclic form of the parasite have also been developed to reflect the natural route of transmission [159].

5.3.2 Th1 and Th2 Cytokines

The majority of people infected with visceralising *Leishmania* maintain an asymptomatic infection, but the mechanisms that mediate effective control of the disease are relatively unknown. The Th1/Th2 dichotomy which influences CL outcomes does not appear to have a clear role in determining the resistance/susceptibility profiles in human infection or in experimental models of VL [160]. A strong cytokine response is induced during VL, and the production of IFN-γ appears crucial for the control of parasites and the development of resistance to infection [161]. Active infection is associated with the presence of both Th1 and Th2 cytokines. Multiple cytokines and chemokines are produced in response to VL infection with elevated levels of IFN-γ, TNF, IL-6, IL-8, IL-10, IL-12, IL-15, IL-18, IL-33, IP-10 and MIG observed in the serum of VL patients [162, 163]. Immune correlates of protection using whole blood assays have shown that subclinical VL infections and cured VL patients display a strong Th1 response with significantly elevated levels of IFNγ [158]. IL-10 was only elicited from patients with active VL disease, supporting the view that IL-10 is a key immunosuppressive cytokine in VL patients that contributes to host susceptibility [164].

Whilst clinical studies using samples from the peripheral blood of patients are informative, they may not necessarily reflect the events or immune mechanisms occurring in infected visceral organs. Studies in experimental rodent models demonstrate that organ-specific immune responses play a significant role in host defence of VL with defined patterns of tissue tropism and differential responses developing in the liver and spleen [165].

5.3.3 Adaptive Immune System: Contributions of B and T Cells to VL

B cells are not considered to play a significant protective role during *Leishmania* infection and have been implicated in exacerbating VL clinical disease [166, 167]. In contrast, T cells are critical for effective antileishmanial host responses. Immunocompromised mice lacking functional T cells, such as nude mice [168], severe combined immunodeficiency mice (SCID) [169] and recombinase-activating gene (RAG) knockout mice [170] all show enhanced susceptibility to *L. (L.) donovani* infection, which can be overcome via reconstitution of T cell populations. Effector $CD4^+$ T cells are responsible for the production of cytokines that are critical for the activation of macrophages and the initiation of effective host protective responses. Cytotoxic $CD8^+$ T cells play a host protective role and are required for effective clearance of parasites [168] and the generation of memory responses [171]. Antigen-specific $CD4^+$ and $CD8^+$ cells are activated during infection in both humans [172] and mice [173] and are required for optimal host response to infection. Administration of antigen-specific $CD8^+$ T cells to *L. (L.) donovani*-infected mice significantly decreased parasite burdens in the liver and spleen [174], and the induction of $CD8^+$ T cell responses is being explored as a therapeutic intervention [175]. Interestingly, in an intradermal model of VL, the clearance of parasites from the skin correlated with the infiltration and activation of both $CD4^+$ and $CD8^+$ T cells, analogous to the initiation of inflammatory responses and resolution observed in cutaneous infection [159]. Defective and anergic $CD8^+$ T cell responses may impair host responses to infection. $CD8^+$ T cells driven to exhaustion during human VL show upregulation of immune checkpoint inhibitors PD-1 and CTLA-4 with a high IL-10 expression and limited IFNγ production [176]. This may open up new therapeutic pathways as treatment with CTLA-blocking mAbs reduced the level of parasite burdens in a VL mouse model [177].

5.3.4 Immune Responses in the Liver

The hallmark clinical manifestation observed in almost all VL patients is a gross enlargement of the abdomen due to splenomegaly and hepatomegaly. In experimental mice models, hepatosplenomegaly is also a feature and is associated with parasite infection of these tissues. Infection of the liver is evident at 1 week following *L. (L.) donovani* inoculation, peaking at 3–4 weeks postinfection and then resolving with minimal damage to the tissue [174]. This acute resolving infection of the liver is associated with initial dominant reactive oxygen intermediate (ROI) and iNOS responses [178]. Macrophages use both reactive oxygen and nitrogen intermediates in the initial effort to limit *L. (L.) donovani* replication in the liver, with reactive nitrogen intermediates playing a more critical role in the resolution of liver parasite burdens [179].

5.3.4.1 The Granuloma Response

Effective immune responses to VL in the liver are critically dependent on the formation of granuloma structures, which serve to co-ordinate and deliver cellular and soluble host defence factors to the infected tissue. The granuloma environment produces a focus for antileishmanial immune mechanisms in terms of activating and sustaining appropriate parasite killing. During human VL, the presence of granulomas in the liver correlates with the ability to control and maintain infection at a subclinical level. In experimental models of VL, liver granulomas increase in number and size in the first month, leading to the clearance of parasites and the resolution of infection during the second month of infection [180]. Whilst the role of granulomas in generating memory responses is not clear, they may play a role in immunity to reinfection as a focal point for immune responses [181]. Whilst the majority of parasites are cleared from the liver, sterile cure is never achieved, though the liver is resistant to reinfection. The induction of immunosuppression can reactivate infection, which has been observed in the case of HIV patients [182] or people receiving immunosuppressive therapies following organ transplant [183].

The core of the liver granuloma develops from tissue-resident Kupffer cells which are recruited from the sinusoids during the acute phase of the inflammatory response [184]. Kupffer cells are the major phagocytic population within the liver and the prime target for *Leishmania* infection. The generation of antileishmanial responses in the infected Kupffer cell is dependent on granuloma formation to provide the microenvironment for intracellular *L. (L.) donovani* killing [185]. Infected Kupffer cells fuse with other mononuclear phagocytic cells to form the core of the granuloma, resulting in the secretion of chemokines and the infiltration and recruitment of leukocytes. Monocytes and neutrophils migrate to the liver within the first few days of infection and form a cellular mantle around the infected Kupffer cells in the developing granuloma. Bystander Kupffer cells also play a role in granuloma formation, as noninfected Kupffer cells are activated and initiate protective immune responses during experimental VL [186].

Experiments using depleting monoclonal antibodies towards monocytes and neutrophils delay the maturation of hepatic granulomas, indicating that these cell types are essential for parasite killing [187]. The arrival of mononuclear cells leads to the recruitment of $CD4^+$ and $CD8^+$ T cells, which are also essential for intact granuloma responses [188]. B cells accumulate in granulomas over time in an antigen-independent manner and engage in long-lasting interactions with T cells [189]. Interestingly histological analysis of liver tissue shows that the formation and maturation of granulomas are asynchronous with mature granulomas possessing complete mononuclear cell cuffing observed alongside infected Kupffer cells that have failed to initiate granuloma formation. The mechanisms regarding the differential timing of granuloma formation and the inability of some infected Kupffer cells to induce appropriate host defence responses is not well understood. Upon resolution of infection, empty or sterile granulomas are evident in the mouse model which then undergo an involution phase, restoring normal liver tissue function [185].

Chemokines and chemokine receptors have an important role in the development of protective immune responses in the liver due to their ability to attract Th1 cytokine producing cells. Increased production of CCL3 (MIP1a), CCL2 (MCP-1) and CXCL10 (IP-10) occurs in the liver early during infection, and these factors are most likely produced by the infected Kupffer cell [190]. The central role of chemokines in granuloma formation is highlighted by experiments demonstrating that administration of CCL2, CCL3 or IP-10 during experimental VL infection results in accelerated granuloma maturation in the liver and reduced parasite burdens [191]. Furthermore, mice lacking CCL3 or its receptor CCR5 show enhanced susceptibility to *L. (L.) donovani* infection [192]. Initial chemokine production and cell recruitment to the granuloma are T-cell independent, but sustained chemokine production and granuloma maturation require the presence of infiltrating T cells.

5.3.4.2 T Cells

Both $CD4^+$ and $CD8^+$ T cells are critical for granuloma formation, and the increase in $CD4^+$ and $CD8^+$ T cell numbers in the liver during VL infection may reflect expansion of resident populations as well as recruitment from the spleen [193]. During *L. (L.) donovani* infection T cells undergo high rates of apoptosis [170] suggesting immune responses are continually generated throughout the course of infection, rather than being governed by long-lived effector T cell populations. In animals that lack T cells, the absence of sustained chemokine production results in a failure of granuloma formation and the uncontrolled growth of parasites in the liver [194]. $CD8^+$ T cells contribute to the control of liver parasite burdens through their role in granuloma formation [168, 180] and are essential for control in the liver during rechallenge experiments [195]. The activity of $CD8^+$ T cells may involve perforin and FasL-dependent lysis of parasitised macrophages as well as the secretion of pro-inflammatory cytokines and chemokines [196]. The dynamics of $CD8^+$ effector T cells in the liver during *L. (L.) donovani* infection have been visualised using intravital 2-photon microscopy, and $CD8^+$ T cells were observed to accumulate in granulomas in an antigen-specific manner [184]. This study also demonstrated that infected Kupffer cells are the main antigen-presenting cells for $CD8^+$ T cells in the liver and suggested that a sustained interaction with antigen-specific $CD8^+$ T cells may instigate lysis of the infected host cell [184]. However, *Leishmania* parasites have been shown to evade protective immune responses by inducing functional $CD8^+$ T cell exhaustion, driving $CD8^+$ T cell anergy and cell death during experimental [173] and human VL [176].

5.3.4.3 Th1-Type Cytokines

The predominant host protective role of $CD4^+$ T cells during VL is the production of cytokines and chemokines that supports granuloma formation and parasite killing.

Host defence in the liver is critically mediated by pro-inflammatory Th1-type cytokines, including IL-2 [188], IL-12 [197], IFN-γ [161], TNF [198], lymphotoxin (LT) [199] and granulocyte/macrophage colony-stimulating factor (GM-CSF) [200]. IL-2 is a potent T cell growth factor, which enhances granuloma tissue reactions and parasite clearance during experimental *L. (L.) donovani* infection largely through the induction of IFN-γ [201]. Production of IFN-γ by T cells to generate protective responses in the liver is also dependent on IL-12. Control of parasitaemia is lost in the absence of IL-12 and is associated with reduced IFN-γ production and arrested granuloma formation [202]. IL-12 may also exert antileishmanial effects independently of IFN-γ, as administration of IL-12 to IFN-γ knockout mice still resulted in parasite killing [203]. IL-12 is thought to play an important role in the regulation of the cellular immune responses in human VL. PBMCs from patients with active VL are unable to produce IFN-γ in response to *Leishmania* antigens in vitro; however, the addition of IL-12 is restored in vitro IFN-γ production [197].

During experimental VL, IFN-γ plays a critical role in the early immune responses that induce tissue granuloma formation and effectively control parasite replication. The neutralisation of IFN-γ during infection results in poor cellular assembly of granulomas and an increased parasite burden in the liver [161]. Impaired granuloma formation was also observed in mice deficient in IFN-γ and was associated with an inability of infected Kupffer cells to recruit monocytes and T cells to the liver [203]. Therapeutic administration of IFN-γ can activate macrophages in vivo but requires the presence of T cells for antileishmanial activity [204]. The administration of IFN-γ increased the efficacy of antimony chemotherapy in rodent models [205], and IFN-γ has been used as an adjunct therapy for severe or refractory cases of clinical VL [206]. Whilst IFN-γ plays a crucial role in the initiation of the granulomatous response early in infection, mice deficient in IFN-γ are capable of reducing liver parasite burdens in the later stages of infection. An early IFN-γ response leads to the induction of IL-12 and the expression of TNF, and it appears that the late-developing IFN-γ-independent antileishmanial mechanism is mediated by TNF [203].

TNF is essential for the formation and maturation of the hepatic granuloma response [207]. *L. (L.) donovani* infection is fatal in mice lacking TNF with accelerated parasite growth in the liver, impaired hepatic granuloma formation and an enhanced inflammatory response [207]. Neutralisation of TNF during *L. (L.) donovani* infection promotes parasite persistence in the liver indicating TNF is required for hepatic resolution [198]. TNF is produced by infected Kupffer cells throughout the time course of infection [194] and is essential for leukocyte recruitment. LTα, a member of the TNF superfamily of cytokines, is also required for the control of parasite growth in the liver. LTα plays a key role in granuloma formation, facilitating the trafficking of lymphocytes from the perivascular areas of the liver to the infected Kupffer cells [199]. The role of LTα in the liver is distinct from that of TNF, as $CD4^+$ T cells that express both TNF and LTα are needed for efficient killing of parasites within assembled granulomas. Other members of the TNF superfamily, such as CD95L, also contribute to host protective immune responses during VL [170].

5.3.4.4 Th2-Type Cytokines

Whilst the emphasis on liver immune defence is generally focussed on the production of Th1 cytokines, the co-expression of Th2 cytokines may also contribute to host protective responses. For example, the induction of IL-4 is essential for the formation of mature granulomas and for effective parasite killing [208]. The suppressive effect of immunoregulatory cytokines may limit inflammatory tissue damage in the liver, but generally these cytokines downregulate critical antileishmanial responses, particularly those dependent on IFN-γ. The production of TGF-β [209], IL-6 [210] Il-10 [90], IL-27 [211] and IL-33 [212] impair effective control of parasite growth in the liver. Mice deficient in IL-6 showed an enhanced ability to control infection with earlier, and more rapid, parasite killing associated with increased levels of circulating IFN-γ and accelerated granuloma formation [210]. Expression of IL-33, an IL-1 family member, is increased in the liver during human VL, and patients have increased IL-33 serum levels. Lower liver parasite burdens were observed in mice in the absence of IL-33 signalling mice, which was associated with a strong induction of IFN-γ and IL-12 [212]. Investigations into the role of Th17 cells during experimental VL infection have shown that IL-17A acts synergistically with IFN-γ to induce macrophage activation, increasing NO production and promoting the control of parasite replication [213].

IL-10 is a key immunosuppressive cytokine that inhibits resistance to VL and promotes disease progression. Human VL disease is strongly associated with increased production of IL-10 in a variety of clinical settings, and elevated IL-10 levels correlate with the development of pathology [214]. The absence of IL-10 leads to enhanced resistance to experimental VL infections in mice [90]. IL-10 has multiple effects on the immune system and suppresses the production of key cytokines, IL-12 and IFN-γ [90, 215]. Regulation of cellular immune responses by IL-10 includes the suppression of macrophage activation [216] and impaired intracellular killing of *Leishmania* [217]. Whilst there are multiple cellular sources of IL-10 during VL infection, a population of Th1-like $CD4^+$ T cells that make IL-10 have been associated with disease progression [218]. Conventional dendritic cells that make both IL-10 and IL-27 can induce the production of IL-10 from effector Th1-like $CD4^+$ T cells and enhance immunopathology [219]. The neutralisation of IL-10 enhances the production of IFNγ in cells from VL patients [158], and clinical interventions targeting the inhibition of IL-10 have been proposed as an immunotherapy in combination with chemotherapy [164].

5.3.4.5 NK and NKT Cells

NK cells and NKT cells participate in the early innate immune responses in the liver and contribute to the control of parasitaemia [220, 221]. CD1d-dependent activation of NKT cells occurs during *L. (L.) donovani* infection, and these cells also respond with a rapid production of IFN-γ. CD1d-deficient mice show an increased susceptibility to parasitism [222]. During infection, Kupffer cells can activate invariant NKT (iNKT) cells by engagement of CD47 [223], and iNKT cells are essential for regulating chemokines, such as CXCL10 [220]. There is an increasing interest in the development of therapies that enhance iNKT cell function during VL infection, but the benefits of iNKT cell activation may depend on the antigenic stimulation. Use of the glycosphingophospholipid (GSPL) antigen of *L. (L.) donovani* parasites to activate iNKT cells stimulated IFN-γ and IL-17A and led to the clearance of organ parasite burdens [224]. However, iNKT cell activation using glycolipid antigen α-galactosylceramide (α-GalCer) hindered disease resolution in the liver, exacerbating disease [225]. Further investigation may reveal the utility and effectiveness of therapies aimed at modulating NKT cell function to establish a protective host response.

5.3.5 VL and the Spleen: Suppression and Susceptibility

The spleen is a major organ for the induction of immune responses to infection and also a site for the killing of parasites during VL. However, prevalent clinical features of human VL include splenomegaly and a suppression of antigen-specific immune responses [226]. This immunopathology is recapitulated in experimental murine models where splenomegaly is associated with the persistence of parasites and remodelling of the lymphoid tissue [227]. The kinetics of experimental VL infection display distinct organ-specific pattern: the liver displays an acute resolving infection, attributed to effective granuloma tissue responses, whilst VL parasites persist in the spleen resulting in a chronic, unresolved state of infection.

5.3.5.1 Acute Immune Responses

The spleen is a highly organised secondary lymphoid organ, consisting of a specialised marginal zone (MZ), which separates the red pulp and white pulp region. The macrophages in the MZ, the marginal metallophilic macrophages (MMM) and the marginal zone macrophages (MZM), are the main phagocytic cell populations responsible for the clearance of parasites during experimental *L. (L.) donovani* infection. The antileishmanial activity of these specialised splenic macrophages is dependent on interferon regulatory factor-7 (IRF-7) [228].

Acute immune responses generated in the spleen play a key role in the control of *L. (L.) donovani* parasites in the liver during the early phase of infection. The spleen is an important site for DC priming, and DCs are the critical source of early IL-12 following VL infection [25]. A transient and rapid burst in IL-12 has been observed as early as 5 hours postinfection [229] and is a crucial event for the generation of effective antiparasitic immunity [230]. Vascular cell adhesion molecule-1 (VCAM-1) and its ligand very late antigen-4 (VLA-4) are involved in the initiation of early IL-12 secretion from DCs. Blockade of VCAM-1 or VLA-4 suppressed the production of IL-12 by splenic DCs and reduced parasite-specific T cell responses in the spleen. This was also associated with lower levels of IFN-γ, TNF and NO production in the liver and significantly higher liver parasite burdens [231]. Migratory DCs may directly phagocytose parasites; however, it is most likely that splenic DCs acquire antigen and are activated by infected macrophages in the marginal zone. Upon activation DCs migrate to the T cell areas in the periarteriolar lymphoid sheets (PALS), and IL-12-producing DCs are observed in the T cell area of the spleen during VL infection. The production of IL-12 by DCs is essential for the activation of effector T cell populations, and the total $CD4^+$ T cell population in the spleen is expanded during experimental infection [227]. T cells are the dominant leukocyte population in the spleen of VL patients, as compared to normal healthy control aspirates that show a predominance of B cells [232].

5.3.5.2 Chemokines Mediate DC and T Cell Interactions

Chemokine-dependent encounters between DCs and T cells in the spleen are crucial for effective responses to *L. (L.) donovani* infection. Mice deficient in CCL19 and CCL21 show impaired DC migration in the spleen and a decreased production of IL-12 during *L. (L.) donovani* infection. These defects in early DC activation in the spleen were associated with a reduced migration of effector T cells to the liver and impaired granuloma formation [229]. Exogenous administration of IP-10 restores T cell proliferative capacity, leading to decreased parasite burdens in the liver and spleen. IP-10 treatment during experimental VL induced strong expression of iNOS2 and mediated parasitic killing through increased NO synthesis [233]. Together the data demonstrate the importance of chemokines in promoting early DC and $CD4^+$ T cell interactions in the spleen and inducing protective immunity against *L. (L.) donovani*. The infection of DCs impairs their ability to prime $CD4^+$ T cells as noninfected bystander DCs are capable of inducing immune protective $CD4^+$ T cells, whilst infected DCs induce a nonprotective $CD4^+$ T cell responses [234].

5.3.5.3 Neutrophils

Neutrophils may also play a protective role in the acute response in the spleen, as the absence of neutrophils results in a decrease in IFN-γ-producing $CD4^+$ and $CD8^+$ T cells and an enhanced parasite burden in the spleen. This antileishmanial effect appeared to be specific to the spleen as the absence of neutrophils had only minor effects on parasite growth in the liver [235]. Neutrophils do not appear to play a significant role in the chronic stage of infection as long-term administration of neutrophil depleting antibody does not significantly increased parasite burdens in either organ [236].

5.3.5.4 Chronic Infection of the Spleen: IL-10 and TNF

During experimental *L. (L.) donovani* infection in mice, no resolution of infection occurs in the spleen, and animals maintain chronic parasite burdens in this tissue. There is evidence of profound immune dysfunction in the spleen with an impairment of antigen-specific T cell responses, increased T cell apoptosis [170] and the production of regulatory cytokines, such as IL-10 [218] and TGF-β [237]. NK cells are negative regulators of cell-mediated immunity in the spleen and show enhanced secretion of IL-10 in the chronic phase of infection [238]. Marginal zone B cells in the spleen have been shown to suppress antigen-specific CD8 and CD4 T cell responses during the early stages of VL [239]. B cells also suppress NK cells and inhibit the generation of effector memory CD8 T cells after *L. (L.) donovani* infection [239].

Whilst the production of TNF is crucial for the induction and maintenance of host protective responses in the liver, TNF is a key mediator of pathology in the chronically infected spleen. During the latter stages of VL, high numbers of TNF producing cells are present in the spleen, and TNF production is observed in both the red and white pulp regions [230]. TNF is the principal cytokine responsible for the breakdown of splenic architecture following experimental *L. (L.) donovani* infection, contributing to remodelling of the MZ [240] and the loss of stromal cells from the PALS [241]. Infection-induced remodelling of the MZ is associated with a dramatic and rapid loss of MZMs, whilst MMMs undergo repositioning within the sinus. In mice lacking TNF or mice treated with TNF neutralising monoclonal antibodies, MZMs were preserved, indicating that the loss of MZMs is a TNF-dependent process [240]. Evidence for the role of TNF in disease pathogenesis in human VL arises from studies of TNF polymorphisms. A study in Northern Brazil examined polymorphism in the *TNFA* promoter (TNF1 and TNF2 alleles) in neighbourhoods with ongoing transmission. The presence of the TNF2 allele was more frequent in individuals with progressive disease, whilst the TNF1 allele was associated with asymptomatic infection. The presence of the TNF2 susceptibility allele was associated with higher levels of serum TNF as compared to the TNF1 allele, suggesting that increased TNF is involved in the progression of human VL [242].

5.3.5.5 B Cells

The activation of B cells is a key clinical indicator of VL infection with patients displaying polyclonal hypergammaglobulinaemia [243], polyclonal B cell activation and increased circulating immune complexes. The role of immunoglobulins during VL is controversial, as large amounts of immunoglobulins to both parasite-specific and non-specific antigens are produced during infection, including autoantibodies [167]. These immunoglobulins are not thought to be protective as elevated levels of total antibody correlates with disease pathology [244] and have been implicated in the development of anaemia [245] and autoimmunity [167]. Experimental models of VL using B cell-deficient mice have demonstrated that B cells are not required for the control of parasite burdens. Additionally, the reconstitution of mice with immunoglobulin leads to disease exacerbation through complement activation and signalling [246]. However, B cells may have some regulatory role to play in suppressing immunopathology, as the absence of B cells leads to sustained neutrophil-mediated pathology of the liver [74].

5.3.5.6 Dysregulation of Immune Function

Follicular DCs (FDC), a resident stromal cell population, play a key role in the organisation of lymphoid follicles in the spleen and facilitate the germinal centre (GC) reaction. FDCs are involved in B cell activation, proliferation and maturation through presentation of antigen and production of regulatory signals such as chemokines. During the chronic stage of *L. (L.) donovani* infection, the FDC network is destroyed, and there is a concomitant loss of GC [247]. The complete absence of FDCs is associated with the infiltration of heavily parasitised macrophages into the splenic white pulp regions. It has been hypothesised that the B cell function may become dysregulated in the absence of FDCs, and thus the loss of FDCs may contribute to the hypergammaglobulinaemia observed during VL.

5.3.5.7 Alterations to Splenic Architecture

Impaired DC migration plays a major role in the pathogenesis of VL, and alterations to stromal cell populations directly contribute to immunosuppression during the chronic stage of *L. (L.) donovani* infection. Splenic DCs increase in number during the chronic phase of infection but fail to migrate from the MZ to the PALS. This impaired migration is due to a disruption in the fibroblastic reticular cell (FRC) network that guides T cell and DC migration in the T cell zone of the spleen. The changes to the splenic FRC network are due to a TNF-dependent loss of podoplanin $(gp38)^+$ stromal cells [241]. Downregulation of CCR7 from the DC cell surface also impairs DC migration in the spleen during VL. TNF is also implicated in this process, as enhanced levels of TNF increase IL-10 production, and IL-10 directly induces the loss of CCR7 expression on the DC surface [241]. A potential therapeutic role for DCs has been proposed, as adoptive transfer experiments show that administration of in vitro activated DCs can reduce parasite burdens in the spleen. The efficacy of DC therapy relies on both IL-12 and IL-6, with IL-6 thought to suppress the expansion of IL-10 producing T cells [218]. However recent studies demonstrate that some populations of DCs may contribute to splenic pathology, as targeted deletion of DCs during the established phase of infection improved disease resolution [219].

Interventions that preserve splenic structure during VL have been shown to improve the host response to chemotherapy by enhancing parasite killing. Treatment of experimental VL with receptor tyrosine kinase inhibitors reduced splenomegaly, prevented vascular remodelling and restored the integrity of the microarchitecture of the spleen. Importantly, the maintenance of splenic architecture during infection improved the host response to drug treatment, with a tenfold reduction in the amount of antimony required to clear infection [248].

5.4 Conclusions

Leishmania parasites activate the innate and adaptive arms of the immune system, and it is clear that a co-ordinated network of responses is required for effective immune-mediated parasite clearance. The timing of key chemokine and cytokine responses is essential and involves a tight regulation of cellular populations of the immune system. However, *Leishmania* parasites have developed numerous mechanisms to prevent development of immunity and promote resistance (Fig. 5.1). These include induction of immunosuppressive cytokines, interruption of signalling pathways in macrophages and dendritic cells and induction of regulatory T cells. Resistance to infection is also enhanced by the negative regulatory role of NK cells in chronic disease and the presence of Th2 cell-attracting chemokines in

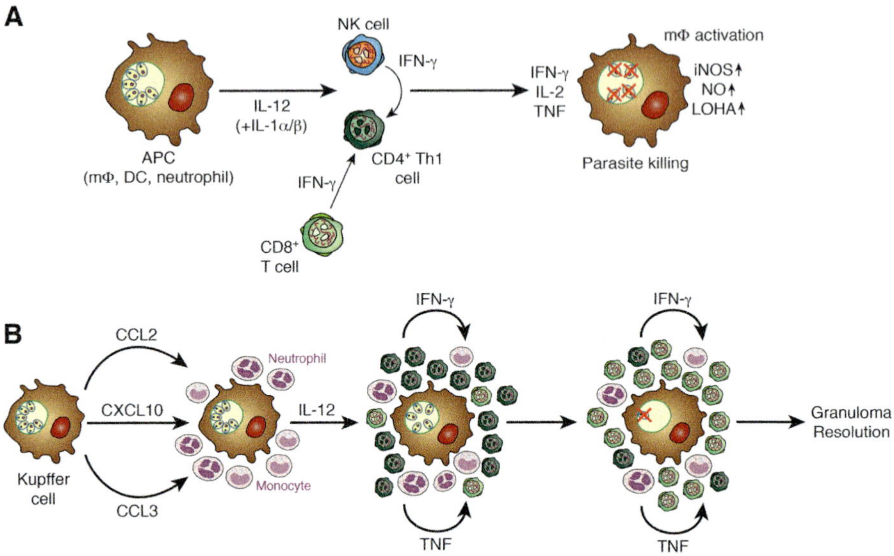

Fig. 5.1 The role of the immune system in resistance to infection. (**a**) Immune responses during *L. (L.) major* infection leading to CL clearance. (**b**) Granuloma formation in the liver leads to clearance of VL infection. *L. (L.) donovani* infection of Kupffer cells leads to the recruitment of immune cells, creating an inflammatory environment that promotes parasite killing (adapted from [193])

lesions. All these mechanisms assist the parasite in avoiding immune clearance and increase the chances of successful transmission of *Leishmania* parasites to a new host. Understanding the complexity of immune responses involved in *Leishmania* disease pathogenesis and protection offers hope for development of effective vaccines and immunotherapeutic interventions. The host immune system also supports the actions of chemotherapy, and understanding which immune modulation interventions will work synergistically with new drug therapies will enhance the clinical treatment of *Leishmania* patients.

References

1. Basu MK, Ray M. Macrophage and *Leishmania*: an unacceptable coexistence. Crit Rev Microbiol. 2005;31(3):145–54.
2. Liew FY, Li Y, Millott S. Tumor necrosis factor-alpha synergizes with IFN-gamma in mediating killing of *Leishmania major* through the induction of nitric oxide. J Immunol. 1990;145(12):4306–10.
3. Iniesta V, Gomez-Nieto LC, Corraliza I. The inhibition of arginase by N(omega)-hydroxy-l-arginine controls the growth of *Leishmania* inside macrophages. J Exp Med. 2001;193(6):777–84.
4. Wei XQ, Charles IG, Smith A, Ure J, et al. Altered immune responses in mice lacking inducible nitric oxide synthase. Nature. 1995;375(6530):408–11.

5. Mukbel RM, Patten C Jr, Gibson K, Ghosh M, et al. Macrophage killing of *Leishmania amazonensis* amastigotes requires both nitric oxide and superoxide. Am J Trop Med Hyg. 2007;76(4):669–75.
6. Trinchieri G. Interleukin-12: a cytokine at the interface of inflammation and immunity. Adv Immunol. 1998;70:83–243.
7. Ricardo-Carter C, Favila M, Polando RE, Cotton RN, et al. *Leishmania major* inhibits IL-12 in macrophages by signalling through CR3 (CD11b/CD18) and down-regulation of ETS-mediated transcription. Parasite Immunol. 2013;35(12):409–20.
8. Belkaid Y, Mendez S, Lira R, Kadambi N, et al. A natural model of *Leishmania major* infection reveals a prolonged "silent" phase of parasite amplification in the skin before the onset of lesion formation and immunity. J Immunol. 2000;165(2):969–77.
9. Peters NC, Sacks DL. The impact of vector-mediated neutrophil recruitment on cutaneous leishmaniasis. Cell Microbiol. 2009;11(9):1290–6.
10. Peters NC, Egen JG, Secundino N, Debrabant A, et al. In vivo imaging reveals an essential role for neutrophils in leishmaniasis transmitted by sand flies. Science. 2008;321(5891):970–4.
11. Ribeiro-Gomes FL, Peters NC, Debrabant A, Sacks DL. Efficient capture of infected neutrophils by dendritic cells in the skin inhibits the early anti-leishmania response. PLoS Pathog. 2012;8(2):e1002536.
12. Ribeiro-Gomes FL, Romano A, Lee S, Roffe E, et al. Apoptotic cell clearance of *Leishmania major*-infected neutrophils by dendritic cells inhibits CD8(+) T-cell priming in vitro by Mer tyrosine kinase-dependent signaling. Cell Death Dis. 2015;6:e2018.
13. Guimaraes-Costa AB, Nascimento MT, Froment GS, Soares RP, et al. *Leishmania amazonensis* promastigotes induce and are killed by neutrophil extracellular traps. Proc Natl Acad Sci USA. 2009;106(16):6748–53.
14. Carlsen ED, Liang Y, Shelite TR, Walker DH, et al. Permissive and protective roles for neutrophils in leishmaniasis. Clin Exp Immunol. 2015;182(2):109–18.
15. Hurrell BP, Schuster S, Grun E, Coutaz M, et al. Rapid sequestration of *Leishmania mexicana* by neutrophils contributes to the development of chronic lesion. PLoS Pathog. 2015;11(5):e1004929.
16. Reiner SL, Locksley RM. The regulation of immunity to *Leishmania major*. Annu Rev Immunol. 1995;13:151–77.
17. Banchereau J, Steinman RM. Dendritic cells and the control of immunity. Nature. 1998;392(6673):245–52.
18. Von Stebut E. Immunology of cutaneous leishmaniasis: the role of mast cells, phagocytes and dendritic cells for protective immunity. Eur J Dermatol. 2007;17(2):115–22.
19. Ritter U, Meissner A, Scheidig C, Korner H. CD8 alpha- and Langerin-negative dendritic cells, but not Langerhans cells, act as principal antigen-presenting cells in leishmaniasis. Eur J Immunol. 2004;34(6):1542–50.
20. Ng LG, Hsu A, Mandell MA, Roediger B, et al. Migratory dermal dendritic cells act as rapid sensors of protozoan parasites. PLoS Pathog. 2008;4(11):e1000222.
21. Woelbing F, Kostka SL, Moelle K, Belkaid Y, et al. Uptake of *Leishmania major* by dendritic cells is mediated by Fcgamma receptors and facilitates acquisition of protective immunity. J Exp Med. 2006;203(1):177–88.
22. Bajenoff M, Breart B, Huang AY, Qi H, et al. Natural killer cell behavior in lymph nodes revealed by static and real-time imaging. J Exp Med. 2006;203(3):619–31.
23. Feijo D, Tiburcio R, Ampuero M, Brodskyn C, et al. Dendritic cells and *Leishmania* infection: adding layers of complexity to a complex disease. J Immunol Res. 2016;2016:3967436.
24. Soong L. Modulation of dendritic cell function by *Leishmania* parasites. J Immunol. 2008;180(7):4355–60.
25. Gorak PM, Engwerda CR, Kaye PM. Dendritic cells, but not macrophages, produce IL-12 immediately following *Leishmania donovani* infection. Eur J Immunol. 1998;28(2):687–95.
26. von Stebut E, Belkaid Y, Jakob T, Sacks DL, et al. Uptake of *Leishmania major* amastigotes results in activation and interleukin 12 release from murine skin-derived dendritic cells: implications for the initiation of anti-*Leishmania* immunity. J Exp Med. 1998;188(8):1547–52.

27. Bennett CL, Misslitz A, Colledge L, Aebischer T, et al. Silent infection of bone marrow-derived dendritic cells by *Leishmania mexicana* amastigotes. Eur J Immunol. 2001;31 (3):876–83.
28. Xin L, Li K, Soong L. Down-regulation of dendritic cell signaling pathways by *Leishmania amazonensis* amastigotes. Mol Immunol. 2008;45(12):3371–82.
29. Silveira FT, Lainson R, Gomes CM, Laurenti MD, et al. Reviewing the role of the dendritic Langerhans cells in the immunopathogenesis of American cutaneous leishmaniasis. Trans R Soc Trop Med Hyg. 2008;102(11):1075–80.
30. Belkaid Y, Butcher B, Sacks DL. Analysis of cytokine production by inflammatory mouse macrophages at the single-cell level: selective impairment of IL-12 induction in *Leishmania*-infected cells. Eur J Immunol. 1998;28(4):1389–400.
31. Flohe S, Lang T, Moll H. Synthesis, stability, and subcellular distribution of major histocompatibility complex class II molecules in Langerhans cells infected with *Leishmania major*. Infect Immun. 1997;65(8):3444–50.
32. Martinez-Lopez M, Iborra S, Conde-Garrosa R, Sancho D. Batf3-dependent CD103+ dendritic cells are major producers of IL-12 that drive local Th1 immunity against *Leishmania major* infection in mice. Eur J Immunol. 2015;45(1):119–29.
33. Akbari M, Honma K, Kimura D, Miyakoda M, et al. IRF4 in dendritic cells inhibits IL-12 production and controls Th1 immune responses against *Leishmania major*. J Immunol. 2014;192(5):2271–9.
34. Bogdan C, Rollinghoff M, Diefenbach A. The role of nitric oxide in innate immunity. Immunol Rev. 2000;173:17–26.
35. Scott P. The role of TH1 and TH2 cells in experimental cutaneous leishmaniasis. Exp Parasitol. 1989;68(3):369–72.
36. Iezzi G, Karjalainen K, Lanzavecchia A. The duration of antigenic stimulation determines the fate of naive and effector T cells. Immunity. 1998;8(1):89–95.
37. Steinman RM, Hemmi H. Dendritic cells: translating innate to adaptive immunity. Curr Top Microbiol Immunol. 2006;311:17–58.
38. O'Garra A. Cytokines induce the development of functionally heterogeneous T helper cell subsets. Immunity. 1998;8(3):275–83.
39. Bogdan C, Moll H, Solbach W, Rollinghoff M. Tumor necrosis factor-alpha in combination with interferon-gamma, but not with interleukin 4 activates murine macrophages for elimination of *Leishmania major* amastigotes. Eur J Immunol. 1990;20(5):1131–5.
40. Darrah PA, Patel DT, De Luca PM, Lindsay RW, et al. Multifunctional TH1 cells define a correlate of vaccine-mediated protection against *Leishmania major*. Nat Med. 2007;13 (7):843–50.
41. Sacks D, Noben-Trauth N. The immunology of susceptibility and resistance to *Leishmania major* in mice. Nat Rev Immunol. 2002;2(11):845–58.
42. Scott P, Eaton A, Gause WC, di Zhou X, et al. Early IL-4 production does not predict susceptibility to *Leishmania major*. Exp Parasitol. 1996;84(2):178–87.
43. Uzonna JE, Spath GF, Beverley SM, Scott P. Vaccination with phosphoglycan-deficient *Leishmania major* protects highly susceptible mice from virulent challenge without inducing a strong Th1 response. J Immunol. 2004;172(6):3793–7.
44. Kedzierski L, Curtis JM, Doherty PC, Handman E, et al. Decreased IL-10 and IL-13 production and increased CD44hi T cell recruitment contribute to *Leishmania major* immunity induced by non-persistent parasites. Eur J Immunol. 2008;38(11):3090–100.
45. Stober CB, Lange UG, Roberts MT, Alcami A, et al. IL-10 from regulatory T cells determines vaccine efficacy in murine *Leishmania major* infection. J Immunol. 2005;175(4):2517–24.
46. Ehrchen JM, Roebrock K, Foell D, Nippe N, et al. Keratinocytes determine Th1 immunity during early experimental leishmaniasis. PLoS Pathog. 2010;6(4):e1000871.
47. Sunderkotter C, Kunz M, Steinbrink K, Meinardus-Hager G, et al. Resistance of mice to experimental leishmaniasis is associated with more rapid appearance of mature macrophages in vitro and in vivo. J Immunol. 1993;151(9):4891–901.

48. Himmelrich H, Launois P, Maillard I, Biedermann T, et al. In BALB/c mice, IL-4 production during the initial phase of infection with *Leishmania major* is necessary and sufficient to instruct Th2 cell development resulting in progressive disease. J Immunol. 2000;164(9):4819–25.
49. Biedermann T, Zimmermann S, Himmelrich H, Gumy A, et al. IL-4 instructs TH1 responses and resistance to *Leishmania major* in susceptible BALB/c mice. Nat Immunol. 2001;2(11):1054–60.
50. Belkaid Y, Rouse BT. Natural regulatory T cells in infectious disease. Nat Immunol. 2005;6(4):353–60.
51. Belkaid Y, Piccirillo CA, Mendez S, Shevach EM, et al. CD4+CD25+ regulatory T cells control *Leishmania major* persistence and immunity. Nature. 2002;420(6915):502–7.
52. Belkaid Y, Hoffmann KF, Mendez S, Kamhawi S, et al. The role of interleukin (IL)-10 in the persistence of *Leishmania major* in the skin after healing and the therapeutic potential of anti-IL-10 receptor antibody for sterile cure. J Exp Med. 2001;194(10):1497–506.
53. Campanelli AP, Roselino AM, Cavassani KA, Pereira MS, et al. CD4+CD25+ T cells in skin lesions of patients with cutaneous leishmaniasis exhibit phenotypic and functional characteristics of natural regulatory T cells. J Infect Dis. 2006;193(9):1313–22.
54. Bourreau E, Ronet C, Darcissac E, Lise MC, et al. Intralesional regulatory T-cell suppressive function during human acute and chronic cutaneous leishmaniasis due to *Leishmania guyanensis*. Infect Immun. 2009;77(4):1465–74.
55. Hoseini SG, Javanmard SH, Zarkesh SH, Khamesipour A, et al. Regulatory T-cell profile in early and late lesions of cutaneous leishmaniasis due to *Leishmania major*. J Res Med Sci. 2012;17(6):513–8.
56. Okwor I, Liu D, Beverley SM, Uzonna JE. Inoculation of killed *Leishmania major* into immune mice rapidly disrupts immunity to a secondary challenge via IL-10-mediated process. Proc Natl Acad Sci USA. 2009;106(33):13951–6.
57. Huber M, Timms E, Mak TW, Rollinghoff M, et al. Effective and long-lasting immunity against the parasite *Leishmania major* in CD8-deficient mice. Infect Immun. 1998;66(8):3968–70.
58. Muller I, Kropf P, Etges RJ, Louis JA. Gamma interferon response in secondary *Leishmania major* infection: role of CD8+ T cells. Infect Immun. 1993;61(9):3730–8.
59. Belkaid Y, Von Stebut E, Mendez S, Lira R, et al. CD8+ T cells are required for primary immunity in C57BL/6 mice following low-dose, intradermal challenge with *Leishmania major*. J Immunol. 2002b;168(8):3992–4000.
60. Uzonna JE, Joyce KL, Scott P. Low dose *Leishmania major* promotes a transient T helper cell type 2 response that is down-regulated by interferon gamma-producing CD8+ T cells. J Exp Med. 2004;199(11):1559–66.
61. Ruiz JH, Becker I. CD8 cytotoxic T cells in cutaneous leishmaniasis. Parasite Immunol. 2007;29(12):671–8.
62. Novais FO, Carvalho LP, Graff JW, Beiting DP, et al. Cytotoxic T cells mediate pathology and metastasis in cutaneous leishmaniasis. PLoS Pathog. 2013;9(7):e1003504.
63. Novais FO, Carvalho AM, Clark ML, Carvalho LP, et al. CD8+ T cell cytotoxicity mediates pathology in the skin by inflammasome activation and IL-1beta production. PLoS Pathog. 2017;13(2):e1006196.
64. Faria DR, Souza PE, Duraes FV, Carvalho EM, et al. Recruitment of CD8(+) T cells expressing granzyme A is associated with lesion progression in human cutaneous leishmaniasis. Parasite Immunol. 2009;31(8):432–9.
65. Santos Cda S, Boaventura V, Ribeiro Cardoso C, Tavares N, et al. CD8(+) granzyme B(+)-mediated tissue injury vs. CD4(+)IFNgamma(+)-mediated parasite killing in human cutaneous leishmaniasis. J Invest Dermatol. 2013;133(6):1533–40.
66. Cardoso TM, Machado A, Costa DL, Carvalho LP, et al. Protective and pathological functions of CD8+ T cells in *Leishmania braziliensis* infection. Infect Immun. 2015;83(3):898–906.

67. Bertholet S, Goldszmid R, Morrot A, Debrabant A, et al. *Leishmania* antigens are presented to CD8+ T cells by a transporter associated with antigen processing-independent pathway in vitro and in vivo. J Immunol. 2006;177(6):3525–33.
68. Houde M, Bertholet S, Gagnon E, Brunet S, et al. Phagosomes are competent organelles for antigen cross-presentation. Nature. 2003;425(6956):402–6.
69. Rodriguez A, Regnault A, Kleijmeer M, Ricciardi-Castagnoli P, et al. Selective transport of internalized antigens to the cytosol for MHC class I presentation in dendritic cells. Nat Cell Biol. 1999;1(6):362–8.
70. Matheoud D, Moradin N, Bellemare-Pelletier A, Shio MT, et al. *Leishmania* evades host immunity by inhibiting antigen cross-presentation through direct cleavage of the SNARE VAMP8. Cell Host Microbe. 2013;14(1):15–25.
71. Ashok D, Schuster S, Ronet C, Rosa M, et al. Cross-presenting dendritic cells are required for control of *Leishmania major* infection. Eur J Immunol. 2014;44(5):1422–32.
72. Sacks DL, Scott PA, Asofsky R, Sher FA. Cutaneous leishmaniasis in anti-IgM-treated mice: enhanced resistance due to functional depletion of a B cell-dependent T cell involved in the suppressor pathway. J Immunol. 1984;132(4):2072–7.
73. Hoerauf A, Solbach W, Rollinghoff M, Gessner A. Effect of IL-7 treatment on *Leishmania major*-infected BALB.Xid mice: enhanced lymphopoiesis with sustained lack of B1 cells and clinical aggravation of disease. Int Immunol. 1995;7(11):1879–84.
74. Smelt SC, Cotterell SE, Engwerda CR, Kaye PM. B cell-deficient mice are highly resistant to *Leishmania donovani* infection, but develop neutrophil-mediated tissue pathology. J Immunol. 2000;164(7):3681–8.
75. Ronet C, Voigt H, Himmelrich H, Doucey MA, et al. *Leishmania major*-specific B cells are necessary for Th2 cell development and susceptibility to *L. major* LV39 in BALB/c mice. J Immunol. 2008;180(7):4825–35.
76. Ronet C, Hauyon-La Torre Y, Revaz-Breton M, Mastelic B, et al. Regulatory B cells shape the development of Th2 immune responses in BALB/c mice infected with *Leishmania major* through IL-10 production. J Immunol. 2010;184(2):886–94.
77. Rodriguez-Pinto D, Saravia NG, McMahon-Pratt D. CD4 T cell activation by B cells in human *Leishmania (Viannia)* infection. BMC Infect Dis. 2014;14:108.
78. Gollob KJ, Antonelli LR, Faria DR, Keesen TS, et al. Immunoregulatory mechanisms and CD4-CD8- (double negative) T cell subpopulations in human cutaneous leishmaniasis: a balancing act between protection and pathology. Int Immunopharmacol. 2008;8(10):1338–43.
79. Liese J, Schleicher U, Bogdan C. The innate immune response against *Leishmania* parasites. Immunobiology. 2008;213(3-4):377–87.
80. Rodriguez NE, Wilson ME. Eosinophils and mast cells in leishmaniasis. Immunol Res. 2014;59(1–3):129–41.
81. Kopf M, Brombacher F, Kohler G, Kienzle G, et al. IL-4-deficient Balb/c mice resist infection with *Leishmania major*. J Exp Med. 1996;184(3):1127–36.
82. Radwanska M, Cutler AJ, Hoving JC, Magez S, et al. Deletion of IL-4Ralpha on CD4 T cells renders BALB/c mice resistant to *Leishmania major* infection. PLoS Pathog. 2007;3(5):e68.
83. Noben-Trauth N, Kropf P, Muller I. Susceptibility to *Leishmania major* infection in interleukin-4-deficient mice. Science. 1996;271(5251):987–90.
84. Alexander J, Brombacher F, McGachy HA, McKenzie AN, et al. An essential role for IL-13 in maintaining a non-healing response following *Leishmania mexicana* infection. Eur J Immunol. 2002;32(10):2923–33.
85. Lazarski CA, Ford J, Katzman SD, Rosenberg AF, et al. IL-4 attenuates Th1-associated chemokine expression and Th1 trafficking to inflamed tissues and limits pathogen clearance. PLoS One. 2013;8(8):e71949.
86. Hurdayal R, Nieuwenhuizen NE, Revaz-Breton M, Smith L, et al. Deletion of IL-4 receptor alpha on dendritic cells renders BALB/c mice hypersusceptible to *Leishmania major* infection. PLoS Pathog. 2013;9(10):e1003699.

87. Brombacher F. The role of interleukin-13 in infectious diseases and allergy. BioEssays. 2000;22(7):646–56.
88. Matthews DJ, Emson CL, McKenzie GJ, Jolin HE, et al. IL-13 is a susceptibility factor for *Leishmania major* infection. J Immunol. 2000;164(3):1458–62.
89. Bourreau E, Prevot G, Pradinaud R, Launois P. Interleukin (IL)-13 is the predominant Th2 cytokine in localized cutaneous leishmaniasis lesions and renders specific CD4+ T cells unresponsive to IL-12. J Infect Dis. 2001;183(6):953–9.
90. Murphy ML, Wille U, Villegas EN, Hunter CA, et al. IL-10 mediates susceptibility to *Leishmania donovani* infection. Eur J Immunol. 2001;31(10):2848–56.
91. O'Garra A, Vieira P. T(H)1 cells control themselves by producing interleukin-10. Nat Rev Immunol. 2007;7(6):425–8.
92. Moore KW, de Waal Malefyt R, Coffman RL, O'Garra A. Interleukin-10 and the interleukin-10 receptor. Annu Rev Immunol. 2001;19:683–765.
93. Costa DL, Cardoso TM, Queiroz A, Milanezi CM, et al. Tr-1-like CD4+CD25-CD127-/lowFOXP3- cells are the main source of interleukin 10 in patients with cutaneous leishmaniasis due to *Leishmania braziliensis*. J Infect Dis. 2015;211(5):708–18.
94. Silvestre R, Cordeiro-Da-Silva A, Santarem N, Vergnes B, et al. SIR2-deficient *Leishmania infantum* induces a defined IFN-gamma/IL-10 pattern that correlates with protection. J Immunol. 2007;179(5):3161–70.
95. Kane MM, Mosser DM. The role of IL-10 in promoting disease progression in leishmaniasis. J Immunol. 2001;166(2):1141–7.
96. Groux H, Cottrez F, Rouleau M, Mauze S, et al. A transgenic model to analyze the immunoregulatory role of IL-10 secreted by antigen-presenting cells. J Immunol. 1999;162(3):1723–9.
97. Padigel UM, Alexander J, Farrell JP. The role of interleukin-10 in susceptibility of BALB/c mice to infection with *Leishmania mexicana* and *Leishmania amazonensis*. J Immunol. 2003;171(7):3705–10.
98. Darrah PA, Hegde ST, Patel DT, Lindsay RW, et al. IL-10 production differentially influences the magnitude, quality, and protective capacity of Th1 responses depending on the vaccine platform. J Exp Med. 2010;207(7):1421–33.
99. Anderson CF, Oukka M, Kuchroo VJ, Sacks D. CD4(+)CD25(-)Foxp3(-) Th1 cells are the source of IL-10-mediated immune suppression in chronic cutaneous leishmaniasis. J Exp Med. 2007;204(2):285–97.
100. Schwarz T, Remer KA, Nahrendorf W, Masic A, et al. T cell-derived IL-10 determines leishmaniasis disease outcome and is suppressed by a dendritic cell based vaccine. PLoS Pathog. 2013;9(6):e1003476.
101. Gimblet C, Loesche MA, Carvalho L, Carvalho EM, et al. IL-22 protects against tissue damage during cutaneous leishmaniasis. PLoS One. 2015;10(8):e0134698.
102. Hezarjaribi HZ, Ghaffarifar F, Dalimi A, Sharifi Z. Evaluation of protective effect of IL-22 and IL-12 on cutaneous leishmaniasis in BALB/c mice. Asian Pac J Trop Med. 2014;7(12):940–5.
103. Demoulin JB, Renauld JC. Interleukin 9 and its receptor: an overview of structure and function. Int Rev Immunol. 1998;16(3–4):345–64.
104. Kopf M, Le Gros G, Bachmann M, Lamers MC, et al. Disruption of the murine IL-4 gene blocks Th2 cytokine responses. Nature. 1993;362(6417):245–8.
105. Monteyne P, Renauld JC, Van Broeck J, Dunne DW, et al. IL-4-independent regulation of in vivo IL-9 expression. J Immunol. 1997;159(6):2616–23.
106. Gessner A, Blum H, Rollinghoff M. Differential regulation of IL-9-expression after infection with *Leishmania major* in susceptible and resistant mice. Immunobiology. 1993;189(5):419–35.
107. Arendse B, Van Snick J, Brombacher F. IL-9 is a susceptibility factor in Leishmania major infection by promoting detrimental Th2/type 2 responses. J Immunol. 2005;174(4):2205–11.
108. Li MO, Wan YY, Sanjabi S, Robertson AK, et al. Transforming growth factor-beta regulation of immune responses. Annu Rev Immunol. 2006;24:99–146.

109. Mougneau E, Bihl F, Glaichenhaus N. Cell biology and immunology of *Leishmania*. Immunol Rev. 2011;240(1):286–96.
110. Barral-Netto M, Barral A, Brownell CE, Skeiky YA, et al. Transforming growth factor-beta in leishmanial infection: a parasite escape mechanism. Science. 1992;257(5069):545–8.
111. Li J, Hunter CA, Farrell JP. Anti-TGF-beta treatment promotes rapid healing of *Leishmania major* infection in mice by enhancing in vivo nitric oxide production. J Immunol. 1999;162 (2):974–9.
112. Heinzel FP, Rerko RM, Hatam F, Locksley RM. IL-2 is necessary for the progression of leishmaniasis in susceptible murine hosts. J Immunol. 1993;150(9):3924–31.
113. Mattner F, Magram J, Ferrante J, Launois P, et al. Genetically resistant mice lacking interleukin-12 are susceptible to infection with *Leishmania major* and mount a polarized Th2 cell response. Eur J Immunol. 1996;26(7):1553–9.
114. Scott P, Artis D, Uzonna J, Zaph C. The development of effector and memory T cells in cutaneous leishmaniasis: the implications for vaccine development. Immunol Rev. 2004;201:318–38.
115. Pakpour N, Zaph C, Scott P. The central memory CD4+ T cell population generated during *Leishmania major* infection requires IL-12 to produce IFN-gamma. J Immunol. 2008;180 (12):8299–305.
116. von Stebut E, Udey MC. Requirements for Th1-dependent immunity against infection with *Leishmania major*. Microbes Infect. 2004;6(12):1102–9.
117. Carrera L, Gazzinelli RT, Badolato R, Hieny S, et al. *Leishmania* promastigotes selectively inhibit interleukin 12 induction in bone marrow-derived macrophages from susceptible and resistant mice. J Exp Med. 1996;183(2):515–26.
118. Von Stebut E, Ehrchen JM, Belkaid Y, Kostka SL, et al. Interleukin 1 alpha promotes Th1 differentiation and inhibits disease progression in *Leishmania major*-susceptible BALB/c mice. J Exp Med. 2003;198(2):191–9.
119. Wang ZE, Reiner SL, Zheng S, Dalton DK, et al. CD4+ effector cells default to the Th2 pathway in interferon gamma-deficient mice infected with *Leishmania major*. J Exp Med. 1994;179(4):1367–71.
120. Scharton TM, Scott P. Natural killer cells are a source of interferon gamma that drives differentiation of CD4+ T cell subsets and induces early resistance to *Leishmania major* in mice. J Exp Med. 1993;178(2):567–77.
121. Satoskar AR, Stamm LM, Zhang X, Satoskar AA, et al. Mice lacking NK cells develop an efficient Th1 response and control cutaneous *Leishmania major* infection. J Immunol. 1999;162(11):6747–54.
122. Nacy CA, Meierovics AI, Belosevic M, Green SJ. Tumor necrosis factor-alpha: central regulatory cytokine in the induction of macrophage antimicrobial activities. Pathobiology. 1991;59(3):182–4.
123. Gessner A, Vieth M, Will A, Schroppel K, et al. Interleukin-7 enhances antimicrobial activity against Leishmania major in murine macrophages. Infect Immun. 1993;61(9):4008–12.
124. Swihart K, Fruth U, Messmer N, Hug K, et al. Mice from a genetically resistant background lacking the interferon gamma receptor are susceptible to infection with *Leishmania major* but mount a polarized T helper cell 1-type CD4+ T cell response. J Exp Med. 1995;181 (3):961–71.
125. Sadick MD, Heinzel FP, Holaday BJ, Pu RT, et al. Cure of murine leishmaniasis with anti-interleukin 4 monoclonal antibody. Evidence for a T cell-dependent, interferon gamma-independent mechanism. J Exp Med. 1990;171(1):115–27.
126. Anderson CF, Mendez S, Sacks DL. Nonhealing infection despite Th1 polarization produced by a strain of *Leishmania major* in C57BL/6 mice. J Immunol. 2005;174(5):2934–41.
127. Havell EA. Evidence that tumor necrosis factor has an important role in antibacterial resistance. J Immunol. 1989;143(9):2894–9.
128. Titus RG, Sherry B, Cerami A. Tumor necrosis factor plays a protective role in experimental murine cutaneous leishmaniasis. J Exp Med. 1989;170(6):2097–104.

129. Bacellar O, Faria D, Nascimento M, Cardoso TM, et al. Interleukin 17 production among patients with American cutaneous leishmaniasis. J Infect Dis. 2009;200(1):75–8.
130. Gonzalez-Lombana C, Gimblet C, Bacellar O, Oliveira WW, et al. IL-17 mediates immunopathology in the absence of IL-10 following *Leishmania major* infection. PLoS Pathog. 2013;9(3):e1003243.
131. Moafi M, Rezvan H, Sherkat R, Taleban R, et al. Comparison of pro-inflammatory cytokines of non-healing and healing cutaneous leishmaniasis. Scand J Immunol. 2017;85(4):291–9.
132. Boulay JL, O'Shea JJ, Paul WE. Molecular phylogeny within type I cytokines and their cognate receptors. Immunity. 2003;19(2):159–63.
133. Chen Q, Ghilardi N, Wang H, Baker T, et al. Development of Th1-type immune responses requires the type I cytokine receptor TCCR. Nature. 2000;407(6806):916–20.
134. Yoshida H, Hamano S, Senaldi G, Covey T, et al. WSX-1 is required for the initiation of Th1 responses and resistance to *L. major* infection. Immunity. 2001;15(4):569–78.
135. Artis D, Johnson LM, Joyce K, Saris C, et al. Cutting edge: early IL-4 production governs the requirement for IL-27-WSX-1 signaling in the development of protective Th1 cytokine responses following *Leishmania major* infection. J Immunol. 2004;172(8):4672–5.
136. Oppmann B, Lesley R, Blom B, Timans JC, et al. Novel p19 protein engages IL-12p40 to form a cytokine, IL-23, with biological activities similar as well as distinct from IL-12. Immunity. 2000;13(5):715–25.
137. Tan ZY, Bealgey KW, Fang Y, Gong YM, et al. Interleukin-23: immunological roles and clinical implications. Int J Biochem Cell Biol. 2009;41(4):733–5.
138. Langrish CL, McKenzie BS, Wilson NJ, de Waal Malefyt R, et al. IL-12 and IL-23: master regulators of innate and adaptive immunity. Immunol Rev. 2004;202:96–105.
139. Tolouei S, Ghaedi K, Khamesipour A, Akbari M, et al. IL-23 and IL-27 levels in macrophages collected from peripheral blood of patients with healing vs non-healing form of cutaneous leishmaniasis. Iran J Parasitol. 2012;7(1):18–25.
140. Diefenbach A, Schindler H, Donhauser N, Lorenz E, et al. Type 1 interferon (IFNalpha/beta) and type 2 nitric oxide synthase regulate the innate immune response to a protozoan parasite. Immunity. 1998;8(1):77–87.
141. Mattner J, Wandersee-Steinhauser A, Pahl A, Rollinghoff M, et al. Protection against progressive leishmaniasis by IFN-beta. J Immunol. 2004;172(12):7574–82.
142. Jaramillo M, Gomez MA, Larsson O, Shio MT, et al. *Leishmania* repression of host translation through mTOR cleavage is required for parasite survival and infection. Cell Host Microbe. 2011;9(4):331–41.
143. Fernandez-Figueroa EA, Rangel-Escareno C, Espinosa-Mateos V, Carrillo-Sanchez K, et al. Disease severity in patients infected with *Leishmania mexicana* relates to IL-1beta. PLoS Negl Trop Dis. 2012;6(5):e1533.
144. Lima-Junior DS, Costa DL, Carregaro V, Cunha LD, et al. Inflammasome-derived IL-1beta production induces nitric oxide-mediated resistance to *Leishmania*. Nat Med. 2013;19(7):909–15.
145. Gurung P, Karki R, Vogel P, Watanabe M, et al. An NLRP3 inflammasome-triggered Th2-biased adaptive immune response promotes leishmaniasis. J Clin Invest. 2015;125(3):1329–38.
146. Charmoy M, Hurrell BP, Romano A, Lee SH, et al. The Nlrp3 inflammasome, IL-1beta, and neutrophil recruitment are required for susceptibility to a nonhealing strain of *Leishmania major* in C57BL/6 mice. Eur J Immunol. 2016;46(4):897–911.
147. Gannavaram S, Bhattacharya P, Ismail N, Kaul A, et al. Modulation of innate immune mechanisms to enhance *Leishmania* vaccine-induced immunity: role of coinhibitory molecules. Front Immunol. 2016;7:187.
148. Oghumu S, Lezama-Davila CM, Isaac-Marquez AP, Satoskar AR. Role of chemokines in regulation of immunity against leishmaniasis. Exp Parasitol. 2010;126(3):389–96.

149. Ajdary S, Alimohammadian MH, Eslami MB, Kemp K, et al. Comparison of the immune profile of nonhealing cutaneous leishmaniasis patients with those with active lesions and those who have recovered from infection. Infect Immun. 2000;68(4):1760–4.
150. Khalil EA, Ayed NB, Musa AM, Ibrahim ME, et al. Dichotomy of protective cellular immune responses to human visceral leishmaniasis. Clin Exp Immunol. 2005;140(2):349–53.
151. Leclercq V, Lebastard M, Belkaid Y, Louis J, et al. The outcome of the parasitic process initiated by *Leishmania infantum* in laboratory mice: a tissue-dependent pattern controlled by the Lsh and MHC loci. J Immunol. 1996;157(10):4537–45.
152. Ali A. Leishmaniases and HIV/AIDS co-infections: review of common features and management experiences. Ethiop Med J. 2002;40(Suppl 1):37–49.
153. Vidal S, Tremblay ML, Govoni G, Gauthier S, et al. The Ity/Lsh/Bcg locus: natural resistance to infection with intracellular parasites is abrogated by disruption of the Nramp1 gene. J Exp Med. 1995;182(3):655–66.
154. Blackwell JM, Goswami T, Evans CA, Sibthorpe D, et al. SLC11A1 (formerly NRAMP1) and disease resistance. Cell Microbiol. 2001;3(12):773–84.
155. Stober CB, Brode S, White JK, Popoff JF, et al. Slc11a1, formerly Nramp1, is expressed in dendritic cells and influences major histocompatibility complex class II expression and antigen-presenting cell function. Infect Immun. 2007;75(10):5059–67.
156. Blackwell JM, Barton CH, White JK, Searle S, et al. Genomic organization and sequence of the human NRAMP gene: identification and mapping of a promoter region polymorphism. Mol Med. 1995;1(2):194–205.
157. Goswami T, Bhattacharjee A, Babal P, Searle S, et al. Natural-resistance-associated macrophage protein 1 is an H+/bivalent cation antiporter. Biochem J. 2001;354(Pt 3):511–9.
158. Singh OP, Gidwani K, Kumar R, Nylen S, et al. Reassessment of immune correlates in human visceral leishmaniasis as defined by cytokine release in whole blood. Clin Vaccine Immunol. 2012;19(6):961–6.
159. Ahmed S, Colmenares M, Soong L, Goldsmith-Pestana K, et al. Intradermal infection model for pathogenesis and vaccine studies of murine visceral leishmaniasis. Infect Immun. 2003;71 (1):401–10.
160. Miralles GD, Stoeckle MY, McDermott DF, Finkelman FD, et al. Th1 and Th2 cell-associated cytokines in experimental visceral leishmaniasis. Infect Immun. 1994;62(3):1058–63.
161. Squires KE, Schreiber RD, McElrath MJ, Rubin BY, et al. Experimental visceral leishmaniasis: role of endogenous IFN-gamma in host defense and tissue granulomatous response. J Immunol. 1989;143(12):4244–9.
162. Kurkjian KM, Mahmutovic AJ, Kellar KL, Haque R, et al. Multiplex analysis of circulating cytokines in the sera of patients with different clinical forms of visceral leishmaniasis. Cytometry A. 2006;69(5):353–8.
163. Sundar S, Reed SG, Sharma S, Mehrotra A, et al. Circulating T helper 1 (Th1) cell- and Th2 cell-associated cytokines in Indian patients with visceral leishmaniasis. Am J Trop Med Hyg. 1997;56(5):522–5.
164. Singh OP, Sundar S. Immunotherapy and targeted therapies in treatment of visceral leishmaniasis: current status and future prospects. Front Immunol. 2014;5:296.
165. Engwerda CR, Kaye PM. Organ-specific immune responses associated with infectious disease. Immunol Today. 2000;21(2):73–8.
166. Bohme MW, Evans DA, Miles MA, Holborow EJ. Occurrence of autoantibodies to intermediate filament proteins in human visceral leishmaniasis and their induction by experimental polyclonal B-cell activation. Immunology. 1986;59(4):583–8.
167. Galvao-Castro B, Sa Ferreira JA, Marzochi KF, Marzochi MC, et al. Polyclonal B cell activation, circulating immune complexes and autoimmunity in human American visceral leishmaniasis. Clin Exp Immunol. 1984;56(1):58–66.
168. Stern JJ, Oca MJ, Rubin BY, Anderson SL, et al. Role of L3T4+ and LyT-2+ cells in experimental visceral leishmaniasis. J Immunol. 1988;140(11):3971–7.

169. Kaye PM, Bancroft GJ. *Leishmania donovani* infection in scid mice: lack of tissue response and in vivo macrophage activation correlates with failure to trigger natural killer cell-derived gamma interferon production in vitro. Infect Immun. 1992;60(10):4335–42.
170. Alexander CE, Kaye PM, Engwerda CR. CD95 is required for the early control of parasite burden in the liver of *Leishmania donovani*-infected mice. Eur J Immunol. 2001;31 (4):1199–210.
171. Stager S, Smith DF, Kaye PM. Immunization with a recombinant stage-regulated surface protein from *Leishmania donovani* induces protection against visceral leishmaniasis. J Immunol. 2000;165(12):7064–71.
172. Mary C, Auriault V, Faugere B, Dessein AJ. Control of *Leishmania infantum* infection is associated with CD8(+) and gamma interferon- and interleukin-5-producing CD4(+) antigen-specific T cells. Infect Immun. 1999;67(11):5559–66.
173. Joshi T, Rodriguez S, Perovic V, Cockburn IA, et al. B7-H1 blockade increases survival of dysfunctional CD8(+) T cells and confers protection against *Leishmania donovani* infections. PLoS Pathog. 2009;5(5):e1000431.
174. Polley R, Stager S, Prickett S, Maroof A, et al. Adoptive immunotherapy against experimental visceral leishmaniasis with CD8+ T cells requires the presence of cognate antigen. Infect Immun. 2006;74(1):773–6.
175. Maroof A, Brown N, Smith B, Hodgkinson MR, et al. Therapeutic vaccination with recombinant adenovirus reduces splenic parasite burden in experimental visceral leishmaniasis. J Infect Dis. 2012;205(5):853–63.
176. Gautam S, Kumar R, Singh N, Singh AK, et al. CD8 T cell exhaustion in human visceral leishmaniasis. J Infect Dis. 2014;209(2):290–9.
177. Murphy ML, Cotterell SE, Gorak PM, Engwerda CR, et al. Blockade of CTLA-4 enhances host resistance to the intracellular pathogen, *Leishmania donovani*. J Immunol. 1998;161 (8):4153–60.
178. Murray HW, Nathan CF. Macrophage microbicidal mechanisms in vivo: reactive nitrogen versus oxygen intermediates in the killing of intracellular visceral *Leishmania donovani*. J Exp Med. 1999;189(4):741–6.
179. Murray HW, Cartelli DM. Killing of intracellular *Leishmania donovani* by human mononuclear phagocytes. Evidence for oxygen-dependent and -independent leishmanicidal activity. J Clin Invest. 1983;72(1):32–44.
180. McElrath MJ, Murray HW, Cohn ZA. The dynamics of granuloma formation in experimental visceral leishmaniasis. J Exp Med. 1988;167(6):1927–37.
181. Moore JW, Moyo D, Beattie L, Andrews PS, et al. Functional complexity of the *Leishmania* granuloma and the potential of in silico modeling. Front Immunol. 2013;4:35.
182. Lachaud L, Bourgeois N, Plourde M, Leprohon P, et al. Parasite susceptibility to amphotericin B in failures of treatment for visceral leishmaniasis in patients coinfected with HIV type 1 and *Leishmania infantum*. Clin Infect Dis. 2009;48(2):e16–22.
183. Antinori S, Cascio A, Parravicini C, Bianchi R, et al. Leishmaniasis among organ transplant recipients. Lancet Infect Dis. 2008;8(3):191–9.
184. Beattie L, Peltan A, Maroof A, Kirby A, et al. Dynamic imaging of experimental *Leishmania donovani*-induced hepatic granulomas detects Kupffer cell-restricted antigen presentation to antigen-specific CD8 T cells. PLoS Pathog. 2010;6(3):e1000805.
185. Murray HW. Tissue granuloma structure-function in experimental visceral leishmaniasis. Int J Exp Pathol. 2001;82(5):249–67.
186. Beattie L, d'El-Rei Hermida M, Moore JW, Maroof A, et al. A transcriptomic network identified in uninfected macrophages responding to inflammation controls intracellular pathogen survival. Cell Host Microbe. 2013;14(3):357–68.
187. Cervia JS, Rosen H, Murray HW. Effector role of blood monocytes in experimental visceral leishmaniasis. Infect Immun. 1993;61(4):1330–3.

188. Murray HW, Stern JJ, Welte K, Rubin BY, et al. Experimental visceral leishmaniasis: production of interleukin 2 and interferon-gamma, tissue immune reaction, and response to treatment with interleukin 2 and interferon-gamma. J Immunol. 1987;138(7):2290–7.
189. Moore JW, Beattie L, Dalton JE, Owens BM, et al. B cell: T cell interactions occur within hepatic granulomas during experimental visceral leishmaniasis. PLoS One. 2012;7(3):e34143.
190. Cotterell SE, Engwerda CR, Kaye PM. *Leishmania donovani* infection initiates T cell-independent chemokine responses, which are subsequently amplified in a T cell-dependent manner. Eur J Immunol. 1999;29(1):203–14.
191. Dey R, Majumder N, Bhattacharyya Majumdar S, Bhattacharjee S, et al. Induction of host protective Th1 immune response by chemokines in *Leishmania donovani*-infected BALB/c mice. Scand J Immunol. 2007;66(6):671–83.
192. Sato N, Kuziel WA, Melby PC, Reddick RL, et al. Defects in the generation of IFN-gamma are overcome to control infection with *Leishmania donovani* in CC chemokine receptor (CCR) 5-, macrophage inflammatory protein-1 alpha-, or CCR2-deficient mice. J Immunol. 1999;163 (10):5519–25.
193. Stanley AC, Engwerda CR. Balancing immunity and pathology in visceral leishmaniasis. Immunol Cell Biol. 2007;85(2):138–47.
194. Engwerda CR, Smelt SC, Kaye PM. An in vivo analysis of cytokine production during *Leishmania donovani* infection in scid mice. Exp Parasitol. 1996;84(2):195–202.
195. Murray HW, Squires KE, Miralles CD, Stoeckle MY, et al. Acquired resistance and granuloma formation in experimental visceral leishmaniasis. Differential T cell and lymphokine roles in initial versus established immunity. J Immunol. 1992;148(6):1858–63.
196. Tsagozis P, Karagouni E, Dotsika E. CD8(+) T cells with parasite-specific cytotoxic activity and a Tc1 profile of cytokine and chemokine secretion develop in experimental visceral leishmaniasis. Parasite Immunol. 2003;25(11–12):569–79.
197. Ghalib HW, Whittle JA, Kubin M, Hashim FA, et al. IL-12 enhances Th1-type responses in human *Leishmania donovani* infections. J Immunol. 1995;154(9):4623–9.
198. Tumang MC, Keogh C, Moldawer LL, Helfgott DC, et al. Role and effect of TNF-alpha in experimental visceral leishmaniasis. J Immunol. 1994;153(2):768–75.
199. Engwerda CR, Ato M, Stager S, Alexander CE, et al. Distinct roles for lymphotoxin-alpha and tumor necrosis factor in the control of *Leishmania donovani* infection. Am J Pathol. 2004;165 (6):2123–33.
200. Murray HW, Cervia JS, Hariprashad J, Taylor AP, et al. Effect of granulocyte-macrophage colony-stimulating factor in experimental visceral leishmaniasis. J Clin Invest. 1995;95 (3):1183–92.
201. Murray HW, Miralles GD, Stoeckle MY, McDermott DF. Role and effect of IL-2 in experimental visceral leishmaniasis. J Immunol. 1993;151(2):929–38.
202. Murray HW. Endogenous interleukin-12 regulates acquired resistance in experimental visceral leishmaniasis. J Infect Dis. 1997;175(6):1477–9.
203. Taylor AP, Murray HW. Intracellular antimicrobial activity in the absence of interferon-gamma: effect of interleukin-12 in experimental visceral leishmaniasis in interferon-gamma gene-disrupted mice. J Exp Med. 1997;185(7):1231–9.
204. Murray HW, Hariprashad J, Aguero B, Arakawa T, et al. Antimicrobial response of a T cell-deficient host to cytokine therapy: effect of interferon-gamma in experimental visceral leishmaniasis in nude mice. J Infect Dis. 1995;171(5):1309–16.
205. Murray HW. Effect of continuous administration of interferon-gamma in experimental visceral leishmaniasis. J Infect Dis. 1990;161(5):992–4.
206. Badaro R, Johnson WD Jr. The role of interferon-gamma in the treatment of visceral and diffuse cutaneous leishmaniasis. J Infect Dis. 1993;167(Suppl 1):S13–7.
207. Murray HW, Jungbluth A, Ritter E, Montelibano C, et al. Visceral leishmaniasis in mice devoid of tumor necrosis factor and response to treatment. Infect Immun. 2000;68 (11):6289–93.
208. Stager S, Alexander J, Carter KC, Brombacher F, et al. Both interleukin-4 (IL-4) and IL-4 receptor alpha signaling contribute to the development of hepatic granulomas with optimal antileishmanial activity. Infect Immun. 2003;71(8):4804–7.

209. Wilson ME, Young BM, Davidson BL, Mente KA, et al. The importance of TGF-beta in murine visceral leishmaniasis. J Immunol. 1998;161(11):6148–55.
210. Murray HW. Accelerated control of visceral *Leishmania donovani* infection in interleukin-6-deficient mice. Infect Immun. 2008;76(9):4088–91.
211. Rosas LE, Satoskar AA, Roth KM, Keiser TL, et al. Interleukin-27R (WSX-1/T-cell cytokine receptor) gene-deficient mice display enhanced resistance to *Leishmania donovani* infection but develop severe liver immunopathology. Am J Pathol. 2006;168(1):158–69.
212. Rostan O, Gangneux JP, Piquet-Pellorce C, Manuel C, et al. The IL-33/ST2 axis is associated with human visceral leishmaniasis and suppresses Th1 responses in the livers of BALB/c mice infected with *Leishmania donovani*. MBio. 2013;4(5):e00383–13.
213. Nascimento MS, Carregaro V, Lima-Junior DS, Costa DL, et al. Interleukin 17A acts synergistically with interferon gamma to promote protection against *Leishmania infantum* infection. J Infect Dis. 2015;211(6):1015–26.
214. Ghalib HW, Piuvezam MR, Skeiky YA, Siddig M, et al. Interleukin 10 production correlates with pathology in human *Leishmania donovani* infections. J Clin Invest. 1993;92(1):324–9.
215. Murray HW, Moreira AL, Lu CM, DeVecchio JL, et al. Determinants of response to interleukin-10 receptor blockade immunotherapy in experimental visceral leishmaniasis. J Infect Dis. 2003;188(3):458–64.
216. Bogdan C, Vodovotz Y, Nathan C. Macrophage deactivation by interleukin 10. J Exp Med. 1991;174(6):1549–55.
217. Bhattacharyya S, Ghosh S, Jhonson PL, Bhattacharya SK, et al. Immunomodulatory role of interleukin-10 in visceral leishmaniasis: defective activation of protein kinase C-mediated signal transduction events. Infect Immun. 2001;69(3):1499–507.
218. Stager S, Maroof A, Zubairi S, Sanos SL, et al. Distinct roles for IL-6 and IL-12p40 in mediating protection against *Leishmania donovani* and the expansion of IL-10+ CD4+ T cells. Eur J Immunol. 2006;36(7):1764–71.
219. Owens BM, Beattie L, Moore JW, Brown N, et al. IL-10-producing Th1 cells and disease progression are regulated by distinct CD11c(+) cell populations during visceral leishmaniasis. PLoS Pathog. 2012;8(7):e1002827.
220. Svensson M, Zubairi S, Maroof A, Kazi F, et al. Invariant NKT cells are essential for the regulation of hepatic CXCL10 gene expression during *Leishmania donovani* infection. Infect Immun. 2005;73(11):7541–7.
221. Kirkpatrick CE, Farrell JP, Warner JF, Denner G. Participation of natural killer cells in the recovery of mice from visceral leishmaniasis. Cell Immunol. 1985;92(1):163–71.
222. Amprey JL, Im JS, Turco SJ, Murray HW, et al. A subset of liver NK T cells is activated during *Leishmania donovani* infection by CD1d-bound lipophosphoglycan. J Exp Med. 2004;200(7):895–904.
223. Beattie L, Svensson M, Bune A, Brown N, et al. *Leishmania donovani*-induced expression of signal regulatory protein alpha on Kupffer cells enhances hepatic invariant NKT-cell activation. Eur J Immunol. 2010;40(1):117–23.
224. Karmakar S, Bhaumik SK, Paul J, De T. TLR4 and NKT cell synergy in immunotherapy against visceral leishmaniasis. PLoS Pathog. 2012;8(4):e1002646.
225. Stanley AC, Zhou Y, Amante FH, Randall LM, et al. Activation of invariant NKT cells exacerbates experimental visceral leishmaniasis. PLoS Pathog. 2008;4(2):e1000028.
226. Zijlstra EE, el-Hassan AM. Leishmaniasis in Sudan. Visceral leishmaniasis. Trans R Soc Trop Med Hyg. 2001;95(Suppl 1):S27–58.
227. Polley R, Zubairi S, Kaye PM. The fate of heterologous CD4+ T cells during *Leishmania donovani* infection. Eur J Immunol. 2005;35(2):498–504.
228. Phillips R, Svensson M, Aziz N, Maroof A, et al. Innate killing of *Leishmania donovani* by macrophages of the splenic marginal zone requires IRF-7. PLoS Pathog. 2010;6(3):e1000813.
229. Ato M, Maroof A, Zubairi S, Nakano H, et al. Loss of dendritic cell migration and impaired resistance to *Leishmania donovani* infection in mice deficient in CCL19 and CCL21. J Immunol. 2006;176(9):5486–93.

230. Engwerda CR, Murphy ML, Cotterell SE, Smelt SC, et al. Neutralization of IL-12 demonstrates the existence of discrete organ-specific phases in the control of *Leishmania donovani*. Eur J Immunol. 1998;28(2):669–80.
231. Stanley AC, Dalton JE, Rossotti SH, MacDonald KP, et al. VCAM-1 and VLA-4 modulate dendritic cell IL-12p40 production in experimental visceral leishmaniasis. PLoS Pathog. 2008;4(9):e1000158.
232. Nylen S, Maurya R, Eidsmo L, Manandhar KD, et al. Splenic accumulation of IL-10 mRNA in T cells distinct from CD4+CD25+ (Foxp3) regulatory T cells in human visceral leishmaniasis. J Exp Med. 2007;204(4):805–17.
233. Gupta G, Bhattacharjee S, Bhattacharyya S, Bhattacharya P, et al. CXC chemokine-mediated protection against visceral leishmaniasis: involvement of the proinflammatory response. J Infect Dis. 2009;200(8):1300–10.
234. Resende M, Moreira D, Augusto J, Cunha J, et al. *Leishmania*-infected MHC class II high dendritic cells polarize CD4+ T cells toward a nonprotective T-bet+ IFN-gamma+ IL-10+ phenotype. J Immunol. 2013;191(1):262–73.
235. McFarlane E, Perez C, Charmoy M, Allenbach C, et al. Neutrophils contribute to development of a protective immune response during onset of infection with *Leishmania donovani*. Infect Immun. 2008;76(2):532–41.
236. Rousseau D, Demartino S, Ferrua B, Michiels JF, et al. In vivo involvement of polymorphonuclear neutrophils in *Leishmania infantum* infection. BMC Microbiol. 2001;1:17.
237. Wilson ME, Recker TJ, Rodriguez NE, Young BM, et al. The TGF-beta response to *Leishmania chagasi* in the absence of IL-12. Eur J Immunol. 2002;32(12):3556–65.
238. Maroof A, Beattie L, Zubairi S, Svensson M, et al. Posttranscriptional regulation of Il10 gene expression allows natural killer cells to express immunoregulatory function. Immunity. 2008;29(2):295–305.
239. Bankoti R, Gupta K, Levchenko A, Stager S. Marginal zone B cells regulate antigen-specific T cell responses during infection. J Immunol. 2012;188(8):3961–71.
240. Engwerda CR, Ato M, Cotterell SE, Mynott TL, et al. A role for tumor necrosis factor-alpha in remodeling the splenic marginal zone during *Leishmania donovani* infection. Am J Pathol. 2002;161(2):429–37.
241. Ato M, Stager S, Engwerda CR, Kaye PM. Defective CCR7 expression on dendritic cells contributes to the development of visceral leishmaniasis. Nat Immunol. 2002;3(12):1185–91.
242. Karplus TM, Jeronimo SM, Chang H, Helms BK, et al. Association between the tumor necrosis factor locus and the clinical outcome of *Leishmania chagasi* infection. Infect Immun. 2002;70(12):6919–25.
243. Ghose AC, Haldar JP, Pal SC, Mishra BP, et al. Serological investigations on Indian kala-azar. Clin Exp Immunol. 1980;40(2):318–26.
244. Anam K, Afrin F, Banerjee D, Pramanik N, et al. Differential decline in Leishmania membrane antigen-specific immunoglobulin G (IgG), IgM, IgE, and IgG subclass antibodies in Indian kala-azar patients after chemotherapy. Infect Immun. 1999;67(12):6663–9.
245. Pontes De Carvalho LC, Badaro R, Carvalho EM, Lannes-Vieira J, et al. Nature and incidence of erythrocyte-bound IgG and some aspects of the physiopathogenesis of anaemia in American visceral leishmaniasis. Clin Exp Immunol. 1986;64(3):495–502.
246. Deak E, Jayakumar A, Cho KW, Goldsmith-Pestana K, et al. Murine visceral leishmaniasis: IgM and polyclonal B-cell activation lead to disease exacerbation. Eur J Immunol. 2010;40(5):1355–68.
247. Smelt SC, Engwerda CR, McCrossen M, Kaye PM. Destruction of follicular dendritic cells during chronic visceral leishmaniasis. J Immunol. 1997;158(8):3813–21.
248. Dalton JE, Maroof A, Owens BM, Narang P, et al. Inhibition of receptor tyrosine kinases restores immunocompetence and improves immune-dependent chemotherapy against experimental leishmaniasis in mice. J Clin Invest. 2010;120(4):1204–16.

Part II

Challenges in the Diagnosis, Treatment and Control of Leishmaniasis in Times of Drug Resistance

Co-infection with HIV

6

Margriet den Boer and Jorge Alvar

Abstract

This chapter describes the epidemiology, current spread, and clinical aspects of HIV/*Leishmania* co-infection and highlights the recently released guidelines of WHO on their management. It discusses the development of resistant *Leishmania* strains for existing anti-*Leishmania* drugs and the complexity of chemotherapy for *Leishmania*/HIV co-infection, which relies on the same drugs that are used in uncomplicated *Leishmania*. Additionally, prospects for future chemotherapeutic alternatives that target *Leishmania* and HIV and tackle both infections simultaneously are summarized.

6.1 Introduction

HIV/*Leishmania* co-infection was first reported in 1985, and since then, it has been reported in 35 countries with a prevalence ranging between 1 and 30% of cases of leishmaniasis, depending on the analyzed geographical areas. It is an expanding but significantly underestimated problem, as it mostly affects neglected populations. Two comprehensive reviews on epidemiology, immunology, and clinical features of HIV-*Leishmania* co-infection published with a decade in between permit a comparison of its progression and knowledge thereof [1, 2].

In 2009, the human immunodeficiency virus (HIV) affected 33.3 million people worldwide and caused 1.8 million deaths (see Fig. 6.1). Currently, 22.5 million of infected people live in sub-Saharan Africa which is where 69% of the 2.6 million new HIV infections in 2009 occurred. However, there are clear indications that

M. den Boer (✉)
Medecins Sans Frontieres, London, UK

J. Alvar
Drugs for Neglected Diseases Initiative, Geneva, Switzerland

A. Ponte-Sucre, M. Padrón-Nieves (eds.), *Drug Resistance in Leishmania Parasites*,
https://doi.org/10.1007/978-3-319-74186-4_6

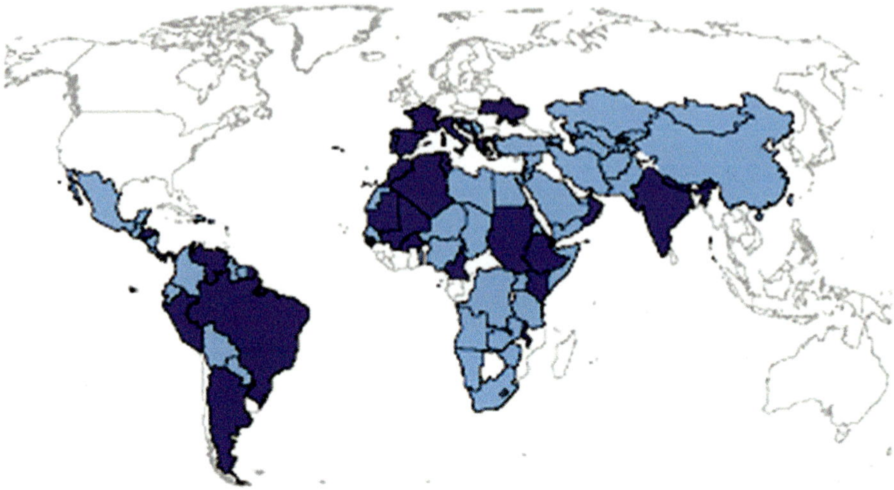

Fig. 6.1 Countries with endemic leishmaniasis and with *Leishmania*-HIV co-infection. Dark blue: countries reporting HIV/Leishmaniasis co-infection. Light blue: leishmaniasis endemic countries. Source: http://www.who.int/leishmaniasis/burden/hiv_coinfection/burden_hiv_coinfection/en/index.html, accessed at 17/5/2011

suggest that the HIV epidemic in Africa and worldwide is stabilizing with 0.5 million less new infections in 2009 than at the peak of the epidemic 12 years ago. Nevertheless, HIV is concentrating and expanding within urban areas (http://www.unaids.org/documents/20101123_GlobalReport_Chap2_em.pdf).

Leishmaniasis is a hypoendemic disease in Southern Europe with less than 0.3 cases per 100,000 inhabitants. Co-infection was first reported in Spain, with most of the cases among HIV-positive intravenous drug users, some of them as an activation of asymptomatic infection when becoming immunosuppressed and others as a new infection when sharing *Leishmania*-infected needles [3]. After the introduction of antiretroviral therapy (ART) at the end of the 1990s, the number of new co-infected cases declined rapidly in all European countries [2, 4].

Both visceral leishmaniasis (VL) and HIV are highly prevalent in East Africa, but VL is a disease of very isolated, remote areas in Ethiopia, Kenya, Somalia, Uganda, and Sudan where the prevalence of HIV is low. Migration and its consequences of malnutrition and poor housing have been identified as major factors in transmission of leishmaniasis [5]. In contrast with Europe, in Africa the lack of access to ART remains a major challenge, although patient coverage rose from 7% in 2003 to 42% in 2008 and in Eastern and Southern Africa to 48%. The prospects for co-infected patients with no access to ART are grim, as they will relapse after leishmaniasis treatment and eventually become unresponsive to leishmaniasis drugs.

Nowadays, Ethiopia has by far the highest prevalence of HIV/VL worldwide (15–30% of VL cases). Most cases occur in a selective group of male young workers that migrate every year from the Highland territories to the fertile lands in the

Northwest of the country (Humera) in order to harvest sesame and sorghum [6]. This region borders Eritrea and Sudan, both areas with a large presence of deployed soldiers, prostitution and HIV transmission, and also highly endemic for VL [7]. It has been shown that infected migrants disseminate leishmaniasis to non-endemic areas when returning home [8, 9].

In Southern Sudan, the number of HIV/VL co-infected patients rose sharply after the peace agreement was signed in 2005 and large-scale migration took place toward Jonglei and the Upper Nile states, well-known areas of leishmaniasis transmission. In 2008, a prevalence of 25% of co-infection among VL patients was found in a specific area of Southern Sudan [10]. The situation is expected to worsen due to the current VL epidemic in Southern Sudan, with more than 10,000 cases since September 2010 (http://www.who.int/leishmaniasis/Upsurge_kalaazar_Southern_Sudan.pdf) and almost 200,000 refugees that recently returned from North Sudan [11]. An additional 800,000 people are expected to return in the coming year after the outcome of the recent referendum for the independence of Southern Sudan. A great majority of these are expected to settle in the two abovementioned endemic states. A VL outbreak that occurred in the early 1990s claimed 100,000 lives in the same area [12].

In the Indian subcontinent (ISC), harboring 75% of the total burden of VL in the world, the number of co-infections is lower than in Africa, with reported figures of less than 1% of all VL cases, although this is disputed by specific studies that estimate an increase in prevalence not only in India but also in Nepal [13–15]. The reasons underlying this discrepancy may be related to a different pattern of transmission; while for HIV an urban pattern was shown, and confined to the South of India, *Leishmania* transmission is mostly rural and the areas with higher endemism are located in the Northern states (Bihar, Jharkhand, Uttar Pradesh, and West Bengal). Bangladesh and Nepal share this dual epidemiological pattern, and consequently the percentage of co-infection has remained low.

In South America, co-infected cases are only reported in Brazil at a low rate of 1:10,000, again, with two different transmission patterns that maintain the rate of co-infected cases at 2% of the total of infected VL patients [16].

CL-HIV co-infection has spread to a much lower extent than VL-HIV (i.e., 0.1% of the total CL cases [16]).

6.2 Clinical Manifestation of HIV/*Leishmania* Co-infection

VL as an opportunistic infection of HIV manifests as an uncontrolled infection with a very high parasite burden. Both HIV and *Leishmania* not only contribute separately to the impairment of the immune response targeting the same cells (macrophages) but also exert a synergistic deleterious effect on the host cells, increasing both virus replication and parasite multiplication [17] and favoring progression of the disease into AIDS [18]. Parasite distribution appears frequently not to be confined exclusively to the typically affected organs in immunocompetent patients but also disseminated into peripheral locations, such as the skin, gut, lungs, peripheral blood, peritoneal fluid, etc. [19, 20]. This distribution may represent a challenge for current chemotherapy. Furthermore, the abundance of parasites in peripheral blood in these patients may increase the chances for transmission via sand flies [21], therefore contributing to the spread of drug-resistant strains, especially via anthroponotic transmission cycles in *Leishmania (L.) donovani.*

When compared with VL-HIV, clinical impairment of leishmaniasis in CL-HIV is much less severe; nevertheless, in an outbreak of *L. (L.) major* in Burkina Faso reported in 2000, CL-HIV patients showed more polymorphic lesions and required longer treatment [22].

Without an adequate immune response, drugs lose, at least partially, their efficacy against *Leishmania* infection; even those compounds previously considered to be effective regardless of the strength of the immune response such as amphotericin-B (AMB). Co-infected patients relapse repeatedly after each treatment course and finally become unresponsive to all drugs used. Prognosis of VL-HIV is poor, although significantly better in patients (1) with a high $CD4^+$ count, (2) maintained under ART, and (3) having achieved parasitological or clinical cure after an initial episode of VL [23]. A drawback is the increased toxicity of antileishmanial drugs in co-infected patients, which negatively impairs prognosis, especially in case of pentavalent antimonials (Sb^V) [2, 24].

6.3 Risk for Drug Resistance in Co-infection

Resistance to antileishmanial drugs has only rarely been documented, except for resistance to Sb^V, widespread in the ISC due to their prolonged misuse [25, 26]. A detailed description of this situation can be found in Chap. 7. Resistance develops experimentally for all drugs, although in practical terms, miltefosine (MIL) and paromomycin are likely to be more prone to the development of resistant strains than AMB, not only because of their mechanism of interaction with the parasite but also because of the requirement for relatively long treatments, increasing the risk of low compliance [27, 28]. Indeed, after a decade of uncontrolled use of MIL in India and Nepal, the total failure rate for MIL reached up to 22% in a 12-month follow-up [29]. Whether this lack of response is due to resistant strains or not has yet to be determined, but this flags a new concern for the use of MIL which is thoroughly described in Chap. 4. On the other hand, although AMB-resistant strains have been described in vitro [30] and a decreased efficacy has been observed in co-infected patients after several treatment cycles [31, 32], no resistant AMB strains were found in these patients, and there is a nil record of resistant strains in the literature despite its constant use in leishmaniasis for many years. AMB resistance has been described for fungal infections in immune-suppressed patients [33].

In the ISC, combination therapy of two antileishmanial drugs in regimen with reduced dosages and duration was proven effective, and in theory, this is the most promising alternative to thwart the increasing trend of resistance [34]. However, for this strategy to be successful, adherence to therapy should be ensured at the primary healthcare level. This is a difficult task in practical terms during massive control campaigns fueled by the need for decentralization of the treatment without proper funds to ensure directly observed treatment (DOT). Poor treatment compliance is another problem and may be worse in patients with a low education level. With no guaranteed compliance, the risk of developing resistant strains cannot be ruled out. For this reason and to expand the life span of the few existing medicines against leishmaniasis, it is highly recommended to use, in the ISC, an alternative regimen consisting of one single iv infusion of 10 mg/kg total dose of liposomal amphotericin-B (L-AMB) with a proven efficacy of >96% in India and an ascertained 100% compliance [35].

In co-infected patients, relapses predispose to the selection of resistant infectious strains. In foci where the source of infection consists of *Leishmania*-contaminated syringes, or those with anthroponotic transmission like East Africa and the ISC, there is a major risk for the spread of these resistant strains to other patients. Resistance can in theory easily appear in immune-compromised patients; a decreased susceptibility of parasite isolates to pentavalent antimonials has been demonstrated in a canine leishmaniasis model after only one treatment [36].

6.4 New WHO Recommended Treatment Guidelines for the Treatment of *Leishmania*/HIV Co-infected Patients

Considering that there are only few published clinical studies on the efficacy of treatments for HIV/VL co-infection outside the Mediterranean area, the Expert Committee on Leishmaniasis provided the following guidance on patient management [37].

Due to their efficacy, safety, and the absence of resistant strains until now, liposomal AMB formulations (L-AMB) constitute the first choice in the treatment of co-infected patients at a dose of 3–5 mg/kg infusions, daily or intermittently for a 10-dose schedule at days 1–5, 10, 17, 24, 31, and 38, up to a 40 mg/kg total dose. Sb^V are more toxic for co-infected patients than for non-co-infected VL patients and require careful monitoring for pancreatitis and cardiotoxicity. Sb^V should therefore only be used in areas where their efficacy is not yet decreased and liposomal AMB formulations are not available. MIL may be used as an alternative to antimonials as it was shown to be safer than antimonials and reasonably effective in co-infected patients [24].

Secondary prophylaxis has shown to prolong survival by reducing the number and severity of relapses in co-infected patients, especially in those with $CD4^+$ counts lower than 200 cells/μL. It also reduces the possibility of transmission of resistant parasites. In zoonotic VL, *Leishmania* parasites are transmitted by the sand fly, from patients only to dogs, and not to humans, meaning that secondary prophylaxis can be completed with any drug, as there is no risk of spread of resistant strains. Based on the experience collected for zoonotic leishmaniasis in the Mediterranean basin, WHO-recommended prophylaxes include L-AMB (3–5 mg/kg/day) administered once every 3 weeks for 12 months and Sb^V (20 mg Sb^V/kg/day every 3–4 weeks) or pentamidine (4 mg/kg/day [300 mg for an adult] every 3–4 weeks).

In anthroponotic foci, where resistant parasites may be transmitted in absence of any animal reservoir within the cycle, it is strongly recommended not to use secondary prophylaxis with medicines used in mainstream therapy regimes for primary attacks [2, 38]. This protocol reduces the options to pentamidine, which is not used anymore for treating primary VL. However, the efficacy of secondary prophylaxis has not yet been ascertained in any anthroponotic foci.

Drug resistance may appear in *Leishmania*/HIV co-infected patients after consecutive relapses despite maintenance therapy with ART and secondary prophylaxis. Combination regimens are not yet studied in co-infected patients. All these data suggest that it is extremely urgent to invest in research into new options for treatment and prophylaxis.

6.5 Perspectives in HIV-*Leishmania* Chemotherapy

No doubt, combination of ART with classical leishmanicidal drugs with minimal euthymic character, that is, as independent as possible of the immune status of the host, like liposomal formulations of AMB, is the golden standard for the next medium-range future. An educated guess for the future, taking into account the current status of the chemotherapy pipeline, is that improvement in chemotherapy will likely come from improvement of current leads or from better formulations that will enable drugs to reach the anatomical locations that harbor *Leishmania* amastigotes in HIV patients. Furthermore, independent advances for both therapies will have a real and positive impact on infection when used in combination.

Perusing the literature, an appealing approach seems to be the development of drugs active on both HIV and *Leishmania*, not necessarily addressing the same or homologous target. Their optimization may be problematic in terms of preserving their activity on both microorganisms.

Although scarce, there are several examples and early proofs of concept for this approach. Leishmanicidal activity of specifically designed HIV drugs, like inhibitors of HIV aspartyl proteinase, has been tested, following a chemotherapeutical "piggy-back" approach, and new molecules with antileishmania and antiviral activities have been discovered by high-throughput screening. Examples for these two new trends ensue.

6.5.1 Inhibitors of Aspartyl Proteinases

The HIV aspartyl proteinases involved in the maturation of viral proteins are inhibited by specific inhibitors (HIV-PIs) and act in combination with viral reverse transcriptase inhibitors in ART. Their application has led to a tremendous reduction in the severity and incidence of AIDS, including co-infections with *Leishmania* [39–41].

The leishmanicidal effects of HIV-PI's were first reported by Savoia et al [42]. The rationale for their use is the inhibition of some proteasomal activities by HIV-PI, together with the leishmanicidal activity described for other human proteasomal inhibitors [42].

Although incomplete, there is a growing awareness of the activity of HIV-PIs on different *Leishmania* developmental stages, compiled in Table 6.1.

The following conclusions can be inferred from this table:

Table 6.1 Leishmanicidal activity of HIV-proteinase inhibitor (HIV-PI)

Ref	HIV-PI[a]	*Leishmania* system and HIV-PI inhibition				Comments
		Species (strain)	Stage assayed[b]			
			Promastigote	Axenic amastigote	Intracellular amastigote[c]	
[42]	IDV	*L. (L.) major*	IC_{50} = 8.3 ± 0.9 μM			
	SQV	LRC-L137	IC_{50} = 7.0 ± 0.7 μM			
	IDV		70% at 50 μM			
	SQV	*L. (L.) infantum*	67% (50 μM)			
		MHOM/TN/80/IPT1				
[44]	NFV	*L. (L.) infantum*	<5% (25 μM)	77% (25 μM)	79.9% (25 μM)	Data for MDM[d] amastigote infection
	RTV	MHOM/MA/67/ITMAP-263	<5% (25 μM)	83% (25 μM)	43.7% (25 μM)	
	SQV		<5% (25 μM)	0% (25 μM)	61.5% (25 μM)	Strain resistant to Sb^V
		L. (L.) donovani			92.4% (25 μM)	
	NFV	(9518)	<5% (25 μM)		52.6% (25 μM)	
	RTV		<5% (25 μM)		50.1% (25 μM)	
	SQV		<5% (25 μM)			
[46]	NFV	*L. (L.) amazonensis*	IC_{50} = 15.1 ± 1.1 μM		86% at 50 μM	IND, SQV IC_{50}s > 50 μM.
	LPV	MHOM/BR/77/LTB0016	IC_{50} = 16.4 ± 0.8 μM		80% at 50 μM	
	APV		IC_{50} = 16.4 ± 0.8 μM			
[45]		*L. (L.) donovani*		66% (12.5 μM)		
		(9518)				
[43]	NFV	*L. (L.) infantum*	IC_{50} = 14.1 ± 0.2 μM		64% (10.5 μM)	
	SQV	(MCAN/VE/98/IBO-78)	IC_{50} = 55.1 ± 6.5 μM		34% (10 μM)	
	NFV	*L. (L.) donovani*	IC_{50} = 14.1 ± 3.9 μM			
	SQV	MHOM/IN/80/DD	IC_{50} = 51.9 ± 3.4 μM			
	NFV	*L. (L.) mexicana*	IC_{50} = 9.9 ± 0.5 μM		74% (10.5 μM)	
	SQV	MHOM/VE/80/NR	IC_{50} = 42.1 ± 7.3 μM		43% (10 μM)	
	NFV	*L. (L.) amazonensis*	IC_{50} = 13.4 ± 3.0 μM			
	SQV	IFLA/BR/67/PH8	IC_{50} = 40 ± 1.2 μM			
	NFV	*L. (V.) braziliensis*	IC_{50} = 14.6 ± 0.4 μM			
	SQV	MHOM/BR/75/M2903	IC_{50} = 36 ± 0.35 μM			
	NFV	*L. (L.) major*	IC_{50} = 13.4 ± 2.5 μM			
	SQV	MHOM/SU/73/5-ASKH	IC_{50} = 46.9 ± 1.5 μM			
	NFV	*L. (L.) pifanoi*		IC_{50} = 9.9 ± 1.4 μM		
	SQV	MHOM/VE/60Ltrod		IC_{50} = 15.2 ± 2.7 μM		

[a] Abbreviations for HIV-PI: IDV.- Indinavir, LPV.- Loponavir , NFV.- Nefinavir, RTV.- Ritonavir, SQV.- Saquinavir.

[b].-Percentage of inhibition of the expressed parameter at (HIV-PI concentration)

[c].- Expressed as inhibition percentage for macrophage:parasite association index.

[d].- MDM.- monocyte derived macrophage

1. There is a strong variation in leishmanicidal activities depending both on the HIV-PI and the species of *Leishmania* tested [42, 43].
2. When a given HIV-PI was tested in parallel on different *Leishmania* species, the efficacy for those causative of CL was scarcely higher than for those producing VL [42, 43].
3. Within a given *Leishmania* species, variation of HIV-PI among different strains is low [42, 43], including those resistant to Sb^V [44].
4. IC_{50}s were higher for *L. (L.) infantum* strains isolated from patients with previous ART therapy [43]; in fact nelfinavir (NFV) resistance is induced by growing the parasites under drug pressure [45].
5. Efficacy of HIV-PIs on macrophages infected with *Leishmania* is maintained regardless of HIV co-infection [43];
6. HIV-PIs kill *Leishmania* at much higher concentrations (micromolar range) than those required for inhibition of viral replication (nanomolar range).

Thus, a real impact of HIV-PIs on the *Leishmania* burden with their current dosing scheme, aside from improvement due to HIV recession, can only be explained if the macrophage may concentrate HIV-PIs up to toxic levels for intracellular parasites. In fact both axenic and intracellular parasites are more susceptible to HIV-PIs than promastigotes [44].

Leishmanicidal targets for HIV-PIs. At first sight, the logical mechanism for HIV-PIs is the inhibition of aspartic proteinase activities in *Leishmania.* Using typical substrates and conditions, this activity and its inhibition by NFV have been evidenced in lysates of *L. (L.) mexicana* and *L. (L.) infantum* [43, 46]. Furthermore, characterization of this aspartic proteinase activity was carried out for *L. (L.) mexicana* [47]. Additional targets, perhaps as a consequence of a prior proteinase inhibition, are suggested by (1) inhibition of karyokinesis by NFV in bi- and polynuclear *L. (L.) mexicana* promastigotes [43] and (2) appearance of plasma membrane blebbings and mitochondria swelling assessed on parasites treated with HIV-PIs at their respective IC_{50} [46]. This last observation seems to be related to an apoptosis-like process induced by NFV on *L. (L.) donovani* axenic amastigotes, evidenced by mitochondrial depolarization and release of endonuclease G, together with induction of oxidative stress [45].

The use of HIV-PIs as leishmanicidal agents in the absence of *Leishmania*/HIV co-infection is questionable; first, there is a large gap in active concentrations for anti-HIV and anti-*Leishmania* effects; second, HIV-PIs are not exempt from toxic side effects, especially at HIV-PI concentrations required for leishmanicidal activity setup in vitro, and *Leishmania* resistance can be easily induced [43, 45]. Finally, oxidative stress induced by NFV is mostly precluded by episomal overexpression of the *gsh1* gene [45], encoding for γ-glutamylcysteinyl synthase, the enzyme responsible for the limiting step in the synthesis of glutathione, immediate precursor of trypanothione, the ultimate responsible for thiol redox in the metabolism in *Leishmania.* As such, inhibition of glutathione synthesis reverts Sb^V resistance [48], so possible cross-resistance between antimonials and HIV-PIs may occur; against this pessimistic statement, we must pinpoint that NFV was active on a *L. (L.) donovani*

Sb^V-resistant strain [44] and, secondly, discrepancy between mechanisms of Sb^V resistance raised in vitro with those from clinical field isolates is not unusual: inhibition of glutathione biosynthesis did not improve Sb^V susceptibility in field isolates of *L. (V.) panamensis* resistant to Sb^V [49]; in the same trend, in transcriptomics for *L. (L.) donovani* strains resistant to Sb^V in Nepal, mRNA levels for γ-glutamylcysteinyl synthase were decreased [50].

Another important issue is the higher expression of virulence factors in parasites treated with sublethal concentrations of HIV-PI, as leishmaniolysin or cysteine proteinase b reported for *L. (L.) amazonensis* [46].

Altogether, HIV-PIs may have a side-lethal activity on *Leishmania*. Nevertheless, there are several concerns. Apparently, there is a risk of easy induction of resistance, toxic side effects, and induction of virulence factors. Additional studies are needed in order to highlight the clinical relevance of this approach and balance its advantages and disadvantages; furthermore, it will be worthwhile to test novel HIV-PIs for their leishmanicidal activity. In conclusion, an educated guess is that the intrinsic leishmanicidal effect of HIV-PIs in patients is much less relevant than the effect caused by improvement in their immune response caused by the inhibition of HIV proliferation. As such, their usefulness as straightforward new leishmanicidal agents ranks much lower than that of current leishmanicidal drugs in non-HIV co-infected *Leishmania* patients.

6.6 High-Throughput Screening for New Anti-HIV and Anti-*Leishmania* Leads

Medium- and high-throughput screening of compounds produced by combinatorial chemistry [51, 52], massive screening of natural products [53–55], or new leads produced by academic groups constitute an important source for promising antileishmanial drugs. The screening of the same series of compounds for anti-pathogenic protozoa and anti-HIV activities nowadays is not infrequent, although the number of groups that specifically focus on a co-treatment philosophy is, in contrast, rather scarce [56].

In many cases for a single drug endowed with both leishmanicidal and antiviral activities, the concentration required for effectiveness on both infections is beyond the threshold of patient cytotoxicity, precluding their use as a single drug for co-therapy; in a series of acrinidone derivatives, 2-(benzothiazol-2-ylamino)-10H-acridin-9-one showed an IC_{50} against *Leishmania* of 3 μM; nevertheless, the anti-HIV activity was higher (IC_{50} = 27.9 μM) and quite close to cytotoxic values for mammalian cells [57]. A reduced number of compounds with anti-*Leishmania* and anti-HIV activities have gone upstream in the pipeline and gone past the stage of initial in vitro tests. For example, the group of Figadère in the Université de Paris-Sud has synthesized more than 200 2-substituted quinolines, and some have both anti-*Leishmania* and anti-HIV activities [56, 58]. A major advantage of these compounds is their druggability including possible oral administration. These compounds have been successfully tested in murine models for CL and VL [59, 60], but not for anti-HIV activity.

Marine products are an endless and mostly untapped source for anti-HIV and anti-*Leishmania* compounds [54, 61–64], and a reduced number are active in both diseases, such as the semisynthetic derivatives of curcuphenol, a sesquiterpene isolated from the sponge *Myrmekioderma styx* [65], which has better leishmanicidal than anti-HIV activity, but both in the micromolar concentration range. Manzamine A and 8–hydroxymanzamine, belonging to the growing family of β-carboline alkaloids, were isolated from sponges from the *Acanthostrongylophora* genus and display remarkable anti-*Leishmania* and anti-HIV activities [66–68].

Very often, the complexity of natural products impairs their chemical synthesis; in such cases, improvement of the antiviral and leishmanicidal activities can be achieved through semisynthetic methods, modifying the natural structure of the compound instead of synthesizing it from scratch. An example of this methodology is illustrated by isoaaptamine, a molecule isolated from sponges of the genus *Hymeniacidon*. Its 9-O-4-ethylbenzoyl derivative showed a sixfold improved anti-*Leishmania* activity compared to the non-acylated natural form while preserving its anti-HIV activity [69].

Anti-HIV and anti-*Leishmania* activities have also been described for marine peptides. Mollamides are cyclic hexapeptides containing a thiazoline group isolated from the tunicate *Didemnum molle* [70]; mollamide B showed a moderate anti-HIV activity, whereas its leishmanicidal effect is threefold higher on a molar basis. Animal antimicrobial peptides and their artificial surrogates may act simultaneously on both pathogens, suggesting their putative future use in co-infections, but this is now only at its very first stage of development.

A caveat for lead optimization is that in many cases, mechanism of actions and targets of anti-HIV and anti-*Leishmania* activity may differ greatly; therefore, it will be unlikely that their optimization will lead to parallel benefits for both targeted microorganisms. An exception will be those modifications not affecting drug-target interaction but the pharmacokinetics or pharmacology of the drug.

6.7 Concluding Remarks

Leishmania chemotherapy in HIV co-infected patients is much more complex than chemotherapy for uncomplicated *Leishmania* infections alone and relies mostly on the same drugs. The major determining factor on outcome is the reduction of the HIV burden by antiretroviral chemotherapy. Due to the reciprocal detriment effect of both infections on the immune system, the use of parasiticidal and highly effective liposomal AMB appears to be the most reliable treatment for VL/HIVE co-infected patients. There are prospects for a single drug tackling both infections simultaneously, but research in this direction is in a very early stage and hampered by a lack of financial support or capacity to assay the same compound for both anti-HIV and anti-*Leishmania* activity. In order to develop and optimize leads and create a chemotherapeutic alternative for co-infected patients, a strong research effort will have to be made.

Acknowledgment The Spanish Agency for International Cooperation for Development supported the WHO Leishmaniasis program and was focused among other activities on the treatment of HIV-*Leishmania* co-infected patients in Ethiopia. LR is supported by grants from EU HEALTH-2007-223414 and Fondo de Investigación Sanitaria RETICS RD06-0021-06 and PS09/01928.

References

1. Alvar J, Canavate C, Gutierrez-Solar B, Jimenez M, et al. *Leishmania* and human immunodeficiency virus coinfection: the first 10 years. Clin Microbiol Rev. 1997;10(2):298–319.
2. Alvar J, Aparicio P, Aseffa A, den Boer M, et al. The relationship between Leishmaniasis and AIDS: the second 10 years. Clin Microbiol Rev. 2008;21(2):334–59.
3. Cruz I, Morales MA, Noguer I, Rodriguez A, et al. *Leishmania* in discarded syringes from intravenous drug users. Lancet. 2002;359(9312):1124–5.
4. Lopez-Velez R. The impact of highly active antiretroviral therapy (HAART) on visceral leishmaniasis in Spanish patients who are co-infected with HIV. Ann Trop Med Parasitol. 2003;97(Suppl 1):143–7.
5. Aagaard-Hansen J, Nombela N, Alvar J. Population movement: a key factor in the epidemiology of neglected tropical diseases. Trop Med Int Health. 2010;15(11):1281–8.
6. Mengesha B, Abuhoy M. Kala-azar among labour migrants in Metema-Humera region of Ethiopia. Trop Geogr Med. 1978;30(2):199–206.
7. Lyons S, Veeken H, Long J. Visceral leishmaniasis and HIV in Tigray, Ethiopia. Trop Med Int Health. 2003;8(8):733–9.
8. Alvar J, Bashaye S, Argaw D, Cruz I, et al. Kala-azar outbreak in Libo Kemkem, Ethiopia: epidemiologic and parasitologic assessment. Am J Trop Med Hyg. 2007;77(2):275–82.
9. Bashaye S, Nombela N, Argaw D, Mulugeta A, et al. Risk factors for visceral leishmaniasis in a new epidemic site in Amhara Region, Ethiopia. Am J Trop Med Hyg. 2009;81(1):34–9.
10. Gorski S, Collin SM, Ritmeijer K, Keus K, et al. Visceral leishmaniasis relapse in Southern Sudan (1999–2007): a retrospective study of risk factors and trends. PLoS Negl Trop Dis. 2010;4(6):e705. https://doi.org/10.1371/journal.pntd.0000705.
11. Moszynski P. Kala-azar outbreak is symptomatic of humanitarian crisis facing southern Sudan. BMJ. 2010;341:c7276. https://doi.org/10.1136/bmj.c7276.
12. Seaman J, Mercer AJ, Sondorp HE, Herwaldt BL. Epidemic visceral leishmaniasis in southern Sudan: treatment of severely debilitated patients under wartime conditions and with limited resources [see comments]. Ann Intern Med. 1996;124(7):664–72.
13. Redhu NS, Dey A, Balooni V, Singh S. *Leishmania*-HIV co-infection: an emerging problem in India. Aids. 2006;20(8):1213–5.
14. Mathur P, Samantaray JC, Vajpayee M, Samanta P. Visceral leishmaniasis/human immunodeficiency virus co-infection in India: the focus of two epidemics. J Med Microbiol. 2006;55 (Pt 7):919–22.
15. Gurubacharya RL, Gurubacharya SM, Gurubacharya DL, Quinkel J, et al. Prevalence of visceral leishmaniasis & HIV co-infection in Nepal. Indian J Med Res. 2006;123(3):473–5.
16. Elkhoury EA. Co-infeccao leishmaniose visceral e AIDS no Brasil. Rev Soc Bras Med Trop. 2007;40(124)
17. Bernier R, Turco SJ, Olivier M, Tremblay M. Activation of human immunodeficiency virus type 1 in monocytoid cells by the protozoan parasite *Leishmania donovani*. J Virol. 1995;69 (11):7282–5.
18. Bentwich Z. Concurrent infections that rise the HIV viral load. J HIV Ther. 2003;8(3):72–5.

19. Rosatelli JB, Souza CS, Soares FA, Foss NT, et al. Generalized cutaneous leishmaniasis in acquired immunodeficiency syndrome. J Eur Acad Dermatol Venereol. 1998;10(3):229–32.
20. Russo R, Laguna F, Lopez-Velez R, Medrano FJ, et al. Visceral leishmaniasis in those infected with HIV: clinical aspects and other opportunistic infections. Ann Trop Med Parasitol. 2003;97 (Suppl 1):99–105.
21. Molina R, Lohse JM, Pulido F, Laguna F, et al. Infection of sand flies by humans coinfected with *Leishmania infantum* and human immunodeficiency virus. Am J Trop Med Hyg. 1999;60 (1):51–3.
22. Guiguemde RT, Sawadogo OS, Bories C, Traore KL, et al. *Leishmania major* and HIV co-infection in Burkina Faso. Trans R Soc Trop Med Hyg. 2003;97(2):168–9.
23. Pintado V, Martin-Rabadan P, Rivera ML, Moreno S, et al. Visceral leishmaniasis in human immunodeficiency virus (HIV)-infected and non-HIV-infected patients. A comparative study. Medicine (Baltimore). 2001;80(1):54–73.
24. Ritmeijer K, Dejenie A, Assefa Y, Hundie TB, et al. A comparison of miltefosine and sodium stibogluconate for treatment of visceral leishmaniasis in an Ethiopian population with high prevalence of HIV infection. Clin Infect Dis. 2006;43(3):357–64.
25. Sundar S. Drug resistance in Indian visceral leishmaniasis. Trop Med Int Health. 2001;6 (11):849–54.
26. Rijal S, Yardley V, Chappuis F, Decuypere S, et al. Antimonial treatment of visceral leishmaniasis: are current in vitro susceptibility assays adequate for prognosis of *in vivo* therapy outcome? Microbes Infect. 2007;9(4):529–35. https://doi.org/10.1016/j.micinf.2007.01.009.
27. Saint-Pierre-Chazalet M, Ben Brahim M, Le Moyec L, Bories C, et al. Membrane sterol depletion impairs miltefosine action in wild-type and miltefosine-resistant *Leishmania donovani* promastigotes. J Antimicrob Chemother. 2009;64(5):993–1001. https://doi.org/10.1093/jac/dkp321.
28. Maarouf M, Adeline MT, Solignac M, Vautrin D, et al. Development and characterization of paromomycin-resistant *Leishmania donovani* promastigotes. Parasite. 1998;5(2):167–73.
29. Bart Ostyn PM, Surendra U, Rudra Pratap S, Shri Prakash S, et al. (2010) Challenges for the implementation of new tools to monitor treatment outcome in Miltefosine-treated Kala-azar Patients in India and Nepal. Kaladrug meeting, Antwerp, 2010
30. Al-Mohammed HI, Chance ML, Bates PA. Production and characterization of stable amphotericin-resistant amastigotes and promastigotes of *Leishmania mexicana*. Antimicrob Agents Chemother. 2005;49(8):3274–80. https://doi.org/10.1128/AAC.49.8.3274-3280.2005.
31. Durand R, Paul M, Pratlong F, Rivollet D, et al. *Leishmania infantum*: lack of parasite resistance to amphotericin B in a clinically resistant visceral leishmaniasis. Antimicrob Agents Chemother. 1998;42(8):2141–3.
32. Lachaud L, Bourgeois N, Plourde M, Leprohon P, et al. Parasite susceptibility to amphotericin B in failures of treatment for visceral leishmaniasis in patients coinfected with HIV type 1 and *Leishmania infantum*. Clin Infect Dis. 2009;48(2):e16–22. https://doi.org/10.1086/595710.
33. Pfaller MA, Diekema DJ. Epidemiology of invasive candidiasis: a persistent public health problem. Clin Microbiol Rev. 2007;20(1):133–63. https://doi.org/10.1128/CMR.00029-06.
34. Sundar S, Sinha PK, Rai M, Verma DK, et al. Comparison of short-course multidrug treatment with standard therapy for visceral leishmaniasis in India: an open-label, non-inferiority, randomised controlled trial. Lancet. 2011;377(9764):477–86. https://doi.org/10.1016/S0140-6736(10)62050-8.
35. Matlashewski GBA, Kroeger A, Battacharya S, Sundar S, et al. Visceral leishmaniasis: elimination with existing interventions. Lancet Infect Dis. 2011;11(4):322–5.
36. Gramiccia M, Gradoni L, Orsini S. Decreased sensitivity to meglumine antimoniate (Glucantime) of *Leishmania infantum* isolated from dogs after several courses of drug treatment. Ann Trop Med Parasitol. 1992;86(6):613–20.
37. WHO. WHO Technical Report Series 949. 2010.

38. World Health Organization Report of the 5th Consultative Meeting on *Leishmania*/HIV Coinfection. WHO Technical Report Series WHO/CDS/NTD/IDM/2007.5. In Addis Ababa, Ethiopia, 20–22 March 2007.
39. de La Rosa R, Pineda JA, Delgado J, Macias J, et al. Incidence of and risk factors for symptomatic visceral leishmaniasis among human immunodeficiency virus type 1-infected patients from Spain in the era of highly active antiretroviral therapy. J Clin Microbiol. 2002;40(3):762–7.
40. del Giudice P, Mary-Krause M, Pradier C, Grabar S, et al. Impact of highly active antiretroviral therapy on the incidence of visceral leishmaniasis in a French cohort of patients infected with human immunodeficiency virus. J Infect Dis. 2002;186(9):1366–70.
41. Lopez-Velez R, Perez-Molina JA, Guerrero A, Baquero F, et al. Clinicoepidemiologic characteristics, prognostic factors, and survival analysis of patients coinfected with human immunodeficiency virus and *Leishmania* in an area of Madrid, Spain. Am J Trop Med Hyg. 1998;58(4):436–43.
42. Savoia D, Allice T, Tovo PA. Antileishmanial activity of HIV protease inhibitors. Int J Antimicrob Agents. 2005;26(1):92–4.
43. Valdivieso E, Rangel A, Moreno J, Saugar JM, et al. Effects of HIV aspartyl-proteinase inhibitors on *Leishmania* sp. Exp Parasitol. 2010;126(4):557–63. https://doi.org/10.1016/j.exppara.2010.06.002.
44. Trudel N, Garg R, Messier N, Sundar S, et al. Intracellular survival of *Leishmania* species that cause visceral leishmaniasis is significantly reduced by HIV-1 protease inhibitors. J Infect Dis. 2008;198(9):1292–9. https://doi.org/10.1086/592280.
45. Kumar P, Lodge R, Trudel N, Ouellete M, et al. Nelfinavir, an HIV-1 protease inhibitor, induces oxidative stress-mediated, caspase-independent apoptosis in *Leishmania* amastigotes. PLoS Negl Trop Dis. 2010;4(3):e642. https://doi.org/10.1371/journal.pntd.0000642.
46. Santos LO, Marinho FA, Altoe EF, Vitorio BS, et al. HIV aspartyl peptidase inhibitors interfere with cellular proliferation, ultrastructure and macrophage infection of *Leishmania amazonensis*. PLoS One. 2009;4(3):e4918. https://doi.org/10.1371/journal.pone.0004918.
47. Valdivieso E, Dagger F, Rascon A. *Leishmania mexicana*: identification and characterization of an aspartyl proteinase activity. Exp Parasitol. 2007;116(1):77–82. https://doi.org/10.1016/j.exppara.2006.10.006.
48. Carter KC, Sundar S, Spickett C, Pereira OC, et al. The *in vivo* susceptibility of *Leishmania donovani* to sodium stibogluconate is drug specific and can be reversed by inhibiting glutathione biosynthesis. Antimicrob Agents Chemother. 2003;47(5):1529–35.
49. Goyeneche-Patino DA, Valderrama L, Walker J, Saravia NG. Antimony resistance and trypanothione in experimentally selected and clinical strains of *Leishmania panamensis*. Antimicrob Agents Chemother. 2008;52(12):4503–6. https://doi.org/10.1128/AAC.01075-08.
50. Decuypere S, Rijal S, Yardley V, De Doncker S, et al. Gene expression analysis of the mechanism of natural Sb(V) resistance in *Leishmania donovani* isolates from Nepal. Antimicrob Agents Chemother. 2005;49(11):4616–21. https://doi.org/10.1128/AAC.49.11.4616-4621.2005.
51. Cipolla L, La Ferla B, Gregori M. Combinatorial approaches to iminosugars as glycosidase and glycosyltransferase inhibitors. Comb Chem High Throughput Screen. 2006;9(8):571–82.
52. Pettersson S, Clotet-Codina I, Este JA, Borrell JI, et al. Recent advances in combinatorial chemistry applied to development of anti-HIV drugs. Mini Rev Med Chem. 2006;6(1):91–108.
53. Balunas MJ, Kinghorn AD. Drug discovery from medicinal plants. Life Sci. 2005;78(5):431–41.
54. Sagar S, Kaur M, Minneman KP. Antiviral lead compounds from marine sponges. Mar Drugs. 2010;8(10):2619–38.
55. Yu D, Morris-Natschke SL, Lee KH. New developments in natural products-based anti-AIDS research. Med Res Rev. 2007;27(1):108–32.

56. Fakhfakh MA, Fournet A, Prina E, Mouscadet JF, et al. Synthesis and biological evaluation of substituted quinolines: potential treatment of protozoal and retroviral co-infections. Bioorg Med Chem. 2003;11(23):5013–23.
57. Delmas F, Avellaneda A, Di Giorgio C, Robin M, et al. Synthesis and antileishmanial activity of (1,3-benzothiazol-2-yl) amino-9-(10H)-acridinone derivatives. Eur J Med Chem. 2004;39 (8):685–90.
58. Grassi F, Guimaraes Correa AB, Mascarenhas RE, Galvao B, et al. Quinoline compounds decrease in vitro spontaneous proliferation of peripheral blood mononuclear cells (PBMC) from human T-cell lymphotropic virus (HTLV) type-1-infected patients. Biomed Pharmacother. 2008;62(7):430–5.
59. Vieira NC, Herrenknecht C, Vacus J, Fournet A, et al. Selection of the most promising 2-substituted quinoline as antileishmanial candidate for clinical trials. Biomed Pharmacother. 2008;62(10):684–9.
60. Nakayama H, Loiseau PM, Bories C, Torres de Ortiz S, et al. Efficacy of orally administered 2-substituted quinolines in experimental murine cutaneous and visceral leishmaniases. Antimicrob Agents Chemother. 2005;49(12):4950–6.
61. Laport MS, Santos OC, Muricy G. Marine sponges: potential sources of new antimicrobial drugs. Curr pharm biotechnol. 2009;10(1):86–105.
62. Donia M, Hamann MT. Marine natural products and their potential applications as anti-infective agents. Lancet Infect Dis. 2003;3(6):338–48.
63. Tziveleka LA, Vagias C, Roussis V. Natural products with anti-HIV activity from marine organisms. Curr Top Med Chem. 2003;3(13):1512–35.
64. Watts KR, Tenney K, Crews P. The structural diversity and promise of antiparasitic marine invertebrate-derived small molecules. Curr Opin Biotechnol. 2010;21(6):808–18.
65. Gul W, Hammond NL, Yousaf M, Peng J, et al. Chemical transformation and biological studies of marine sesquiterpene (S)-(+)-curcuphenol and its analogs. Biochimica et biophysica acta. 2007;1770(11):1513–9.
66. Rao KV, Donia MS, Peng J, Garcia-Palomero E, et al. Manzamine B and E and ircinal A related alkaloids from an Indonesian Acanthostrongylophora sponge and their activity against infectious, tropical parasitic, and Alzheimer's diseases. J Nat Prod. 2006;69(7):1034–40.
67. Rao KV, Kasanah N, Wahyuono S, Tekwani BL, et al. Three new manzamine alkaloids from a common Indonesian sponge and their activity against infectious and tropical parasitic diseases. J Nat Prod. 2004;67(8):1314–8.
68. Rao KV, Santarsiero BD, Mesecar AD, Schinazi RF, et al. New manzamine alkaloids with activity against infectious and tropical parasitic diseases from an Indonesian sponge. J Nat Prod. 2003;66(6):823–8.
69. Gul W, Hammond NL, Yousaf M, Bowling JJ, et al. Modification at the C9 position of the marine natural product isoaaptamine and the impact on HIV-1, mycobacterial, and tumor cell activity. Bioorg Med Chem. 2006;14(24):8495–505.
70. Donia MS, Wang B, Dunbar DC, Desai PV, et al. Mollamides B and C, Cyclic hexapeptides from the Indonesian tunicate Didemnum molle. J Nat Prod. 2008;71(6):941–5.

Visceral Leishmaniasis

7

Shyam Sundar and Jaya Chakravarty

Abstract

Pentavalent antimonials (Sb^V) have been the sheet anchor of therapy for leishmaniasis for >75 years. In the early 1980s, it was realized that a significant subset of patients with visceral leishmaniasis were not responding to Sb^V in the state of Bihar, India. Revised recommendation using ten times more drug provided a transient reprieve; however, a large proportion of patients in India and to some extent in Nepal remained unresponsive to Sb^V. Diverse studies have suggested emergence of Sb^V refractory strains in India. Attempts to find a marker of unresponsiveness have failed so far. Alternative therapeutic options include conventional amphotericin-B or its lipid formulations, oral miltefosine, and paromomycin and short course multidrug therapy. In the Indian subcontinent, the only recommended monotherapy is a single dose of liposomal amphotericin-B (L-AMB, dose 10 mg/kg) which is efficacious, safe, and ensures complete compliance. Multidrug therapy has high efficacy, short course, less toxicity, and prevents development of resistance. If these scarce antileishmanial drugs are to be protected from going down the lane of Sb^V, multidrug, short course, affordable treatment of VL should be evolved with access to all.

7.1 Introduction

Visceral leishmaniasis (VL) (Fig. 7.1), also known as "Kala-azar," is typically caused by parasites belonging to the *L. (L.) donovani* complex, which includes two species: *Leishmania (L.) donovani*, the causative organism of VL in the Indian subcontinent (ISC) and Africa, and *Leishmania (L.) infantum* [(*L. (L.) chagasi*)], which causes VL in the Mediterranean basin and Central and South

S. Sundar (✉) · J. Chakravarty
Department of Medicine, Institute of Medical Sciences, Banaras Hindu University, Varanasi, India

A. Ponte-Sucre, M. Padrón-Nieves (eds.), *Drug Resistance in Leishmania Parasites*,
https://doi.org/10.1007/978-3-319-74186-4_7

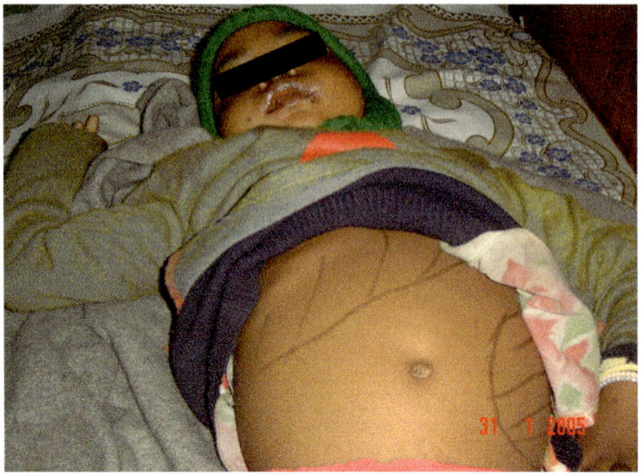

Fig. 7.1 A child with visceral leishmaniasis with hepatosplenomegaly

America. Approximately 0.2–0.4 million VL cases and 0.7–1.2 million CL cases occur each year. In 2015 more than 90% of global VL cases occurred in seven countries: India, Bangladesh, Sudan, South Sudan, Kenya, Brazil, and Ethiopia [1].

The number of VL cases is highest in the ISC. The World Health Organization (WHO) has targeted VL for elimination from this region as a public health problem by 2020. A memorandum of understanding was signed by India, Bangladesh, Nepal, and later Bhutan and Thailand to eliminate Kala-azar from this region. Elimination has been defined as bringing the annual incidence of Kala-azar to less than one case per 10,000 population at block PHC (Primary Health Centre) level in India and Bangladesh and district level in Nepal and Bhutan. Currently, Nepal has eliminated the disease at district level and sustained the situation for the past 2 years. Bangladesh has achieved the elimination target in 90% of endemic upazilas. India has achieved the target in more than two thirds of endemic blocks [2].

VL is the systemic and most severe form of leishmaniasis, characterized by prolonged fever, splenomegaly, lymphadenopathy, hepatomegaly, pancytopenia, progressive anemia, and weight loss. If untreated, VL is uniformly fatal. Some patients with VL may develop a chronic form of dermal leishmaniasis characterized by indurated nodules or depigmented macules, which is called post-Kala-azar dermal leishmaniasis (PKDL). PKDL is quite common (occurring in >50% patients with VL) in Sudan, where it may occur concurrently with VL and heals spontaneously in most patients [3]. In the ISC, it affects only a small proportion of patients, 6 months to several years after an episode of VL, and treatment is necessary [4]. Patients with PKDL serve as an important reservoir of infection.

Natural transmission of leishmaniasis is carried out by female sand flies. In South Asia and the Horn of Africa, the predominant mode of transmission is anthroponotic and patients with Kala-azar or post-Kala-azar dermal leishmaniasis (Fig. 7.2) and those with asymptomatic infection may be the reservoirs for driving transmission [5–8]. In the Mediterranean, the Middle East, and Brazil, the disease is zoonotic, with the domestic dog as the most important reservoir host sustaining transmission [6].

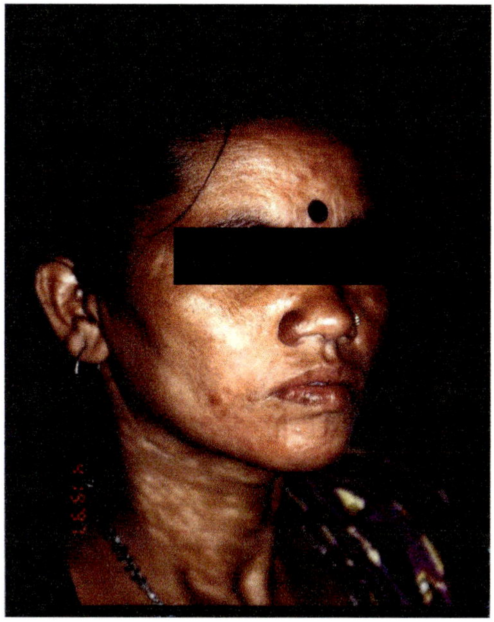

Fig. 7.2 A patient with post-Kala-azar dermal leishmaniasis with multiple nodules on the face

7.2 Visceral Leishmaniasis: The Challenges

7.2.1 Increase in the Risk Factors for Leishmaniasis

Environmental changes like deforestation, urbanization, and migration of nonimmune people to endemic areas have led to the increase in the incidence of leishmaniasis. Migration from nonendemic to endemic areas is a major risk factor for the spread of VL as these people, on their return, can spread the disease in a nonimmune population. This issue is exemplified by the severe epidemic in Southern Sudan which led to the death of 100,000 patients [9]. In the ISC, VL is associated with low socioeconomic status. Even when free drugs are available, patients cannot afford the costs of transportation to the hospital and of hospitalization. Thus, untreated VL and PKDL cases harbor the parasite and disseminate it [10].

The HIV/AIDS pandemic has modified the natural history of leishmaniasis [11] (see Chap. 6). Both diseases exert a synergistic detrimental effect on the cellular immune response because they target similar immune cells [12, 13].

HIV infection increases the risk of developing VL in areas of endemicity, reduces the likelihood of a therapeutic response, and greatly increases the probability of relapse [14–17]. At the same time, VL promotes the clinical progression of HIV disease and the development of AIDS-defining conditions. These factors make HIV/VL co-infected patients a potential source for spreading drug-resistant parasites [11, 18]. Furthermore, transmission of the infection via needle sharing in HIV/VL co-infected patients in southern Europe threatens to convert an apparently zoonotic disease into the anthroponotic form [11, 19, 20]. HIV-VL co-infection has been reported from more than 35 countries. Initially, most of these cases were from southwestern Europe, but the number of cases is increasing in sub-Saharan Africa especially Ethiopia, Brazil, and South Asia [19, 21, 22]. In the hyperendemic region of Bihar, India, 1.8–4.5% of VL patients were HIV-positive [23, 24]. There was an increase in the incidence of VL/HIV co-infection from 0.32/100,000 in 2007 to 1.08/100,000 in 2010 in northern Brazil [25]. In Ethiopia HIV co-infection ranged from 10.4% to 40% among VL patients from different centers [26, 27].

Most people with leishmanial infection do not develop into clinical disease. These asymptomatic infections are defined differently in studies as either a positive serological test, polymerase chain reaction (PCR), or leishmanin skin test (LST) in individuals who are otherwise in a healthy condition. In prospective studies, the ratio of incident infection to clinical disease varies from 1:2.4 in Sudan [28], 4.1–5.6:1 in Kenya [29] and Ethiopia [30], 4.1–8.9:1 in the ISC [31–33], 18:1 in Brazil [34], to 50:1 in Spain [35]. A mathematical modeling study based on data from the ISC has shown that transmission of *L. (L.) donovani* is predominantly driven by asymptomatically infected hosts [36]. A detailed description can be found in Chap. 4. Thus, in the era of elimination of VL in the Indian subcontinent, the current challenge is to find out which subset of asymptomatics have the highest risk of developing into clinical VL and sustaining transmission.

7.2.2 Challenges in the Diagnosis of VL

The diagnosis of VL is complicated by the fact that its clinical features are shared by a number of commonly occurring diseases like malaria, typhoid fever, tuberculosis, etc. The sequestration of parasites in the spleen, bone marrow, or lymph nodes is a challenge, and demonstration of parasites necessitates embarking upon invasive procedures which are difficult to perform in the prevailing field conditions (Fig. 7.3). Additional details on challenges in VL diagnosis can be found in Chap. 4.

Molecular techniques such as PCR can be used for the diagnosis of VL, but these techniques remain restricted to referral hospitals and research centers, despite efforts to simplify them.

Antigen-based tests like the latex agglutination test detecting a heat-stable, low-molecular-weight carbohydrate antigen in the urine of VL patients have demonstrated a good specificity but only low to moderate (48–87%) sensitivity in East Africa and the ISC [37–40].

Antibody-based tests, though widely used, have drawbacks. Antibodies remain detectable up to several years after cure; therefore, VL relapse cannot be diagnosed by serology [41, 42]. In endemic areas, a significant proportion of healthy individuals with no history of VL are positive for antileishmanial antibodies owing to a group of patients with asymptomatic infections. The seroprevalence in healthy populations varies from <10% in low to moderate endemic areas [29, 43, 44] to >30% in high-transmission foci or areas where household contacts are common

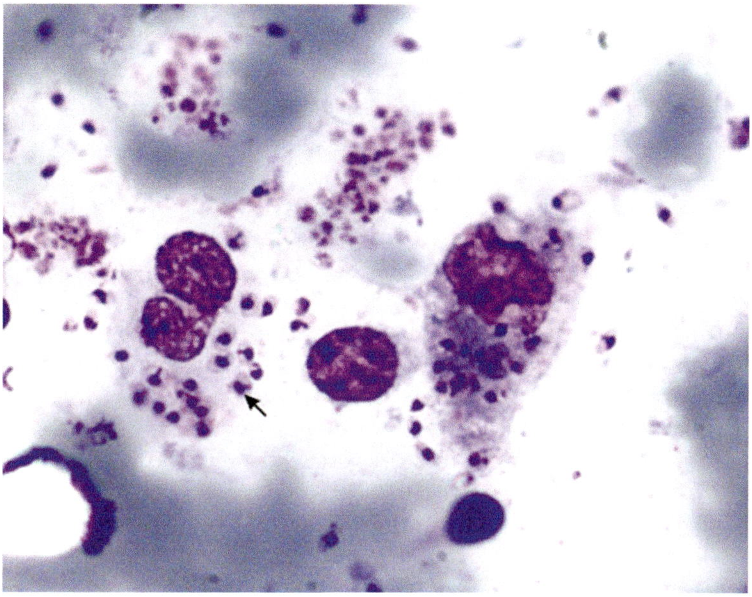

Fig. 7.3 Microphotograph showing two infected macrophages with multiple amastigotes

[45–47]. Another drawback is that over 40% of HIV co-infected individuals have no detectable specific antibody levels against *Leishmania* [11].

The direct agglutination test (DAT) and the rK39-based immunochromatographic test (ICT) are the two serological tests which have been extensively validated in the field. In a meta-analysis performed by [39], DAT had a demonstrated sensitivity of 94.8% (95% confidence intervals (CI), 92.7–96.4) and specificity of 97.1% (95% CI, 93.9–98.7), respectively. The performance of DAT was not influenced by region or by species of *Leishmania*. Its main drawbacks are cumbersome procedure, the regular quality control of antigen, the need for the storage of the antigen at 2–8 °C once it has been dissolved, and the prolonged incubation time needed for performing the assay. rK39 is a 39-amino acid repeat that is part of a kinesin-related protein in *L. (L.) chagasi* and is conserved within the *L. (L.). donovani* complex [48]. Immunochromatographic strip tests (ICTs) based on rK39 are easy to perform, rapid, and cheap and yield reproducible results. A meta-analysis that included 13 validation studies of the rK39 immunochromatographic test (ICT) showed sensitivity and specificity estimates of 93.9% (95% CI, 87.7–97.1) and 95.3% (95% CI, 88.8–98.1), respectively [39]. However, this test shows regional variation and has been shown to be less accurate in East Africa [49–51]. Another format of rK39 ICT has been reported with higher sensitivity and specificity in Africa [52]. There is an urgent need to develop a sensitive, easy-to-use, noninvasive antigen-detection test for the diagnosis of primary VL (particularly in HIV co-infected patients), which would also diagnose relapses.

7.2.3 Challenges in VL Treatment

Over the years there have been many challenges in the treatment of VL. The number of antileishmanials is small. All of them except miltefosine (MIL) have to be administered parenterally. The duration of treatment is long, drugs are toxic, and hospitalization is required for monitoring. As new therapies have been developed for VL, e.g., L-AMB, oral MIL, and paromomycin, the standard pentavalent antimonials (Sb^V) have been rendered obsolete in some regions.

7.2.3.1 Antimonials

First indications of drug resistance came from unconfirmed reports from the four most affected districts in North Bihar of about 30% patients not responding to the prevailing regimen of Sb^V [53]. An expert committee of the Government of Bihar recommended that Sb^V should be used in two 10-day courses with a 10-day interval [54]. Aikat et al. [55] followed these recommendations and described only 1% patients' refractory to Sb^V therapy. However, only a few years later, Thakur et al. [56] randomized patients to receive Sb^V 20 mg/kg (maximum 600 mg) either for 20 days or longer in case of partial or delayed response and demonstrated that 86% of patients were cured in the former group.

Surprisingly, the cure rate with 10 mg/kg for 20 days was much lower compared with earlier results. In the same year, the WHO [57] expert committee recommended Sb^V to be used in doses of 20 mg/kg up to a maximum of 850 mg for 20 days and a repetition of the same regimen for 20 days in cases of treatment failures. Four years later, [58] again reviewed the WHO recommendations and published a report of a clinical trial in which Sb^V at 20 mg/kg (max. 850 mg) for 40 days cured 97% of patients, while 20-day treatment at the same doses cured only 81% of patients. Three years later, the same group reported a further decline in cure rate to 71% after 20 days of treatment at the same doses [59]. Furthermore, by the early 1990s, extending the therapy to 30 days could cure only 64% of patients in a hyperendemic district of Bihar [60]. Five years later, in a bigger study, 156 patients were randomized in three groups for treatment either with (a) Sb^V alone for 30 days or (b) Sb^V plus interferon-γ (IFN-γ) for 15 days or (c) Sb^V plus IFN-γ 30 for days. Only 36% of patients were cured with Sb^V alone, and addition of IFN-γ improved the cure rate to 42% and 49% in groups b and c, respectively [61].

Between 1994 and 1997, a study was conducted to document the level of Sb^V resistance in the hyperendemic region of Bihar and to determine whether therapeutic failure had spread to the neighboring state of Uttar Pradesh (UP). At Bihar and UP sites, 209 and 111 patients were treated, respectively. The results demonstrated that only 35% of patients could be cured at Bihar, and of these, primary unresponsiveness was seen in 52% patients, whereas another 8% relapsed after an initial cure. In UP, on the other hand, 98% were cured initially and one (1%) relapsed.

Thus, it was apparent from the study that Sb^V continued to be effective in the state of UP, but in North Bihar, where most of the disease occurred, it was ineffective in most patients [62]. There were reports of antimony resistance spreading to the Terai regions of Nepal, especially from the district adjoining the hyperendemic areas of

Bihar, where up to 30% of the patients were unresponsive, though in eastern Nepal a 90% cure rate had been reported [63]. These studies confirmed that a high level of antimony resistance existed in Bihar, whereas it was still effective in surrounding areas.

There had been speculations whether Indian *L. (L.) donovani* had become truly refractory to Sb^V or resistance occurred because of the inadequate doses being used in Bihar. In a study to determine whether acquired drug resistance was present in Bihar, *L. (L.) donovani* isolates were taken from responders and nonresponders. In vitro amastigote-macrophage assay showed that isolates from patients who did respond to sodium stibogluconate treatment were threefold more sensitive, with 50% effective doses (ED_{50}) ~2.5 mg Sb^V/mL compared to isolates from patients who did not respond (ED_{50} ~7.5 mg Sb^V/mL) [64]. The significant differences in amastigote sensitivity supported the concept of acquired resistance in Bihar.

The reasons behind the appearance of resistance were that (a) Sb^V was freely available and (b) both qualified medical practitioners and unqualified quacks prescribed the drug. This unrestricted availability of the drug led to widespread misuse. Most patients (73%) consulted unqualified practitioners first [65]. It was a common practice to start with a small dose and gradually build up to the full dose over a week; it was also advocated to have drug-free periods to minimize the toxicity, especially renal toxicity. It was common for physicians to split the daily dose in two injections to be given twice a day. These practices resulted in the buildup of a subtherapeutic blood level and increased tolerance of parasites to Sb^V. In a study to detect the factors leading to antimony resistance in Indian VL, it was observed that only 26% of the patients were treated according to the WHO guidelines, 42% did not take the drug regularly, and 36% stopped the drug on their own initiative. Almost half of the patients, receiving pentamidine as a second-line drug, had not received adequate antimony treatment before being labeled as refractory to Sb^V. These facts indicate large-scale misuse of antileishmanial drugs in Bihar, contributing to development of drug resistance [65]. Moreover, there were several manufacturers of Sb^V in India, and not all produced consistent quality products, resulting in occasional batches being substandard and toxic, adding to the problems associated with Sb^V therapy and serious toxicity and deaths related to the drug [66].

Another reason for the increasing frequency of *Leishmania* resistant to Sb^V in India while parasites still remained sensitive in the rest of the world could be that transmission in Bihar is anthroponotic. In this type of life cycle, once Sb^V resistance gets established, it spreads exponentially through the population and organisms, is sensitive to the drug, and gets eliminated quickly, whereas drug-resistant parasites continue to circulate in the community.

7.2.3.2 Other Antileishmanial Drugs

Pentamidine was the first drug to be used in patients, refractory to Sb^V, and, initially, high cure rates were reported [67]. But its efficacy declined over the years, and a decade later, it cured only approximately 70% of patients [59, 68, 69]. Its use in VL was ultimately abandoned due to its decreased efficacy and serious toxicities.

AMB-B is a polyene antibiotic used predominantly as an antifungal drug, but it also has excellent antileishmanial activity. Due to the high affinity of AMB-B for 24-substituted sterols, aqueous pores are formed in the plasma membrane leading to increased membrane permeability and killing of *Leishmania*. In Sb^V refractory regions in India, it has been used extensively with excellent results [70, 71]. AMB-B has excellent cure rates (~100%) at a dose of 0.75–1 mg/kg for 15–20 daily or alternate days intravenous infusions; however, most of the patients experience infusion reactions (e.g., fever, chills, and thrombophlebitis) and, occasionally, serious toxic episodes (e.g., hypokalemia, nephrotoxicity, myocarditis, and even death). It was recommended as a first-line drug by the Indian National Expert Committee for Sb^V refractory regions [72] (NVBDCP). The need for infusions, hospitalization for prolonged periods, high cost of the drug, requirement for close monitoring, and high incidence of adverse events (occasionally serious) constitute important drawbacks that prevented its implementation at the primary health-care level in Bihar. Clinical resistance to AMB-B is rare.

Lipid-associated amphotericin-B (L-AMB) preparations are as effective as conventional AMB-B and have negligible adverse reactions. The dose requirement of L-AMB varies in different geographical regions; while for patients in the ISC a small dose induces high cure rates, a higher dose is needed for patients from the Mediterranean region and Brazil [73–75]. It is possible to administer high doses of L-AMB over a short period with high cure rates [76]. Although its high price precluded its use in the developing countries, it was the drug of choice for VL in Mediterranean. However, a preferential pricing agreement with WHO (agreement between Gilead and WHO of 14 March 2007) reduced the price of L-AMB (AmBisome®) for endemic regions to $20 (now $18) per 50-mg vial [77]. The preferential pricing made L-AMB a feasible option for the treatment of VL in the endemic region.

MIL, an alkyl phospholipid, is the first oral agent approved for the treatment of leishmaniasis. At the recommended doses (100 mg daily for patients weighing 25 kg and 50 mg daily for those weighing <25 kg for 4 weeks), cure rates were 94% for VL [78]. Its limitations are high cost, need for monitoring for gastrointestinal side effects, and occasional hepatic toxicity and nephrotoxicity. As it is teratogenic, women of child-bearing potential have to observe contraception measures for the duration of treatment and an additional 3 months. Furthermore, it has a long-terminal half-life, which ranges between 150 and 200 h. About four half-lives are required to reach more than 90% clearance of the plateau levels (at steady state). Thus, subtherapeutic levels may remain for several weeks after a standard course of treatment. This fact may lead to the quick emergence of resistance. Free availability and quick recovery (within 10 days, most patients feel better) coupled with the high cost of the drug may motivate patients to prematurely discontinue treatment, and suboptimal compliance will ultimately lead to the emergence of parasite resistance [79]. Due to its oral advantage, this drug was chosen for the elimination program in India, Nepal, and Bangladesh [77, 80]. However, after a decade of use of the drug in the ISC, the relapse rate doubled and its efficacy appeared to have declined [81]. Another recent study from India revealed a cure rate of only 92.6% at 12 months [82]. While in Nepal the results were worse, with relapse rate of 10.8% at 6 and of 20.0% at

12 months [83]. In Bangladesh, a phase IV study showed a cure rate of only 85% [84]. Its efficacy was low in a study from Ethiopia where the final cure among non-HIV-infected patients 6 months after treatment in the MIL group was only 75.6% [85]. The dwindling efficacy of MIL monotherapy in the ISC is a matter of great concern, and it has been replaced with other therapies for the elimination initiative. A complementary explanation of this situation is given in Chap. 4.

Paromomycin, an aminoglycoside-aminocyclitol antibiotic, has been used for the treatment of VL in a parenteral formulation and CL in both topical and parenteral formulations. In a phase III trial in the ISC, paromomycin was shown to be non-inferior to AMB-B and was approved by the Indian government in August 2006 for the treatment of patients with VL [86]. Clinical resistance with this drug in VL has not been reported.

However, following a 60-day parenteral course for treatment of CL in two *L. (L.) aethiopica* cases, isolates taken from relapsed patients were three- to fivefold less susceptible to the drug—after treatment—than isolates taken before treatment in an amastigote-macrophage assay [87]. The advantages of this agent include its cost, approximately US $10 per patient [88]. The disadvantages are the need for intramuscular injection, monitoring of serum transaminases, and the existence of inadequate data regarding its use in pregnancy.

7.3 Control of Visceral Leishmaniasis

7.3.1 Free Distribution of Drugs

The high cost of the antileishmanial drugs coupled with their easy, over-the-counter availability often leads to underdosing and incomplete treatment. This has been the major factor for antimony resistance, and this reason could lead to resistance to another drug like MIL too. Considering that majority of the population cannot afford to purchase and complete a full course of treatment, it is recommended that antileishmanials should be made available free of cost to be distributed through public and/or private health-care providers like antitubercular and antiretroviral drugs, and antileishmanial drugs should be withdrawn from the open market.

7.3.2 Monitoring Therapy

The appearance of Sb^V resistance in the anthroponotic cycle in Bihar suggests that resistance could also expand to other antileishmanial drugs as well. A similar potential for resistance to originate exists in East Africa, another anthroponotic focus of VL with intense transmission, where poverty, illiteracy, and poor health-care facilities are common.

A recent study demonstrated that even in 2008, critical flaws remained in VL case management in the primary health-care services in Bihar, like obsolete use of antimonials with high failure rates and long patient delay. After reviewing the visceral leishmaniasis 191 records of all 150 patients sampled and interviewing 139 patients or their guardian, it was concluded that 81% of patients had first presented themselves to unqualified practitioners, the median delay before reaching the appropriate primary health-care facility was 40 days (IQR 31–59 days), and 48% of VL patients were still being treated with Sb^V out of which 40% needed a second treatment course [89]. Similar concerns were raised for MIL when in a phase IV trial in India, involving domiciliary treatment with MIL and weekly supervision, showed doubling of the relapse rates in one of the clinical centers [79]. These findings suggest that monitoring therapy is imperative to prevent emergence of resistance. The directly observed treatment strategy (DOTS) for tuberculosis has been a big success, and either a parallel or integrated with DOTS system could be organized for leishmaniasis. This will lead to better compliance, completion of the treatment course, and ultimately, prevent resistance.

7.3.3 Combination Therapy

The growing resistance of the parasite to antileishmanial drugs suggests that the currently used monotherapy needs to be reviewed. Multidrug combination therapy has been used successfully in tuberculosis, leprosy, and malaria. The rationale behind combination therapy is increased activity through use of compounds with synergistic or additive activity, preventing the emergence of drug resistance; lower dose requirement, thereby reducing chances of toxic side effects and cost; and increased spectrum of activity.

A randomized, noncomparative, group sequential, triangular design study assigned 181 subjects to treatment with 5 mg/kg of L-AMB alone (group A; 45 subjects), 5 mg/kg of L-AMB followed by MIL for 10 days (group B; 46 subjects) or 14 days (group C; 45 subjects), or 3.75 mg/kg of L-AMB followed by MIL for 14 days (group D; 45 subjects). When it became apparent that all regimens were effective, 45 additional, nonrandomized patients were assigned to receive 5 mg/kg of L-AMB followed by MIL for 7 days (group E). All 226 subjects had initial apparent cure responses. Nine months after treatment, final cure rates were high (>95%) and similar in all multidrug groups. These results suggest that single infusion of L-AMB (in most instances, administered in an outpatient setting) followed by a brief self-administered course of MIL could be an excellent option against Indian Kala-azar [90]. The preferential pricing opened the prospect of combining lower total doses of L-AMB in other combination regimens [77]. In another study in the ISC, three-drug combinations (single injection of 5 mg/kg L-AMB and 7-day oral MIL or 10-day 11 mg/kg intramuscular paromomycin or 10 days each of MIL and paromomycin) were used. All the combinations showed an excellent cure rate and were non-inferior to the standard treatment [91].

Combination therapy provides shorter duration treatment with much improved compliance that will prevent the emergence of resistance. Since the pipeline for the antileishmanial drugs is nearly empty, it is imperative to protect and prolong the effective life of the existing drugs. In the recent guidelines published by the WHO, this combination therapy has been made one of the preferred treatment for VL in the ISC [92].

7.3.4 Novel Therapy

Liposomal AMB is one of the safest and most efficacious among antileishmanials. With the decrease in the price of L-AMB (AmBisome®) for endemic regions [77], an open-label study in India comparing the efficacy of single-dose L-AMB at a dose of 10 mg per kilogram of body weight to conventional AMB, at 1 mg per kilogram, given every other day for 15 doses was conducted. Cure rates at 6 months were similar in the two groups: 95.7% (95% CI, 93.4–97.9) in the liposomal-therapy group and 96.3% (95% CI, 92.6–99.9) in the conventional-therapy group [93]. The low-dose requirement, preferential pricing, excellent efficacy, a single-day hospitalization, no safety concerns, and monitoring requirement make a single infusion of the liposomal preparation an excellent option for the ISC. All these factors led WHO to recommend this treatment as one of the best option for this region [92]. The single dose ensures 100% compliance and therefore decreases the chances of resistance. To test the feasibility in primary health centers, a study was done in Bangladesh where the cure rate at 6 months was 97% [94].

Encouraged by the success of the single-dose L-AMB therapy in the ISC, a randomized controlled trial was done to compare the efficacy and safety of single dose of L-AMB 7.5–10 mg/kg body weight or multiple doses, 7 times 3 mg/kg on days 1–5, 14, and 21 in East Africa. However, the trial was terminated after the third interim analysis because of low efficacy of all the regimens [95].

7.3.5 Monitoring Drug Resistance

Ideally, parasite resistance should be monitored, rather than relapses or unresponsiveness.

It will also permit the identification of key intracellular targets and parasite defense mechanisms, which can then be exploited to rationally develop analogues of existing drugs that would not be affected by the most common defenses. Analysis of genetic markers that determine high antileishmanial resistance, performed systematically for every parasite isolate that shows low antileishmanial sensitivity, would facilitate the tracking of the level of resistance in affected populations. At present, there are no molecular markers of resistance available for the currently used antileishmanial drugs, and the only reliable method for monitoring resistance of isolates is the technically demanding in vitro amastigote-macrophage model. Development of drug resistance markers and tools easy to use in the field should be encouraged. See Chaps. 4 and 15 for a detailed discussion of this topic.

7.3.6 Management of PKDL

Patients with PKDL serve as an important reservoir of infection, and in ISC, treatment is essential. In India, AMB-B 60–80 doses over 4 months or MIL for 12 weeks are the recommended regimens. However, the inordinately long regimens especially for patients without any physical handicap lead to frequent noncompliance. Better and shorter and acceptable options need to be developed [96].

7.3.7 Management of HIV/VL Co-infection

Another potential source for the emergence of drug resistance is the HIV/VL co-infected patients. These patients have high parasite burden and a weak immune response, respond poorly to treatment, and have a high relapse rate. Therefore, they are the ideal candidates to harbor drug-resistant parasites. All antileishmanial therapies are less effective in HIV-positive patients. There is a high mortality rate due to concurrent illness, complications, and drug toxicity. Pentavalent antimonials (Sb^V) and AMB-B are more toxic to HIV patients, who require close monitoring for pancreatitis, cardiotoxicity, and nephrotoxicity [22]. In Ethiopia, MIL was found to be less effective than Sb^V in co-infected patients, and side effects were worse in these patients [52]. The best option for these patients is L-AMB. Secondary prophylaxis to prevent relapses has been reported in several publications, but more evidence from clinical trials is needed to establish a beneficial effect [22]. Initiation of HAART (highly active antiretroviral therapy) dramatically decreases the incidence of VL co-infection. Therefore, HAART in combination with antileishmanials should be advocated strictly in these patients. A detailed description of this topic can be found in Chap. 6.

7.4 Vector Control

Vector control is an important strategy for decreasing the spread of VL. Residual insecticide spraying of houses and animal shelters was shown to be efficacious in India [97], where the vector (*Phlebotomus argentipes*) is restricted to areas in and around the home. However, in Sudan and other endemic countries in East Africa, transmission occurs mainly outside villages [98]. Therefore, indoor residual spraying for disease control is unlikely to be as efficient in this region. Case-control studies conducted in Bangladesh and Nepal demonstrated that sleeping under a nonimpregnated bed net during the warm months was a protective factor against VL [99, 100]. The mass distribution of insecticide-treated nets (ITNs) in Sudan was accompanied by a 27% reduction in the incidence of VL in an observational study [101]. A recent study showed that VL was associated with housing conditions like living in a thatched house or in a house with damp floors, which suggests that improving living conditions could decrease the incidence of VL [102].

7.5 Conclusion

Inventory of antileishmanial agents is very small; emergence of drug resistance and decreased efficacy of some drugs is further complicating the control of leishmaniasis. A better understanding of the mechanisms of action of the drugs and unraveling the puzzle of drug resistance mechanisms with easy-to-use markers of resistance may pave the way for more rational use of drugs. Directly observed therapy given free, in treatment centers manned by trained personnel, will go a long way in controlling the disease as well as drug resistance.

Combination chemotherapy is rapidly emerging as the norm for treating several infective disorders like malaria, tuberculosis, HIV, etc., and its application is strongly advocated for VL. Novel therapy like single-dose L-AmB which ensures complete compliance has revolutionized the treatment of VL in the ISC. Strict monitoring of these novel drug therapies is required to ensure their efficacy at field level.

References

1. World Health Organization. Leishmaniasis: situation and trends. 2015. http://www.who.int/gho/neglected_diseases/leishmaniasis/en/
2. World Health Organization. Leishmaniasis. Kala azar elimination Program. Report of a WHO consultation of a partners. Geneva, Switzerland. 10–11 February 2015. http://apps.who.int/iris/bitstream/10665/185042/1/9789241509497_eng.pdf
3. Zijlstra EE, Musa AM, Khalil EA, el-Hassan IM, et al. Post-kala-azar dermal leishmaniasis. Lancet Infect Dis. 2003;3:87–98.
4. Thakur CP, Kumar K. Post kala-azar dermal leishmaniasis: a neglected aspect of kala-azar control programmes. Ann Trop Med Parasitol. 1992;86:355–9.
5. Magill AJ. Epidemiology of the leishmaniases. Dermatol Clin. 1995;13:505–23.
6. Pearson RD, Jeronimo SMB, de Queiroz SA. Leishmaniasis. In: Guerrant RL, Walker DH, Weller PF, editors. Tropical infectious diseases: principles, pathogens and practice. Philadelphia: Churchill Livingstone; 1999. p. 797–813.
7. Mary C, Faraut F, Drogoul MP, Xeridat B, et al. Reference values for *Leishmania infantum* parasitemia in different clinical presentations: quantitative polymerase chain reaction for therapeutic monitoring and patient follow-up. Am J Trop Med Hyg. 2006;75:858–63.
8. Bhattarai NR, Van der Auwera G, Khanal B, De Doncker S, et al. PCR and direct agglutination as *Leishmania* infection markers among healthy Nepalese subjects living in areas endemic for kala-azar. Trop Med Int Health. 2009;14:404–11.
9. Seaman J, Mercer AJ, Sondorp E. The epidemic of visceral leishmaniasis in western Upper Nile, southern Sudan: course and impact from 1984 to 1994. Int J Epidemiol. 1996;25:862–71.
10. Desjeux P. The increase in risk factors for leishmaniasis worldwide. Trans R Soc Trop Med Hyg. 2001;95:239–43.
11. Alvar J, Cañavate C, Gutiérrez-Solar B, Jiménez M, et al. *Leishmania* and human immunodeficiency virus co-infection: the first 10 years. Clin Microbiol Rev. 1997;10:298–319.
12. Tremblay M, Olivier M, Bernier R. *Leishmania* and the pathogenesis of HIV infection. Parasitol Today. 1996;12:257–61.
13. Olivier M, Badaro R, Medrano FJ, Moreno J. The pathogenesis of *Leishmania*/HIV co-infection: cellular and immunological mechanisms. Ann Trop Med Parasitol. 2003;97:79–98.

14. Rosenthal E, Marty P, Poizot-Martin I, Reynes J, et al. Visceral leishmaniasis and HIV-1 co-infection in southern France. Trans R Soc Trop Med Hyg. 1995;89:159–62.
15. Gradoni L, Scalone A, Gramiccia M, Troiani M. Epidemiological surveillance of leishmaniasis in HIV-1-infected individuals in Italy. AIDS. 1996;10:785–91.
16. Lopez-Velez R, Perez-Molina JA, Guerrero A, Baquero F, et al. Clinicoepidemiologic characteristics, prognostic factors, and survival analysis of patients co-infected with human immunodeficiency virus and *Leishmania* in an area of Madrid, Spain. Am J Trop Med Hyg. 1998;58:436–43.
17. Pintado V, Martín-Rabadán P, Rivera ML, Moreno S, et al. Visceral leishmaniasis in human immunodeficiency virus (HIV)-infected and non-HIV-infected patients. A comparative study. Medicine (Baltimore). 2001;80:54–73.
18. Laguna F, Videla S, Jiménez-Mejías ME, Sirera G, et al. Amphotericin-B lipid complex versus meglumine antimoniate in the treatment of visceral leishmaniasis in patients infected with HIV: a randomized pilot study. J Antimicrob Chemother. 2003;52:464–8.
19. Alvar J, Gutiérrez-Solar B, Pachón I, Calbacho E, et al. AIDS and *Leishmania infantum*. New approaches for a new epidemiological problem. Clin Dermatol. 1996;14:541–6.
20. Molina R, Gradoni L, Alvar J. HIV and the transmission of *Leishmania*. Ann Trop Med Parasitol. 2003;97:29–45.
21. Desjeux P, Alvar J. *Leishmania* HIV co-infections: epidemiology in Europe. Ann Trop Med Parasitol. 2003;97:3–15.
22. Alvar J, Aparicio P, Aseffa A, Den Boer M, et al. The relationship between leishmaniasis and AIDS: the second 10 years. Clin Microbiol Rev. 2008;21:334–59.
23. Burza S, Mahajan R, Sanz MG, Sunyoto T, et al. HIV and visceral leishmaniasis coinfection in Bihar, India: an underrecognized and underdiagnosed threat against elimination. Clin Infect Dis. 2014;59:552–5.
24. Burza S, Sinha PK, Mahajan R, Lima MA, et al. Five-year field results and long-term effectiveness of 20 mg/kg liposomal amphotericin B (Ambisome) for visceral leishmaniasis in Bihar, India. PLoS Negl Trop Dis. 2014;8:e2603.
25. Albuquerque LC, Mendonça IR, Cardoso PN, Baldaçara LR, et al. HIV/AIDS-related visceral leishmaniasis: a clinical and epidemiological description of visceral leishmaniasis in northern Brazil. Rev Soc Bras Med Trop. 2014;47:38–46.
26. Mengesha B, Endris M, Takele Y, Mekonnen K, et al. Prevalence of malnutrition and associated risk factors among adult visceral leishmaniasis patients in Northwest Ethiopia: a cross sectional study. BMC Res Notes. 2014;7:75.
27. Mengistu G, Ayele B. Visceral Leishmaniasis and HIV co-infection in patients admitted to Gondar university hospital, northwest Ethiopia. Ethiop J Health Dev. 2007;21:53–60.
28. Zijlstra EE, el-Hassan AM, Ismael A, Ghalib HW. Endemic kala-azar in eastern Sudan: a longitudinal study on the incidence of clinical and subclinical infection and post-kala-azar dermal leishmaniasis. Am J Trop Med Hyg. 1994;51:826–36.
29. Schaefer KU, Kurtzhals JA, Gachihi GS, Muller AS, et al. A prospective seroepidemiological study of visceral leishmaniasis in Baringo district, rift valley province, Kenya. Trans R Soc Trop Med Hyg. 1995;89:471–5.
30. Ali A, Ashford RW. Visceral leishmaniasis in Ethiopia. IV. Prevalence, incidence and relation of infection to disease in an endemic area. Ann Trop Med Parasitol. 1994;88:289–93.
31. Bern C, Haque R, Chowdhury R, Ali M, et al. The epidemiology of visceral leishmaniasis and asymptomatic leishmanial infection in a highly endemic Bangladeshi village. Am J Trop Med Hyg. 2007;76:909–14.
32. Ostyn B, Gidwani K, Khanal B, Picado A, et al. Incidence of symptomatic and asymptomatic *Leishmania donovani* infections in high-endemic foci in India and Nepal: a prospective study. PLoS Negl Trop Dis. 2011;5:e1284.
33. Topno RK, Das VN, Ranjan A, Pandey K, et al. Asymptomatic infection with visceral leishmaniasis in a disease-endemic area in Bihar, India. Am J Trop Med Hyg. 2010;83:502–6.

34. Evans TG, Teixeira MJ, McAuliffe IT, Vasconcelos I, et al. Epidemiology of visceral leishmaniasis in northeast Brazil. J Infect Dis. 1992;166:1124–32.
35. Moral L, Rubio EM, Moya M. A leishmanin skin test survey in the human population of l'Alacantí region (Spain): implications for the epidemiology of *Leishmania infantum* infection in southern Europe. Trans R Soc Trop Med Hyg. 2002;96:129–32.
36. Stauch A, Sarkar RR, Picado A, Ostyn B, et al. Visceral leishmaniasis in the Indian subcontinent: modelling epidemiology and control. PLoS Negl Trop Dis. 2011;5:e1405.
37. Rijal S, Boelaert M, Regmi S, Karki BMS, et al. Evaluation of a urinary antigen-based latex agglutination test in the diagnosis of kala-azar in eastern Nepal. Trop Med Int Health. 2004;9:724–9.
38. Sundar S, Agrawal S, Pai K, Chance M, et al. Detection of leishmanial antigen in the urine of patients with visceral leishmaniasis by a latex agglutination test. Am J Trop Med Hyg. 2005;73:269–71.
39. Chappuis F, Rijal S, Singh R, Acharya P, et al. A meta-analysis of the diagnostic performance of the direct agglutination test and rK39 dipstick for visceral leishmaniasis. Br Med J. 2006;333:723.
40. Sundar S, Jha TK, Thakur CP, Sinha PK, et al. Injectable paromomycin for visceral leishmaniasis in India. N Engl J Med. 2007;356:2571–81.
41. Hailu A. Pre- and post-treatment antibody levels in visceral leishmaniasis. Trans R Soc Trop Med Hyg. 1990;84:673–5.
42. De Almeida SL, Romero HD, Prata A, Costa RT, et al. Immunologic tests in patients after clinical cure of visceral leishmaniasis. Am J Trop Med Hyg. 2006;75:739–43.
43. Koirala S, Karki P, Das ML, Parija SC, et al. Epidemiological study of kala-azar by direct agglutination test in two rural communities of eastern Nepal. Trop Med Int Health. 2004;9:533–7.
44. Schenkel K, Rijal S, Koirala S, Koirala S, et al. Visceral leishmaniasis in southeastern Nepal: a cross-sectional survey on *Leishmania donovani* infection and its risk factors. Trop Med Int Health. 2006;11:1792–9.
45. Ibrahim ME, Lambson B, Yousif AO, Deifalla NS, et al. Kala-azar in a high transmission focus: an ethnic and geographic dimension. Am J Trop Med Hyg. 1999;61:941–4.
46. Sundar S, Maurya R, Singh RK, Bharti K, et al. Rapid, noninvasive diagnosis of visceral leishmaniasis in India: comparison of two immunochromatographic strip tests for detection of anti-K39 antibody. J Clin Microbiol. 2006;44:251–3.
47. Sundar S, Singh RK, Maurya R, Kumar B, et al. Serological diagnosis of Indian visceral leishmaniasis: direct agglutination test versus rK39 strip test. Trans R Soc Trop Med Hyg. 2006;100:533–7.
48. Burns JM Jr, Shreffler WG, Benson DR, Ghalib HW, et al. Molecular characterization of a kinesin-related antigen of *Leishmania chagasi* that detects specific antibody in African and American visceral leishmaniasis. Proc Natl Acad Sci USA. 1993;90:775–9.
49. Veeken H, Ritmeijer K, Seaman J, Davidson R. Comparison of an rK39 dipstick rapid test with direct agglutination test and splenic aspiration for the diagnosis of kala-azar in Sudan. Trop Med Int Health. 2003;8:164–7.
50. Diro E, Techane Y, Tefera T, Assefa Y, et al. Field evaluation of FD-DAT, rk-39 dipstick and KATEX (urine latex agglutination) for diagnosis of visceral leishmaniasis in northwest Ethiopia. Trans R Soc Trop Med Hyg. 2007;101:908–14.
51. Boelaert M, El-Safi S, Hailu A, Mukhtar M, et al. Diagnostic tests for kala-azar: a multi-centre study of the freeze-dried DAT, rK39 strip test and KAtex in East-Africa and the Indian subcontinent. Trans R Soc Trop Med Hyg. 2008;102:32–40.
52. Ritmeijer K, Melaku Y, Mueller M, Kipngetich S, et al. Evaluation of a new recombinant K39 rapid diagnostic test for Sudanese visceral leishmaniasis. Am J Trop Med Hyg. 2006;74:76–80.
53. Peters W. The treatment of kala-azar. New approach to an old problem. Indian J Med Res. 1981;73:1–18.

54. Anonymous. Proceedings of the meeting of an expert group on kala-azar held at Indian Council of Medical Research Headquarters. Indian Council of Medical Research, New Delhi, 9 Sept 1977.
55. Aikat BK, Sahaya S, Pathania AG, Bhattacharya PK, et al. Clinical profile of cases of kala-azar in Bihar. Indian J Med Res. 1979;70:563–70.
56. Thakur CP, Kumar M, Singh SK, Sharma D, et al. Comparison of regimens of treatment with sodium stibogluconate in kalaazar. Br Med J (Clin Res Ed). 1984;288:895–7.
57. World Health Organization. The leishmaniases: report of a WHO expert committee. WHO Tech Rep Ser. 1984;701:99–108.
58. Thakur CP, Kumar M, Kumar P, Mishra BN, et al. Rationalisation of regimens of treatment of kala-azar with sodium stibogluconate in India: a randomised study. Br Med J (Clin Res Ed). 1988;296:1557–61.
59. Thakur CP, Kumar M, Pandey AK. Evaluation of efficacy of longer duration of therapy of fresh cases of kala-azar with sodium stibogluconate. Indian J Med Res. 1991;93:103–10.
60. Jha TK, Singh NK, Jha SN. Therapeutic use of sodium stibogluconate in kala-azar from some hyperendemic districts of N. Bihar, India (Abstract). J Assoc Physicians India. 1992;40:868.
61. Sundar S, Agrawal NK, Sinha PR, Horwith GS, et al. Short-course, low dose Amphotericin-B lipid complex therapy for visceral leishmaniasis unresponsive to antimony. Ann Intern Med. 1997;127:133–7.
62. Sundar S, More DK, Singh MK, Singh VP, et al. Failure of pentavalent antimony in visceral leishmaniasis in India: report from the center of the Indian epidemic. Clin Infect Dis. 2000;31:1104–7.
63. Rijal S, Chappuis F, Singh R, Bovier PA, et al. Treatment of visceral leishmaniasis in south-eastern Nepal: decreasing efficacy of sodium stibogluconate and need for a policy to limit further decline. Trans R Soc Trop Med Hyg. 2003;97:350–4.
64. Lira R, Sundar S, Makharia A, Kenney R, et al. Evidence that the high incidence of treatment failures in Indian kala-azar is due to the emergence of antimony-resistant strains of *Leishmania donovani*. J Infect Dis. 1999;180:564–7.
65. Sundar S, Thakur BB, Tandon AK, Agrawal NR, et al. Clinico-epidemiological study of drug resistance in Indian kala-azar. Br Med J. 1994;308:307.
66. Sundar S, Sinha PR, Agrawal NK, Srivastava R, et al. A cluster of severe antimony cardiotoxicity in kala-azar due to high osmolality lot. Am J Trop Med Hyg. 1998;59:139–43.
67. Jha TK. Evaluation of diamidine compounds (pentamidine isethionate) in the treatment of resistant cases of kala-azar occurring in North Bihar, India. Trans R Soc Trop Med Hyg. 1983;77:167–70.
68. Jha SN, Singh NK, Jha TK. Changing response to diamidine compounds in cases of kalaazar unresponsive to antimonials. J Assoc Phys India. 1991;39:314–6.
69. Thakur CP, Kumar M, Pandey AK. Comparison of regimens of treatment of antimony resistant kala-azar patients: a randomized study. Am J Trop Med Hyg. 1991;45:435–41.
70. Thakur CP, Singh RK, Hassan SM, Kumar R, et al. Amphotericin-B deoxycholate treatment of visceral leishmaniasis with newer modes of administration and precautions: a study of 938 cases. Trans R Soc Trop Med Hyg. 1999;93:9–23.
71. Mishra M, Biswas UK, Jha DN, Khan AB. Amphotericin versus pentamidine in antimony-unresponsive kala-azar. Lancet. 1992;340:1256–7.
72. National Vector Borne Disease Control Programme (NVBDCP). Guidelines for Vector Borne Disease http://nvbdcp.gov.in/kal8.html. Accessed 20 Sept 2010.
73. Sundar S, Singh VP, Sharma S, Makharia M, et al. Response to interferon-gamma plus pentavalent antimony in Indian visceral leishmaniasis. J Infect Dis. 1997;176:1117–9.
74. Sundar S, Jha TK, Thakur CP, Mishra M, et al. Low-dose liposomal Amphotericin-B in refractory Indian visceral leishmaniasis: a multicenter study. Am J Trop Med Hyg. 2002;66:143–6.
75. Berman JD. DS Food and Drug administration approval of AmBisome (liposomal Amphotericin-B) for treatment of visceral leishmaniasis. Clin Infect Dis. 1999;28:49–51.

76. Sundar S, Jha TK, Thakur CP, Mishra M, et al. Single dose liposomal Amphotericin-B in the treatment of visceral leishmaniasis in India: a multicenter study. Clin Infect Dis. 2003;37:800–4.
77. Olliaro P, Sundar S. Anthropometrically derived dosing and drug costing calculations for treating visceral leishmaniasis in Bihar, India. Trop Med Int Health. 2009;14:88–92.
78. Sundar S, Jha TK, Thakur CP, Engel J, et al. Oral Miltefosine for Indian visceral leishmaniasis. N Engl J Med. 2002;347:1739–46.
79. Sundar S, Murray HW. Availability of Miltefosine for the treatment of kala-azar in India. Bull World Health Organ. 2005;83:394–5.
80. Sundar S, Mondal D, Rijal S, Bhattacharya S, et al. Implementation research to support the initiative on the elimination of kala azar from Bangladesh, India and Nepal--the challenges for diagnosis and treatment. Trop Med Int Health. 2008;13:2–5.
81. Sundar S, Singh A, Rai M, Prajapati VK, et al. Efficacy of miltefosine in the treatment of visceral leishmaniasis in India after a decade of use. Clin Infect Dis. 2012;55:543–50.
82. Burza S, Nabi E, Mahajan R, Mitra G, et al. One-year follow-up of immunocompetent male patients treated with miltefosine for primary visceral leishmaniasis in Bihar. India Clin Infect Dis. 2013;57:1363–4.
83. Rijal S, Ostyn B, Uranw S, Rai K, et al. Increasing failure of miltefosine in the treatment of Kala-azar in Nepal and the potential role of parasite drug resistance, reinfection, or noncompliance. Clin Infect Dis. 2013;56:1530–8.
84. Rahman M, Ahmed BN, Faiz MA, Chowdhury MZ, et al. Phase IV trial of miltefosine in adults and children for treatment of visceral leishmaniasis (kala-azar) in Bangladesh. Am J Trop Med Hyg. 2011;85:66–9.
85. Ritmeijer K, Dejenie A, Assefa Y, Hundie TB, et al. A comparison of Miltefosine and sodium stibogluconate for treatment of visceral leishmaniasis in an Ethiopian population with high prevalence of HIV infection. Clin Infect Dis. 2006;43:357–64.
86. Sundar S, Singh RK, Bimal SK, Gidwani K, et al. Comparative evaluation of parasitology and serological tests in the diagnosis of visceral leishmaniasis in India: a phase III diagnostic accuracy study. Trop Med Int Health. 2007;12:284–9.
87. Teklemariam S, Hiwot AG, Frommel D, Miko TL, et al. Aminosidine and its combination with sodium stibogluconate in the treatment of diffuse cutaneous leishmaniasis caused by *Leishmania aethiopica*. Trans R Soc Trop Med Hyg. 1994;88:334–9.
88. Sundar S, Rai M. Treatment of visceral leishmaniasis. Expert Opin Pharmacother. 2005;6:2821–9.
89. Hasker E, Singh SP, Malaviya P, Singh RP, et al. Management of visceral leishmaniasis in rural primary health care services in Bihar, India. Trop Med Int Health. 2010;15:55–62.
90. Sundar S, Rai M, Chakravarty J, Agarwal D, et al. New treatment approach in Indian visceral leishmaniasis: single-dose liposomal Amphotericin-B followed by short-course oral Miltefosine. Clin Infect Dis. 2008;47:1000–6.
91. Sundar S, Sinha PK, Rai M, Verma DK, et al. Comparison of short-course multidrug treatment with standard therapy for visceral leishmaniasis in India: an open-label, non-inferiority, randomised controlled trial. Lancet. 2011;5:477–86.
92. World Health Organization. Control of leishmaniasis: report of a meeting of the WHO Expert Committee on the Control of Leishmaniases, Geneva, 22–26 March, 2010. http://www.whqlibdoc.WHO.int/trs/WHO_TRS_949_eng.pdf. Accessed 20 Mar 2011.
93. Sundar S, Chakravarty J, Agarwal D, Rai M, et al. Single-dose liposomal Amphotericin-B for visceral leishmaniasis in India. N Engl J Med. 2010;362:504–12.
94. Mondal DAJ. Efficacy and safety of single-dose liposomal amphotericin B for visceral leishmaniasis in a rural public hospital in Bangladesh: A feasibility study. Lancet Glob Health. 2014;2:e51–e7.
95. Khalil EA, Weldegebreal T, Younis BM, Omollo R, et al. Safety and efficacy of single dose versus multiple doses of AmBisome for treatment of visceral leishmaniasis in eastern Africa: a randomised trial. PLoS Negl Trop Dis. 2014;8:e2613.

96. Sundar S, Chakravarty J. An update on pharmacotherapy for leishmaniasis. Expert Opin Pharmacother. 2015;16:237–52.
97. Kaul SM, Sharma RS, Dey KP, Rai RN, et al. Impact of DDT indoor residual spraying on Phlebotomus argentipes in a kala-azar endemic village in eastern Uttar Pradesh. Bull World Health Organ. 1994;72:79–81.
98. Hassan MM, Elraba'a FMA, Ward RD, Maingon R, et al. Detection of high rates of in-village transmission of *Leishmania donovani* in eastern Sudan. Acta Trop. 2004;92:77–82.
99. Bern C, Joshi AB, Jha SN, Das ML, et al. Factors associated with visceral leishmaniasis in Nepal: bed-net use is strongly protective. Am J Trop Med Hyg. 2000;63:184–8.
100. Bern C, Hightower AW, Chowdhury R, Ali M, et al. Risk factors for kala-azar in Bangladesh. Emerg Infect Dis. 2005;11:655–62.
101. Ritmeijer K, Davies C, van Zorge R, Wang SJ, et al. Evaluation of a mass distribution programme for fine-mesh impregnated bednets against visceral leishmaniasis in eastern Sudan. Trop Med Int Health. 2007;12:404–14.
102. Singh SP, Hasker E, Picado A, Gidwani K, et al. Risk factors for visceral leishmaniasis in India: further evidence on the role of domestic animals. Trop Med Int Health. 2010;15:29–35.

American Tegumentary Leishmaniasis

8

Olga Zerpa, Maritza Padrón-Nieves, and Alicia Ponte-Sucre

Abstract

American tegumentary leishmaniasis is an endemic anthropozoonosis undergoing expansion on the American continent. The disease is caused by several *Leishmania* species and thus there are intraspecific parasitological dissimilarities that may generate different pathologies. Furthermore, in America *Leishmania* spp. has diverse reservoirs (that may change continuously) and can use various vectors to infect humans and mammals. Antimonials are the drugs of choice for the treatment of American tegumentary leishmaniasis; however, their efficacy is not predictable, and this may be linked to parasite drug resistance. This is further complicated by the fact that the etiological parasitic species in America belong to both the *Leishmania* and the *Viannia* subgeni. For all these reasons, the identification of the etiological infectious agent—up to the species level—is fundamental for precise clinical diagnosis, treatment, and prognosis and for control of the disease. The present chapter offers a description of American tegumentary leishmaniasis, a fundamental piece of knowledge for the comprehension of the challenges we face for leishmaniasis in times of drug resistance. As a way to better understand the unique scenario that America offers for leishmaniasis, some data related to the figures present in the Old World will be presented.

O. Zerpa
Emerson Clinical Research Institute, Washington, DC, USA

M. Padrón-Nieves · A. Ponte-Sucre (✉)
Laboratorio de Fisiología Molecular, Instituto de Medicina Experimental, Escuela Luis Razetti, Universidad Central de Venezuela, Caracas, Venezuela

A. Ponte-Sucre, M. Padrón-Nieves (eds.), *Drug Resistance in Leishmania Parasites*,
https://doi.org/10.1007/978-3-319-74186-4_8

8.1 Introduction

Leishmaniasis is an infectious disease caused by flagellate protozoa of the genus *Leishmania* (*L.*). The disease is transmitted to humans through the bite of an insect vector, the sand fly. Depending on the vertebrate reservoir found in a specific geographic zone, the disease is classified as zoonotic or anthroponotic. Leishmaniasis is characterized by a spectrum of clinical, histopathological, and immunological features linked to the pathogenicity of the infecting parasite as well as to the immunological response of the host. As stated in various chapters of the present volume, clinical manifestations of leishmaniasis include lesions in the skin and/or the mucous membranes or invasion of visceral organs [1].

Leishmania infections range in severity from asymptomatic lesions to disfiguring tegumentary leishmaniasis and fatal visceral leishmaniasis. Dermotrophic strains cause American tegumentary leishmaniasis (ATL) characterized by a spectrum of clinical manifestations including localized cutaneous (LCL), diffuse cutaneous (DCL), disseminated (DL), and mucocutaneous (MCL) leishmaniasis [1, 2].

The main species (in 86–98% of cases) causing the limited clinical manifestations (characteristic ulcerative lesion) of leishmaniasis in the Old World are *L.* (*L.*) *major*, *L.* (*L.*) *tropica*, *L.* (*L.*) *aethiopica*, and some zymodemes of *L.* (*L.*) *infantum*. The lesions usually heal spontaneously in periods from 3 months to 2 years or may evolve to a relapsing lesion known as leishmaniasis *recidiva cutis* (*L.* (*L.*) *tropica*) or to DCL (*L.* (*L.*) *aethiopica*) [3]. *L.* (*L.*) *donovani* visceral infections may develop into post-Kala-azar dermic leishmaniasis (PKDL) [4]. A detailed description of the clinical manifestations of Old World cutaneous leishmaniasis is found in a recently published comprehensive review [3].

Most of the species that cause tegumentary leishmaniasis occur in the New World. They are numerous and belong to both the *Leishmania* (*L.*) and *Viannia* (*V.*) subgeni, being *L.* (*V.*) *braziliensis* the most prevalent species, followed by *L.* (*L.*) *amazonensis* and *L.* (*L.*) *mexicana*, *L.* (*V.*) *guyanensis*, and *L.* (*V.*) *panamensis*. Other types of *Leishmania* that may also produce the disease in America are *L.* (*L.*) *pifanoi*, *L.* (*L.*) *shawi*, *L.* (*L.*) *venezuelensis*, *L. lainsoni*, and *L.* (*V.*) *peruviana* [2]. This variety in the species responsible for New World ATL determines the diverse clinical manifestations of the disease, including the aggressive and destructive MCL [1]. The comprehensive review written by Goto and Lindoso [1] offers a detailed description of the clinical manifestations of New World cutaneous leishmaniasis.

8.2 Epidemiology, Classification of *Leishmania*, Vectors

Tegumentary leishmaniasis is endemic in 82 countries all over the world, with approximately 1.5 million cases per year. Africa hosts most of the reported cases, then comes cases found in the Middle East, and finally those found in Latin America, being Chile the only country, which has not reported cases [1, 5]. Around 70–75% of global incidence occur only in ten countries: Afghanistan, Algeria, Brazil,

Colombia, Costa Rica, Ethiopia, the Islamic Republic of Iran, Peru, Sudan, and the Syrian Arab Republic. [6, 7]. Approximately 35,000 cases of mucosal leishmaniasis occur annually, mainly in Brazil, Peru, and Bolivia [8], and for CL high-burden countries, the population at risk of CL varies between 14% and 100%, and all together 399 million people are at risk of CL [7].

The disease is a dermatological syndrome. It is diagnosed in 3.3% of the skin-related infections present in tourists that visit Latin America [9]. Cases predominate among agricultural workers, followed by students and finally housewives and children; its incidence in males is higher, possibly due to their greater risk of vector exposure caused by their type of (outdoors) work.

Risk factors to emergence and spread of tegumentary leishmaniasis include environmental factors (temperature and water storage, irrigation habits, deforestation, climate changes), immunosuppression (HIV or organ transplant), the use of immunosuppression therapy, and appearance of drug resistance. There is also an increased incidence in leishmaniasis in traveling people. Finally, war, people displacement by geopolitical problems, poor socioeconomic status, and low-level household also contribute to spread the disease [10].

The *Leishmania* species as well as the immune status of the host determine the clinical features of ATL. The initial lesion appears at the site where the insect bites. The incubation period lasts 2 weeks to 3 months. The initial lesion is a small, itchy, erythematous papule or nodule that eventually results in the enlargement of the draining lymph node. This initial wound may heal spontaneously; alternatively, it may evolve after several weeks, to patent disease with different clinical features [11].

Due to the diversity of the species that may co-exist in some geographical areas, correlation between clinical features of the disease and the infecting species of *Leishmania* is not straightforward; this is further complicated by the fact that the laboratory procedures needed for species identification are complex and sophisticated. These facts, as well as the range of drug sensitivities expressed by New World *Leishmania*, constitute a challenge for the prognosis of ATL [1].

As clearly described in the introduction of this volume, the *Leishmania* parasite alternates between two extreme environments to which the parasite must adapt, i.e., the mammalian host (amastigotes, without flagellum) and the insect vector (promastigote, flagellar form). Successful transmission occurs when the parasitized vector sucks blood from a vertebrate and inoculates promastigotes present in the proboscis. As the parasite enters the vertebrate circulation, the parasite is phagocytosed by macrophages. Although phlebotomine sand flies (*Phlebotomus* and *Lutzomyia*) transmit the disease, only anthropophilic *Lutzomyia* (~30 species), distributed all over America, can potentially function as vectors for *Leishmania* [5].

Regarding reservoirs, it is fundamental to differentiate zoonotic leishmaniasis, in which the reservoirs are wild or domestic animals, and anthroponotic leishmaniasis, when humans constitute the main host. This latter form of transmission is typical but not exclusive for the VL produced by *L.* (*L.*) *donovani*, and the LCL caused by *L.* (*L.*) *tropica*, in the Old World but is not common in America [5].

In America, many vertebrates have been identified as reservoirs: the sloth (*Choloepus* (*C.*) *didactylus*) for *L (V.). guyanensis* and *C. Hoffmani* for *L. (V.)*

panamensis, the opossum (*Didelphis marsupialis*) for *L. (V.) guyanensis*, the rice rats (*Oryzomis capito*), and the agouti (*Dasyprocta Nectomys*) for *L. (L.) amazonensis*. Also, several rodents' species function as reservoirs for *L. (L.) mexicana* and the rat (*Rattus rattus*) for *L. (V.) guyanensis* [12–16].

8.3 Clinical Spectrum, Immune and Pathologic Consequences of ATL

ATL may occur in three general forms with a range of clinical, histological, and immunological features that differ among them. LCL is located at one end of the spectrum and occurs in immune-competent patients. It is characterized by one or a few usually ulcerated lesions. The anergic DCL is located at the other end of the spectrum and is characterized by the clinical expression of numerous nodules, non-ulcerated papules, and plaques. Mucocutaneous lesions are located in the intermediate area of the spectrum, with extensive lesions prone to relapse [1, 17].

8.3.1 Localized Cutaneous Leishmaniasis (LCL)

LCL (Fig. 8.1a) is the most prevalent form of the disease and is caused by dermotropic *Leishmania* species [18]. Both *Vianna* and *Leishmania* subgeni produce it. The lesions, varying in number from one to ten, appear in an exposed area of the body surface. The established lesion is a well-delimited round, painless ulcer, with raised edges and a central crust, sometimes hemorrhagic. It starts as an erythematous papule after the bite of the vector. It grows and, in a few weeks, develops into an ulcer with little secretion but purulent if a secondary infection builds up. It may occur also as papules that surround the primary ulcer and may be accompanied by inflamed lymphatic tracts and nodes. The ulcers may heal spontaneously, leaving a hypopigmented, smooth, thin scar. The host–parasite balance, as well as other undefined factors, determines the evolution to other forms of the disease [3, 19, 20].

The ulcer differentiates to a typical epithelioid granuloma with a mixed pattern of Th1 and Th0 cytokines and a predominance of a Th1 response. Nodules and plaques on the skin may be flat; in the ulcers, the skin is abruptly lost producing epidermal hyperplasia. A macrophage infiltrate with epithelioid differentiation occupies the dermis, and a variable number of lymphoid cells and plasma cells (including a moderate number of Langerhans type giant cells) surround and/or invade the macrophage infiltration. The patients are normally immune-competent and develop a positive Montenegro test [1, 21]. For differential diagnosis, the following diseases should be considered: piodermitis, sporotrichosis, chromomycosis, skin cancer, cutaneous tuberculosis, and varicose ulcers and traumatic ulcers.

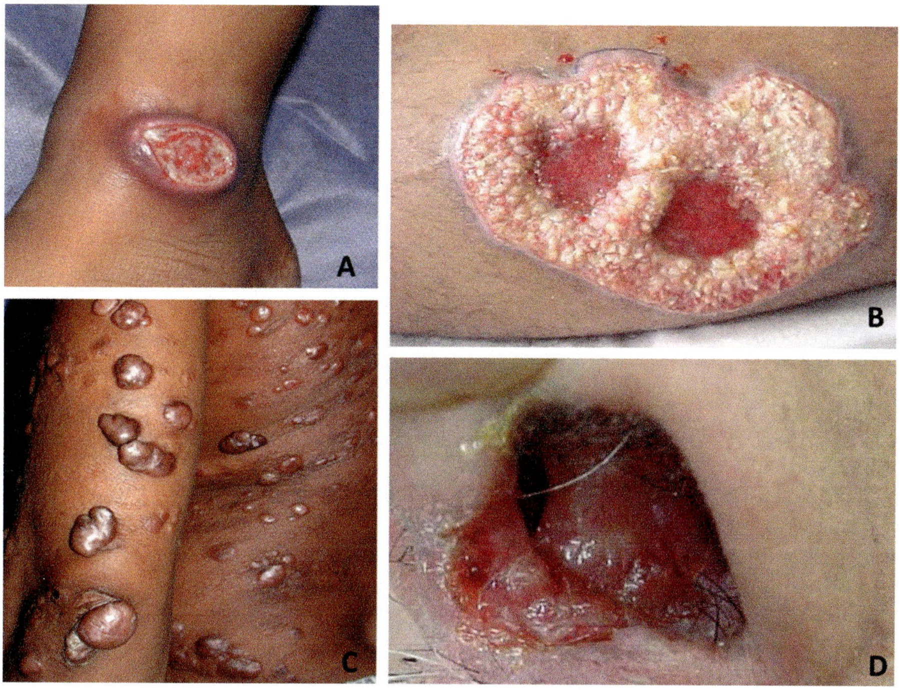

Fig. 8.1 Clinical forms of tegumentary leishmaniasis. (**a**) Localized cutaneous leishmaniasis. (**b**) Leishmaniasis recidiva cutis. (**c**) Diffuse cutaneous leishmaniasis. (**d**) Disseminated cutaneous leishmaniasis.

8.3.2 Leishmaniasis Recidiva Cutis (LRC)

LRC (Fig. 8.1b) is rare in the New World and in the Old World is associated with infections produced by *L.* (*L.*) *tropica*. Characteristic papular and vesicular lesions appear, in or around the healed scar. Most of the identified parasites that produce this form of the disease in the New World belong to the subgeni *Viannia* [22], but *L.* (*L.*) *amazonensis* in Brazil [23] and *L.* (*V.*) *panamensis* in Ecuador [24] can produce it [1].

8.3.3 Diffuse Cutaneous Leishmaniasis (DCL)

DCL is a true anergic form of tegumentary leishmaniasis characterized by the presence of nodular lesions that do not ulcerate (Fig. 8.1c) [25, 26]. This uncommon (described in Brazil, Mexico, Venezuela, the Dominican Republic, and Colombia) presentation of leishmaniasis is characterized by a lack of a cell-mediated immune response, although it may produce protective antibodies. It is caused by parasites of the subgeni *Leishmania*, i.e., *L.* (*L.*) *mexicana*, and *L.* (*L.*) *amazonensis* in the New World and by *L.* (*L.*) *aethiopica* in the Old World.

DCL seems to eclose mainly in childhood, beginning the early manifestations before the age of 15. It is believed that this predisposition is related to genetic and metabolic individual factors [27]. Some authors refer that an initial LCL lesion may be the origin of the spread of parasites by lymphatic and hematic means and that the subsequent inhibition of specific cellular immunity may lead to DCL appearance [28].

In early stages, the disease is characterized by the presence of papules, plaques, or erythematous nodules generally in localized skin areas. These lesions (full of parasites probably due to the Th2 immune response) may be asymmetrical, affecting a single extremity, or may be symmetrical but limited only to the upper or lower limbs [19, 25]. The lesions ulcerate if they suffer trauma, and invasion of the nasal mucosa occurs once the clinical disease becomes severe. This form of the disease is not accompanied by a strong inflammatory reaction. In DCL the initial sores relapse with the formation of nodules on the edge of the scar that remains with little changes over months or years and abruptly spread through the body surface.

Histological sections demonstrate atrophy of the epidermis, with dermo-epidermal boundary rectification. A dense macrophage infiltration invades the dermis, accompanied by a moderate amount of vacuolated lymphoid and plasma cells. The inflammation reaches the subcutaneous tissue, and vacuolated macrophages contain a large number of parasites [1].

Cytokines and accessory signals on the skin decline; this situation compromises the function of antigen-presenting cells and induces a parasite-specific anergy. The granuloma is characterized by a predominantly Th2 response, with a high percentage of naive T cells that react against the parasite. The Montenegro test is negative [21]. In rare occasions, the initial diagnosis is positive but then becomes negative [19, 29, 30]. The titers of anti-*Leishmania* antibodies are high but decrease after treatment, a response that does not reveal a protective activity [31]. For differential diagnosis, the following diseases should be considered: lepromatous leprosy, cutaneous neurofibromatosis, lymphomas, and xanthomatosis.

8.3.4 Disseminated Leishmaniasis (DL)

DL (Fig. 8.1d) is characterized by the presence of multiple (10–300) pleomorphic small lesions, mainly acneiform and papular, in two noncontiguous areas of the body [20]. In 29% of cases, at least a mucocutaneous lesion is found. The clinical outcome includes a verrucous plaque, sarcoid, chronic ulcers with poor response to treatment and relapse with extensive lesions with a variable immunological response. It is produced by parasites of the subgeni *Leishmania* and *Viannia*. However, there are areas in northeast Brazil where *L.* (*V.*) *braziliensis* has been the only species found in infected patients [32].

As for DCL, some authors refer that an initial LCL lesion may be the origin of the spread of parasites by lymphatic and hematic means and that the subsequent inhibition of specific cellular immunity may lead to DL appearance [26, 28]. The lesions develop transformations similar to those found in LCL, and the epithelioid differentiation of the epidermis concurs with epithelial proliferation, hyperkeratosis, parakeratosis, and scale-crusts. The parasites appear in varying numbers and must be sought within macrophages. Also, similar to LCL and in contrast to DCL, the infection is not age related, and it is mainly a result of the exposure of the host to the infected vector and to the immune response of the patients [28, 31].

DL pathogenesis is not still fully dilucidated; however, the absence of a cell-mediated immune response, with decreased $CD4^+$ T cell titers in peripheral blood, and a poor response by these cells to the *Leishmania* antigen seem to be a common feature. In DL patients, epidermal Langerhans cells are not frequent, and the granuloma has a mixed pattern of Th1 and Th2 cytokines. The Montenegro test has been reported to be negative depending on the geographical area where the patient lives; thus it has been claimed to be negative in Brazil and positive in Venezuela [21]. For the differential diagnosis, the following diseases should be considered: skin tuberculosis, chromomycosis, sporotrichosis, sarcoidosis, and leprosy.

8.3.5 Mucocutaneous Leishmaniasis (MCL)

One of the most severe forms of damage that occur in leishmaniasis involves the upper respiratory tract mucosa. It includes metastases by way of blood vessels or lymphatic system or by expansion of a face LCL [1, 31]. MCL appears years after the onset of cutaneous leishmaniasis and is characterized by the destruction of the walls of oral–nasal and pharyngeal cavities, potentially evolving to disfiguring lesions. The initial symptoms are mild and include nasal inflammation and stuffiness; ulceration and perforation of the nose septum could slowly ensue. The lesion may extend to the face, the soft palate, the pharynx, or the larynx. A cutaneous lesion can accompany the mucocutaneous lesion. *L.* (*V.*) *braziliensis* is the etiological agent in most cases, but species like *L.* (*V.*) *panamensis*, *L.* (*V.*) *guyanensis*, *L.* (*L.*) *amazonensis*, and *L.* (*L.*) *major* may also cause MCL [20].

The epidemiological data demonstrate that 5–7% of patients with LCL develop MCL [31, 33]. However, the frequency of MCL varies according to geographical location: In Brazil, it varies from 0.4% in the south [34, 35] to 1.4% in the central region [20] and to 2.7% in the northeast [36]. In the Andean countries, MCL may represent 7.1% of the registered cases of leishmaniasis [37]; Bolivia exhibits a high frequency of 20%, Ecuador a medium frequency of 7.7% [38], Colombia a low frequency of 2.3%, and Venezuela a very low frequency of 0.4% [37]. Most patients are over 40 years of age, although this form of the disease may also affect children [1].

The clinical manifestations begin with nasal obstruction, rhinorrhea, mucocutaneous bleeding, and shedding of serous crusts, impaired olfaction, and cacosmia. Physical examination at the beginning of the disease demonstrates erythema and infiltration in the nasal mucosa, mainly in the septum and inferior turbinate. If the disease develops without diagnosis and treatment, it progresses to an ulcer with serous crusts, surrounded by diffuse infiltrations of the mucosa (because of a poor definition of the granuloma); it may compromise the cartilaginous septum and produce drilling and deformation and even the total destruction of the septum giving the appearance of "tapir nose." The discharge of the nose can occasionally be purulent, due to bacterial infections and polypoid degeneration of the nasal mucosa.

These features are accompanied with significant shrinkage of the nasal wing and collapse of the corresponding nostril. Sometimes the acute inflammatory processes that occur around the nasal vestibule produce severe pain that could compromise the maxillary region of the affected side [17, 33, 39, 40]. At advanced stages of the disease, a destruction of the midface may occur.

In some cases, invasion of the nose and palate occurs; the patients report a feeling of "fullness" in the mouth, toothache, teeth loss, and spontaneous bleeding of the gums. These lesions grow profusely and may compromise the upper lip; they may also produce indurations, infiltration, and ulceration of the hard palate, amputation of the uvula, and lesions of the soft palate. Additionally, dysphagia, open rhinolalia, and regurgitation of food, as well as damage of the laryngeal structures such as epiglottis, ventricular bands, and vocal cords, may occur. Finally, the upper airway may also be compromised due to the tension produced by the formation of a granuloma in the mucosa and subsequent fibrosis; some cases may even require tracheotomy. In severe cases, there is deterioration of the patient's general condition and even death if the compromise of the respiratory tract is serious [1].

Histological sections support a diffuse mixed infiltrate [1]. The macrophage infiltrate differentiates into an epithelioid tissue with low densities of parasites [17, 30]). Langerhans cells ($CD1a^+$) and $CD83^+$ cells cannot be found in the epithelium [30, 41]. This situation might reflect the migration of Langerhans cells to the lymph node, or the action of the parasites on Langerhans cells during the chronic phase of the disease, circumstances that may cause an inadequate and deficient transduction of the signals necessary for an adequate immune response. In the epidermis, there is a strong expression of major histocompatibility complex (MHC)-II and intercellular adhesion molecule 1 (ICAM)-1, which confirms the state of hypersensitivity of this clinical form of leishmaniasis. The MCL granuloma expresses a mixed pattern of cytokine production (Th1/Th2, and a high CD4/CD8 ratio) [42, 43].

The Montenegro test reaction is strongly positive (Restrepo 1980). *Leishmania* antibody levels are variable and correlate with the extent of the patient's clinical profile [19, 31]. For differential diagnosis, the following diseases should be considered: in the nasal area, trauma, bacterial infections, syphilis, cocaine use, chromium poisoning, half-facial malignant granuloma, paracoccidioidomycosis, nasal polyps, rhinosporidiosis, leprosy, and squamous and basal cell carcinoma and in the palate and larynx carcinoma, paracoccidioidomycosis, and tuberculosis.

The number of diseases with which MCL should be differentially diagnosed is high; therefore, it is fundamental to carry out further examinations. These tests must include fungal serology, intradermal tests, mycological studies, mycobacteria, chest X-ray, nose and paranasal sinuses tomography, and histopathological analysis. Additionally, there may be complications such as conjunctival lesions with distortion of the palpebral fissure and, in rare cases, loss of the eyesight. Moreover, healing processes can lead to a decreased size of the mouth and airways that hinder feeding and breathing. Finally, extension of the lesion at the base of the skull with bacterial infection can cause meningitis or osteomyelitis.

8.3.6 Tegumentary Leishmaniasis in HIV-Infected Patients

HIV/*Leishmania* co-infection has been reported in 35 countries. In the Old World, there are reports of PKDL in HIV-infected patients [5, 44]. In the New World, the manifestations can be similar to those found in non-immunosuppressed patients with no signs of aggravation, but they can be quite unusual. A full description of this problem is covered in Chaps. 5 and 6, this same volume; therefore the theme will not be discussed in detail herein.

8.4 Diagnosis and Treatment

Diagnosis of ATL is relatively simple, and in most cases the demonstration of the parasite by direct methods after clinical suspicion is sufficient to establish the treatment. The diagnosis cannot be intuitive but has to be confirmatory of the parasite (etiological agent) or its antigen(s) in the lesion. These forms of diagnosis are called direct, while those immunological tests used if the direct approaches fail are called indirect parameters of diagnosis [45].

The sensitivity of the direct examination tests is low (50–70% in the Old World, 15–30%, in the New World, where chronic cases and MCL are frequent). The detection level is higher, reaching 44–58% by culturing the biopsies and 38–52% by injection into hamsters [1, 46–48].

On the other hand, serodiagnosis includes a set of indirect methods seldom used for the diagnosis of LCL in the Old World because the results may be variable, the sensitivity of the tests is low, and there may be cross-reactivity with other infections. Unfortunately, the sensitivities of these methods are not better for New World leishmaniasis. However, still they are in use. The most commonly used assays for ATL serodiagnosis are thus the indirect immunofluorescence assay (IIFA) and the enzyme-linked immunosorbent assay (ELISA) [1, 46–49]. In ATL, the anti-*Leishmania* antibody levels do not remain high after treatment; this means that positive results of serologic diagnostic method generally indicate current infection.

Excluding direct microscopic examination of biopsies, the additional diagnostic methods require a complex laboratory structure and technical skills, as well as longer times to obtain the results [1]. Furthermore, the approaches to detect the etiological agent have low sensitivity and do not always identify the *Leishmania* species. Recent efforts aim to develop assays to detect the parasite DNA in the patients [5].

Among the variety of molecular approaches developed for the diagnosis of leishmaniasis and the identification of the etiological agent, the polymerase chain reaction (PCR) assay is considered one of the best methods. It is based on the complementarity that exists between the two strands of DNA. The method relies on cycles of repeated heating and cooling of DNA melting and its enzymatic replication in the presence of primers, which are short DNA fragments containing sequences complementary to the target region. This cycling enables selective and repeated amplification and eventually the identification of the infecting *Leishmania* species [1, 50, 51].

Finally, the anti-*Leishmania* delayed-type hypersensitivity or Montenegro skin test diagnoses *Leishmania* infection, and therefore is used in epidemiological studies to determine infection prevalence. The test does not distinguish between present and past infections, and thus its importance as a diagnostic tool is questionable for people living in endemic areas. The test is positive in patients with more than 19 months of treatment [48, 52, 53] and in 75% of non-infected individuals, with no disease manifestation in the past, but living in an endemic area [52]. This test may be useful, however, for the diagnosis in travelers that do not normally live in endemic areas.

The treatment of leishmaniasis must include the thorough cleaning of the lesions with topical antiseptics and the treatment of secondary bacterial infections with topical and/or mouth antibiotics. Afterward, the patient should be treated with the adequate chemotherapy to kill the parasite. Alternatively, attempts to develop an immunotherapy against leishmaniasis have been performed in many laboratories and places including Venezuela [54, 55]. The data suggest that immunotherapy might be an excellent therapy for LCL, with few side effects and low-cost administration. However, further studies are needed to confirm the results. Finally the surgical reconstruction of the sequelae in nasal pyramid and portion of the upper lip skin is advisable to do it after confirming that there is no active disease for a period of 1 year or longer.

8.5 Challenges of ATL in the Era of Drug Resistance

ATL is a serious public health problem in America both in rural and urban areas; its incidence has dramatically increased in the last two decades. ATL affects zones considered endemic for leishmaniasis, but it is also increasing in travelers living in non-endemic parts who have visited endemic areas [1]. Furthermore, co-infection is an additional concern because of its increasing rates, either by HIV, by additional parasites like *T. cruzi* or helminths, or the special case of co-infection represented by *Leishmania* RNA viruses, or LRV, which are endosymbionts reported so far essentially in Latin America and frequently associated with treatment failure. These issues are thoroughly described in Chaps. 4 and 6 from the present volume. This means that fighting against leishmaniasis must be among priority programs related to endemic and epidemic diseases that must integrate other pathogens and monitoring conditions and must also incorporate public and private institutions, scientific societies, and affected communities.

Diagnosis seems to be a dilemma due to the variety of *Leishmania* species that produce ATL. This is especially true for *L. (V.) braziliensis* in LCL and LMC patients as the parasite is scarce in the tissues. For this reason, main goals to be reached must include the use of homogeneous protocols for *Leishmania* antigen purification according to validated protocols with quality control analysis; additionally, the cutoff determination of the diagnosis method for leishmanina must be performed in order to homogenize the criteria of positive and negative readings. When talking about direct microscopy and PCR, a lot of discussion still exist. Microscopy on a smear is more frequently used since it means a speedy (<1 h) result. Molecular diagnosis is much more sensitive than microscopy. However, specificity depends on the performance of each laboratory, the selected target, and the selected protocol, many of them in house protocols with an intrinsical variability evidenced when the protocol is transferred from one lab to another, highlighting the lack of consensus that exist.

Tegumentary leishmaniasis therapy in America is mostly restricted to the use of antimonials (Sb^V) and more recently miltefosine (MIL) for some types of LCL. However, in Latin America, the efficacy of this medicament is rather unpredictable with 7% treatment failure in Bolivia, 16% in Brazil, 23.9% in Peru, and up to 39% in Colombia [1]. Furthermore, the guidelines for regional implementation are unfortunately not homogeneous [1, 56]. This all means that therapeutic failure, defined as the clinical phenotype in which the patient does not improve at the end of a treatment (absence of response), or in which the clinical symptoms reappear after the initial cure (relapse), is a real challenge that should be clearly differentiated from clinical resistance in order to avoid the ambiguity of both meanings.

Drug resistance represents an intrinsic characteristic of parasites with a significantly lower susceptibility to a drug than that of their susceptible counterparts. Drug resistance is an adaptive trait. Exposure to drugs (e.g., due to external factors like suboptimal doses or poor quality of the medicaments that induces the expression and function of ATP-binding cassette (ABC) transporters and proteins) promotes an increase in the frequency of occurrence of this phenotype, and although it is expressed in the patient, the associated phenotype must be confirmed experimentally evaluated in parasites isolated from the lesion [56–58].

On the other hand, treatment failure is a multifactorial complex phenomenon. Drug, host, and parasite factors may contribute to it. In the case of American field strains of *Leishmania* (but not only, as beautifully described in Chaps. 4 and 15 of this volume), special attention should be paid to the variable intrinsic drug sensitivity usually related to species-specific issues as is the case of the *Viannia* subgenus already described, as well as to epigenetic features that may change different functions in the parasites. This means that the specific contribution of the parasite physiology to treatment failure is difficult to address [59–61]. This is especially true since as has been described in various chapters of this volume (Introduction, Chaps. 4 and 15), the in vitro data is normally obtained using the extracellular form of the parasite (the promastigotes) and seldom using its intracellular form (the amastigotes), and results are infrequently compared to the treatment outcome of patients from whom parasites are isolated.

However, it is fundamental to find easy tools to be used in the common clinical laboratory to evaluate if relapses that occur in patients associate with metabolic changes that might be associated to the fitness of infecting isolates. In such isolates (isolated from patient suffering DCL and refractory to Sb^V), a correlation between glucose uptake and plasma membrane potential has been evaluated. The results were compared with those obtained from reference strains and demonstrated that *Leishmania* parasites (*L. (L.) amazonensis* and *L. (L.) mexicana*) causing DCL incorporate glucose at an efficient rate, albeit without significant changes in the plasma membrane potential as their corresponding reference strains. One isolate did not change its accumulation rate of glucose compared to its reference strain and expressed a less polarized membrane potential insensitive to mitochondrial inhibitors, thus suggesting a metabolic dysfunction in this isolate. Further validation of the concepts herein established and whether or not the third isolate corresponds with a drug-resistant phenotype needs to be demonstrated at the genetic level [62, 63].

In the case of ATL, especially in Latin America, this is further complicated due to the many infecting species of *Leishmania*, including parasites of subgeni *Leishmania* and *Viannia*. In fact, isolates of *L. (V.) braziliensis* with lower susceptibility to Sb^V have been reported even before the start of treatment, although they have probably never been in contact with the drug (s) [64]. It is not clear if this difference is due to an intrinsic unresponsiveness to the drugs, expressed by members of the *Viannia* subgenus, but certainly constitutes an issue that should recall our attention and emphasize that the contribution of the parasite to therapeutic failure could not only correspond to the expression of drug resistance. That is, the existence of additional phenotypes could be determinant for the phenomenon of therapeutic failure. Unfortunately, and again returning to the experimental determination of this phenomena, these phenotypes are not necessarily easy to identify in the available systems and therefore and is fundamental to describe specific cellular markers easy to evaluate in the clinical laboratory, a situation that challenges the classical view of how the factors responsible for that therapeutic failure are evaluated [59–61, 63, 64].

In summary, Old World leishmaniasis has a better therapeutic outcome, except when caused by *L. (L.) aethiopica*, than New World leishmaniasis where therapeutic responses are mixed. This all means that treatment guidelines and protocols have to be reevaluated on a global basis considering the huge differences between Old and New World leishmaniasis [1], that the concept of monotherapy with regard to resistance has to be reevaluated, and that diagnosis and satisfactory treatment are imperative challenges for the adequate outcome in ATL, especially in an era of drug resistance.

Acknowledgments The authors are grateful for the financing support received from the Coordination for Research, Faculty of Medicine, UCV and the Council for Scientific and Humanistic Research (CDCH), Universidad Central de Venezuela. Likewise, they are grateful for the support conferred by the Alexander von Humboldt Foundation and the University of Würzburg through the Siebold-Collegium Institute for Advanced Studies, Germany, to Alicia Ponte-Sucre.

References

1. Goto H, Lindoso JA. Current diagnosis and treatment of cutaneous and mucocutaneous Leishmaniasis. Exp Rev Anti Infect Ther. 2010;8:419–33.

2. Souza AS, Giudice A, Pereira JM, Guimaraes LH, et al. Resistance of *Leishmania (Viannia) braziliensis* to nitric oxide: correlation with antimony therapy and TNF-alpha production. BMC Infect Dis. 2010;10:209.
3. Akilov OE, Khachemoune A, Hasan T. Clinical manifestations and classification of Old World cutaneous leishmaniasis. Int J Dermatol. 2007;46:132–42.
4. Zijlstra EE, Musa AM, Khalil EA, el-Hassan IM, et al. Post-kala-azar dermal leishmaniasis. Lancet Infect Dis. 2003;3:87–98.
5. World Health Organization. Control of the leishmaniasis. Technical report series 949. 2010.
6. Alvar J, Velez ID, Bern C, Herrero M, et al. Leishmaniasis worldwide and global estimates of its incidence. PLoS One. 2012;7(5):e35671.
7. World Health Organization. Wkly Epidemiol Rec. 2016;91:285–96.
8. Savoia D. Recent updates and perspectives on Leishmaniasis. J Infect Dev Ctries. 2015;9 (6):588–96.
9. Gautret P, Schlagenhauf P, Gaudart J, Castelli F, et al. Multicenter EuroTravNet/GeoSentinel study of travel-related infectious diseases in Europe and for the GeoSentinel Surveillance Network. Emerg Infect Dis. 2009;15:1783–90.
10. Oryan A, Akbari M. Worldwide risk factors in leishmaniasis. Asian Pac J Trop Med. 2016;9 (10):925–32.
11. Machado P, Araújo C, Da Silva AT, Almeida RP, et al. Failure of early treatment of cutaneous leishmaniasis in preventing the development of an ulcer. Clin Infect Dis. 2002;34:E69–73.
12. Feliciangeli MD. Vectors of leishmaniasis in Venezuela. Parassitologia. 1991;33:229–36.
13. Feliciangeli MD, Rodriguez N, Bravo A, Arias F, et al. Vectors of cutaneous leishmaniasis in north-central Venezuela. Med Vet Entomol. 1994;8:317–24.
14. Feliciangeli MD, Rabinovich J. Abundance of *Lutzomyia ovallensi* but not *Lu. Gomezi* (Diptera: Psychodidae) correlated with cutaneous leishmaniasis incidence in north-central Venezuela. Med Vet Entomol. 1998;12:121–31.
15. De Lima H, Rodriguez N, De Guglielmo Z, Rodriguez A, et al. Cotton rats and black rats as possible reservoirs of cutaneous leishmaniasis in an endemic area in Lara State, Venezuela. Mem Inst Oswaldo Cruz. 2002;97:169–74.
16. Gramiccia M, Gradoni L. The current status of zoonotic leishmaniases and approaches to disease control. Int J Parasitol. 2005;35(11–12):1169–80.
17. Ridley DS, Marsden PD, Cuba CC, Barreto AC. A histological classification of mucocutaneous leishmaniasis in Brazil and its clinical evaluation. Trans R Soc Trop Med Hyg. 1980;74:508.
18. Scarisbrick JJ, Chiodini PL, Watson J, Moody A, et al. Clinical features and diagnosis of 42 travellers with cutaneous leishmaniasis. Travel Med Infect Dis. 2006;4:14–21.
19. Convit J, Ulrich M, Fernández CT, Tapia FJ, et al. The clinical and immunological spectrum of American cutaneous leishmaniasis. Trans R Soc Trop Med Hyg. 1993;87:444–8.
20. Reithinger R, Dujardin JC, Louzir H, Pirmez C, et al. Cutaneous leishmaniasis. Lancet Infect Dis. 2007;7:581–96.
21. Restrepo Isaza M. La reacción de Montenegro en la epidemiología de la leishmaniasis sudamericana. Bol Of Sanit Panam. 1980;89:130.
22. Oliveira-Neto MP, Mattos M, Souza CS, Fernandes O, et al. Leishmaniasis recidiva cutis in New World cutaneous leishmaniasis. Internat J Dermatol. 1998;37:846–9.
23. Bittencourt AL, Costa JM, Carvalho EM, Barral A. Leishmaniasis recidiva cutis in American cutaneous leishmaniasis. Internat J Dermatol. 1993;32:802–5.
24. Calvopina M, Uezato H, Gomez EA, Korenaga M, et al. Leishmaniasis recidiva cutis due to *Leishmania (Viannia) panamensis* in subtropical Ecuador: isoenzymatic characterization. Internat J Dermatol. 2006;45:116–20.
25. Barral A, Costa JM, Bittencourt AL, Barral-Netto M, et al. Polar and subpolar diffuse cutaneous leishmaniasis in Brazil: clinical and immunopathologic aspects. Internat J Dermatol. 1995;34:474–9.

26. Ortega Moreno ME, Lugo DA, Belizario Ochoa DC, Galindo Martinez WA, et al. Comparación clínica de la Leishmaniasis Cutánea Difusa y Leishmaniasis Diseminada en Venezuela. Dermatol Venez. 2013;51:29–35.
27. Zerpa O, Convit J. Leishmaniasis Cutánea Difusa en Venezuela. Gaz Méd Bahía. 2009;79(Supl. 3):30–4.
28. Jiménez A, Vásquez DA, Albarracín N, Vélez ID. Leishmaniasis Diseminada en Colombia: Reporte de un caso. Dermatol Venez. 2012;50(2):46–9.
29. Tapia FJ, Cáceres-Dittmar G, Sánchez MA, Fernández CT, et al. Adhesion molecules in lesions of American cutaneous leishmaniasis. Exp Dermatol. 1994;3:17–22.
30. Castés M1, Tapia FJ. Immunopathology of American tegumentary leishmaniasis. Acta Cient Venez. 1998;49(1):42–56.
31. Ulrich M, Rodriguez V, Centeno M, Convit J. Differing antibody IgG isotypes in the polar forms of leprosy and cutaneous leishmaniasis characterized by antigen specific T cell anergy. Clin Exp Immunol. 1995;100:54–8.
32. Turetz ML, Machado PR, Ko AI, Alves F, et al. Disseminated leishmaniasis: a new and emerging form of leishmaniasis observed in northeastern Brazil. J Infect Dis. 2002;186:1829–34.
33. el-Hassan AM, Meredith SEO, Yagi HI, Khalil EAG, et al. Sudanese mucocutaneous leishmaniasis: epidemiology, clinical features, diagnosis, immune responses and treatment. Trans R Soc Trop Med Hyg. 1995;89:647–52.
34. Grimaldi G Jr, Tesh RB, McMahon-Pratt D. A review of the geographic distribution and epidemiology of leishmaniasis in the New World. American J Trop Med Hyg. 1989;41:687–725.
35. de Castro EA, Luz E, Telles FQ, Pandey A, et al. Eco-epidemiological survey of *Leishmania (Viannia) braziliensis* American cutaneous and mucocutaneous leishmaniasis in Ribeira Valley River, Parana State, Brazil. Acta Trop. 2005;93:141–9.
36. Jones TC, Johnson WD Jr, Barretto AC, Lago E, et al. Epidemiology of American cutaneous leishmaniasis due to *Leishmania braziliensis braziliensis*. J Infect Dis. 1987;156:73–83.
37. Davies CR, Reithinger R, Campbell-Lendrum D, Feliciangeli D, et al. The epidemiology and control of leishmaniasis in Andean countries. Cad Saude Publica. 2000;16:925–50.
38. Garcia AL, Parrado R, Rojas E, Delgado R, et al. Leishmaniases in Bolivia: comprehensive review and current status. Am J Trop Med Hyg. 2009;80:704–11.
39. Marsden PD. Mucocutaneous leishmaniasis ("Espundia" Escomel, 1911). Trans R Soc Trop Med Hyg. 1986;80:859–76.
40. Marsden PD, Nonata RH. Mucocutaneous leishmaniasis-a review of clinical aspects. Rev Soc Bras Med Trop. 1975;IX:325–6.
41. Sanchez MA, Caceres-Dittmar G, Oriol O, Mosca W, et al. Epidermal Langerhans cells and dendritic epidermal T cells in murine cutaneous leishmaniasis. Immunocytochemical study. Acta Microsc. 1993;2:180–7.
42. Moll H. The role of chemokines and accessory cells in the immunoregulation of cutaneous leishmaniasis. Behring Inst Mitt. 1997;99:73–8.
43. Moll H. The role of dendritic cells at the early stages of *Leishmania* infection. Adv Exp Med Biol. 2000;479:163–73.
44. Puig L, Pradinaud R. *Leishmania* and HIV co-infection: dermatological manifestations. Ann Trop Med Parasitol. 2003;97:107–14.
45. Stuart K, Brun R, Croft S, Fairlamb A, et al. Kinetoplastids: related protozoan pathogens, different diseases. J Clin Invest. 2008;118:1301–10.
46. Cuba CA, Marsden P, Barreto AC, Rocha R, et al. Diagnóstico parasitológico e inmunológico de leishmaniasis tegumentaria Americana. Rev Med Exp. 1980;17:1–4.
47. Weigle KA, De Davalos M, Heredia P, Molineros R, et al. Diagnosis of cutaneous and mucocutaneous leishmaniasis in Colombia: a comparison of seven methods. Am J Trop Med Hyg. 1987;36:489–96.
48. Reed SG. Diagnosis of leishmaniasis. Clin Dermatol. 1996;14:471–8.

49. Bray RS, Lainson R. The immunology and serology of leishmaniasis: the fluorescent antibody staining technique. Trans R Soc Trop Med Hyg. 1965;59:535–44.
50. Osman OF, Oskam L, Zijlstra EE, Kroon NC, et al. Evaluation of PCR for diagnosis of visceral leishmaniasis. J Clin Microbiol. 1997;35(10):2454–7.
51. Rodríguez N, Guzman B, Rodas A, Takiff H, et al. Diagnosis of cutaneous leishmaniasis and species discrimination of parasites by PCR and hybridization. J Clin Microbiol. 1994;32 (9):2246–52.
52. Sassi A, Louzir H, Ben Salah A, Mokni M, et al. Leishmanin skin test lymphoproliferative responses and cytokine production after symptomatic or asymptomatic *Leishmania major* infection in Tunisia. Clin Exp Immunol. 1999;116:127–32.
53. Shaw JJ, Lainson R. Leishmaniasis in Brazil: X. Some observations of intradermal reactions to different trypanosomatid antigens of patients suffering from cutaneous and mucocutaneous leishmaniasis. Trans R Soc Trop Med Hyg. 1975;69:323–35.
54. Convit J, Castellanos PL, Ulrich M, Castes M, et al. Immunotherapy of localized, intermediate, and diffuse forms of American cutaneous leishmaniasis. J Infect Dis. 1989;160:104–15.
55. Convit J. Leishmaniasis immunological and clinical aspects and vaccines in Venezuela. Clin Dermatol. 1996;14:479–87.
56. Yardley V, Ortuno N, Llanos-Cuentas A, Chappuis F, et al. American tegumentary leishmaniasis: is antimonial treatment outcome related to parasite drug susceptibility? J Infect Dis. 2006;194:1168–75.
57. Yardley V, Croft SL, De Doncker S, Dujardin JC, et al. The sensitivity of clinical isolates of *Leishmania* from Peru and Nepal to miltefosine. Am J Trop Med Hyg. 2005;73:272–5.
58. Rijal S, Yardley V, Chappuis F, Decuypere S, et al. Antimonial treatment of visceral leishmaniasis: are current in vitro susceptibility assays adequate for prognosis of in vivo therapy outcome? Microbes Infect. 2007;9:529–35.
59. Ponte-Sucre A. Physiological consequences of drug resistance in *Leishmania* and their relevance for chemotherapy. Kinetoplastid Biol Dis. 2003;2:14.
60. Natera S, Machuca C, Padrón-Nieves M, Romero A, et al. *Leishmania sp.*: proficiency of drug resistant parasites. Int J Antimicrob Agents. 2007;29:637–42.
61. Padrón-Nieves M, Díaz E, Romero A, Machuca C, et al. Valor pronóstico de los cambios fisiológicos asociados a la quimio-resistencia en *Leishmania*. VITAE Academia Biomédica Digital (33). 2007.
62. Padrón-Nieves M, Machuca C, Díaz E, Cotrim P, et al. Correlation between glucose uptake and membrane potential in *Leishmania* parasites isolated from DCL patients with therapeutic failure: a proof of concept. Parasitol Res. 2014;113(6):2121–8.
63. Padrón-Nieves M, Ponte-Sucre A. Marcadores de resistencia en *Leishmania*: susceptibilidad in vitro a drogas leishmanicidas vs. retención de calceina en aislados de pacientes venezolanos con leishmaniasis cutanea difusa. Arch Venez Farmacol y Ter. 2015;32:29–33.
64. Vanaerschot M, Dumetz F, Roy S, Ponte-Sucre A, et al. Treatment failure in leishmaniasis: drug-resistance or another (epi) phenotype? Expert Rev Anti Infect Ther. 2014;12:937–46.

The Challenges of Effective Leishmaniasis Treatment

9

Sarah Hendrickx, Louis Maes, Simon L. Croft, and Guy Caljon

Abstract

During the past decades, visceral leishmaniasis therapy has been faced with the rapid emergence of drug resistance against the pentavalent antimonials which had been used as mainstay of treatment for over 70 years. Even though cutaneous leishmaniasis cannot be linked to development of drug resistance, the huge species- and strain-specific variations in drug susceptibilities severely complicate effective treatment as well. A new challenge in leishmaniasis control has arisen with increasing numbers of treatment failures against all of the currently used anti-leishmanial standard drugs. The exact causes of these treatment failures are still not fully comprehended, but they are most likely a consequence of the complex interplay between parasite, host and drug. In this chapter, the generally accepted underlying factors of treatment failure are discussed along with their consequences for therapy, drug design and other related challenges.

S. Hendrickx · L. Maes · G. Caljon (✉)
Laboratory for Microbiology, Parasitology and Hygiene (LMPH), University of Antwerp, Wilrijk (Antwerp), Belgium
e-mail: Guy.Caljon@uantwerpen.be

S. L. Croft
Department of Immunology and Infection, Faculty of Infectious and Tropical Diseases, London School of Hygiene and Tropical Medicine, London, UK

A. Ponte-Sucre, M. Padrón-Nieves (eds.), *Drug Resistance in Leishmania Parasites*,
https://doi.org/10.1007/978-3-319-74186-4_9

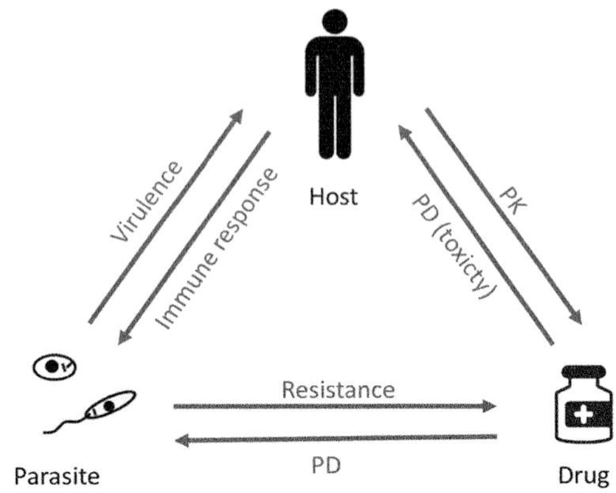

Fig. 9.1 Pharmacodynamic (PD)/pharmacokinetic (PK) relationship between host, drug and pathogen [based on 6]

9.1 Introduction

9.1.1 Treatment Failure

One of the greatest challenges in the control of cutaneous (CL) and visceral leishmaniasis (VL) remains the proportion of treatment failures occurring after drug treatment. The reported failure rates upon both VL and CL treatment with the current anti-leishmanial reference drugs have increased significantly during the past decade [1–5]. Although these treatment failure cases are often falsely linked to refractoriness of the causative *Leishmania* isolate to the drug, it has only been associated with 'acquired' drug resistance in a minority of cases. An unfavourable treatment outcome frequently does not merely result from the parasite's drug susceptibility profile but can be related to the wide variety of factors arising from the complex interplay between drug, parasite and host during chemotherapeutic intervention (Fig. 9.1). Generally, treatment failure (TF; see also Chaps. 4 and 15) is characterized by patients not responding to a given therapy or presenting a relapse within a few months after the completion of treatment. TF is particularly common in disease-endemic areas where repeated drug treatment schedules are needed, hence enhancing the risk of more rapid induction of drug resistance. Although most cases of TF have a multifactorial origin, some of the causes can either be specifically linked to drug-, parasite- or host-related factors, as listed below.

9.2 Direct and Indirect Causes of Treatment Failure

9.2.1 Parasite-Related Factors

One of the most straightforward causes of treatment relapse is a low **intrinsic drug susceptibility** of the *Leishmania* species involved. There is a significant variation in intrinsic susceptibility between the 17 *Leishmania* species that are infectious to man, resulting in the possible co-existence of both 'intrinsic' and 'acquired' drug resistances [7]. While intrinsic resistance generally implies a lack of initial response of the parasite to a drug, acquired drug resistance involves selection of less susceptible or resistant parasites upon drug exposure. Acquired resistance is mostly linked not only to successive treatment cycles with the same drug resulting in relapses but also due to underdosing, incomplete treatment related to poor therapy adherence or non-compliance, poor drug quality or 'single-drug' use [8–13]. It is important to note that when the parasite's pretreatment drug susceptibility is relatively low, it may be more prone to develop acquired resistance upon subsequent drug exposure. Clearly defining drug 'resistance' (DR) in the characterization of clinical isolates and the efficacy of anti-leishmanial drugs remains very difficult mainly because of the lack of validated standard operating procedures and clearly defined species-specific 'breakpoint concentrations' that should be validated on a large number of clinical isolates from primary unresponsive patients [7, 14]. Over the past years, **drug resistance** in relation to the currently used anti-leishmanial reference drugs has increasingly been reported, the most well-known example being the huge rise in the number of antimony (Sb^V) unresponsive cases in the Indian subcontinent (ISC) since the 1970s (see also Chaps. 4 and 7). Nowadays, more than 65% of the patients in India, Nepal or Bangladesh no longer respond to Sb^V therapy [15], which enforced the implementation of other drugs such as miltefosine (MIL) and a liposomal formulation of amphotericin-B (AMB) in the regional Kala-azar elimination programme [16]. In the past 5 years, a few reports on MIL and AMB resistance in clinical isolates have also surfaced [17–20]. However, most isolates that are obtained after treatment relapse still appear to be drug-susceptible in the routine laboratory susceptibility assays [1].

Nowadays, a potential role of the **parasite (epi-)phenotype** in TF is being suggested as well (see also Chap. 15). This (epi-)phenotype comprises all parasite factors other than drug resistance, such as species-specific reduced drug susceptibility, parasite infectivity and aberrant or atypical interactions with the host immune system [21]. For example, MIL TF has been associated with an increased infectivity of relapse isolates [22], while Sb^V resistance was shown to entail a fitness advantage with regard to metacyclogenesis, infectivity and virulence in vitro and in vivo [23–25]. Also, MIL resistance has been linked to changes in parasite fitness, although no consensus has yet been reached on the specific effect on the parasite. Previous research using *L. (L.) infantum* revealed a decrease in virulence associated with MIL resistance [26], while another study on *L. (L.) major* demonstrated a fitness gain [27], suggesting that fitness effects may either be specific to the resistance mechanism that can differ among the species or be dependent on the parasite stage that was used during the experimental selection of resistance. The latter is particularly true for paromomycin (PMM), as the outcome of the resistance selection procedure was clearly stage-specific [28]. Given this definite link between fitness and drug resistance, increased parasite fitness may well be responsible for variations in baseline drug susceptibility and the associated TF.

More recently, it has also been demonstrated in mouse infection models that infecting *Leishmania* populations do not necessarily behave homogeneously in terms of **replication potential and metabolic state**: intracellular amastigotes have been identified showing different rates of division and metabolic activity, of which some can be called 'quiescent' [29–32]. If these latter forms are also present in clinical VL or CL, they could in principle form the basis of a latent infection state and negatively impact on drug efficacy.

9.2.2 Drug-Related Factors

Obviously, the **quality of the drug** plays a determining role for treatment success. There are a lot of counterfeit and substandard drugs being sold in endemic countries that are intentionally mislabelled and often contain subtherapeutic concentrations of the active ingredient or even no active drug at all [10, 11]. This malpractice not only decreases or completely abolishes the overall responsiveness to treatment but also endangers future application given the enhanced risk of emerging drug resistance.

As VL is mostly endemic in subtropical and tropical areas, **drug formulation and stability** in warm and humid zone four environments[1] are very important to retain drug efficacy [33]. For example, the liposomal formulation of AmB (L-AmB; AmBisome®) is highly effective against VL and is currently recommended as first-line therapy in endemic areas with Sb^V resistance [34]. However, the need to ship and store the drug continuously below 25 °C to keep the intravenous formulation stable makes it very challenging to widely use this formulation in tropical areas. At the moment, an oral 'lipid particle' formulation of AmB that is thermally stable and retains excellent efficacy in animal models is being developed [35].

Unlike in antimicrobial research [6], yet another challenge is that the **pharmacokinetic and pharmacodynamic (PK/PD)** properties of most anti-leishmanial compounds have been very poorly documented. Understanding how drugs are working at the target site (PD) and how they behave in animal models and human patients (PK) will certainly help to better define therapeutic efficacy. First of all, fairly little is known about the drug accumulation within the acidic environment of the phagolysosome, while the overall intracellular drug concentration in the macrophage may still be too limited to predict time and concentration kinetics on top of the variety of factors involved in the drug distribution at the level of the patient. Among others, protein binding plays a determining role as it directly influences the availability for macrophage uptake and subsequent activity on the intracellular parasite [36–38]. Since a link between changes in pH and antibacterial activity was already demonstrated for PMM, knowledge on accumulation and/or transport of drugs into the macrophage and the acidic phagolysosome environment becomes essential. However, the drug's pharmacokinetic characteristics still depend too much on measuring the drug concentrations in plasma, while these do not necessarily reflect the actual intracellular drug concentrations to which the parasite is exposed. In spite

[1]Regions which have a mean annual temperature > 22 °C measured in the open air combined with a mean annual partial water vapour pressure > 15–27 hPa.

of these limitations, plasma levels still steer current dosage schemes in vivo. The development of a successful treatment schedule for CL is further burdened by the unknown distribution of the drug to the skin. Most of the currently available drugs are used to treat both VL and CL [39], although generally higher doses and longer treatment courses are needed in CL therapy. This is not surprising given the drug's indirect access to the skin via blood and lymph and the lower rates of blood flow and oxygen tension in the skin compared to the viscera. In this view, the design of topical formulations for CL would be more ideal. Past topical formulation studies were usually faced with the insufficient delivery of the anti-leishmanial reference drugs at the target site [40] and already suggested that a critical exposure phase in the dermis is an obligatory characteristic for an effective topical formulation in vivo [41]. The challenging drug distribution between viscera and the skin is further complicated by the differences observed between uninfected and infected tissues. For example, it was demonstrated that permeation markers such as caffeine and ibuprofen, as well as some anti-leishmanial drugs, have different in vitro permeation properties through normal mouse skin compared to mouse skin removed from a CL-infected nodule [42], indicating the occurrence of some infection-dependent changes in PK properties that could be related to the oedema and the altered immunological profile in the inflamed skin. In experimental models of VL treated with AmBisome®, organ enlargement and other pathophysiological factors also cause differences in drug distribution and elimination between the liver and spleen [43]. Yet another very important factor in PK/PD is whether a drug acts in a time-dependent or concentration-dependent manner, as this will impact on the dosing regimen. While the treatment schedule should be prolonged for drugs with a time-dependent mode of action, the dose for drugs with a concentration-dependent activity should be increased [37]. For example, the dose of PMM was adapted in response to the lower efficacy rates observed in East Africa rather than prolonging the treatment duration [37, 44]. Finally, PD/PK modelling for the current anti-leishmanial reference drugs becomes even more complicated by the various host-related factors. In the past, suboptimal drug exposure, showing plasma concentrations lower than anticipated, has been linked to incorrect dosing of Indian and Nepalese males and children during MIL therapy [45, 46] which endorses why attention must be paid to drug exposure kinetics, either by quantifying the drug plasma levels [47] or by a using the dried blood spot method to quantify MIL concentrations in treated patients [48].

Understandably, **poor adherence or non-compliance** [1, 12], which are often falsely used as synonyms, can also be involved in TF. Although they both lead to suboptimal drug exposure, adherence to drug treatment is linked to the extent to which the patient's drug uptake corresponds with the agreed recommendations by a health-care provider, while compliance rather involves patient inactivity [49].

Other drug-related factors that can facilitate TF are long **treatment schedules** [50, 51] and **intrinsic drug properties** such as a long elimination half-life ($t_{1/2}$) [52], which both may result in parasite exposure to prolonged suboptimal drug concentrations.

9.2.3 Host-Related Factors

Some patient groups are more prone to TF than others with the most important factor being the **host immune system**. Immunocompromised patients, such as HIV-positive people, children and elderly, are more likely to experience a relapse than immunocompetent patients [45, 53]. Moreover, HIV/VL co-infection in humans is often associated with the occurrence of parasites in other tissues [53–55], while the absence of an effective immune response has been shown to decrease drug efficacy in in vivo experiments. In case of HIV/VL co-infection, anti-leishmanial therapy is often combined with highly active antiretroviral therapy (HAART). Analogous to treatment of malaria and tuberculosis where drug–drug interactions have been well characterized (www.hiv-druginteractions.org), understanding interactions between anti-leishmanial and antiretroviral drugs could help in designing more effective treatments.

Some risk factors associated with relapse after VL treatment in immunocompetent patients are **gender, age and specific clinical signs,** e.g. a smaller decrease in splenomegaly at discharge and the time between onset of symptoms and the start of treatment [56]. Additionally, **geographic region** plays a key role. The currently used reference drugs (MIL, PMM and L-AmB) are less effective against African VL (e.g. in Ethiopia and Sudan) compared to Indian VL, resulting in higher rates of treatment relapse [57–59]. For MIL and Sb^V, this has resulted in the implementation of deviating dosing schemes based on the geographical region and the immune status of the host [59].

9.3 Experimental Approaches to Predict Treatment Outcome In Vitro

As the pharmacokinetic, pharmacodynamic and host immune phenomena largely define the response of a pathogen to a drug, predicting treatment outcome merely based on in vitro data only remains virtually impossible. Nonetheless, some in vitro assays may provide useful information which can lead to making early predictions towards the in vivo behaviour.

9.3.1 Drug Susceptibility Determination

9.3.1.1 In Vitro

Drug susceptibility testing of clinical isolates would logically be the most convenient method to predict treatment outcome, as has been established for antibiotics in bacterial and fungal infections [60, 61] and for malaria [62]. As for most anti-leishmanial drugs, the exact mechanism of action and the mechanisms of resistance are not completely elucidated, it still remains difficult to link in vitro drug susceptibility values from isolates from patients that are cured, relapsed or did not respond to treatment to the corresponding treatment outcome. For Sb^V, a positive correlation could be found between the in vitro drug susceptibility profile and the actual treatment outcome of patients ranging from susceptible (S/S) over intermediate (R/S) to resistant (R/R). In this case, R/R cases could be linked to nonresponders or relapse cases, while S/S strains could be linked to cure. The intermediate R/S profile could even be linked to an increased risk for R/R development [63]. A small pilot study with MIL using Nepalese VL strains suggested of a correlation between patient treatment outcome and the in vitro survival under in vitro drug pressure using promastigote back transformation. Further validation on a larger sample set unfortunately failed to corroborate these preliminary results [64].

For strains or species causing mucocutaneous (MCL), diffuse cutaneous leishmaniasis (DCL) or post-Kala-azar dermal leishmaniasis (PKDL), prediction of treatment outcome based on in vitro susceptibility results is even harder to do. Although one study claims to observe differences in promastigote suceptibility to predict Sb^V TF for CL and MCL [65], there is little evidence that clinical isolates from relapse cases have an altered drug susceptibility. As their disease pathology rather originates from the induced host immune response, these forms of the disease are generally more difficult to treat and require long treatment courses and the application of drug combinations [66, 67]. Given the role of host-dependent immune responses in the clinical manifestation and disease pathology, the geographical factor needs to be taken into account with African VL found to be far less responsive to treatment than Indian VL despite a comparable in vitro susceptibility [68].

9.3.1.2 In Vivo

In the past, various laboratory animal studies have been explored to predict PK/PD characteristics of drug candidates [69–71]. Although such studies do consider the host's immunity, there is still no validated model fully representative of human disease. Several CL mouse models have been described which, depending on the mouse species, show a variable susceptibility for infection, a diverse disease progression and an outcome characterized by the induction of different immune cascades [72]. An additional obscuring factor is the variable disease patterns caused by the different species responsible for CL. The same problems also apply for VL where different mouse models are associated with differential progression in various tissues, although they are all linked to a transient and self-curing disease pattern [72]. The Syrian golden hamster model is characterized by a progressive disease and a symptomatology that fairly resembles human VL (heptosplenomegaly, weight loss, fatal outcome). Unfortunately, not much is known about the specific immune responses as only few specific anti-hamster antibodies/reagents are available [73]. Designing an animal model that is able to predict the relationship between the drug concentration in the plasma and in the infected tissues would certainly be very helpful. However, even then one may still seriously question the translational capacities of such animal model to human disease.

9.3.2 Studying Drug-Resistance Mechanisms and Drug-Related Physiological Modifications

Research into the previously mentioned drug-induced alterations in parasite physiology can certainly help in predicting the success of a given therapy. By comparing resistant clinical isolates or generating resistant laboratory mutants, the mechanisms underlying drug resistance can be unravelled and allow to identify putative biomarkers of resistance, avoid the emergence of resistance and define strategies to combat the already existing resistance in the field. Generally, acquired drug resistance is associated with either gene amplification of drug target enzymes, structural and functional modifications of drug target enzymes or transporters decreasing intracellular drug concentrations [74, 75]. Research into the (epi-) phenotypic modifications associated with drug resistance can serve as an attractive tool to predict parasite behaviour under drug pressure [21].

9.3.3 Detection of In Vivo Sanctuary Sites

A few reports already stated that *Leishmania* might reside in specific yet undefined sanctuary sites upon drug treatment [21, 22]. Inside these hidden niches, parasites could be protected from drug exposure, allowing parasite survival during drug exposure and subsequent re-proliferation. Several suggestions have been made concerning the nature of these parasitic safe havens during drug therapy, but research so far failed to unequivocally identify such a 'sanctuary site' [22]. One of the most important problems is the lack of sensitivity of the currently available laboratory techniques to detect very low residual parasite burdens in experimental infection models.

9.3.3.1 Pathology

Gross pathology can be used to assess the parasite burdens inside target and off-target organs and tissues. After fixation of whole organs, residual parasites present in that specific specimen can be detected by microscopy. Although detection limits are usually somewhat lower compared to modern molecular techniques [76, 77], this technique allows evaluation of possible changes in organ architecture upon drug treatment or infection, which can then be linked to drug access and activity [78].

9.3.3.2 Bio-Imaging

In the past, several reporter strains have been developed allowing in vivo visualization of the parasite inside its target organs. Although these are particularly useful in studies on infection, pathology and chemotherapy, these models have not yet been fully exploited in terms of analysing key properties of anti-leishmanial drug action. With high-resolution imaging, far more information could be extracted in the future on the in vivo rate of parasite killing and on differences in drug activity between different target tissues. The conventional luciferase (LUC)- and red-shifted luciferase (PpyRE9) transfected strains represent promising tools in laboratory animals to detect potential parasite reservoirs upon drug treatment as they allow longitudinal follow-up of infection after treatment [79]. The rapid visualization of dividing parasites at specific target sites at the start of treatment relapse could unravel these specific parasite niches and expose the real problem sites under drug pressure. Although for *Trypanosoma* detection limits of around 100 parasites have been reported [80], it will be challenging to obtain similar sensitivities for *Leishmania* within the phagolysosomal compartment of macrophages determined by the intra-parasite expression levels of the reporter gene [81].

9.3.3.3 qPCR

Molecular techniques have gained considerable momentum in *Leishmania* diagnosis and detect parasite DNA or RNA in a very specific way with high sensitivity compared to the conventional microscopic techniques [76, 82]. While several target sequences can be used, the applicability of a kDNA-based assay has been reported most frequently [83–85]. This type of assay uses the conserved region of the *Leishmania* kinetoplast DNA minicircles as a specific target, which is possible in both conventional and quantitative PCR assays. Although the use of kDNA-based PCR assays has already been suggested for predicting relapse or parasite reactivation after treatment [84], some studies are aiming to identify even more sensitive targets to identify the parasite sanctuary sites and predict relapse [85].

9.3.4 Prediction of Drug Dynamics In Vitro

A few novel in vitro assays have been explored to gain insight into the specific aspects of the drug dynamics, more particularly aiming at determining the time-dependent activity of anti-leishmanial reference drugs. The ‘time-to-kill’ of a given compound was defined as the time required to kill 100% of the parasites based upon microscopic assessment of Giemsa-stained drug-treated infected cells combined with a >95% reduction in the promastigote back-transformation assay [86]. A host cell-dependent drug action was already demonstrated in vitro for some anti-leishmanial reference compounds [87] and implies a cellular pharmacokinetic role resulting from differences in drug metabolism and accumulation but still suffers from additional differences between infected and uninfected cells [88]. Although any additional impact of the patient’s immune system is largely ignored in any in vitro system, the required information on the compound’s time-dependent cidal action in vitro could serve as a first step in understanding in vivo action dynamics. Of course, expanding such in vitro assays to make in vivo predictions would definitely be challenging as the amastigote division rate inside in vitro macrophages and animal models still needs to be further evaluated [29, 30]. While previous research in *Mycobacteria* models already demonstrated whether drugs were active against dividing or non-dividing bacteria, the first studies characterizing *Leishmania* spp. survival and multiplication in vitro and in vivo have only just started [29, 30, 89]. A second assay that has been mentioned to predict drug dynamics is looking at the reversibility of action and evaluates whether the drug effect can be alleviated upon its removal after different exposure times, providing complementary information on the drug's time-dependent mechanism of killing [90].

References

1. Rijal S, Ostyn B, Uranw S, Rai K, et al. Increasing failure of miltefosine in the treatment of kala-azar in Nepal and the potential role of parasite drug resistance, reinfection, or noncompliance. Clin Infect Dis. 2013;56(11):1530–8.
2. Mueller M, Ritmeijer K, Balasegaram M, Koummuki Y, et al. Unresponsiveness to AmBisome in some Sudanese patients with kala-azar. Trans R Soc Trop Med Hyg. 2007;101(1):19–24.
3. Salih NA, van Griensven J, Chappuis F, Antierens A, et al. Liposomal amphotericin B for complicated visceral leishmaniasis (kala-azar) in eastern Sudan: how effective is treatment for this neglected disease? Tropical Med Int Health. 2014;19(2):146–52.
4. Sundar S, More DK, Singh MK, Singh VP, et al. Failure of pentavalent antimony in visceral leishmaniasis in India: report from the center of the Indian epidemic. Clin Infect Dis. 2000;31 (4):1104–7.
5. Uliana SR, Trinconi CT, Coelho AC. Chemotherapy of leishmaniasis: present challenges. Parasitology. 2017;20:1–17.
6. Asin-Prieto E, Rodriguez-Gascon A, Isla A. Applications of the pharmacokinetic/pharmacodynamic (PK/PD) analysis of antimicrobial agents. J Infect Chemother. 2015;21(5):319–29.
7. Maes L, Cos P, Croft S. The relevance of susceptibility tests, breakpoints and markers. In: Ponte-Sucre A, Diaz E, Padrón-Nieves M, editors. Drug resistance in *Leishmania* parasites. Vienna: Springer; 2013. p. 407–29.
8. Bryceson A. A policy for leishmaniasis with respect to the prevention and control of drug resistance. Tropical Med Int Health. 2001;6(11):928–34.
9. Senior K. Global health-care implications of substandard medicines. Lancet Infect Dis. 2008;8 (11):666.
10. Dorlo TP, Eggelte TA, de Vries PJ, Beijnen JH. Characterization and identification of suspected counterfeit miltefosine capsules. Analyst. 2012;137(5):1265–74.
11. Dorlo TP, Eggelte TA, Schoone GJ, de Vries PJ, et al. A poor-quality generic drug for the treatment of visceral leishmaniasis: a case report and appeal. PLoS Negl Trop Dis. 2012;6(5): e1544.
12. Uranw S, Ostyn B, Dorlo TP, Hasker E, et al. Adherence to miltefosine treatment for visceral leishmaniasis under routine conditions in Nepal. Tropical Med Int Health. 2013;18(2):179–87.
13. Caudron JM, Ford N, Henkens M, Mace C, et al. Substandard medicines in resource-poor settings: a problem that can no longer be ignored. Tropical Med Int Health. 2008;13 (8):1062–72.
14. Hendrickx S, Guerin PJ, Caljon G, Croft SL, et al. Evaluating drug resistance in visceral leishmaniasis: the challenges. Parasitology. 2016;109:1–11.
15. Frezard F, Demicheli C, Ribeiro RR. Pentavalent antimonials: new perspectives for old drugs. Molecules. 2009;14(7):2317–36.
16. Dhillon GP, Sharma SN, Nair B. Kala-azar elimination programme in India. J Indian Med Assoc. 2008;106(10):664, 6–8.
17. Purkait B, Kumar A, Nandi N, Sardar AH, et al. Mechanism of amphotericin B resistance in clinical isolates of *Leishmania donovani*. Antimicrob Agents Chemother. 2012;56(2):1031–41.
18. Cojean S, Houze S, Haouchine D, Huteau F, et al. *Leishmania* resistance to miltefosine associated with genetic marker. Emerg Infect Dis. 2012;18(4):704–6.
19. Hendrickx S, Boulet G, Mondelaers A, Dujardin JC, et al. Experimental selection of paromomycin and miltefosine resistance in intracellular amastigotes of *Leishmania donovani* and *L. infantum*. Parasitol Res. 2014;113(5):1875–81.

20. Srivastava S, Mishra J, Gupta AK, Singh A, et al. Laboratory confirmed miltefosine resistant cases of visceral leishmaniasis from India. Parasit Vectors. 2017;10(1):49.
21. Vanaerschot M, Dumetz F, Roy S, Ponte-Sucre A, et al. Treatment failure in leishmaniasis: drug-resistance or another (epi-) phenotype? Expert Rev Anti-Infect Ther. 2014;12(8):937–46.
22. Rai K, Cuypers B, Bhattarai NR, Uranw S, et al. Relapse after treatment with miltefosine for visceral leishmaniasis is associated with increased infectivity of the infecting *Leishmania donovani* strain. MBio. 2013;4(5):e00611–e006113.
23. Ouakad M, Vanaerschot M, Rijal S, Sundar S, et al. Increased metacyclogenesis of antimony-resistant *Leishmania donovani* clinical lines. Parasitology. 2011;138(11):1392–9.
24. Vanaerschot M, De Doncker S, Rijal S, Maes L, et al. Antimonial resistance in *Leishmania donovani* is associated with increased *in vivo* parasite burden. PLoS One. 2011;6(8):e23120.
25. Vanaerschot M, Maes I, Ouakad M, Adaui V, et al. Linking *in vitro* and *in vivo* survival of clinical *Leishmania donovani* strains. PLoS One. 2010;5(8):e12211.
26. Hendrickx S, Beyers J, Mondelaers A, Eberhardt E, et al. Evidence of a drug-specific impact of experimentally selected paromomycin and miltefosine resistance on parasite fitness in *Leishmania infantum*. J Antimicrob Chemother. 2016;71(7):1914–21.
27. Turner KG, Vacchina P, Robles-Murguia M, Wadsworth M, et al. Fitness and phenotypic characterization of miltefosine-resistant *Leishmania major*. PLoS Negl Trop Dis. 2015;9(7): e0003948.
28. Hendrickx S, Inocencio da Luz RA, Bhandari V, Kuypers K, et al. Experimental induction of paromomycin resistance in antimony-resistant strains of *L. donovani*: outcome dependent on *in vitro* selection protocol. PLoS Negl Trop Dis. 2012;6(5):e1664.
29. McConville MJ, Saunders EC, Kloehn J, Dagley MJ. *Leishmania* carbon metabolism in the macrophage phagolysosome - feast or famine? F1000Res. 2015;4(F1000 Faculty Rev):938.
30. Saunders EC, Ng WW, Kloehn J, Chambers JM, et al. Induction of a stringent metabolic response in intracellular stages of *Leishmania mexicana* leads to increased dependence on mitochondrial metabolism. PLoS Pathog. 2014;10(1):e1003888.
31. Mandell MA, Beverley SM. Continual renewal and replication of persistent *Leishmania major* parasites in concomitantly immune hosts. Proc Natl Acad Sci U S A. 2017;114(5):E801–10.
32. Kloehn J, Saunders EC, O'Callaghan S, Dagley MJ, et al. Characterization of metabolically quiescent *Leishmania* parasites in murine lesions using heavy water labeling. PLoS Pathog. 2015;11(2):e1004683.
33. Grimm W. Storage conditions for stability testing in the EC, Japan and USA; the most important market for drug products. Drug Dev Ind Pharm. 1993;19(20):2795–830.
34. Sundar S, Mehta H, Suresh AV, Singh SP, et al. Amphotericin B treatment for Indian visceral leishmaniasis: conventional versus lipid formulations. Clin Infect Dis. 2004;38(3):377–83.
35. Wasan KM, Wasan EK, Gershkovich P, Zhu X, et al. Highly effective oral amphotericin B formulation against murine visceral leishmaniasis. J Infect Dis. 2009;200(3):357–60.
36. Dorlo TP, Balasegaram M, Beijnen JH, de Vries PJ. miltefosine: a review of its pharmacology and therapeutic efficacy in the treatment of leishmaniasis. J Antimicrob Chemother. 2012;67 (11):2576–97.
37. Levison ME, Levison JH. Pharmacokinetics and pharmacodynamics of antibacterial agents. Infect Dis Clin N Am. 2009;23(4):791–815.
38. Wijnant GJ, Van Bocxlaer K, Yardley V, Murdan S, et al. Efficacy of paromomycin-chloroquine combination therapy in experimental cutaneous leishmaniasis. Antimicrob Agents Chemother. 2017;61(8). https://doi.org/10.1128/AAC.00358-17. pii: e00358-17.
39. Sundar S, Chakravarty J. An update on pharmacotherapy for leishmaniasis. Expert Opin Pharmacother. 2015;16(2):237–52.
40. Van Bocxlaer K, Yardley V, Murdan S, Croft SL. Topical formulations of miltefosine for cutaneous leishmaniasis in a BALB/c mouse model. J Pharm Pharmacol. 2016;68(7):862–72.
41. Garnier T, Mantyla A, Jarvinen T, Lawrence J, et al. *In vivo* (coloque en cursiva) studies on the antileishmanial activity of buparvaquone and its prodrugs. J Antimicrob Chemother. 2007;60 (4):802–10.

42. Van Bocxlaer K, Yardley V, Murdan S, Croft SL. Drug permeation and barrier damage in *Leishmania*-infected mouse skin. J Antimicrob Chemother. 2016;71(6):1578–85.
43. Voak AA, Harris A, Qaiser Z, Croft SL, et al. Treatment of experimental visceral leishmaniasis with single-dose liposomal amphotericin B – pharmacodynamics and biodistribution at different stages of disease. Antimicrob Agents Chemother. 2017. https://doi.org/10.1128/AAC.00497-17. pii: AAC.00497-17.
44. Hailu A, Musa A, Wasunna M, Balasegaram M, et al. Geographical variation in the response of visceral leishmaniasis to paromomycin in East Africa: a multicentre, open-label, randomized trial. PLoS Negl Trop Dis. 2010;4(10):e709.
45. Ostyn B, Hasker E, Dorlo TP, Rijal S, et al. Failure of miltefosine treatment for visceral leishmaniasis in children and men in South-East Asia. PLoS One. 2014;9(6):e100220.
46. Dorlo TP, Rijal S, Ostyn B, de Vries PJ, et al. Failure of miltefosine in visceral leishmaniasis is associated with low drug exposure. J Infect Dis. 2014;210(1):146–53.
47. Castro MD, Gomez MA, Kip AE, Cossio A, et al. Pharmacokinetics of miltefosine in children and adults with cutaneous leishmaniasis. Antimicrob Agents Chemother. 2017;61(3). pii: e02198-16.
48. Kip AE, Rosing H, Hillebrand MJ, Blesson S, et al. Validation and clinical evaluation of a novel method to measure miltefosine in leishmaniasis patients using dried blood spot sample collection. Antimicrob Agents Chemother. 2016;60(4):2081–9.
49. Burkhart PV, Sabate E. Adherence to long-term therapies: evidence for action. J Nurs Scholarsh. 2003;35(3):207.
50. Geli P, Laxminarayan R, Dunne M, Smith DL. "One-size-fits-all"? Optimizing treatment duration for bacterial infections. PLoS One. 2012;7(1):e29838.
51. Croft SL, Olliaro P. Leishmaniasis chemotherapy—challenges and opportunities. Clin Microbiol Infect. 2011;17(10):1478–83.
52. Hastings IM, Watkins WM, White NJ. The evolution of drug-resistant malaria: the role of drug elimination half-life. Philos Trans R Soc Lond Ser B Biol Sci. 2002;357(1420):505–19.
53. Jarvis JN, Lockwood DN. Clinical aspects of visceral leishmaniasis in HIV infection. Curr Opin Infect Dis. 2013;26(1):1–9.
54. Nissapatorn V, Sawangjaroen N. Parasitic infections in HIV infected individuals: diagnostic & therapeutic challenges. Indian J Med Res. 2011;134(6):878–97.
55. Lindoso JA, Cota GF, da Cruz AM, Goto H, et al. Visceral leishmaniasis and HIV coinfection in Latin America. PLoS Negl Trop Dis. 2014;8(9):e3136.
56. Burza S, Sinha PK, Mahajan R, Lima MA, et al. Risk factors for visceral leishmaniasis relapse in immunocompetent patients following treatment with 20 mg/kg liposomal amphotericin B (Ambisome) in Bihar, India. PLoS Negl Trop Dis. 2014;8(1):e2536.
57. Monge-Maillo B, López-Vélez R. Miltefosine for visceral and cutaneous leishmaniasis: drug characteristics and evidence-based treatment recommendations. Clin Infect Dis. 2015;60(9):1398–404.
58. Musa AM, Younis B, Fadlalla A, Royce C, et al. Paromomycin for the treatment of visceral leishmaniasis in Sudan: a randomized, open-label, dose-finding study. PLoS Negl Trop Dis. 2010;4(10):e855.
59. Sundar S, Chakravarty J. Liposomal amphotericin B and leishmaniasis: dose and response. J Global Infect Dis. 2010;2(2):159–66.
60. Boothe DM. Interpreting culture and susceptibility data in critical care: perks and pitfalls. J Vet Emerg Crit Care (SanAntonio). 2010;20(1):110–31.
61. Rex JH, Goldberger M, Eisenstein BI, Harney C. The evolution of the regulatory framework for antibacterial agents. Ann N Y Acad Sci. 2014;1323:11–21.
62. Duru V, Khim N, Leang R, Kim S, et al. *Plasmodium falciparum* dihydroartemisinin-piperaquine failures in Cambodia are associated with mutant K13 parasites presenting high survival rates in novel piperaquine *in vitro* assays: retrospective and prospective investigations. BMC Med. 2015;13:305.

63. Inocencio da Luz RA, Vermeersch M, Dujardin JC, Cos P, et al. *In vitro* sensitivity testing of *Leishmania* clinical field isolates: preconditioning of promastigotes enhances infectivity for macrophage host cells. Antimicrob Agents Chemother. 2009;53(12):5197–203.
64. Hendrickx S, Eberhardt E, Mondelaers A, Rijal S, et al. Lack of correlation between the promastigote back-transformation assay and miltefosine treatment outcome. J Antimicrob Chemother. 2015;70(11):3023–6.
65. Grogl M, Thomason TN, Franke ED. Drug resistance in leishmaniasis: its implication in systemic chemotherapy of cutaneous and mucocutaneous disease. Am J Trop Med Hyg. 1992;47(1):117–26.
66. Ramesh V, Katara GK, Verma S, Salotra P. Miltefosine as an effective choice in the treatment of post-kala-azar dermal leishmaniasis. Br J Dermatol. 2011;165(2):411–4.
67. Palumbo E. Treatment strategies for mucocutaneous leishmaniasis. J Global Infect Dis. 2010;2 (2):147–50.
68. Moore EM, Lockwood DN. Treatment of visceral leishmaniasis. J Global Infect Dis. 2010;2 (2):151–8.
69. Fortin A, Hendrickx S, Yardley V, Cos P, et al. Efficacy and tolerability of oleylphosphocholine (OlPC) in a laboratory model of visceral leishmaniasis. J Antimicrob Chemother. 2012;67 (11):2707–12.
70. Nettey H, Allotey-Babington GL, Somuah I, Banga NB, et al. Assessment of formulated amodiaquine microparticles in *Leishmania donovani* infected rats. J Microencapsul. 2017;34 (1):21–8.
71. Balaraman K, Vieira NC, Moussa F, Vacus J, et al. *In vitro* and *in vivo* antileishmanial properties of a 2-n-propylquinoline hydroxypropyl beta-cyclodextrin formulation and pharmacokinetics via intravenous route. Biomed Pharmacother. 2015;76:127–33.
72. Loeuillet C, Bañuls AL, Hide M. Study of *Leishmania* pathogenesis in mice: experimental considerations. Parasit Vectors. 2016;9:144.
73. Loría-Cervera EN. Animal models for the study of leishmaniasis. Rev Inst Med Trop Sao Paulo. 2014;56(1):1–11.
74. Ponte-Sucre A. Physiological consequences of drug resistance in *Leishmania* and their relevance for chemotherapy. Kinetoplastid Biol Dis. 2003;2(1):14.
75. Kedzierski L, Sakthianandeswaren A, Curtis JM, Andrews PC, et al. Leishmaniasis: current treatment and prospects for new drugs and vaccines. Curr Med Chem. 2009;16(5):599–614.
76. Eberhardt E, Mondelaers A, Hendrickx S, Van den Kerkhof M, et al. Molecular detection of infection homogeneity and impact of miltefosine treatment in a Syrian golden hamster model of *Leishmania donovani* and *L. infantum* visceral leishmaniasis. Parasitol Res. 2016;115 (10):4061–70.
77. Forestier CL. Imaging host-*Leishmania* interactions: significance in visceral leishmaniasis. Parasite Immunol. 2013;35(9–10):256–66.
78. Yurdakul P, Dalton J, Beattie L, Brown N, et al. Compartment-specific remodeling of splenic micro-architecture during experimental visceral leishmaniasis. Am J Pathol. 2011;179(1):23–9.
79. Rouault E, Lecoeur H, Meriem AB, Minoprio P, et al. Imaging visceral leishmaniasis in real time with golden hamster model: monitoring the parasite burden and hamster transcripts to further characterize the immunological responses of the host. Parasitol Int. 2017;66(1):933–9.
80. Lewis MD, Fortes Francisco A, Taylor MC, Burrell-Saward H, et al. Bioluminescence imaging of chronic *Trypanosoma cruzi* infections reveals tissue-specific parasite dynamics and heart disease in the absence of locally persistent infection. Cell Microbiol. 2014;16(9):1285–12300.
81 Michel G, Ferrua B, Lang T, Maddugoda MP, et al. Luciferase-expressing *Leishmania infantum* allows the monitoring of amastigote population size, *in vivo*, *ex vivo* and *in vitro*. PLoS Negl Trop Dis. 2011;5(9):e1323.
82 Moreira ND, Vitoriano-Souza J, Roatt BM, Vieira PMA, et al. Parasite burden in hamsters infected with two different strains of *Leishmania* (*Leishmania*) *infantum*: "Leishman Donovan units" versus real-time PCR. PLoS One. 2012;7(10):e47907.

83 Satow MM, Yamashiro-Kanashiro EH, Rocha MC, Oyafuso LK, et al. Applicability of kDNA-PCR for routine diagnosis of American tegumentary leishmaniasis in a tertiary reference hospital. Rev Inst Med Trop Sao Paulo. 2013;55(6):393–9.
84 Nicodemo AC, Amato VS, Tuon FF, Souza RM, et al. Usefulness of kDNA PCR in the diagnosis of visceral leishmaniasis reactivation in co-infected patients. Rev Inst Med Trop Sao Paulo. 2013;55(6):429–31.
85 Ceccarelli M, Galluzzi L, Migliazzo A, Magnani M. Detection and characterization of *Leishmania (Leishmania)* and *Leishmania (Viannia)* by SYBR green-based real-time PCR and high resolution melt analysis targeting kinetoplast minicircle DNA. PLoS One. 2014;9(2):e88845.
86 Maes L, Beyers J, Mondelaers A, Van den Kerkhof M, et al. *In vitro* 'time-to-kill' assay to assess the cidal activity dynamics of current reference drugs against *Leishmania donovani* and *Leishmania infantum*. J Antimicrob Chemother. 2017;72(2):428–30.
87 Seifert K, Escobar P, Croft SL. *In vitro* activity of anti-leishmanial drugs against *Leishmania donovani* is host cell dependent. J Antimicrob Chemother. 2010;65(3):508–11.
88 Koniordou M, Patterson S, Wyllie S, Seifert K. Snapshot profiling of the antileishmanial potency of lead compounds and drug candidates against intracellular *Leishmania donovani* amastigotes, with a focus on human-derived host cells. Antimicrob Agents Chemother. 2017;61(3). pii: e01228-16.
89 Tegazzini D, Diaz R, Aguilar F, Pena I, et al. A replicative *in vitro* assay for drug discovery against *Leishmania donovani*. Antimicrob Agents Chemother. 2016;60(6):3524–32.
90 Duenas-Romero AM, Loiseau PM, Saint-Pierre-Chazalet M. Interaction of sitamaquine with membrane lipids of *Leishmania donovani* promastigotes. Biochim Biophys Acta. 2007;1768 (2):246–52.

Part III

Molecular Features of Drug Resistant Leishmania

The Role of Proteomics in the Study of Drug Resistance

10

Leonardo Saboia-Vahia, Jose Batista de Jesus, and Patricia Cuervo

Abstract

The recent completion of the genomic sequencing of three species of *Leishmania*, *L. (L.) major*, *L. (L.) infantum*, and *L. (V.) braziliensis* has enormous relevance to the study of the leishmaniasis pathogenesis. However, since in *Leishmania* the control of gene expression relies on the stability or processing of the mature mRNA, as well as on the posttranslational modifications of proteins, the genomic sequences alone are insufficient to predict protein expression within the parasites. In this scenario, proteomic technologies provide feasible pathways to functional studies of this parasite. With the challenging increase of natural drug resistance by *Leishmania*, the combination of the available genomic resources of these parasites with powerful high-throughput proteomic analysis is urgently needed to shed light on resistance mechanisms and identify new drug targets against *Leishmania*. Diverse proteomic approaches have been used to describe and catalogue global protein profiles of *Leishmania* spp. reveal changes in protein expression during development, determine the subcellular localization of gene products, evaluate host-parasite interactions, and elucidate drug resistance mechanisms. The characterization of these proteins has advanced, although

L. Saboia-Vahia
Laboratório de Pesquisa em Leishmaniose, Instituto Oswaldo Cruz, FIOCRUZ, Rio de Janeiro, RJ, Brasil

Laboratório de Doenças Infecto-Parasitárias, Universidade Federal de São João del Rei, Divinópolis, MG, Brasil

J. B. de Jesus
Departamento de Medicina, Universidade Federal de São João del Rei, São João del Rei, MG, Brasil

P. Cuervo (✉)
Laboratório de Pesquisa em Leishmaniose, Instituto Oswaldo Cruz, FIOCRUZ, Rio de Janeiro, RJ, Brasil
e-mail: patricia.cuervo@fiocruz.br

A. Ponte-Sucre, M. Padrón-Nieves (eds.), *Drug Resistance in Leishmania Parasites*,
https://doi.org/10.1007/978-3-319-74186-4_10

many fundamental questions remain unanswered. Here we discuss the recent proteomic discoveries that have contributed to the understanding of drug resistance mechanisms in *Leishmania* parasites.

10.1 Introduction

Since the publication of the genome data from *L. (L.) major* [1], *L. (L.) infantum*, and *L. (V.) braziliensis* [2], over 14 *Leishmania* spp. genomes have been sequenced to date (http://tritrypdb.org/tritrypdb/). Such achievements, in addition to the accumulation of genomic data from other *Leishmania* species, strains, and clinical isolates (http://www.genedb.org/, http://www.uniprot.org/, [3]), offer the prospects of new drug target identification and/or the exploration of particular metabolic pathways for drug development.

For example, analysis of genomic data from *L. (L.) infantum* and *L. (V.) braziliensis* has revealed the presence of a gene encoding cyclopropane fatty acyl phospholipid synthase [2]. Because this enzyme appears to be involved in the maintenance of the parasite's membrane and is not present in humans, it has been pointed as a putative chemotherapeutic target [2]. Such as this enzyme, many additional targets are probably encoded within the genome of *Leishmania* spp. The discovery of these targets is urgently needed given the increasing treatment failure observed with the mainstay chemotherapy, the pentavalent antimonials (Sb^V) [4], and the emergence of clinically resistant isolates [5–12].

Although all information concerning potential drug targets is contained in sequence databases, the promises of such target identifications are hampered by several factors. First, the limited functional annotation of the genomic sequence data determines that more than 50% of the predicted proteins have unknown functions [1, 2], which presents itself as an attractive challenge. Unfortunately, in 2017, 12 years after the first sequencing of a *Leishmania* genome, this scenario has not changed, and near 50% of the predicted proteins lacks functional annotation. Second, while the complexity of the cell cycle of these parasites would indicate that specific repertoires of genes are expressed in the promastigote and amastigote stages, global microarray genomic analyses have revealed that most *Leishmania* genes are constitutively transcribed [13–15], which is in agreement with the polycistronic organization of this parasite's genome [1]. Third, of the approximately 8000–9000 coding genes found in *Leishmania*, ~6200 are common to all trypanosomatids sequenced thus far, ~1000 are *Leishmania*-specific, and only ~200 genes (including some pseudogenes) are species-specific [1, 2, 16–18]. Hence, the modest differences between the genome sequences of *L. (L.) major*, *L. (L.) infantum*, *L. (V.) braziliensis*, and *L. (L.) mexicana* do not reflect the vast differences among the clinical phenotypes of leishmaniasis that are associated with each of these species. Consequently, the *Leishmania* genome sequences alone are insufficient to predict whole protein expression profiles throughout the life cycle of the parasite or under specific drug pressure or other experimental conditions. Interestingly, genome heterogeneity arisen from large-scale gene copy number variation, and extensive aneuploidy is observed in natural *Leishmania* isolates in response to drug pressure [3, 19–22]. However, it is still unknown whether this variation in gene and chromosome copy number is followed by variation in protein abundance.

Gene regulation in *Leishmania*, as with other trypanosomatids, occurs principally at the posttranscriptional level [23, 24] mainly by mechanisms that involve RNA stabilization and 3′UTR signatures [25–27]. Seminal analyses of RNA transcripts using DNA microarrays in either broad gene expression studies [13, 14, 28–31] or studies limited to specific genes [32–34] have shown that less than 6 % of genes are modulated at the mRNA level during the different stages of the *Leishmania* life cycle. Gene expression studies specifically related to drug resistance in *Leishmania* have shown the same pattern [19, 20, 32, 35]. Changing this scenario, recently, RNA sequencing (RNA-seq) of *L. (L.) mexicana* transcriptome showed that over 3.000 genes (~40%) are differentially expressed between promastigotes and amastigotes [18].

Although mRNA quantification has resulted in the increase of knowledge of several cellular processes of *Leishmania*, the direct analysis of protein levels is advantageous because the relationship between transcript abundance and protein expression levels in this parasite has been shown to be poor [15, 36]. In addition, information concerning cellular localization, posttranslational modifications, or protein interactions cannot be obtained from mRNA data [37]. As aneuploidy and regulation at translational and posttranslational levels make the scenario of protein expression in this parasite more complex [36], high-resolution proteomic approaches have the potential of shedding light on protein patterns that define a clinical phenotype. This may include either a phenotype associated with a specific disease manifestation or one associated with the susceptibility or resistance to a specific drug.

Proteomic analyses, therefore, provide data that are of crucial significance for the description and comprehension of the biology of *Leishmania* parasites, which are not evident from the genome sequence or the mRNA transcripts. The proteome is defined as the set of proteins expressed by a cell or organism under specific conditions and at a given point in time. The field of proteomics intends to provide detailed descriptions and integration of protein data to better ascertain protein function in biological systems. By allowing the characterization of complex systems, proteomic approaches offer the opportunity to identify proteins involved with drug resistance in *Leishmania*, in addition to new drug targets for this parasite.

In general, most proteomic studies of *Leishmania* and other trypanosomatids involve protein fractionation from a protein mixture using SDS-PAGE and/or two-dimensional electrophoresis (2DE), followed by gel excision and enzymatic digestion of protein spots. Peptides are submitted to mass spectrometry (MS) methods that combine soft ionization sources [matrix-assisted laser desorption/ionization (MALDI) or electrospray ionization (ESI)] [38–40] with various mass analyzers. Subsequent protein identification is accomplished by linking mass spectral data to genome sequence databases using bioinformatics tools [41, 42]. Gel-free shotgun liquid chromatography tandem mass spectrometry (LC/MS/MS) analyses have the potential to map more thoughtfully the *Leishmania* proteome under different conditions [43–45]. High-resolution LC/MS/MS for quantitative analysis by isotopic labeling of proteins or even label-free approaches has in fact revealed new aspects of *Leishmania* stage-specific proteomes or specific parasite phenotypes [46–53].

Proteomic studies of *Leishmania* spp. have focused on global proteome profiling [46, 49, 54–65], detailed descriptions of stage-specific protein expression [14, 36, 43, 47, 51, 66–76], posttranslational modifications (PTMs) [44, 52, 59, 73, 77–81], identification of proteins from subcellular proteomes and secretomes [45, 71, 82–91], and determining potential drug targets or proteins involved in drug resistance [50, 53, 56, 64, 92–100], among others.

Proteomic studies for identifying molecules potentially involved in the drug resistance of *Leishmania* spp. can be classified according to the approach carried out using either (1) axenic promastigotes and/or amastigotes that have been selected to drug resistance in vitro or (2) axenic promastigotes and/or amastigotes derived from clinical isolates that are considered naturally drug resistant. The compounds evaluated in these studies include drugs currently used for the treatment of leishmaniasis, such as Sb^V, amphotericin-B (AMB), and miltefosine (MIL), in addition to model drugs for the study of resistance, such as methotrexate (MTX) and arsenite, and drug under development, such as bicyclic nitro drugs [24, 50, 53, 56, 79, 92–99, 101–108] (Table 10.1). This chapter will discuss the applications of proteomic approaches to the study of *Leishmania* drug resistance, focusing on the identified molecules and on the inferred mechanisms of resistance to current medicines used for the treatment of leishmaniasis.

Table 10.1 Drugs analyzed in proteomic studies of *Leishmania* drug resistance

Drug	Drug status	Species analyzed	References
Antimonials	First-line drug	*L. (L.) infantum* *L. (L.) donovani* *L. (V.) panamensis* *L. (L.) braziliensis*	[79, 93–95, 102–106, 124]
Miltefosine	First-line drug	*L. (L.) donovani* *L. (L.) infantum*	[93, 96, 97, 99, 108]
Amphotericin-B	First-line drug	*L. (L.) infantum*	[107]
Methotrexate	Model drug	*L. (L.) major*	[56, 92]
α-Difluoromethylornithine	HAT drug	*L. (L.) donovani*	[50]
Arsenite	Model drug	*L. (L.) donovani*	[101]
Bicyclic nitro-drugs	Under development	*L. (L.) donovani*	[53]

HAT human African trypanosomiasis

10.2 Proteomic Approaches Used for Studying Drug Resistance in *Leishmania* spp.

10.2.1 Protein Expression Mapping by Two-Dimensional Electrophoresis

Proteomic studies of drug resistance in *Leishmania* spp. have traditionally used 2DE and 2D differential in-gel electrophoresis (2D-DIGE) for the comparative analysis of protein expression in drug-resistant and drug-sensitive parasites [56, 79, 92–99, 101–108]. This technique undoubtedly provided the basis for further developments in proteomics and, despite having certain limitations, is still used for protein expression mapping. The separation of complex cellular extracts by 2DE is achieved by coupling two independent electrophoretic separations, using isoelectric focusing in the first dimension and SDS-PAGE in the second [109, 110]. Soon after the first reports of 2DE appeared, this method was widely adopted by researchers around the world in several distinct applications [111–113].

The first works of what can be called the earliest *Leishmania* proteomics, even before the term "proteomics" was coined, came from the early 1980s. In these works, 2DE was used to (1) separate cell lysates of *L. (L.) tropica* for further detection of antigenic proteins using rabbit sera [114] and (2) for the comparative analysis of protein expression patterns from distinct *Leishmania* species that cause American tegumentary leishmaniasis with the aim of detecting species-specific markers [115]. However, issues concerning reproducibility, specifically involving the stability of the pH gradients, discouraged the widespread use of the method. Additionally, the absence of a protein identification system prevented the designation of interesting proteins. Identification was achieved by co-migration with purified proteins or through the use of antibodies. Using these methods, the regulation of tubulin expression during *Leishmania* differentiation was demonstrated [116]. Despite the drawbacks, valuable information was obtained, such as the demonstration that *Leishmania* resistance to MTX, an antiproliferative agent, is mediated by a mutation in the target enzyme, dihydrofolate reductase (DHFR), which alters the physicochemical properties of the protein [117].

The introduction of immobilized pH gradients in the first dimension [118] eliminated the reproducibility issues associated with pH stability. At the same time, N-terminal sequencing using traditional Edman chemistry applied to proteins separated by SDS-PAGE allowed the identification of peptides and proteins, as well as molecular mass determination [119]. However, it was the introduction of soft ionization techniques for peptides and proteins (MALDI and ESI) that allowed the acquisition of mass spectra of these molecules at the subnanomolar level and also changed the paradigm of protein identification [38–40]. Coupling 2DE and MS identification, Drummelsmith et al. observed up to a fourfold increase in the expression of several spots of trypanothione (TRYR) protein in transfected *L. (L.) major* promastigotes overexpressing the TRYR gene [56]. This assay validated the use of 2DE for drug resistance studies. Decades after 2D appeared, various studies started using fluorescent dyes, mainly the Cy dyes, which is the principle of 2D-DIGE, in order to achieve a more reliable quantification of differences among the spots detected [47, 81].

As mentioned above, all proteomic studies on *Leishmania* drug resistance have used the combination of 2DE or 2D-DIGE and MS for protein identification. However, 2DE presents major limitations, including the inability to resolve low abundance proteins and hydrophobic proteins, which represent important sources of information in the case of drug resistance mechanisms. Also in quantitative studies, the protein co-migration is a significant issue when deciding which protein contributed the more for the change in intensity observed in one spot. Despite these limitations, 2DE is a well-characterized technique for protein separation, and it is distinguished by its visual array that allows the detection of posttranslational modified states [120, 121].

10.2.2 Other Proteomic Approaches for Studying Drug Resistance

Although the use of fluorescent dyes has turned 2D-DIGE into a quantitative tool with better sensitivity and reproducibility than 2D, the gel-based approaches are still very limited regarding linearity, dynamic range, and reliability for quantifying differences in protein abundance, being limited to the resolution of soluble and abundant proteins [118, 122]. These limitations have been surpassed by the development of mass spectrometry (MS)-based approaches. In MS-based proteomic approaches, proteins can be identified and quantified by means of detection and quantification of their peptides [123]. The main methods include labeling with stable isotopes or label-free approaches. Labeling methods introduce a mass tag into proteins or peptides, either metabolically, enzymatically, or by chemical means; labeling based on isobaric tags for relative and absolute quantification (iTRAQ) and stable isotope labeling by amino acids in cell culture (SILAC) has been used in several proteomic analysis of *Leishmania* parasites enabling quantification of stage-specific proteins, characterization of posttranslational modifications, and quantification of protein abundance in parasites selected for drug resistance [48, 50, 53, 72, 78, 107, 124]. On the other hand, label-free methods correlate the ion intensity signal of peptide mass spectra or the number of peptide spectral counts with the protein quantity [122, 125, 126]. Shotgun label-free methods for protein quantification in *Leishmania* have been little explored, but there is a nice example of the potential of this approach in the quantitative analysis of the proteome of *L. (L.) mexicana* reported by Paape et al. [43].

Using high-throughput proteomic technologies, pharmacoproteomics allows discovery and validation of novel drug targets and generates information about drug metabolism and transport as well as about drug efficacy, resistance, and toxicity [127, 128]. Successful examples of these applications can be found in cancer research [129, 130]. Pharmacoproteomics has started to be used successfully in the study of drugs under development for leishmaniasis treatment [53].

10.3 Proteomics of Drug Resistance in *Leishmania* spp.

10.3.1 Proteomics of Model Drugs for Understanding Resistance in *Leishmania*: Methotrexate and Arsenite Resistance

Model drugs such as MTX and arsenite have been widely used for the study of molecular mechanisms of drug resistance in *Leishmania* [131–136]. In fact, much of the current knowledge of resistance mechanisms and novel potential drug targets in this genus came from studies using resistant parasites obtained after in vitro selection with these drugs [137–140]. The resistance of *Leishmania* to these compounds includes events such as DNA amplification [131, 141, 142], decreased drug accumulation, and increased drug efflux [98, 143–145], among others. Methotrexate is an anticancer drug that inhibits DHFR which is responsible for the conversion of dihydrofolate to tetrahydrofolate. Derivatives of tetrahydrofolate are essential for the biosynthesis of purines and pyrimidines. Therefore, in the presence of MTX, DNA synthesis is prevented. Although this antifolate is toxic to *Leishmania*, it was found to be much more toxic to mammalian cells than for the parasites, thus preventing its use as a chemotherapeutic agent for leishmaniasis [146].

The first recorded proteomic study on drug resistance in *Leishmania* was conducted using *L. (L.) major* promastigotes that were induced in vitro to MTX resistance [56]. Comparison of MTX-resistant parasites to sensitive parasites using 2DE revealed the overexpression of the pteridine reductase PTR1, a known primary mediator of MTX resistance. It was demonstrated that the PTR1 overexpression was due to several gene amplification events in the resistant parasites [56]. As PTR1 is able to reduce dihydrofolate to tetrahydrofolate to a minimal extent, its overexpression could compensate for the inhibition of DHFR by MTX [56].

In a further study, Drummelsmith et al. [92] observed that *L. (L.) major* MTX-resistant promastigotes exhibited increased expression of proteins involved in stress response, such as chaperonins, heat-shock proteins, and enolase, as well as enzymes such as argininosuccinate synthetase (ARGG), which catalyzes the penultimate step in arginine biosynthesis. As in the case of PTR1, the overexpression of ARGG was also the result of gene amplification events, which was most likely a result of the structural proximity of the PTR1 and ARGG coding genes [92]. Other proteins with less easily predicted roles in drug resistance, such as methionine adenosyltransferase (MAT), were also identified in this study. This enzyme is overexpressed both in sensitive cells shocked with MTX and in mutants resistant to the drug, suggesting that it may play a significant role in the initial cellular responses to MTX in *L. (L.) major*. Unlike other proteins, the overexpression of MAT was not due to gene amplification events [92]. In addition, it was observed that increases in *S*-adenosylmethionine level, which is synthesized by MAT, correlated with the selection and emergence of MTX resistance in *L. major* [92].

Finally, a proteomic analysis of *L. (L.) donovani* induced to arsenite resistance was reported. However, as 2DE gels from wild-type and arsenite-resistant parasites were completely different, comparison of the differential protein expression between the two conditions was precluded [101].

10.3.2 Proteomics of Antimonial Resistance

Pentavalent antimonials in the forms of sodium stibogluconate (Pentostam) and meglumine antimoniate (Glucantime) are first-line drugs for the treatment of distinct forms of leishmaniasis [147]. Despite the fact that Sb^V have been used worldwide for almost 80 years, reports on the clinical resistance and descriptions of resistant parasites started to appear within the last 20 years [4–12]. This situation is particularly alarming in India where widespread failure to Sb^V treatment in previously untreated patients has been reported [see Chap. 4 in this volume; 8, 148, 149]. Recently, it was demonstrated that arsenic contamination of drinking water might have contributed to the development of antimonial resistance in *Leishmania* parasites circulating in the Bihar region [150]. The complexity of the resistance scenario is augmented and sometimes obscured by the variation in the clinical response to Sb^V due to species-specific sensitivity to these drugs [see Chap. 15 in this volume; 4, 151, 152].

The understanding of the mechanism of action of Sb^V drugs and resistance to them has come from laboratory parasites, in which resistance has been selected in vitro by the pressure of the drug. For antileishmanial activity, it is necessary that the Sb^V be reduced to the trivalent form Sb^{III}. Although debatable, reduction of the drug can apparently occur both in the macrophage and in the amastigote [153–156]. Reduction would be accomplished by either an enzymatic mechanism involving a thiol-dependent reductase [157] and/or an arsenate/antimonate reductase [158, 159] or by some nonenzymatic mechanism [160]. Regarding the internalization of the drug, it was demonstrated that AQP1, a transporter of trivalent metalloids [161], mediates the uptake of Sb^{III} in *Leishmania* [162]. The expression level of *aqp1* can correlate to the sensitivity to the drug [163, 164], and a major cluster of *L. (L.) donovani* isolates from the Indian subcontinent (ISC), which are resistant to Sb^V, presents a mutation in the *aqp1* gene that results in a nonfunctional protein and therefore reduced influx of Sb^{III} [3].

The activity of antimoniate seems to center around thiol redox metabolism [165], although early reports pointed to glycolytic and fatty acid β-oxidation pathways [166] or a programmed cell death (PCD) pathway involving DNA fragmentation non-mediated by caspase [167, 168]. On the other hand, increased levels of γ-glutamylcysteine synthetase and ornithine decarboxylase [137, 169], the enzymes involved in the synthesis of glutathione and polyamines, which are precursors of trypanothione, have been observed in parasites selected for resistance to Sb^{III} or arsenite [137, 145]. As a consequence, accumulation of trypanothione and glutathione contributes to the resistant phenotype [165, 170]. In addition, mechanisms for the increased efflux of Sb-thiol complexes [145] and/or decreased drug influx mediated by decreased/nonfunctional AQP1, besides drug sequestration involving a P-glycoprotein member of the ABC transporters (PgpA/MRPA), as well as other transporters (ABCC4, ABCC5, MRP1), might also influence the antimonial resistance [163, 171–177]. Despite some controversies, it seems that resistance to antimony is a multifactorial phenomenon involving various mechanisms such as decreased drug uptake, diminished metal reduction, increased glutathione and trypanothione synthesis, and augmented drug efflux/sequestration [4, 139, 140, 178].

Proteomics of antimonial resistance have provided further evidence for some of the above mechanisms and have added new pieces to the resistance puzzle. Seminal studies comparing between Sb^{III}-sensitive and in vitro selected Sb^{III}-resistant axenic amastigotes of *L. (L.) infantum* using 2DE (pH range, 5–6) and ESI-MS/MS revealed some downregulated proteins in Sb^{III}-resistant parasites, including the LACK receptor, β-tubulin, proteasome pa26 subunit, pyruvate kinase, and the kinetoplastid membrane protein 11 (KMP-11) [94]. Although none of these proteins had previously been associated with antimony resistance, several of them have been further observed in other *Leishmania* species selected for antimony resistance or other drugs [79, 95, 96, 102, 124] (Table 10.2).

Regarding KMP-11, Western blot analysis confirmed that the levels of KMP-11 were lower in Sb^{III}-resistant parasites when compared to the parental wild-type parasites. However, overexpression of the protein did not alter the Sb^{III} susceptibility of parasites. In addition, Northern blot analysis revealed that the downregulation of KMP-11 was not due to a decrease in mRNA levels [94]. Decrease of KMP-11 abundance was also observed in the phosphoproteome analysis of *L. (V.) braziliensis* selected for antimonial resistance [79]. In *L. (L.) donovani*, it was shown that KMP11 could increase the lipid bilayer pressure [179, 180]. Thus, the marked decreased level of KMP-11, probably due to an increased turnover rate of this protein, could alter the interaction of transporters or putative efflux systems, enhancing activity for pumping Sb^{III} out of parasites [94]. On other hand, it has been proposed that the decrease of this protein could be part of a general mechanism of response to the stress caused by the drug pressure [79].

Argininosuccinate synthetase (ARGG) was the only protein identified as overexpressed in the *L. (L.) infantum* drug-resistant mutant. Increased abundance of ARGG was also observed in *L. braziliensis* resistant to antimonial [79] and in MTX-resistant *L. (L.) major* [92]. Both in Sb^{V}-resistant *L. (L.) donovani* and in MTX-resistant *L. (L.) major*, it was observed that the genomic region coding for this gene is amplified [20, 92], supporting the overexpression of ARGG [94]. However, the role that ARGG plays on resistance to Sb^{V} is unknown. The role in resistance of the other identified proteins remains to be established.

Regarding proteomic studies of field isolates, a proteomic analysis recently compared *L. (L.) donovani* parasites obtained from both a Sb^{V}-unresponsive and Sb^{V}-responsive patient to identify proteins involved in antimonial resistance [93]. First, the Sb^{V}-resistant and Sb^{V}-sensitive phenotypes of these isolates were corroborated by in vitro growth inhibition assays. Second, RT-PCR analysis showed that the expression levels of *aqp1*, *gsh1*, and *PgpA* (*mrpa*), which are genes associated with in vitro-induced resistance, were not differentially expressed between the sensitive and resistant clinical isolates. Third, it was shown that the parasites from the Sb^{V}-unresponsive patient were more resistant to Sb^{III}- and Sb^{V}-induced PCD. The PCD features analyzed were the mitochondrial membrane potential ($\Delta\psi m$), DNA fragmentation, and externalization of phosphatidylserine residues followed by membrane permeabilization [93]. Thus, based on previous evidence suggesting that Sb^V kill *Leishmania* by a process involving several features of PCD [167, 168, 181], the soluble proteomes of these isolates naturally resistant and sensitive to Sb^V were analyzed by 2DE (pH range, 4–5) and MS/MS. Enolase, 14-3-3, ATP-dependent RNA helicase, dipeptidyl-peptidase III, 20 S proteasome α5 subunit, small kinetoplastid calpain-related protein SKCRP14.1, and many heat-shock proteins (HSP) were among the identified proteins. From this group, the 14-3-3 protein; the HSP83, whose abundances are increased in resistant parasites; and the SKCRP14.1, which is decreased in resistant parasites, were highlighted as having different roles in PCD. In fact, *Leishmania* HSP83 is an orthologue of the mammalian HSP90, which is involved in mitochondrial apoptotic pathways [182] (Table 10.2).

The genes coding for 14-3-3, HSP83, and SKCRP14.1 were cloned and the 14-3-3 and HSP83 constructs were transfected individually into the sensitive parasites, whereas the SKCRP14.1 construct was transfected into the resistant parasites [93]. While sensitive parasites transfected with the 14-3-3 construct did not show any increase in resistance to Sb^{III} or Sb^V when compared with sensitive control parasites, the sensitive parasites overexpressing HSP83 were more than twofold resistant to Sb^{III} compared with sensitive control parasites. In Sb^{III}-treated parasites, a more intense drug-mediated DNA fragmentation was observed in the control parasites when compared to HSP83-overexpressing parasites. It was also found that after treatment with Sb^{III}, the $\Delta\psi m$ was higher in HSP83-overexpressing parasites than in control cells [93]. Overexpression of SKCRP14.1 increased the sensitivity of resistant parasites to Sb^{III} and Sb^V, in addition to the sensitivity of transfectant parasites to Sb^{III}-induced DNA fragmentation. After treatment with Sb^{III} , no variations in the $\Delta\psi m$ were observed between SKCRP14.1 transfectants and control cells [93]. As will be described below, several of these effects were also observed in MIL-treated parasites [93], which highlights the phenomenon of cross-resistance but also reveals contrasting mechanisms involved in *Leishmania* drug resistance. Hence, although the network of molecules through which HSP83 and SKCRP14.1 interfere with drug-induced PCD pathways in *L. (L.) donovani* remains to be elucidated, this study demonstrated that these proteins modulate drug susceptibility in this parasite. It remains to be established if these phenomena are observed in other *L. (L.) donovani* clinical isolates and in other *Leishmania* species for which resistance has been reported. Interestingly, it recently was demonstrated that antimony-resistant *L. (L.) infantum* exhibited decreased abundance of SKCRP14.1, reinforcing the observation that antimonial resistance is associated with a decrease in cell death-related proteins [106]. In addition, increased abundance of HSP83, as well other HSPs and chaperones, has been further observed in different *Leishmania* species selected for antimonial resistance [79, 95, 103–106, 124], including *L. (L.) infantum*, *L. (V.) panamensis*, and *L. (V.) braziliensis* (Table 10.2). Remarkably, parasites selected for MIL, amphotericin-B, or alpha-difluoromethylornithine resistance also present a significant increase in various heat-shock proteins [50, 96, 107, 108], suggesting that these proteins are part of a general response to the stress caused by the drug pressure. The increased protection against drug-related stress and drug-related programmed cell death may contribute to the resistance phenotype as a whole

Table 10.2 *Leishmania* spp. proteins associated with drug resistance (proteins highlighted by the authors). Sb^{V}, pentavalent antimonial; Sb^{III}, trivalent antimonial

Protein	Species	Natural (N) or selected (S) resistance	Life stage	Increased (I) or decreased (D) abundance	Mechanism	References
Antimonials Sb^{V}-Sb^{III}						
MRPA (ABCC3) ABC-thiol transporter	*L. (L.) infantum*	S	Promastigotes	I	Increased drug sequestration	[106]
S-adenosylmethionine synthetase (SAMS)	*L. (V.) panamensis*	S and N	Promastigotes	I	Increased levels of intracellular T$[SH]_2$—Increased protection against ROS	[95, 104]
	L. (L.) infantum	S				
S-adenosylhomocysteine hydrolase (SAHH)	*L. (L.) donovani*	S	Promastigotes—Amastigotes	I	Increased levels of intracellular T$[SH]_2$—Increased protection against ROS	[104, 124]
	L. (V.) panamensis	S and N				
Rab7 homolog	*L. (V.) panamensis*	S and N	Promastigotes—Amastigotes	I	Transport of vesicles containing sequestered metal–thiol conjugates?	[95, 104]
	L. (L.) infantum	S				
Pteridine reductase (PTR1)	*L. (V.) braziliensis*	S	Promastigotes	I	Increased antioxidant response	[79, 95]
Tryparedoxin peroxidase	*L. (V.) braziliensis*	S	Promastigotes	I	Increased antioxidant response	[79, 95, 106]
Peroxiredoxin	*L. (L.) infantum*	S				

(continued)

Table 10.2 (continued)

Protein	Species	Natural (N) or selected (S) resistance	Life stage	Increased (I) or decreased (D) abundance	Mechanism	References
Heat shock proteins/chaperones						
HSP83, HSP70, HSP60 Protein disulfide isomerase	*L. (L.) donovani*	N and S	Promastigotes—Amastigotes	I	Increased general response to stress/protection against drug-related programmed cell death	[79, 93, 95, 103–106, 124]
	L. (V.) panamensis	S				
	L. (V.) braziliensis	S				
	L. (L.) infantum	S				
Small kinetoplastid calpain-related protein (SKCRP14.1)	*L. (L.) donovani*	N	Promastigotes—Amastigotes	D	Decreased cell death-related proteins	[93, 106]
	L. (L.) infantum	S				
Proliferative cell nuclear antigen (PCNA)	*L. (L.) donovani*	N	Promastigotes	I	Increased DNA repair?	[79, 95, 102]
	L. (V.) braziliensis	S				
Glycolytic enzymes						
Triose phosphate isomerase Fructose-1,6-bisphosphate aldolase Glyceraldehyde 3-phosphate dehydrogenase Hexokinase Glycosomal malate dehydrogenase	*L. (L.) donovani*	N and S	Promastigotes	I	Increased glycolysis—Increased parasite fitness? General stress response mechanism	[93, 95, 102, 103, 106, 124]
	L. (L.) infantum	S				
	L. (V.) panamensis	S				
	L. (V.) braziliensis	S				

Aldehyde dehydrogenase Enolase						
TCA-related enzymes						
Pyruvate dehydrogenase Dihydrolipoamide dehydrogenase Isocitrate dehydrogenase	*L. (L.) infantum*	S	Promastigotes	I or D	General stress response mechanism Increased parasite fitness? (only for I)	[95, 99, 106]
Transcription, translation-related proteins						
Translation initiation factor 1-alpha Elongation factor 1-alpha Elongation factor 2 Elongation factor 5-alpha 60s ribosomal L23a Various ribosomal proteins	*L. (L.) infantum*	S	Promastigotes	I	General stress response mechanism Increased parasite fitness?	[79, 95, 102–104, 106, 124]
	L. (L.) donovani	N and S				
	L. (V.) panamensis	S				
	L. (V.) braziliensis	S				
Oligopeptidase B (OPB)	*L. (V.) panamensis*	S	Promastigotes	I	Increased virulence—Increased parasite fitness?	[104]
Cysteine leucine rich protein	*L. (L.) donovani*	N	Promastigotes	I	Increased virulence—Increased parasite fitness?	[102]
Succinyl-CoA:3-ketoacid-coenzyme A transferase	*L. (L.) infantum*	S	Promastigotes	I	Compensatory production of acetyl-CoA?	[99]
Histone 4	*L. (L.) infantum*	S	Promastigotes	I	General stress response mechanism Increased parasite fitness?	[106]
Cyclophilin-A	*L. (V.) braziliensis*	S	Promastigotes	D	???	[95]
	L. (L.) infantum	S				

(continued)

Table 10.2 (continued)

Protein	Species	Natural (N) or selected (S) resistance	Life stage	Increased (I) or decreased (D) abundance	Mechanism	References
Activated protein kinase c receptor (LACK)	*L. (L.) infantum*	S	Promastigotes	I	Increased signal transduction?	[94]
GP63, Leishmanolysin	*L. (L.) infantum*	S	Promastigotes	D	???	[106]
Amastin, surface protein	*L. (L.) infantum*	S	Promastigotes	D	???	[106]
Argininosuccinate synthetase (ARGG)	*L. (L.) infantum*	S	Amastigotes Promastigotes	I	???	[79, 94]
	L. (V.) braziliensis	S				
Kinetoplastid membrane protein (KMP-11)	*L. (L.) infantum*	S	Amastigotes	D	General stress response mechanism?	[79, 94]
	L. (V.) braziliensis	S	Promastigotes			
Ca^{2+} homeostasis-related proteins						
Hypothetical protein containing an EF-hand calcium binding domain	*L. (L.) infantum*	S	Promastigotes	I	Ca^{2+} homeostasis—Increased protection against ROS	[99]
Calcium binding protein				D		
Tetratricopeptide repeat (TPR) domain protein	*L. (V.) braziliensis*	S	Promastigotes	I	General stress response mechanism	[79]
Calcireticulin	*L. (V.) braziliensis*	S	Promastigotes	I	General stress response mechanism	[79]

Beta-tubulin	*L. (L.) donovani*	N and S	Promastigotes	I	Signal transduction?—General stress response mechanism?	[79, 94, 95, 102, 124]
	L. braziliensis	S				
	L. (L.) infantum	S				
Nucleoside diphosphate kinase b (NDKb)	*L. braziliensis*	S	Promastigotes	I	???	[79]
Succinyl-CoA ligase	*L. braziliensis*	S	Promastigotes	D	???	[79, 95]
RNA helicase	*L. (L.) donovani*	N	Amastigotes-Promastigotes	D	???	[79, 95]
	L. braziliensis	S				
	L. (L.) infantum	S				
Miltefosine (HePC)						
Transcription, translation-related proteins						
Eukaryotic initiation factor 4A (eIF4A)	*L. (L.) donovani*	S	Promastigotes	I	General stress response mechanism	[96, 97]
Elongation factor 1-beta	*L. (L.) infantum*	N			Increased parasite fitness?	
Heat shock proteins						
HSP83, HSP60, stress-induced protein sti1	*L. (L.) infantum*	N	Promastigotes	I	Increased general response to stress/protection against drug-related programmed cell death	[96, 108*]
HSPA9B	*L. (L.) donovani*	S		D		
Peroxiredoxin	*L. (L.) infantum*	N	Promastigotes	I	Increased antioxidant response	[96]
Proliferative cell nuclear antigen (PCNA)	*L. (L.) infantum*	N	Promastigotes	I	Increased DNA repair?	[96]

(continued)

Table 10.2 (continued)

Protein	Species	Natural (N) or selected (S) resistance	Life stage	Increased (I) or decreased (D) abundance	Mechanism	References
Mitochondrial ATPase β-subunit	*L. (L.) infantum*	N	Promastigotes	I	Increased DNA repair?	[96]
TCA-related enzymes						
Pyruvate dehydrogenase Dihydrolipoamide dehydrogenase Isocitrate dehydrogenase	*L. (L.) infantum*	S	Promastigotes	D	General stress response mechanism Increased parasite fitness?	[99]
Ca^{2+} homeostasis-related proteins						
Hypothetical protein containing an EF-hand calcium binding domain	*L. (L.) infantum*	S	Promastigotes	I	Ca^{2+} homeostasis—Increased protection against ROS	[99]
Calcium binding protein				D		
Succinyl-CoA:3-ketoacid-coenzyme A transferase	*L. (L.) infantum*	S	Promastigotes	I	Compensatory production of acetyl-CoA?	[99]
Activated protein kinase c receptor (LACK)	*L. (L.) infantum*	N	Promastigotes	I	Increased signal transduction?	[96]
S-adenosylmethionine synthetase (SAMS)	*L. (L.) infantum*	N	Promastigotes	D	Increased levels of intracellular T$[SH]_2$—Increased protection against ROS	[96]
Amphotericin-B (AmB)						
S-adenosylmethionine synthetase (SAMS)	*L. (L.) infantum*	S	Promastigotes	I	Increased levels of intracellular T$[SH]_2$—Increased protection against ROS	[107]

Glycolytic enzymes						
Fructose-1,6-bisphosphate Aldolase Glyceraldehyde 3-phosphate dehydrogenase Hexokinase Enolase	*L. (L.) infantum*	S	Promastigotes	I	General stress response mechanism Increased parasite fitness?	[107]
TCA-related enzymes						
Aconitase Dihydrolipoamide dehydrogenase Malate dehydrogenase Succinate dehydrogenase flavoprotein	*L. (L.) infantum*	S	Promastigotes	I	General stress response mechanism Increased parasite fitness?	[107]
Trypanothione reductase	*L. (L.) infantum*	S	Promastigotes	I	Increased antioxidant response	[107]
Tryparedoxin peroxidases	*L. (L.) infantum*	S	Promastigotes	I	Increased antioxidant response	[107]
Iron superoxide dismutase	*L. (L.) infantum*	S	Promastigotes	I	Increased antioxidant response	[107]
Heat shock proteins/chaperones						
HSP83, HSP70, HSP60 T-complex protein 1 (several subunits)	*L. (L.) infantum*	S	Promastigotes	I	Increased general response to stress/protection against drug-related programmed cell death	[107]
Various peptidases	*L. (L.) infantum*	S	Promastigotes	I	Increased parasite fitness—virulence?	[107]
Transcription, translation-related proteins						
Various protein synthesis-related proteins Various ribosomal proteins	*L. (L.) infantum*	S	Promastigotes	I	Increased parasite fitness?	[107]

(continued)

Table 10.2 (continued)

Protein	Species	Natural (N) or selected (S) resistance	Life stage	Increased (I) or decreased (D) abundance	Mechanism	References
H1A-2 P-type H^+-ATPase	*L. (L.) infantum*	S	Promastigotes	D	Protection against drug-related programmed cell death	[107]
Methotrexate (MTX)						
Pteridine reductase (PTR1)	*L. (L.) major*	S	Promastigotes	I	Increased antioxidant response	[56, 92]
Beta-tubulin	*L. (L.) major*	S	Promastigotes	I	Signal transduction?—General stress response mechanism	[56]
Methionine adenosyltransferase (MAT)	*L. (L.) major*	S	Promastigotes	I	Increased levels of intracellular T $[SH]_2$—Increased protection against ROS	[92]
S-adenosylmethionine synthetase (SAMS)	*L. (L.) major*	S	Promastigotes	I	Increased levels of intracellular T $[SH]_2$—Increased protection against ROS	[92]
Argininosuccinate synthetase (ARGG)	*L. (L.) major*	S	Promastigotes	I	???	[92]
α-difluoromethylornithine (DFMO)						
S-adenosylmethionine synthetase (SAMS)	*L. (L.) donovani*	S	Promastigotes	I	Increased levels of intracellular T $[SH]_2$—Increased protection against ROS	[50]
Cystathione β-lyase like protein	*L. (L.) donovani*	S	Promastigotes	I	Increased levels of intracellular T $[SH]_2$—Increased protection against ROS	[50]
Dihydrofolate reductase (DHFR-TS)	*L. (L.) donovani*	S	Promastigotes	I	Increased levels of intracellular T $[SH]_2$—Increased protection against ROS	[50]

Nucleoside diphosphate kinase b (NDKb)	*L. (L.) donovani*	S	Promastigotes	I	???	[50]
Heat shock proteins/chaperones						
HSP83, HSP70, HSP60	*L. (L.) donovani*	S	Promastigotes	I	???	[50]
Transcription, translation-related proteins						
Various protein synthesis-related proteins Various ribosomal proteins	*L. (L.) donovani*	S	Promastigotes	I	Increased parasite fitness?	[50]
TCA-related enzymes and fatty acid metabolism						
Isocitrate dehydrogenase 3-ketoacyl-CoA thiolase Oxoglutarate ehydrogenase	*L. (L.) donovani*	S	Promastigotes	I D	General stress response mechanism Increased parasite fitness?	[50]
Tryparedoxin	*L. (L.) donovani*	S	Promastigotes	I	Increased antioxidant response	[50]
Histone 2 Histone 3 Histone 4	*L. (L.) donovani*	S	Promastigotes	I	General stress response mechanism Increased parasite fitness?	[50]
Glycolytic enzymes						
Fructose-1,6-bisphosphate Aldolase Glucose-6-phosphate 1-dehydrogenase, Phosphoenolcarboxykinase	*L. (L.) donovani*	S	Promastigotes	D	General stress response mechanism Increased parasite fitness?	[50]
Various peptidases	*L. (L.) donovani*	S	Promastigotes	I	Increased parasite fitness?	[50]

(continued)

Table 10.2 (continued)

Protein	Species	Natural (N) or selected (S) resistance	Life stage	Increased (I) or decreased (D) abundance	Mechanism	References
Cyclophilins	*L. (L.) donovani*	S	Promastigotes	I	General stress response mechanism Increased parasite fitness?	[50]
Bicyclic nitro-drugs						
Hypothetical NADH:FMN dependent oxidoreductase	*L. (L.) donovani*	S	Promastigotes	D	Reduced drug bio-activation	[53]

?: suggested mechanism with different level of evidences
???: unknown mechanism

[79, 104], resulting in parasites able/adapted to cope with the oxidative stress caused by drug, probably by maintaining functional proteostasis mediated by HSPs.

Cell membrane proteins are very interesting for understanding drug transport in resistant parasites. In this context, Kumar et al. reported a set of overexpressed proteins in membrane- and cytosolic-enriched fractions of an *L. (L.) donovani* clinical isolate resistant to Sb^{V}, when compared to a sensitive isolate [102]. The six proteins identified in the membrane-enriched fraction were two ABC transporters, a fragment of HSP83, a cysteine-leucine-rich protein (CLrP), a GPI transamidase, and a 60S ribosomal protein (L23a). Remarkably, these authors further demonstrated that CLrP is a glycosylated protein with dual localization, in the membrane and nucleolus, whereas the 60s ribosomal L23a protein (60sRL23a) is localized in the cytosol [183, 184]. It was also shown that antimonial-resistant clinical isolates of *L. (L.) donovani* present higher mRNA and protein levels of CLrP and 60sRL23a as compared to antimonial-sensitive parasites [183, 184]. Overexpression of CLrP or 60sRL23a in a sensitive isolate of *L. (L.) donovani* significantly decreased its responsiveness to Sb^{V} and Sb^{III}, in the case of CLrP, and also to MIL and paromomycin, in the case of 60sRL23a. Such reduction on drug sensitivity was followed by increased parasite infectivity to murine macrophages or increased proliferation rate, for CLrP- or 60sRL23a-overexpressing mutants, respectively [183, 184]. Such studies reveal that resistant parasites exhibit a higher fitness than sensitive parasites, showing increased infectivity capability to host cells and increased proliferation rate.

Metabolic isotopic labeling of *L. (L.) infantum* resistant to Sb^{III} followed by comparative proteomic analysis of membrane and cytosolic fractions allowed the observation of increased levels of the ABC transporter MRPA (ABCC3) for the first time in a proteomic study [106]. The increased abundance of MRPA, a well-known protein involved in antimonial sequestration, was accompanied by alterations in the abundance of other transporters such as folate/biopterin transporters that presented diminished abundance in resistant parasites. Interestingly, folate/biopterin transporters have been previously pointed out as potential chemotherapeutic targets in *Leishmania* [185, 186].

It is pertinent to mention that in the proteomic studies that aimed to analyze *Leishmania* membrane proteins associated with drug resistance, the identification of a higher number of membrane proteins, more representative of this fraction, has been hampered possibly by the inherent limitation of 2DE for resolving hydrophobic proteins. Furthermore, the methods hitherto used for sample preparation do not favor the representativeness of such proteins. Such limitations can be overcome using better solubilizing agents and gel-free shotgun proteomic approaches, as demonstrated for other cells and tissues [187–189].

Seminal proteomic studies of antimonial resistance in *L. (L.) donovani* isolates identified β-tubulin, enolase, fructose-1,6-bisphosphate aldolase, the proteasome subunit α5, a carboxypeptidase, a fragment of HSP70, and the proliferative cell nuclear antigen (PCNA) [102]. Interestingly, further expression analyses, by Western blot and qPCR, confirmed that promastigote and amastigotes of resistant parasites exhibit ≥threefold and ~fivefold increased levels of PCNA, respectively, compared to the antimonial-sensitive parasites [190]. Overexpression of PCNA in antimonial-sensitive isolate resulted in significant increase of Sb^V IC_{50}, and parasites overexpressing PCNA exhibited less DNA fragmentation compared to wild-type sensitive parasites upon treatment. In addition, parasites overexpressing PCNA modulated negatively nitric oxide (NO) production in infected macrophages [190].

Because many of the soluble proteins detected in drug resistance studies are among the abundant proteins commonly identified in proteomic studies of *Leishmania* and other trypanosomatids [59, 68, 71, 191], it is difficult to elucidate a clear role in resistance for them. However, as mentioned above, further proteomic analyses of *Leishmania* resistance mechanisms either to Sb^V or to other drugs have corroborated that HSPs, glycolytic enzymes, TCA-related enzymes, transcription-/translation-related proteins, peptidases, as well as DNA repair-related proteins, among other noncanonical resistance proteins, exhibit altered abundance in resistant parasites (Table 10.2) [50, 79, 95, 96, 103–108, 124]. Such findings reinforce the idea that resistant parasites exhibit a better general fitness than sensitive parasites, mediated by the (1) remodeling of their glycolytic metabolism, (2) increasing of virulence factor abundance, (3) and more efficient protein homeostasis and DNA repair, which together result in an increased proliferation and infectivity capability to host cells. Some of these phenotypic traits have been corroborated in a mutant *Leishmania* line that is deficient in glucose transport [100]. A detailed description of the association between fitness and drug resistance in *Leishmania* can be found in Chap. 15, this same volume. Proteomic studies of antimonial resistance have also shown that enzymes that are precursors of trypanothione, such as *S*-adenosylmethionine synthetase (SAMS) and *S*-adenosylhomocysteine hydrolase (SAHH), present increased abundance in *L. (V.) panamensis*, *L. (L.) infantum*, and *L. (L.) donovani* resistant parasites [95, 104, 124]. In addition, proteins involved in redox homeostasis, such as tryparedoxin, peroxiredoxin, and pteridine reductase, are also more abundant in resistant parasites [79, 95, 106]. Together, these findings corroborate the hypothesis that antimonial resistance is closely associated with nitrosative and oxidative stress resistance and remodels the parasite thiol redox metabolism.

10.3.3 Proteomics of Miltefosine Resistance

MIL [hexadecylphosphocholine (HePC]), an alkyl phospholipid compound, is the only oral drug currently available for the treatment of leishmaniasis. Originally intended for breast cancer treatment, MIL proved to be effective against *Leishmania* both in vitro and in animal models [192, 193]. This drug was registered and approved for visceral leishmaniasis (VL) treatment in India in 2002, followed by Germany in 2004. In Colombia in 2005, MIL was approved for the treatment of cutaneous leishmaniasis (CL), where it reached cure rates of over 91% [194, 195]. In 2005, the governments of India, Nepal, and Bangladesh adopted MIL as the first-line treatment for VL elimination [196, 197]. The oral administration of MIL avoids the need of patient hospitalization in VL cases and reduces the inconvenience of injectable drugs, augmenting treatment adherence [195, 198]. Despite the recent approval of MIL for disease treatment, clinical failures during treatment of VL and CL caused by different *Leishmania* species have already been reported [199–201]. MIL is registered for the oral treatment of canine leishmaniasis in several European countries since 2007 (Milteforan®) and was authorized recently (2016) for the treatment of dogs with VL in Brazil, despite studies showing that the improvement in the clinical symptoms was not followed by parasitological clearance [202]. In fact, that study did not recommend the use of MIL for dog treatment, especially in endemic areas of Brazil where dogs have a crucial role in the maintenance and transmission of the parasite [202]. In addition, failure treatment has been reported in naturally infected dogs treated with MIL [203].

Although MIL exhibits in vitro activity against various *Leishmania* species [204], the mechanism of action of this compound is not well understood. However, based on evidence obtained in tumor cell lines, it is known that MIL acts by triggering apoptotic pathways [205]. Evidences of PCD induced by MIL have also been reported for *L. (L.) donovani* promastigotes [206]. This drug appears to affect the integrity of cellular membranes by interfering with lipid metabolism, resulting in the decrease of phosphatidylcholine synthesis [207, 208]. In addition, intracellular drug accumulation seems to be required for the drug's activity. Accumulation involves, among other steps, the translocation of the drug across the cellular membrane, which is accomplished with a recently identified complex of proteins including a P-type ATPase termed *L. (L.) donovani* MIL transporter (LdMT) and its β-subunit, LdRos3 [209]. Interestingly, the expression levels of these proteins are diminished in *L. (V.) braziliensis*, which would help to explain the low sensitivity of this species to the drug [210]. In addition, a common feature of MIL-resistant parasites consists of a decrease in drug accumulation mainly due to either the decreased uptake or increased efflux of the drug [211].

As described above, in *L. (L.) donovani* field isolates, HSP83 and SKCRP14.1 were implicated in the modulation of parasite sensitivity to Sb^V through a mechanism involving features of PCD [93]. In the same study, it was observed that Sb^V-resistant parasites were also cross-resistant to both MIL and AMB when compared with the Sb^V-sensitive parasites. It was also shown that the parasites from the Sb^V-unresponsive patient were more resistant to MIL-induced PCD. Besides being resistant to antimonial, the HSP83-overexpressing parasites were also resistant to MIL and were less sensitive to drug-mediated DNA fragmentation when compared to control parasites. In addition, 10 μM MIL first induced a more rapid hyperpolarization of the mitochondria in HSP83 transfectants when compared to the control cells, followed by a depolarization that took place more slowly in HSP83-overexpressing parasites than in control ones [93]. However, the effect of MIL treatment on SKCRP14.1-overexpressing parasites was the opposite of that observed with antimonial treatment. Resistant parasites transfected with SKCRP14.1 became more resistant to MIL compared with the transfectant control. In addition, SKCRP14.1 overexpression was significantly protected against MIL-induced mitochondrial depolarization and led to resistance against MIL-mediated DNA fragmentation when compared with the control [93]. These results reveal the contrasting roles of the proteins in the resistance mechanisms of *Leishmania* and highlight the importance of setting the individual action scenarios for each drug.

Recently, a study was conducted using a MIL-resistant *L. (L.) donovani* isolate, which was selected in vitro by sequential exposure to the drug [97]. In this study, the total cell extracts of sensitive and resistant promastigotes were analyzed by 2DE (pH range, 4–7), and two differentially expressed spots were identified by LC/MS/MS. The identified spots corresponded to the probable eukaryotic initiation factor 4A (eIF4A), a protein belonging to the DEAD-box subfamily of ATP-dependent helicases. This protein participates in the regulation of translation initiation, and it has been reported that its overexpression confers lithium resistance in *Saccharomyces cerevisiae*, probably by restoring protein synthesis [212].

In a recent study, Carnielli et al. used 2D-DIGE/MS to study the differences in protein abundances between MIL-sensitive and MIL-resistant *L. (L.) infantum* isolates from VL patients with different MIL treatment outcomes [96]. Among 46 spots exhibiting different intensity, 22 proteins were identified. Proteins with increased abundance in MIL-resistant isolates were associated with (1) redox homeostasis, such as peroxiredoxin and *S*-adenosylmethionine synthetase (SAMS); (2) stress response, including several HSPs; (3) DNA repair, such as PCNA and mitochondrial ATPase β-subunit; and (4) glycolytic and TCA-related enzymes, among others. A very similar group of proteins was observed in proteomic studies of *L. (L.) infantum* in vitro selected for amphotericin-B resistance [107] and in *L. (L.) donovani* selected for resistance against DL-α-difluoromethylornithine (DFMO), an inhibitor of ornithine decarboxylase, the first enzyme of the polyamine biosynthetic pathway [50] (Table 10.2). These results corroborate the multifactorial character of drug resistance phenomenon in *Leishmania* and also show that irrespective of the chemotherapy used to select the resistant lines, resistant parasites respond in similar ways to the drug pressure (either in vivo or in vitro) exhibiting increased resistance to oxidative and nitrosative stress, remodeling their glycolytic metabolism and increasing their virulence.

10.4 Proteomic Challenges in the Study of Drug Resistance

Proteomic studies of drug resistance in *Leishmania* have increased over the last decade. The reports reviewed here illustrate the value of proteomic approaches for the identification of proteins and mechanisms involved in resistance phenomenon. Those studies show that proteomic screens are useful in defining new roles for already well-characterized proteins in addition to assigning roles for proteins of unknown function. A summary of the proteins identified from proteomic studies using either resistant parasites selected in vitro or parasites from clinical isolates that are considered naturally drug resistant (proteins highlighted by the authors) is presented in Table 10.2. As can be seen in this table, many proteins identified in these studies have been implicated in *Leishmania* drug resistance using other approaches, but many other proteins are new or even unexpected in the scenario of drug resistance.

Several proteins classically described as being involved in various resistance mechanisms are membrane proteins, and this fraction has not been widely explored in drug resistance studies in *Leishmania*. This fact points to the first challenge: the deep analysis of distinct subcellular fractions of the parasite. Besides the contribution of protein annotation, subcellular proteomic analysis offers the possibility of inferring protein function and elucidating biochemical pathways in drug resistance, which can be exploited for purposes of drug development. In addition, it should be taken into account that the approach used for the proteomic studies revisited here has been 2DE, with the already mentioned limitations, applied to whole cell extracts and analyzing only some pH ranges. As a result, a large part of the *Leishmania* "resistance" proteome remains to be analyzed, which points to the second challenge: the need for a comprehensive proteomic study using better solubilizing detergents for sample preparation and gel-free methods [213] that ensures greater coverage of the proteome. Such an approach will require more powerful and specific bioinformatics tools to cope with the analysis of the enormous quantity of data that would be produced. In fact, data analysis represents a considerable bottleneck in the proteomic studies of parasites, mainly because ~50% of the coding genes do not have a functional annotation, which is why it represents the third challenge.

In very nice example of the exploitation of pharmacoproteomics for the study of drug targets and mechanisms of action, Wyllie et al. [53] studied by proteomic and genomic approaches the effects of bicyclic nitro-compounds on *L. (L.) donovani*. Nitro drugs are being used as part of a combination therapy for human African trypanosomiasis (HAT) [214], and bicyclic nitro-compounds are potential candidates for the treatment of VL (www.dndi.org). Comparing susceptible and drug-resistant parasites, authors identified the hypothetical NADH/FMN-dependent oxidoreductase as the activating nitroreductase (NTR2) and demonstrated that its overexpression rendered parasite hypersensitive to bicyclic nitro-compounds. In addition, it was demonstrated that knockout of NTR2 rendered parasites completely resistant to the compounds [53]. This study shows the potential of pharmacoproteomics to study drug mechanisms and resistance in trypanosomatids.

A common trait of proteomic studies in *Leishmania*, as well as in other organisms, is the recurrent identification of a group of proteins that correspond to the most abundant ones [215]. This precludes the identification of the less abundant proteins and obscures the studied phenomenon. Thus, a dedicated analysis of *Leishmania* most abundant proteins with the subsequent construction of an interactive database containing raw mass data and mass spectra data of these proteins would allow a better exploitation of the proteomic studies, saving time and optimizing resources [216]. In addition, the wide use of transfection models and the potential exploitation of a putative RNA interference (RNAi) pathway, at least in *L.* (*Viannia*) parasites [2, 217], would reinforce and complement the proteomic analysis of changes associated with drug resistance. Finally, as far as we know, proteomic studies of the resistance to other drugs used for leishmaniasis treatment, such as pentamidine, paromomycin, and azoles, have not been reported.

10.5 Concluding Remarks

As the cellular proteome is a dynamic scenario, it should be considered that inter- and intraspecific *Leishmania* genetic variation, in addition to host immune responses and host genetic background, might influence the resistant or sensitive phenotype of the parasites [218]. Thus, despite being rich and detailed, proteomic profiles represent specific patterns that need to be contextualized into a "biological system" level where the complexity must be governed by well-defined mechanisms. The continued advances in proteomic technology development, together with genome data and bioinformatics analysis, could reveal effective therapeutic strategies for species-specific treatments in the future, individualizing the epidemiological settings and valorizing the patients [219]. A large endeavor joining expertise, technologies, facilities, and knowledge would be desirable for obtaining and (re-) interpreting proteomic data of drug resistance in *Leishmania*.

References

1. Ivens AC, Peacock CS, Worthey EA, Murphy L, et al. The genome of the kinetoplastid parasite, *Leishmania major*. Science. 2005;309:436–42.
2. Peacock CS, Seeger K, Harris D, Murphy L, et al. Comparative genomic analysis of three *Leishmania* species that cause diverse human disease. Nat Genet. 2007;39:839–47.
3. Imamura H, Downing T, Van den Broeck F, Sanders MJ, et al. Evolutionary genomics of epidemic visceral leishmaniasis in the Indian subcontinent. Elife. 2016;5:e12613.
4. Croft SL, Sundar S, Fairlamb AH. Drug resistance in leishmaniasis. Clin Microbiol Rev. 2006;19:111–26.
5. Faraut-Gambarelli F, Piarroux R, Deniau M, Giusiano B, et al. *In vitro* and *in vivo* resistance of *Leishmania infantum* to meglumine antimoniate: a study of 37 strains collected from patients with visceral leishmaniasis. Antimicrob Agents Chemother. 1997;41:827–30.
6. Lira R, Sundar S, Makharia A, Kenney R, et al. Evidence that the high incidence of treatment failures in Indian kalaazar is due to the emergence of antimony-resistant strains of *Leishmania donovani*. J Infect Dis. 1999;180:564–7.
7. Palacios R, Osorio LE, Grajalew LF, Ochoa MT. Treatment failure in children in a randomized clinical trial with 10 and 20 days of meglumine antimonate for cutaneous leishmaniasis due to *Leishmania Viannia* species. Am J Trop Med Hyg. 2001;64:187–93.
8. Sundar S. Drug resistance in Indian visceral leishmaniasis. Trop Med Int Health. 2001;6:849–54.
9. Abdo MG, Elamin WM, Khalil EA, Mukhtar MM. Antimony-resistant *Leishmania donovani* in eastern Sudan: incidence and *in vitro* correlation. East Mediterr Health J. 2003;9:837–43.
10. Das VN, Ranjan A, Bimal S, Siddique NA, et al. Magnitude of unresponsiveness to sodium stibogluconate in the treatment of visceral leishmaniasis in Bihar. Natl Med J India. 2005;18:131–3.
11. Hadighi R, Mohebali M, Boucher P, Hajjaran H, et al. Unresponsiveness to Glucantime treatment in Iranian cutaneous leishmaniasis due to drug-resistant *Leishmania tropica* parasites. PLoS Med. 2006;3:e162.
12. Rojas R, Valderrama L, Valderrama M, Varona MX, et al. Resistance to antimony and treatment failure in human *Leishmania* (*Viannia*) infection. J Infect Dis. 2006;193:1375–83.
13. Holzer TR, McMaster WR, Forney JD. Expression profiling by whole-genome interspecies microarray hybridization reveals differential gene expression in procyclic promastigotes,

lesion-derived amastigotes, and axenic amastigotes in *Leishmania mexicana*. Mol Biochem Parasitol. 2006;146:198–218.
14. Leifso K, Cohen-Freue G, Dogra N, Murray A, et al. Genomic and proteomic expression analysis of *Leishmania* promastigote and amastigote life stages: the *Leishmania* genome is constitutively expressed. Mol Biochem Parasitol. 2007;152:35–46.
15. Cohen-Freue G, Holzer TR, Forney JD, McMaster WR. Global gene expression in *Leishmania*. Int J Parasitol. 2007;37:1077–86.
16. El-Sayed NM, Myler PJ, Blandin G, Berriman M, et al. Comparative genomics of trypanosomatid parasitic protozoa. Science. 2005;309:404–9.
17. Rogers MB, Hilley JD, Dickens NJ, Wilkes J, et al. Chromosome and gene copy number variation allow major structural change between species and strains of *Leishmania*. Genome Res. 2011;21:2129–42.
18. Fiebig M, Kelly S, Gluenz E. Comparative life cycle transcriptomics revises *Leishmania mexicana* genome annotation and links a chromosome duplication with parasitism of vertebrates. PLoS Pathog. 2015;11:e1005186.
19. Ubeda JM, Légaré D, Raymond F, Ouameur AA, et al. Modulation of gene expression in drug resistant *Leishmania* is associated with gene amplification, gene deletion and chromosome aneuploidy. Genome Biol. 2008;9:R115.
20. Leprohon P, Légaré D, Raymond F, Hardiman G, et al. Gene expression modulation is associated with gene amplification, supernumerary chromosomes and chromosome loss in antimony-resistant *Leishmania infantum*. Nucleic Acids Res. 2009;37:1387–99.
21. Downing T, Imamura H, Decuypere S, Clark TG, et al. Whole genome sequencing of multiple *Leishmania donovani* clinical isolates provides insights into population structure and mechanisms of drug resistance. Genome Res. 2011;21:2143–56.
22. Mannaert A, Downing T, Imamura H, Dujardin JC. Adaptive mechanisms in pathogens: universal aneuploidy in *Leishmania*. Trends Parasitol. 2012;28:370–6.
23. Clayton C, Shapira M. Post-transcriptional regulation of gene expression in trypanosomes and leishmanias. Mol Biochem Parasitol. 2007;156:93–101.
24. Haile S, Papadopoulou B. Developmental regulation of gene expression in trypanosomatid parasitic protozoa. Curr Opin Microbiol. 2007;10:569–77.
25. Lee MG, Atkinson BL, Giannini SH, Van der Ploeg LH. Structure and expression of the hsp 70 gene family of *Leishmania major*. Nucleic Acids Res. 1988;16:9567–85.
26. Quijada L, Soto M, Alonso C, Requena JM. Analysis of post-transcriptional regulation operating on transcription products of the tandemly linked *Leishmania infantum* hsp70 genes. J Biol Chem. 1997;272:4493–9.
27. Holzer TR, Mishra KK, LeBowitz JH, Forney JD. Coordinate regulation of a family of promastigote-enriched mRNAs by the 3′UTR PRE element in *Leishmania mexicana*. Mol Biochem Parasitol. 2008;157:54–64.
28. Saxena A, Lahav T, Holland N, Aggarwal G, et al. Analysis of the *Leishmania donovani* transcriptome reveals an ordered progression of transient and permanent changes in gene expression during differentiation. Mol Biochem Parasitol. 2007;152:53–65.
29. Almeida R, Gilmartin BJ, McCann SH, Norrish A, et al. Expression profiling of the *Leishmania* life cycle: cDNA arrays identify developmentally regulated genes present but not annotated in the genome. Mol Biochem Parasitol. 2004;136:87–100.
30. Akopyants NS, Matlib RS, Bukanova EN, Smeds MR, et al. Expression profiling using random genomic DNA microarrays identifies differentially expressed genes associated with three major developmental stages of the protozoan parasite *Leishmania major*. Mol Biochem Parasitol. 2004;136:71–86.
31. Rochette A, Raymond F, Ubeda JM, Smith M, et al. Genome-wide gene expression profiling analysis of *Leishmania major* and *Leishmania infantum* developmental stages reveals substantial differences between the two species. BMC Genomics. 2008;9:255.

32. Guimond C, Trudel N, Brochu C, Marquis N, et al. Modulation of gene expression in *Leishmania* drug resistant mutants as determined by targeted DNA microarrays. Nucleic Acids Res. 2003;31:5886–96.
33. Quijada L, Soto M, Requena JM. Genomic DNA macroarrays as a tool for analysis of gene expression in *Leishmania*. Exp Parasitol. 2005;111:64–70.
34. Depledge DP, Evans KJ, Ivens AC, Aziz N, et al. Comparative expression profiling of *Leishmania*: modulation in gene expression between species and in different host genetic backgrounds. PLoS Negl Trop Dis. 2009;3:e476.
35. Leprohon P, Légaré D, Girard I, Papadopoulou B, et al. Modulation of *Leishmania* ABC protein gene expression through life stages and among drug-resistant parasites. Eukaryot Cell. 2006;5:1713–25.
36. McNicoll F, Drummelsmith J, Müller M, Madore E, et al. A combined proteomic and transcriptomic approach to the study of stage differentiation in *Leishmania infantum*. Proteomics. 2006;6:3567–81.
37. Walther TC, Mann M. Mass spectrometry-based proteomics in cell biology. J Cell Biol. 2010;190:491–500.
38. Karas M, Hillemkamp F. Laser desorption ionization of proteins with molecular masses exceeding 10000 kDa. Anal Chem. 1988;60:2299–301.
39. Tanaka K, Waki H, Ido Y, Akita S, et al. Protein and polymer analyses up to m/z 100000 by laser ionization time-of-flight mass spectrometry. Rapid Commun Mass Spectrom. 1988;2:151–3.
40. Fenn J, Mann M, Meng CK, Wong SF, et al. Electrospray ionization for mass spectrometry of large biomolecules. Science. 1989;246:64–71.
41. Cuervo P, Domont GB, De Jesus JB. Proteomics of trypanosomatids of human medical importance. J Proteomics. 2010;73:845–67.
42. Paape D, Aebischer T. Contribution of proteomics of *Leishmania* spp. to the understanding of differentiation, drug resistance mechanisms, vaccine and drug development. J Proteomics. 2011;74:1614–24.
43. Paape D, Barrios-Llerena ME, Le BT, Mackay L, et al. Gel free analysis of the proteome of intracellular *Leishmania mexicana*. Mol Biochem Parasitol. 2010;169:108–14.
44. Tsigankov P, Gherardini PF, Helmer-Citterich M, Späth GF, et al. Phosphoproteomic analysis of differentiating *Leishmania* parasites reveals a unique stage-specific phosphorylation motif. J Proteome Res. 2013;12:3405–12.
45. Braga MS, Neves LX, Campos JM, Roatt BM, et al. Shotgun proteomics to unravel the complexity of the *Leishmania infantum* exoproteome and the relative abundance of its constituents. Mol Biochem Parasitol. 2014;195:43–53.
46. Pawar H, Sahasrabuddhe NA, Renuse S, Keerthikumar S, et al. A proteogenomic approach to map the proteome of an unsequenced pathogen - *Leishmania donovani*. Proteomics. 2012;12:832–44.
47. Pescher P, Blisnick T, Bastin P, Spath GF. Quantitative proteome profiling informs on phenotypic traits that adapt *Leishmania donovani* for axenic and intracellular proliferation. Cell Microbiol. 2011;13:978–91.
48. Biyani N, Madhubala R. Quantitative proteomic profiling of the promastigotes and the intracellular amastigotes of *Leishmania donovani* isolates identifies novel proteins having a role in *Leishmania* differentiation and intracellular survival. Biochim Biophys Acta. 2012;1824:1342–50.
49. Sardar AH, Kumar S, Kumar A, Purkait B, et al. Proteome changes associated with *Leishmania donovani* promastigote adaptation to oxidative and nitrosative stresses. J Proteomics. 2013;81:185–99.
50. Singh AK, Roberts S, Ullman B, Madhubala R. A quantitative proteomic screen to identify potential drug resistance mechanism in alpha-difluoromethylornithine (DFMO) resistant *Leishmania donovani*. J Proteomics. 2014;102:44–59.

51. Tsigankov P, Gherardini PF, Helmer-Citterich M, Späth GF, et al. Regulation dynamics of *Leishmania* differentiation: deconvoluting signals and identifying phosphorylation trends. Mol Cell Proteomics. 2014;13:1787–99.
52. Zilberstein D. Proteomic analysis of posttranslational modifications using iTRAQ in *Leishmania*. Methods Mol Biol. 2015;1201:261–8.
53. Wyllie S, Roberts AJ, Norval S, Patterson S, et al. Activation of bicyclic nitro-drugs by a novel nitroreductase (NTR2) in *Leishmania*. PLoS Pathog. 2016;12:e1005971.
54. Acestor N, Masina S, Walker J, Saravia NG, et al. Establishing two-dimensional gels for the analysis of *Leishmania* proteomes. Proteomics. 2002;2:877–9.
55. Góngora R, Acestor N, Quadroni M, Fasel N, et al. Mapping the proteome of *Leishmania Viannia* parasites using two-dimensional polyacrylamide gel electrophoresis and associated technologies. Biomédica. 2003;23:153–60.
56. Drummelsmith J, Brochu V, Girard I, Messier N, et al. Proteome mapping of the protozoan parasite *Leishmania* and application to the study of drug targets and resistance mechanisms. Mol Cell Proteomics. 2003;2:146–55.
57. Brobey RK, Mei FC, Cheng X, Soong L. Comparative two-dimensional gel electrophoresis maps for promastigotes of *Leishmania amazonensis* and *Leishmania major*. Braz J Infect Dis. 2006;10:1–6.
58. Brobey RK, Soong L. Establishing a liquid-phase IEF in combination with 2-DE for the analysis of *Leishmania* proteins. Proteomics. 2007;7:116–20.
59. Cuervo P, de Jesus JB, Junqueira M, Mendonça-Lima L, et al. Proteome analysis of *Leishmania (Viannia) braziliensis* by two-dimensional gel electrophoresis and mass spectrometry. Mol Biochem Parasitol. 2007;154:6–21.
60. Costa MM, Andrade HM, Bartholomeu DC, Freitas LM, et al. Analysis of *Leishmania chagasi* by 2-D difference gel electrophoresis (2-D DIGE) and immunoproteomic: identification of novel candidate antigens for diagnostic tests and vaccine. J Proteome Res. 2011;10:2172–84.
61. Aebischer T. *Leishmania* spp. proteome data sets: a comprehensive resource for vaccine development to target visceral leishmaniasis. Front Immunol. 2014;5:260.
62. da Fonseca Pires S, Fialho LC Jr, Silva SO, Melo MN, et al. Identification of virulence factors in *Leishmania infantum* strains by a proteomic approach. J Proteome Res. 2014;13:1860–72.
63. McCall LI, Zhang WW, Dejgaard K, Atayde VD, et al. Adaptation of *Leishmania donovani* to cutaneous and visceral environments: in vivo selection and proteomic analysis. J Proteome Res. 2015;14:1033–59.
64. Alcolea PJ, Tuñón GI, Alonso A, García-Tabares F, et al. Differential protein abundance in promastigotes of nitric oxide-sensitive and resistant *Leishmania chagasi* strains. Proteomics Clin Appl. 2016;10:1132–46.
65. Yau WL, Lambertz U, Colineau L, Pescher P, et al. Phenotypic characterization of a *Leishmania donovani* cyclophilin 40 null mutant. J Eukaryot Microbiol. 2016;63:823–33.
66. Thiel M, Bruchhaus I. Comparative proteome analysis of *Leishmania donovani* at different stages of transformation from promastigotes to amastigotes. Med Microbiol Immunol. 2001;190:33–6.
67. El Fakhry Y, Ouellette M, Papadopoulou B. A proteomic approach to identify developmentally regulated proteins in *Leishmania infantum*. Proteomics. 2002;2:1007–17.
68. Bente M, Harder S, Wiesgigl M, Heukeshoven J, et al. Developmentally induced changes of the proteome in the protozoan parasite *Leishmania donovani*. Proteomics. 2003;3:1811–29.
69. Nugent PG, Karsani SA, Wait R, Tempero J, et al. Proteomic analysis of *Leishmania mexicana* differentiation. Mol Biochem Parasitol. 2004;136:51–62.
70. Walker J, Vasquez JJ, Gomez MA, Drummelsmith J, et al. Identification of developmentally-regulated proteins in *Leishmania panamensis* by proteome profiling of promastigotes and axenic amastigotes. Mol Biochem Parasitol. 2006;147:64–73.
71. Foucher AL, Papadopoulou B, Ouellette M. Prefractionation by digitonin extraction increases representation of the cytosolic and intracellular proteome of *Leishmania infantum*. J Proteome Res. 2006;5:1741–50.

72. Rosenzweig D, Smith D, Opperdoes F, Stern S, et al. Retooling *Leishmania* metabolism: from sand fly gut to human macrophage. FASEB J. 2008a;22:590–602.
73. Morales MA, Watanabe R, Laurent C, Lenormand P, et al. Phosphoproteomic analysis of *Leishmania donovani* pro- and amastigote stages. Proteomics. 2008;8:350–63.
74. Mojtahedi Z, Clos J, Kamali-Sarvestani E. *Leishmania major*: identification of developmentally regulated proteins in procyclic and metacyclic promastigotes. Exp Parasitol. 2008;119:422–9.
75. Paape D, Lippuner C, Schmid M, Ackermann R, et al. Transgenic, fluorescent *Leishmania mexicana* allow direct analysis of the proteome of intracellular amastigotes. Mol Cell Proteomics. 2008;7:1688–701.
76. Nirujogi RS, Pawar H, Renuse S, Kumar P, et al. Moving from unsequenced to sequenced genome: reanalysis of the proteome of *Leishmania donovani*. J Proteomics. 2014;97:48–61.
77. Rosenzweig D, Smith D, Myler PJ, Olafson RW, et al. Post-translational modification of cellular proteins during *Leishmania donovani* differentiation. Proteomics. 2008b;8:1843–50.
78. Hem S, Gherardini PF, Osorio y Fortéa J, Hourdel V, et al. Identification of *Leishmania*-specific protein phosphorylation sites by LC-ESI-MS/MS and comparative genomics analyses. Proteomics. 2010;10:3868–83.
79. Moreira D de S, Pescher P, Laurent C, Lenormand P, et al. Phosphoproteomic analysis of wild-type and antimony-resistant *Leishmania braziliensis* lines by 2D-DIGE technology. Proteomics. 2015;15:2999–3019.
80. Bachmaier S, Witztum R, Tsigankov P, Koren R, et al. Protein kinase A signaling during bidirectional axenic differentiation in *Leishmania*. Int J Parasitol. 2016;46:75–82.
81. Morales MA, Watanabe R, Dacher M, Chafey P, et al. Phosphoproteome dynamics reveal heat-shock protein complexes specific to the *Leishmania donovani* infectious stage. Proc Natl Acad Sci USA. 2010;107:8381–6.
82. de Oliveira AH, Ruiz JC, Cruz AK, Greene LJ, et al. Subproteomic analysis of soluble proteins of the microsomal fraction from two *Leishmania* species. Comp Biochem Physiol Part D Genomic Proteomics. 2006;1:300–8.
83. Hide M, Ritleng AS, Brizard JP, Monte-Allegre A, et al. *Leishmania infantum*: tuning digitonin fractionation for comparative proteomic of the mitochondrial protein content. Parasitol Res. 2008;103:989–92.
84. Silverman JM, Chan SK, Robinson DP, Dwyer DM, et al. Proteomic analysis of the secretome of *Leishmania donovani*. Genome Biol. 2008;9:R35.
85. Cuervo P, De Jesus JB, Saboia-Vahia L, Mendonça-Lima L, et al. Proteomic characterization of the released/secreted proteins of *Leishmania* (*Viannia*) *braziliensis* promastigotes. J Proteomics. 2009;73:79–92.
86. Brotherton MC, Racine G, Ouameur AA, Leprohon P, et al. Analysis of membrane-enriched and high molecular weight proteins in *Leishmania infantum* promastigotes and axenic amastigotes. J Proteome Res. 2012;11:3974–85.
87. Lynn MA, Marr AK, McMaster WR. Differential quantitative proteomic profiling of *Leishmania infantum* and *Leishmania mexicana* density gradient separated membranous fractions. J Proteomics. 2013;82:179–92.
88. Santarém N, Racine G, Silvestre R, Cordeiro-da-Silva A, et al. Exoproteome dynamics in *Leishmania infantum*. J Proteomics. 2013;84:106–18.
89. Atayde VD, Aslan H, Townsend S, Hassani K, et al. Exosome secretion by the parasitic protozoan *Leishmania* within the sand fly midgut. Cell Rep. 2015;13:957–67.
90. Kumar A, Misra P, Sisodia B, Shasany AK, et al. Proteomic analyses of membrane enriched proteins of *Leishmania donovani* Indian clinical isolate by mass spectrometry. Parasitol Int. 2015;64:36–42.
91. Lima BS, Fialho LC Jr, Pires SF, Tafuri WL, et al. Immunoproteomic and bioinformatic approaches to identify secreted *Leishmania amazonensis*, *L. braziliensis*, and *L. infantum* proteins with specific reactivity using canine serum. Vet Parasitol. 2016;223:115–9.

92. Drummelsmith J, Girard I, Trudel N, Ouellette M. Differential protein expression analysis of *Leishmania major* reveals novel roles for methionine adenosyltransferase and S-adenosylmethionine in methotrexate resistance. J Biol Chem. 2004;279:33273–80.
93. Vergnes B, Gourbal B, Girard I, Sundar S, et al. A proteomics screen implicates HSP83 and a small kinetoplastid calpain-related protein in drug resistance in *Leishmania donovani* clinical field isolates by modulating drug-induced programmed cell death. Mol Cell Proteomics. 2007;6:88–101.
94. El Fadili K, Drummelsmith J, Roy G, Jardim A, et al. Down regulation of KMP-11 in *Leishmania infantum* axenic antimony resistant amastigotes as revealed by a proteomic screen. Exp Parasitol. 2009;123:51–7.
95. Matrangolo FS, Liarte DB, Andrade LC, de Melo MF, et al. Comparative proteomic analysis of antimony-resistant and -susceptible *Leishmania braziliensis* and *Leishmania infantum chagasi* lines. Mol Biochem Parasitol. 2013;190:63–75.
96. Carnielli JB, de Andrade HM, Pires SF, Chapeaurouge AD, et al. Proteomic analysis of the soluble proteomes of miltefosine-sensitive and -resistant *Leishmania infantum chagasi* isolates obtained from Brazilian patients with different treatment outcomes. J Proteomics. 2014;108:198–208.
97. Singh G, Chavan HD, Dey CS. Proteomic analysis of miltefosine-resistant *Leishmania* reveals the possible involvement of eukaryotic initiation factor 4A (eIF4A). Int J Antimicrob Agents. 2008a;31:584–6.
98. Singh G, Jayanarayan KG, Dey CS. Arsenite resistance in *Leishmania* and possible drug targets. Adv Exp Med Biol. 2008b;625:1–8.
99. Vincent IM, Racine G, Légaré D, Ouellette M. Mitochondrial proteomics of antimony and miltefosine resistant *Leishmania infantum*. Proteomes. 2015;3:328–46.
100. Akpunarlieva S, Weidt S, Lamasudin D, Naula C, et al. Integration of proteomics and metabolomics to elucidate metabolic adaptation in *Leishmania*. J Proteomics. 2017;155: 85–98.
101. Sharma S, Singh G, Chavan HD, Dey CS. Proteomic analysis of wild type and arsenite-resistant *Leishmania donovani*. Exp Parasitol. 2009;123:369–76.
102. Kumar A, Sisodia B, Misra P, Sundar S, et al. Proteome mapping of overexpressed membrane-enriched and cytosolic proteins in sodium antimony gluconate (SAG) resistant clinical isolate of *Leishmania donovani*. Br J Clin Pharmacol. 2010;70:609–17.
103. Peláez RG, Muskus CE, Cuervo P, Marín-Villa M. Differential expression of proteins in *Leishmania (Viannia) panamensis* associated with mechanisms of resistance to meglumine antimoniate. Biomedica. 2012;32:418–29.
104. Walker J, Gongora R, Vasquez JJ, Drummelsmith J, et al. Discovery of factors linked to antimony resistance in *Leishmania panamensis* through differential proteome analysis. Mol Biochem Parasitol. 2012;183:166–76.
105. Messaritakis I, Christodoulou V, Mazeris A, Koutala E, et al. Drug resistance in natural isolates of *Leishmania donovani* s.l. promastigotes is dependent of Pgp170 expression. PLoS One. 2013;8:e65467.
106. Brotherton MC, Bourassa S, Leprohon P, Légaré D, et al. Proteomic and genomic analyses of antimony resistant *Leishmania infantum* mutant. PLoS One. 2013;8:e81899.
107. Brotherton MC, Bourassa S, Légaré D, Poirier GG, et al. Quantitative proteomic analysis of amphotericin B resistance in *Leishmania infantum*. Int J Parasitol Drugs Drug Resist. 2014;4:126–32.
108. Vacchina P, Norris-Mullins B, Carlson ES, Morales MA. A mitochondrial HSP70 (HSPA9B) is linked to miltefosine resistance and stress response in *Leishmania donovani*. Parasit Vectors. 2016;9:621.
109. MacGillivray AJ, Rickwood D. The heterogeneity of mouse-chromatin nonhistone proteins as evidenced by two-dimensional polyacrylamide-gel electrophoresis and ion-exchange chromatography. Eur J Biochem. 1974;41:181–90.

110. O'Farrell PH. High resolution two-dimensional electrophoresis of proteins. J Biol Chem. 1975;250:4007–21.
111. Anderson L, Anderson NG. High resolution two-dimensional electrophoresis of human plasma proteins. Proc Natl Acad Sci USA. 1977;74:5421–5.
112. Bravo R, Celis JE. A search for differential polypeptide synthesis throughout the cell cycle of HeLa cells. J Cell Biol. 1980;84:795–802.
113. Taylor J, Anderson NL, Scandora AE Jr, Willard KE, et al. Design and implementation of a prototype human protein index. Clin Chem. 1982;28:861–6.
114. Handman E, Mitchell GF, Goding JW. Identification and characterization of protein antigens of *Leishmania tropica* isolates. J Immunol. 1981;126:508–12.
115. Saravia NG, Gemmell MA, Nance SL, Anderson NL. Two-dimensional electrophoresis used to differentiate the causal agents of American tegumentary leishmaniasis. Clin Chem. 1984;30:2048–52.
116. Fong D, Chang KP. Tubulin biosynthesis in the developmental cycle of a parasitic protozoan, *Leishmania mexicana*: changes during differentiation of motile and nonmotile stages. Proc Natl Acad Sci USA. 1981;78:7624–8.
117. Arrebola R, Olmo A, Reche P, Garvey EP, et al. Isolation and characterization of a mutant dihydrofolate reductase-thymidylate synthase from methotrexate-resistant *Leishmania* cells. J Biol Chem. 1994;269:10590–6.
118. Görg A, Postel W, Günther S. The current state of two-dimensional electrophoresis with immobilized pH gradients. Electrophoresis. 1988;9:531–46.
119. Matsudaira PT. Sequence from picomole quantities of proteins electroblotted onto polyvinylidene difluoride membranes. J Biol Chem. 1987;262:10035–8.
120. Carrette O, Burkhard PR, Sanchez JC, Hochstrasser DF. State-of-the-art two-dimensional gel electrophoresis: a key tool of proteomics research. Nat Protoc. 2006;1:812–23.
121. Rabilloud T, Chevallet M, Luche S, Lelong C. Two-dimensional gel electrophoresis in proteomics: past, present and future. J Proteomics. 2010;73:2064–77.
122. Bantscheff M, Lemeer S, Savitski MM, Kuster B. Quantitative mass spectrometry in proteomics: critical review update from 2007 to the present. Anal Bioanal Chem. 2012;404:939–65.
123. Duncan MW, Aebersold R, Caprioli RM. The pros and cons of peptide-centric proteomics. Nat Biotechnol. 2010;28:659–64.
124. Biyani N, Singh AK, Mandal S, Chawla B. Differential expression of proteins in antimony-susceptible and -resistant isolates of *Leishmania donovani*. Mol Biochem Parasitol. 2011;179:91–9.
125. Altelaar AF, Munoz J, Heck AJ. Next-generation proteomics: towards an integrative view of proteome dynamics. Nat Rev Genet. 2013;14:35–48.
126. Wiśniewski JR, Hein M, Cox J, Mann M. A "Proteomic Ruler" for protein copy number and concentration estimation without spike-in standards. Mol Cell Proteomics. 2014;13:3497–506.
127. Reddy PJ, Jain R, Paik YK, Downey R, et al. Personalized medicine in the age of pharmacoproteomics: a close up on India and need for social science engagement for responsible innovation in post-proteomic biology. Curr Pharmacogenomics Person Med. 2011;9: 67–75.
128. Matthews H, Hanison J, Nirmalan N. "Omics"-informed drug and biomarker discovery: opportunities, challenges and future perspectives. Proteomes. 2016;4:E28.
129. Goldstein RL, Yang SN, Taldone T, Chang B, et al. Pharmacoproteomics identifies combinatorial therapy targets for diffuse large B cell lymphoma. J Clin Invest. 2015;125:4559–71.
130. Shu S, Lin CY, He HH, Witwicki RM, et al. Response and resistance to BET bromodomain inhibitors in triple-negative breast cancer. Nature. 2016;529:413–7.
131. Detke S, Katakura K, Chang KP. DNA amplification in arsenite resistant *Leishmania*. Exp Cell Res. 1989;180:161–70.
132. Ouellette M, Hettema E, Wust D, Fase-Fowler F, et al. Direct and inverted DNA repeats associated with P-glycoprotein gene amplification in drug resistant *Leishmania*. EMBO J. 1991;10:1009–16.

133. Callahan HL, Beverley SM. Heavy metal resistance: A new role for P-glycoproteins in *Leishmania*. J Biol Chem. 1991;266:18427–30.
134. Bello AR, Nare B, Freedman D, Hardy L, et al. PTR1: a reductase mediating salvage of oxidized pteridines and methotrexate resistance in the protozoan parasite *Leishmania major*. Proc Natl Acad Sci USA. 1994;91:11442–6.
135. Prasad V, Kaur J, Dey CS. Arsenite-resistant *Leishmania donovani* promastigotes express an enhanced membrane P-type adenosine triphosphatase activity that is sensitive to verapamil treatment. Parasitol Res. 2000;86:661–4.
136. Richard D, Kündig C, Ouellette M. A new type of high affinity folic acid transporter in the protozoan parasite *Leishmania* and deletion of its gene in methotrexate-resistant cells. J Biol Chem. 2002;277:29460–7.
137. HaimeurA GC, Pilote S, Mukhopadhyay R, Rosen BP, et al. Elevated levels of polyamines and trypanothione resulting from overexpression of the ornithine decarboxylase gene in arsenite-resistant *Leishmania*. Mol Microbiol. 1999;34:726–35.
138. Brochu C, Wang J, Roy G, Messier N, et al. Antimony uptake systems in the protozoan parasite *Leishmania* and accumulation differences in antimony-resistant parasites. Antimicrob Agents Chemother. 2003;47:3073–9.
139. Ouellette M, Drummelsmith J, Papadopoulou B. Leishmaniasis: drugs in the clinic, resistance and new developments. Drug Resist Updat. 2004;7:257–66.
140. Ashutosh SS, Goyal N. Molecular mechanisms of antimony resistance in *Leishmania*. J Med Microbiol. 2007;56:143–53.
141. Ouellette M, Borst P. Drug resistance and P-glycoprotein gene amplification in the protozoan parasite *Leishmania*. Res Microbiol. 1991;142:737–46.
142. Beverley SM. Gene amplification in *Leishmania*. Annu Rev Microbiol. 1991;45:417–44.
143. Dey S, Papadopoulou B, Haimeur A, Roy G, et al. High level arsenite resistance in *Leishmania* tarentolae is mediated by an active extrusion system. Mol Biochem Parasitol. 1994;67:49–57.
144. Dey S, Ouellette M, Lightbody J, Papadopoulou B, Rosen BP. An ATP-dependent As(III)-glutathione transport system in membrane vesicles of *Leishmania tarentolae*. Proc Natl Acad Sci USA. 1996;93:2192–7.
145. Mukhopadhyay R, Dey S, Xu N, Gage D, et al. Trypanothione overproduction and resistance to antimonials and arsenicals in *Leishmania*. Proc Natl Acad Sci USA. 1996;93:10383–7.
146. Coderre JA, Beverley SM, Schimke RT, Santi DV. Overproduction of a bifunctional thymidylate synthetase-dihydrofolate reductase and DNA amplification in methotrexate-resistant *Leishmania tropica*. Proc Natl Acad Sci USA. 1983;80:2132–6.
147. Croft SL, Coombs GH. Leishmaniasis—current chemotherapy and recent advances in the search for novel drugs. Trends Parasitol. 2003;19:502–8.
148. Thakur CP, Sinha GP, Pandey AK, Kumar N, et al. Do the diminishing efficacy and increasing toxicity of sodium stibogluconate in the treatment of visceral leishmaniasis in Bihar, India, justify its continued use as a first-line drug? An observational study of 80 cases. Ann Trop Med Parasitol. 1998;92:561–9.
149. Sundar S, More DK, Singh MK, Singh VP, et al. Failure of pentavalent antimony in visceral leishmaniasis in India: report from the center of the Indian epidemic. Clin Infect Dis. 2000;31: 1104–1107s.
150. Perry MR, Wyllie S, Prajapati VK, Feldmann J, et al. Visceral leishmaniasis and arsenic: an ancient poison contributing to antimonial treatment failure in the Indian subcontinent? PLoS Negl Trop Dis. 2011;5:e1227.
151. Berman JD, Chulay JD, Hendricks LD, Oster CN. Susceptibility of clinically sensitive and resistant *Leishmania* to pentavalent antimony *in vitro*. Am J Trop Med Hyg. 1982;31:459–65.
152. Navin TR, Arana BA, Arana FE, Berman JD, et al. Placebo-controlled clinical trial of sodium stibogluconate (Pentostam) versus ketoconazole for treating cutaneous leishmaniasis in Guatemala. J Infect Dis. 1992;165:528–34.
153. Burguera J, Burguera M, Petit de Pena Y, Lugo A, et al. Selective determination of antimony (III) and antimony(V) in serum and urine and of total antimony in skin biopsies of patients with

cutaneous leishmaniasis treated with meglumine antimoniate. Trace Elem Med. 1993;10: 66–70.
154. Callahan HL, Portal AC, Devereaux R, Grogl M. An axenic amastigote system for drug screening. Antimicrob Agents Chemother. 1997;41:818–22.
155. Ephros M, Bitnun A, Shaked P, Waldman E, Zilberstein D. Stage-specific activity of pentavalent antimony against *Leishmania donovani* axenic amastigotes. Antimicrob Agents Chemother. 1999;43:278–82.
156. Shaked-Mishan P, Ulrich N, Ephros M, Zilberstein D. Novel intracellular SbV reducing activity correlates with antimony susceptibility in *Leishmania donovani*. J Biol Chem. 2001; 276:3971–6.
157. Denton H, McGregor JC, Coombs GH. Reduction of anti-leishmanial pentavalent antimonial drugs by a parasite-specific thiol-dependent reductase, TDR1. Biochem J. 2004;381:405–12.
158. Zhou Y, Messier N, Ouellette M, Rosen BP, et al. *Leishmania major* LmACR2 is a pentavalent antimony reductase that confers sensitivity to the drug pentostam. J Biol Chem. 2004; 279:37445–51.
159. Mukhopadhyay R, Bisacchi D, Zhou Y, Armirotti A, et al. Structural characterization of the As/Sb reductase LmACR2 from *Leishmania major*. J Mol Biol. 2009;386:1229–39.
160. Frézard F, Demicheli C, Ferreira CS, Costa MA. Glutathione-induced conversion of pentavalent antimony to trivalent antimony in meglumine antimoniate. Antimicrob Agents Chemother. 2001;45:913–6.
161. Tsukaguchi H, Shayakul C, Berger UV, Mackenzie B, et al. Molecular characterization of a broad selectivity neutral solute channel. J Biol Chem. 1998;273:24737–43.
162. Gourbal B, Sonuc N, Bhattacharjee H, Legare D, et al. Drug uptake and modulation of drug resistance in *Leishmania* by an aquaglyceroporin. J Biol Chem. 2004;279:31010–7.
163. Decuypere S, Rijal S, Yardley V, De Doncker S, et al. Gene expression analysis of the mechanism of natural Sb(V) resistance in *Leishmania donovani* isolates from Nepal. Antimicrob Agents Chemother. 2005;49:4616–21.
164. Marquis N, Gourbal B, Rosen BP, Mukhopadhyay R. Modulation in aquaglyceroporin AQP1 gene transcript levels in drug-resistant *Leishmania*. Mol Microbiol. 2005;57:1690–9.
165. Wyllie S, Cunningham ML, Fairlamb AH. Dual action of antimonial drugs on thiol redox metabolism in the human pathogen *Leishmania donovani*. J Biol Chem. 2004;279:39925–32.
166. Berman JD, Waddell D, Hanson BD. Biochemical mechanisms of the antileishmanial activity of sodium stibogluconate. Antimicrob Agents Chemother. 1985;27:916–20.
167. Sereno D, Holzmuller P, Mangot I, Cuny G. Antimonial-mediated DNA fragmentation in *Leishmania infantum* amastigotes. Antimicrob Agents Chemother. 2001;45:2064–9.
168. Sudhandiran G, Shaha C. Antimonial-induced increase in intracellular Ca2+ through non-selective cation channels in the host and the parasite is responsible for apoptosis of intracellular *Leishmania donovani* amastigotes. J Biol Chem. 2003;278:25120–32.
169. Grondin K, Haimeur A, Mukhopadhyay R, Rosen BP, et al. Co-amplification of the gamma-glutamylcysteine synthetase gene gsh1 and of the ABC transporter gene pgpA in arsenite-resistant *Leishmania* tarentolae. EMBO J. 1997;16:3057–65.
170. Haimeur A, Brochu C, Genest P, Papadopoulou B, et al. Amplification of the ABC transporter gene PGPA and increased trypanothione levels in potassium antimonyl tartrate (SbIII) resistant *Leishmania tarentolae*. Mol Biochem Parasitol. 2000;108:131–5.
171. Callahan HL, Roberts WL, Rainey PM, Beverley SM. The PGPA gene of *Leishmania major* mediates antimony (SbIII) resistance by decreasing influx and not by increasing efflux. Mol Biochem Parasitol. 1994;68:145–9.
172. Légaré D, Richard D, Mukhopadhyay R, Stierhof YD, et al. The *Leishmania* ATP-binding cassette protein PGPA is an intracellular metal-thiol transporter ATPase. J Biol Chem. 2001; 276:26301–7.
173. Mittal MK, Rai S, Ravinder GS, Sundar S, et al. Characterization of natural antimony resistance in *Leishmania donovani* isolates. Am J Trop Med Hyg. 2007;76:681–8.

174. Goyeneche-Patino DA, Valderrama L, Walker J, Saravia NG. Antimony resistance and trypanothione in experimentally selected and clinical strains of *Leishmania panamensis*. Antimicrob Agents Chemother. 2008;52:4503–6.
175. Mukherjee A, Padmanabhan PK, Singh S, Roy G, et al. Role of ABC transporter MRPA, gamma-glutamylcysteine synthetase and ornithine decarboxylase in natural antimony-resistant isolates of *Leishmania donovani*. J Antimicrob Chemother. 2007;59:204–11.
176. Mandal G, Sarkar A, Saha P, Singh N, et al. Functionality of drug efflux pumps in antimonial resistant *Leishmania donovani* field isolates. Indian J Biochem Biophys. 2009;46:86–92.
177. Monte-Neto R, Laffitte MC, Leprohon P, Reis P, et al. Intrachromosomal amplification, locus deletion and point mutation in the aquaglyceroporin AQP1 gene in antimony resistant *Leishmania (Viannia) guyanensis*. PLoS Negl Trop Dis. 2015;9:e0003476.
178. Hefnawy A, Berg M, Dujardin JC, De Muylder G. Exploiting knowledge on *Leishmania* drug resistance to support the quest for new drugs. Trends Parasitol. 2017;33:162–74.
179. Jardim A, Hanson S, Ullman B, McCubbin WD, et al. Cloning and structure-function analysis of the *Leishmania donovani* kinetoplastid membrane protein-11. Biochem J. 1995;305:315–20.
180. Fuertes MA, Berberich C, Lozano RM, Gimenez-Gallego G, et al. Folding stability of the kinetoplastid membrane protein-11 (KMP-11) from *Leishmania infantum*. Eur J Biochem. 1999;260:559–67.
181. Lee N, Bertholet S, Debrabant A, Muller J, et al. Programmed cell death in the unicellular protozoan parasite *Leishmania*. Cell Death Differ. 2002;9:53–64.
182. Cohen-Saidon C, Carmi I, Keren A, Razin E. Antiapoptotic function of Bcl-2 in mast cells is dependent on its association with heat shock protein 90. Blood. 2006;107:1413–20.
183. Das S, Shah P, Tandon R, Yadav NK, et al. Over-expression of cysteine leucine rich protein is related to SAG resistance in clinical isolates of *Leishmania donovani*. PLoS Negl Trop Dis. 2015;9:e0003992.
184. Das S, Shah P, Baharia RK, Tandon R, et al. Over-expression of 60s ribosomal L23a is associated with cellular proliferation in SAG resistant clinical isolates of *Leishmania donovani*. PLoS Negl Trop Dis. 2013;7:e2527.
185. Dridi L, Ahmed Ouameur A, Ouellette M. High affinity S-Adenosylmethionine plasma membrane transporter of *Leishmania* is a member of the folate biopterin transporter (FBT) family. J Biol Chem. 2010;285:19767–75.
186. Vickers TJ, Beverley SM. Folate metabolic pathways in *Leishmania*. Essays Biochem. 2011;51:63–80.
187. Wiśniewski JR, Zougman A, Mann M. Combination of FASP and StageTip-based fractionation allows in-depth analysis of the hippocampal membrane proteome. J Proteome Res. 2009a;8:5674–8.
188. Wiśniewski JR, Duś-Szachniewicz K, Ostasiewicz P, Ziółkowski P, et al. Absolute proteome analysis of colorectal mucosa, adenoma, and cancer reveals drastic changes in fatty acid metabolism and plasma membrane transporters. J Proteome Res. 2015;14:4005–18.
189. Vildhede A, Wiśniewski JR, Norén A, Karlgren M, et al. Comparative proteomic analysis of human liver tissue and isolated hepatocytes with a focus on proteins determining drug exposure. J Proteome Res. 2015;14:3305–14.
190. Tandon R, Chandra S, Baharia RK, Das S, et al. Characterization of the proliferating cell nuclear antigen of *Leishmania donovani* clinical isolates and its association with antimony resistance. Antimicrob Agents Chemother. 2014;58:2997–3007.
191. Parodi-Talice A, Durán R, Arrambide N, Prieto V, et al. Proteome analysis of the causative agent of Chagas disease: *Trypanosoma cruzi*. Int J Parasitol. 2004;34:881–6.
192. Croft SL, Neal RA, Pendergast W, Chan JH. The activity of alkyl phosphorylcholines and related derivatives against *Leishmania donovani*. Biochem Pharmacol. 1987;36:2633–6.
193. Kuhlencord A, Maniera T, Eibl H, Unger C. Hexadecylphosphocholine: oral treatment of visceral leishmaniasis in mice. Antimicrob Agents Chemother. 1992;36:1630–4.

194. Sundar S, Jha TK, Thakur CP, Engel J, et al. Oral miltefosine for Indian visceral leishmaniasis. N Engl J Med. 2002;347:1739–46.
195. Soto J, Soto P. Miltefosine: oral treatment of leishmaniasis. Expert Rev Anti Infect Ther. 2006;4:177–85.
196. Sundar S, Mondal D, Rijal S, Bhattacharya S, et al. Implementation research to support the initiative on the elimination of kala azar from Bangladesh, India and Nepal–the challenges for diagnosis and treatment. Trop Med Int Health. 2008;13:2–5.
197. World Health Organization. Regional strategic framework for elimination of Kala-azar from the South-East Asia region (2005–2015). New Delhi: WHO Regional Office for South-East Asia; 2005.
198. Murray HW, Berman JD, Davies CR, Saravia NG. Advances in leishmaniasis. Lancet. 2005; 366:1561–77.
199. Calvopina M, Gomez EA, Sindermann H, Cooper PJ, et al. Relapse of new world diffuse cutaneous leishmaniasis caused by *Leishmania* (*Leishmania*) *mexicana* after miltefosine treatment. Am J Trop Med Hyg. 2006;75:1074–7.
200. Zerpa O, Ulrich M, Blanco B, Polegre M, et al. Diffuse cutaneous leishmaniasis responds to miltefosine but then relapses. Br J Dermatol. 2007;156:1328–35.
201. Pandey BD, Pandey K, Kaneko O, Yanagi T, et al. Relapse of visceral leishmaniasis after miltefosine treatment in a Nepalese patient. Am J Trop Med Hyg. 2009;80:580–2.
202. Andrade HM, Toledo VP, Pinheiro MB, Guimarães TM, et al. Evaluation of miltefosine for the treatment of dogs naturally infected with *L. infantum* (= *L. chagasi*) in Brazil. Vet Parasitol. 2011;181:83–90.
203. Proverbio D, Spada E, Bagnagatti De Giorgi G, Perego R. Failure of miltefosine treatment in two dogs with natural *Leishmania infantum* infection. Case Rep Vet Med. 2014;640151. https://doi.org/10.1155/2014/640151
204. Escobar P, Matu S, Marques C, Croft SL. Sensitivities of *Leishmania* species to hexadecylphosphocholine (miltefosine), ET-18-OCH(3) (edelfosine) and amphotericin B. Acta Trop. 2002;81:151–7.
205. van Blitterswijk WJ, Verheij M. Anticancer alkylphospholipids: mechanisms of action, cellular sensitivity and resistance, and clinical prospects. Curr Pharm Des. 2008;14:2061–74.
206. Paris C, Loiseau PM, Bories C, Bréard J. Miltefosine induces apoptosis-like death in *Leishmania donovani* promastigotes. Antimicrob Agents Chemother. 2004;48:852–9.
207. Croft SL, Seifert K, Duchêne M. Antiprotozoal activities of phospholipid analogues. Mol Biochem Parasitol. 2003;126:165–72.
208. Rakotomanga M, Blanc S, Gaudin K, Chaminade P, et al. Miltefosine affects lipid metabolism in *Leishmania donovani* promastigotes. Antimicrob Agents Chemother. 2007;51:1425–30.
209. Pérez-Victoria FJ, Sánchez-Cañete MP, Castanys S, Gamarro F. Phospholipid translocation and miltefosine potency require both *L. donovani* miltefosine transporter and the new protein LdRos3 in *Leishmania* parasites. J Biol Chem. 2006a;281:23766–75.
210. Sánchez-Cañete MP, Carvalho L, Pérez-Victoria FJ, Gamarro F, et al. Low plasma membrane expression of the miltefosine transport complex renders *Leishmania braziliensis* refractory to the drug. Antimicrob Agents Chemother. 2009;53:1305–13.
211. Pérez-Victoria FJ, Sánchez-Cañete MP, Seifert K, Croft SL, et al. Mechanisms of experimental resistance of *Leishmania* to miltefosine: Implications for clinical use. Drug Resist Updat. 2006b;9:26–39.
212. Montero-Lomelí M, Morais BL, Figueiredo DL, Neto DC, et al. The initiation factor eIF4A is involved in the response to lithium stress in *Saccharomyces cerevisiae*. J Biol Chem. 2002; 277:21542–8.
213. Wiśniewski JR, Zougman A, Nagaraj N, Mann M. Universal sample preparation method for proteome analysis. Nat Methods. 2009b;6:359–62.
214. Priotto G, Kasparian S, Mutombo W, Ngouama D, et al. Nifurtimox-eflornithine combination therapy for second-stage African *Trypanosoma brucei gambiense* trypanosomiasis: a multicentre, randomised, phase III, non-inferiority trial. Lancet. 2009;374:56–64.

215. Gygi SP, Corthals GL, Zhang Y, Rochon Y, et al. Evaluation of two-dimensional gel electrophoresis-based proteome analysis technology. Proc Natl Acad Sci USA. 2000;97: 9390–5.
216. Junqueira M, Spirin V, Santana Balbuena T, Waridel P, et al. Separating the wheat from the chaff: unbiased filtering of background tandem mass spectra improves protein identification. J Proteome Res. 2008;7:3382–95.
217. Lye LF, Owens K, Shi H, Murta SM, et al. Retention and loss of RNA interference pathways in trypanosomatid protozoans. PLoS Pathog. 2010;6:e1001161.
218. Carter KC, Hutchison S, Henriquez FL, Légaré D, et al. Resistance of *Leishmania donovani* to sodium stibogluconate is related to the expression of host and parasite gamma-glutamylcysteine synthetase. Antimicrob Agents Chemother. 2006;50:88–95.
219. Araujo RP, Liotta LA, Petricoin EF. Proteins, drug targets and the mechanisms they control: the simple truth about complex networks. Nat Rev Drug Discov. 2007;6:871–80.

The Role of ABC Transporters in Drug-Resistant *Leishmania* 11

Adriano C. Coelho and Paulo C. Cotrim

Abstract

The ATP-binding cassette (ABC) transporters belong to the largest family of transmembrane proteins found in living organisms. These proteins are present in prokaryotes and eukaryotes and are mainly involved in the transport of a variety of molecules across cellular membranes, whereas others are involved in biological processes unrelated to transport. The genome sequencing of several *Leishmania* species confirmed the presence of members for all eight different subfamilies of ABC transporters (ABCA to ABCH), according to their specific functional and molecular characteristics. These proteins have recently been characterized in *Leishmania*; some of them associated with drug resistance, which is a significant field in leishmaniasis chemotherapy, a disease still lacking effective treatment, with increasing daily reports of therapeutic failure. In this chapter, we focus our discussion on the association of these proteins with drug resistance in leishmaniasis and its fundamental role in the pathology and pharmacology of this medically important protozoan parasite that currently infects around 12 million people in the world.

A. C. Coelho
Departamento de Biologia Animal, Instituto de Biologia, Universidade Estadual de Campinas, Campinas, SP, Brazil

P. C. Cotrim (✉)
Departamento de Moléstias Infecciosas e Parasitárias, Instituto de Medicina Tropical, Faculdade de Medicina, Universidade de São Paulo, São Paulo, SP, Brazil
e-mail: pccotrim@usp.br

A. Ponte-Sucre, M. Padrón-Nieves (eds.), *Drug Resistance in Leishmania Parasites*,
https://doi.org/10.1007/978-3-319-74186-4_11

11.1 Introduction

The ABC proteins belong to the largest family of transmembrane proteins found in all life kingdoms from bacteria to humans [1, 2]. According to Saurin et al. [3], ABC proteins are responsible for the export and import molecules in prokaryotes, while in eukaryotes these transporters have only export functions. These proteins are mainly involved in the transport across cellular membranes of a variety of molecules like ions, peptides, sugars, lipids, or even large molecules like polypeptides and polysaccharides. Based on the huge variety of molecules involved in the transport of these proteins, ABC transporters can be classified as members of the permeome of a given organism that are all proteins responsible for membrane permeability and that encompasses the full set of transporters and channels. Moreover, the ABC proteins can be also involved in biological processes unrelated to transport like DNA repair, DNA translation, or even gene expression [4]. The first identification of an ABC transporter in *Leishmania* was associated with drug resistance [5], indicating that these proteins may play similar roles to those described for other parasitic protozoa [6] and cancer cells [7].

The ABC transporters have a highly conserved structure that includes an ATPase domain (Fig. 11.1). The ATP-binding cassette, or NBD (nucleotide-binding domain), is composed of three major conserved motifs: the Walker A and B motifs and the signature motif C, a specific sequence, characteristic of all ABC transporters, located just upstream of the Walker B motif [2, 8]. The motif C is a unique sequence of ABC transporters that distinguishes them from other proteins containing the NBD (Fig. 11.1). The general structure of these transporters consists of four structural

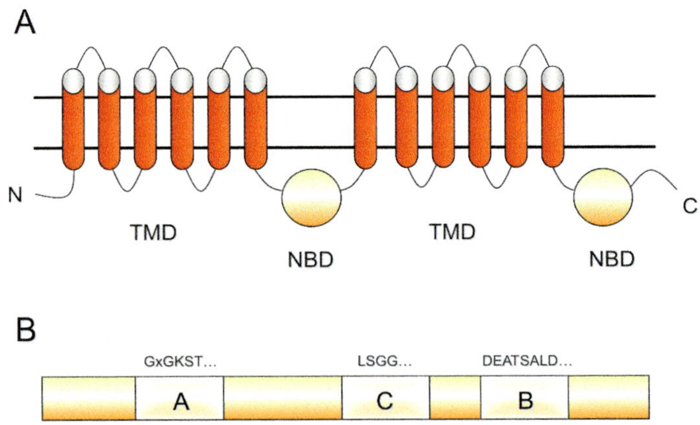

Fig. 11.1 General protein structure of a typical ABC transporter (**A**) and its nucleotide-binding domain (NBD) (**B**). (**A**) The ABC transporter consists of two halves of the protein, each half containing a transmembrane domain (TMD) (in red) and a NBD (in yellow). (**B**) The NBD is constituted by the Walker A and B motifs found in all ATP-binding proteins. In addition, in the NBD, there is the signature, or C motif. These domains are represented in the figure and the most common amino acids found in these motifs are shown indicated above the diagram. In general, the subfamilies contain characteristic residues in these and other regions

domains in the same polypeptide chain, containing two hydrophobic transmembrane domains (TMD) with multiple α-helices and two nucleotide-binding domains (NBD) responsible for ATP binding and hydrolysis, to catalyze the transport of metabolites and drugs across membranes [2, 9, 10] (Fig. 11.1A). Besides full transporters containing two TMDs and two NBDs represented as $(TM\text{-}NBD)_2$, eukaryotic ABC transporters can be organized as half-transporters with one TMD and one NBD with a structural organization TMD-NBD or NBD-TMD. Moreover, there are some intracellular ABC proteins with no TMD and two NBDs fused in the same molecule. The diversity of ABC family in *Leishmania* has been revealed by the genome sequencing of several species of the parasite that contain members of all eight subfamilies (ABCA to ABCH subfamilies) [11] (Table 11.1). Differently, not all subfamilies are present in other eukaryotes, as for example, ABCH subfamily that is not present in the genome of humans and yeast and in the plant *Arabidopsis thaliana* [13] (Table 11.1). Despite this diversity, most of functions described for the ABC transporters in *Leishmania* to date are restricted to drug resistance and traffic of phospholipids [5, 14–18]. Nevertheless, it is probable that cellular functions not yet elucidated will clarify the functional importance of this huge gene family that represents 0.5% of the entire *Leishmania* genome.

11.2 ABC Proteins in *Leishmania*: Functions and Their Role in Drug Resistance

11.2.1 ABC Genes in *Leishmania*: Organization and Distribution

Leishmania parasites contain a full set of ABC proteins with a variety of structures envisaged by the genome database. The use of next-generation sequencing technologies in recent years has become available several high-quality draft and finished genomes of several species, strains, and clinical isolates of the parasite [19–21], most of them available at TritrypDB (www.tritrypdb.org). According to the TritrypDB, the genome of *Leishmania (Leishmania) major*, *L. (L.) mexicana*, *L. (L.) infantum*, and *L. (Viannia) braziliensis*, species responsible for cutaneous (CL), diffuse cutaneous (DCL), visceral (VL), and mucocutaneous (MCL) leishmaniasis respectively, contains around 8000 genes with different numbers of ABC genes present in each genome. A systematic BLAST analysis using the NBD consensus demonstrated the presence of 42 genes in the genome of *L. (L.) major* and *L. (L.) mexicana* and 43 and 39 genes in the genome of *L. (L.) infantum* and *L (V.) braziliensis* respectively [11, 12] (Table 11.1).

In silico analyses have demonstrated that at least three known genes are absent in *L. (V.) braziliensis* (ABCB4, ABCC1, and ABCG3), when compared with those present in other three *Leishmania* species (Table 11.1). Unfortunately, the role of these ABC proteins in the biology of the parasites is still unknown. There is also an extra gene present in the genome of *L. (L.) infantum,* named ABCC9. It is considered the most divergent member of the ABCC subfamily [12] (Table 11.1). It is absent in

Table 11.1 Number of ABC genes in trypanosomatid parasites and other eukaryotes

Subfamily	*L. (L.) major*	*L. (L.) mexicana*	*L. (L.) infantum*	*L. (V.) braziliensis*	*T. brucei*	*T. cruzi*	*A. thaliana*	Yeast	*Drosophila*	Human
ABCA	10	10	10	10	2	5	16	0	19	12
ABCB	4	4	4	3[a]	2	2	27	4	10	11
ABCC	8	8	9	7[a]	3	4	14	7	12	12
ABCD	3	3	3	3	3	3	2	2	2	4
ABCE	1	1	1	1	1	1	2	0	1	1
ABCF	3	3	3	3	3	3	5	5	3	3
ABCG	6	6	6	5[a]	4	4	40	10	15	5
ABCH	3	3	3	3	1	3	0	0	3	0
Other	4	4	4	4	3	3	0	3	0	0
Total	42[a]	42[a]	43[a]	39[a]	22	28	106	32	56	48

[a]In *L. (L.) major* and *L. (L.) mexicana*, one member of the subfamily ABCC is not present and it is present as a pseudogene in *L. (V.) braziliensis* while in *L. (L.) infantum* this gene is functional (ABCC9) [12]. Other three members present in *L. (L.) major, L. (L.) mexicana* and *L. (L.) infantum* are absent in *L. (V.) braziliensis*: ABCB4, ABCC1, and ABCG3 according to the last version of the TritrypDB (www.tritrypdb.org)

the genome of *L. (L.) major* and *L. (L.) mexicana* and its orthologue in *L. (L.) braziliensis* is a pseudogene [12] (Table 11.1).

Two other trypanosomatids related to *Leishmania*, *Trypanosoma brucei* and *T. cruzi*, have less members of ABC proteins (28 and 22 ABC proteins, respectively), when compared to *Leishmania*, indicating that the *ABC* gene family is expanded in *Leishmania* genus [4] (Table 11.1). This expansion occurs because there are higher number of representatives of the subfamilies ABCA, ABCC, and ABCG that are in clusters of two or even three genes in the genome [11, 22]. One example is the occurrence of *ABCC* genes in tandem in chromosome 23 and 31 of *L. (L.) major* [11, 23]. The synteny of these genes is well conserved among *Leishmania* species, indicating that the duplication process for ABCA, ABCC, and ABCG subfamilies occurred before speciation of the *Leishmania* genus, despite over 15 million years of proposed divergence [24, 25]. However, the duplication events are restricted to some subfamilies of ABC proteins; gene members of ABCD, ABCE, and ABCF subfamilies have no duplication in the genome of *Leishmania* species, and their orthologues are present in the genome of *T. brucei* and *T. cruzi* [4, 11] (Table 11.1).

The pattern of gene expression of ABC genes was initially studied using customized DNA microarrays. The stage-specific expression in *Leishmania* showed that a restricted number of *ABC* genes are regulated throughout the parasite life cycle [11]. The genes *ABCA3* and *ABCG3* were detected as preferentially expressed in the amastigote stage, while the gene *ABCF3* was increased in the promastigote form. No other variation in the expression of these genes was observed [11]. On other hand, changes in the expression level of ABC proteins from subfamilies ABCA, ABCB, ABCC, and ABCG were observed in promastigotes and axenic amastigotes of *L. (L.) infantum* by proteomic analyses [26]. These results confirm the importance in assessing stage-specific protein expression by proteomic analysis in trypanosomatid parasites that control the gene expression almost exclusively at the post-transcriptional level [27].

Studies about the role of ABC proteins in the biology, metabolism, transmission and infection of leishmaniases are limited and even unknown. So far, some members of the ABCA subfamily were described as being related to the traffic of phospholipids, while some ABCB, ABCC, and ABCG members are associated with drug resistance [11], an important challenge in the disease chemotherapy. These ABC proteins may be associated with drug resistance by two main mechanisms: (1) increase in the ABC protein levels, due to gene amplification or overexpression of the respective gene, and/or (2) gene mutations capable to change the biochemical properties of the ABC transporter and thus affecting the drug transport capacity [28]. To facilitate the explanation, we will subdivide this chapter according to the different ABC subfamilies.

11.2.2 ABCA Subfamily's Proteins

The ABCA subfamily contains some of the largest members of all ABC transporters, with more than 200 kDa of predicted molecular weight. They share a high degree of sequence conservation and have been mostly related to lipid trafficking and drug resistance in cancer cells [29]. In *Leishmania* this subfamily contains ten members that are conserved in the genome of all species of the parasite (Table 11.1) [11, 30]. These ten members are homologous to the mammalian ABCA subfamily (ABCA1–ABCA10) presenting the typical structure of an ABC transporter $(TM\text{-}NBD)_2$. Some of ABCA subfamily members are duplicated in the *Leishmania* genus. The duplication process occurred during the evolution of the genus and suggests that these members may have acquired different functions in the biology of the parasite when compared to the other trypanosomatids that have less ABCA members (Table 11.1).

In *Leishmania*, only two ABCA proteins have been characterized (ABCA4 and ABCA8), although none of them has been associated with drug resistance [14, 18]. On the contrary, in human cells, at least two members of this subfamily (ABCA2 and ABCA3) were associated with drug resistance in cancer chemotherapy [29, 31, 32]. One of the main mechanisms involved is based in the subcellular drug sequestration to the lysosomes that afterwards export them out of the cell.

ABCA4 and ABCA8 were characterized in *L. (L.) tropica* and originally termed as LtrABCA2 and LtrABCA1.1, respectively [14, 18]. The ABCA8 was the first member of this subfamily described in unicellular eukaryotes, and it corresponds to a protein of around 200 kDa, localized in the flagellar pocket and in the plasma membrane [18]. ABCA4 is not only localized in the flagellar pocket but also in internal vesicles [14]. Transfectants of *L. (L.) tropica* overexpressing both genes have a decreased retention of the phospholipids, phosphatidylcholine, phosphatidylserine, and phosphatidylethanolamine, suggesting that these ABCA proteins play a role in lipid movements across the plasma membrane [14, 18]. According to these authors [14], the differences in their cellular localization may occur due to the high divergence in the NH2-terminus sequence of these ABCA proteins. Interestingly, the overexpression of these ABCA members (ABCA4 and ABCA8) in *L. (L.) tropica* reduced the in vitro infectivity into macrophages [14, 18]. Overexpression of these two ABCA transporters in *Leishmania* demonstrated no resistance to several compounds, like amphotericin-B, miltefosine, and edelfosine, or to other known substrates of ABC transporters [14, 18].

In trypanosomes, a restricted number of ABCA members are present in their genome, with only two in *T. brucei* and five in *T. cruzi* [11] (Table 11.1). Only one member has been characterized in *T. cruzi*, TcABC-1, a protein with approximately 41% amino acid identity with the *Leishmania* ABCA4 and ABCA8 proteins [33]. The role of TcABC-1 in *T. cruzi* seems to be associated with endocytosis and vesicular trafficking. TcABC-1 is also located in the plasma membrane and flagellar pocket and it is present in all the stages of the parasite, except in the trypomastigote stage [33]. There is no report correlating drug resistance with TcABC-1 in the literature, corroborating up to date the data described for *Leishmania* ABCA proteins.

Customized DNA microarrays have been used in the analysis of ABC transporter genes in antimonial resistance to determine whether they are involved in drug resistance [11]. ABCA3 and ABCC3 were found to be overexpressed in a *L. (L.)*

infantum strain resistant to trivalent antimony (Sb^{III}), compared with an antimony-sensitive *L. (L.) infantum* promastigote parasite [11]. However, transfection of both genes in wild-type parasites did not confer higher levels of Sb^{III} resistance than with the overexpression of ABCC3 alone, excluding a possible role in drug resistance for ABCA3 protein.

The role of the other eight members of this subfamily remains to be described and even including if they are involved in drug resistance in *Leishmania*.

11.2.3 ABCB Subfamily's Proteins

The ABCB subfamily contains four members in *Leishmania* (Table 11.1). Unlike the ABCA subfamily with all its members classified as "full" transporters with a duplication of the TM and NBD structure $(TM\text{-}NBD)_2$, the ABCB subfamily has two "full" and two "half" transporters [11]. The main functions described in the literature related to these proteins are drug resistance, transport of peptides, iron/sulfur clusters biogenesis, and association with the RNA interference process in *Caenorhabditis elegans* [34–38].

ABCB4 was the first ABCB transporter described in *L. (L.) donovani*, originally known as LdMDR1 or MDR1 (multidrug resistance-1) [34, 36, 39–42]. Together with ABCB2 (or MDR2), these full transporters of the ABCB subfamily were described in several species of *Leishmania* as proteins involved in resistance to different drugs [34, 36, 39–42] (Table 11.2). Members of this subfamily have also been associated with drug resistance in cancer cells, named as ABCB4 (MDR1) and ABCB2 (MDR2) [9]. The MDR phenotype is characterized by the presence of cross-resistance to a well-defined spectrum of drugs. The general resistance mechanism consists in an increase of drug efflux from the cell that reduces the concentration of the drug inside it [9]. The overexpression and function of these transporters is measured by the decreased signal produced by rhodamine-123, a fluorescent marker of ABCB proteins also known as P-glycoproteins [43, 56]. In vinblastine resistant *L. (L.) amazonensis*, a reduced accumulation of this dye has been observed as a consequence of its increased efflux through the transporters [43].

The ABCB4 transporter has been described in several species of *Leishmania* (Table 11.2), as associated with drug resistance to vinblastine and daunorubicin, as well as to other unrelated hydrophobic drugs such as puromycin, adriamycin, and doxorubicin, none of them used for leishmaniasis chemotherapy [34, 36, 39, 40, 42–44]. Curiously, it was demonstrated that ABCB4 overexpression could lead to miltefosine/edelfosine cross-resistance in a *Leishmania* mutant selected for daunomycin [57]. No gene amplification or overexpression was observed in *L. (L.) donovani* resistant to miltefosine, suggesting that the mechanism of resistance is not due the *ABCB4* gene amplification [58].

Miltefosine, a hexadecylphosphocholine, is the first effective oral drug for the treatment of visceral leishmaniasis [59, 60]. The mode of action of this drug is not well known but is associated with changes in alkyl-lipid metabolism and phospholipid biosynthesis [61]. Phospholipids were previously demonstrated to be substrates

Table 11.2 ABC proteins involved in drug resistance in *Leishmania* spp.

ABC protein (Alias)	*Leishmania* sp.	Drug(s)	Cellular localization	Reference (s)
ABCB2 (MDR2)	*L. (L.) amazonensis*	5-Fluorouracil	Flagellar pocket and multivesicular tubule lysosome	[41]
ABCB4 (MDR1)	*L. (L.) amazonensis, L. (Mundinia) enrietti, L. (L.) donovani, L. (L.) tropica*	Vinblastine, purommycin daunomycin, miltefosine, edelfosine	Golgi apparatus, endoplasmic reticulum, multivesicular tubule lysosome, and mitochondria	[34, 36, 39, 43–45]
ABCC3 (PGPA/ MRPA)	*L. (L.) major, L. (S.) tarentolae, L. (L.) infantum*	Sb^{V}, Sb^{III}, and As^{III}	Vesicles between the nucleus and flagellar pocket	[17, 46–49]
ABCC4	*L. (L.) infantum*	Sb^{III}	Tubular compartment oriented along the longitudinal axis of the parasite	[12]
ABCC5	*L. (L.) infantum*	Sb^{III}	Tubular compartment oriented along the longitudinal axis of the parasite	[12]
ABCC7 (PRP1)	*L. (L.) major*	Pentamidine and Sb^{III}	Intracellular vesicles	[16, 50, 51]
ABCG2	*L. (L.) major*	Sb^{III}	Intracellular vesicles and partially located in the plasma membrane	[52]
ABCG4	*L. (L.) infantum*	Miltefosine, edelfosine, sitamaquine	Plasma membrane and flagellar pocket	[15]
ABCG6	*L. (L.) donovani, L. (L.) infantum*	Camptothecin, miltefosine, chloroquine, and sitamaquine.	Plasma membrane and flagellar pocket	[53, 54]
ABCI4	*L. (L.) major*	Sb^{V}, Sb^{III}, As^{III}, cadmium (Cd^{II}), and some toxic porphyrins	Mitochondria and plasma membrane	[55]

for ABCB transporters [62]; however, if ABCB4 is associated with miltefosine treatment, failure against *Leishmania* is still controversial. The main resistance mechanism is related with two proteins present in the parasite plasma membrane responsible for the accumulation of phospholipids [63–65] (see chapter "Mechanisms of Miltefosine Resistance in *Leishmania*" in the previous edition of this book). Localization of ABCB4 in *Leishmania* is intracellular, in endocytic and secretory compartments including the Golgi apparatus, the endoplasmic reticulum (ER), and the multivesicular tubule (MVT) lysosome [45]. The subcellular location of ABCB4 indicates that the mechanism of drug resistance mediated by this protein occurs in two steps: first, the drug is accumulated in intracellular compartments and subsequently eliminated through exocytosis, a mechanism not associated with drug efflux across the plasma membrane, as it has been described for its mammals MDR1 orthologues [9]. Finally, the fact that the ABCB4 member is not present in the *L. (V.) braziliensis* genome (Table 11.1) suggests that it must not play a role in miltefosine resistance in this species [65]. As in *L. (V.) braziliensis*, no other trypanosome contains this orthologue in its genome [11].

Pentamidine-resistant mutants of *L. (L.) mexicana*, *L. (L.) donovani* and *L. (L.) amazonensis* have a reduced accumulation of the drug in the mitochondria [66–68] and this phenotype could be due an ABCB4's decreased import activity. Although these observations are not directly associated with this ABC transporter, the use of classical inhibitors of ABC transporters such as verapamil (a calcium channel blocker known to reverse multidrug resistance in *Plasmodium falciparum* and mammalian cells) [10, 69] was able to reverse pentamidine resistance in *Leishmania* cells [50, 66], indicating that an ABC transporter is involved in this process. Moreover, it was demonstrated that iron levels could have a modulatory effect on the ABCB4's capacity to transport pentamidine to an intracellular organelle connected to the mitochondria. In this model, iron can potentiate pentamidine accumulation, while its deprivation causes the reduction of the drug inside the parasite [70]. These authors also demonstrated an inversed correlation between *ABCB4* gene copy number and pentamidine resistance. Additional studies are needed to clarify if this hypothesis is correct or not.

The second full transporter that belongs to the ABCB subfamily in *Leishmania* is ABCB2 also known as MDR2 (Table 11.2). This member was initially characterized in *L. (L.) amazonensis* and its expression increases in 5-fluorouracil-resistant parasites, although it has not been related to any other ABCB4 substrate. LaMDR2 exhibits 47% amino acid identity to its most closely related protein, LaMDR1 [41]. In mammalian cells, MDR2 transports phospholipids [32]; however, no cross-resistance to miltefosine was observed in parasites overexpressing this ABC transporter [41]. The subcellular localization of ABCB2 overlaps with the MVT lysosome and the flagellar pocket, suggesting that ABCB2 and ABCB4 in *Leishmania* have a similar localization (Table 11.2).

Besides these two ABCB proteins in *Leishmania*, the two other members, ABCB1 and ABCB3, are half-transporters (TM-NBD) also present in other trypanosomatids [11] (Table 11.1). There is still no information about the role of ABCB1 in the biology of *Leishmania*, while the role of ABCB3 was recently reported. In human cells, these half-transporters are involved in important biological processes such as transport of peptides into the endoplasmic reticulum, iron metabolism and transport of iron/sulfur protein precursors for the ABCB proteins located in the mitochondria [32]. They are intracellular and located in the membrane of mitochondria, endoplasmic reticulum and lysosomes.

Leishmania is auxotrophic for heme and must acquire porphyrins from its host. The ABCB3 is an orthologue of the yeast ATM1 protein and the human ABCB6 and ABCB7 proteins. These ABC proteins, located in the inner mitochondrial membrane, are involved in the transport of porphyrins and in the biogenesis of mitochondrial heme and in cytosolic iron/sulfur clusters [11, 32, 71, 72]. These findings confirmed the essentiality of ABCB3 protein in *Leishmania* in the survival of the parasite. The inactivation of just one allele of *ABCB3* gene reduced the replication of intracellular amastigotes and the virulence of the parasites in mouse [72]. Interestingly, the resistance mechanism mediated by ABCB4 depends on the iron levels present in *Leishmania* [70] that is regulated by ABCB3 [72].

11.2.4 ABCC Subfamily's Proteins

The ABCC proteins are responsible for the transport of toxic compounds into intracellular compartments, and it is also known that this subfamily is often implicated in the secretion of toxic molecules and in cellular detoxification in eukaryotic cells [2, 73]. These proteins are also involved in the ATP-dependent transport of thiol conjugates. In human tumor cells, the ABCC proteins are also known as MRP (multidrug resistance associated proteins) conferring resistance to multiple drugs and to other compounds conjugated to anionic molecules such as reduced glutathione, glucoronate, sulfate, phosphate, and glutamate [74]. Glutathione (GSH) is a tripeptide that contains an unusual linkage between the amine group of cysteine and the carboxyl group of the glutamate side chain. GSH protects cells from reactive oxygen species such as free radicals and peroxides acting as an antioxidant [75].

The first ABC transporter described in *Leishmania* involved in drug resistance was ABCC3 (also known as PGPA or MRPA) (Table 11.2). This protein is involved in trivalent arsenite (As^{III}), Sb^{V} and Sb^{III} resistance in *Leishmania* spp. ABCC3 does not confer resistance to hydrophobic drugs like vinblastine and puromycin, although it was initially detected in DNA amplicons (or H-circles) of methotrexate-resistant promastigotes [5, 46–48, 76]. Later on, it was demonstrated that methotrexate resistance in *Leishmania* depends on the presence of the pteridine resistance gene (*PTR1*), implicated in the synthesis of reduced folates, and not to the *ABCC3* gene contained in these circles [77, 78]. The H-circles are circular DNA originated through recombination of repetitive sequences [79, 80] with an important role in drug resistance in *Leishmania* [81–83].

Leishmania parasites transfected with the *ABCC3* gene reach high levels of resistance to As^{III} and antimonials; however, these levels of resistance are smaller than the observed for mutants selected with these same drugs [17, 46]. These findings indicate that other genes, not directly related to ABCC3 overexpression, are involved in the resistance mechanism. For example, in three independent mutants of *L. (L.) major* resistant to Sb^{III} characterized by whole genome sequencing and comparative genomic hybridization, the *ABCC3* gene was not amplified in these mutants and the mechanism of resistance was due to a terminal deletion in the polyploidy chromosome 31 that contains the gene of the aquaglyceroporin (*AQP1*) and an intrachromosomal amplification of a subtelomeric *locus* on chromosome 34 containing ascorbate-dependent peroxidase and glucose-6-phosphate dehydrogenase genes [84]. Nevertheless, one should not rule out the relationship of *ABCC3* gene with the mechanism of resistance, since gene disruption of *ABCC3* gene in *Leishmania* results in a greater sensitivity of amastigotes against As^{III} and Sb^{III} [85]. The ABCC3 transporter is located in the intracellular vesicular membrane close to the flagellar pocket and confers resistance by sequestering metal-thiol conjugates to these vesicles [47] (Table 11.2).

Among the thiol conjugates, tripanothione (TSH) is the major reduced thiol in *Leishmania*, a conjugate of GSH with spermidine [86]. The biosynthesis of TSH involves two main enzymes: γ-glutamylcysteine synthetase (GSH1) and ornithine decarboxylase (ODC) [87]. These enzymes are responsible for the synthesis of GSH and spermidine respectively, and can be found in increased levels in mutants of *Leishmania* resistant to metals and also in clinical isolates resistant to Sb^{V} [86, 88–91]. However, in some metal-resistant mutants of *Leishmania*, these genes are not amplified, as it was described in *L. (L.) mexicana*, *L. (L.) tropica*, and *L. (L.) major* mutants resistant to Sb^{III} [84, 92, 93].

Co-transfection experiments demonstrated a strong synergy between *ABCC3* and *GSH1* when both genes are expressed in a revertant strain of *L. (Sauroleishmania) tarentolae* (an antimony-resistant mutant grown in the absence of drug), but not when they are overexpressed in a wild-type background [88, 89]. This evidence suggests that additional factors besides the increased expression of ABC proteins are associated with antimony resistance in *Leishmania*. Furthermore, the *ABCC3* gene does not amplify in response to antimonials in *L. (V.) braziliensis* [94], probably due to the RNAi activity recently described in parasites of subgenus *Viannia* [95]. Extra-chromosomal circular DNAs are transcribed in both strands in *Leishmania* [96, 97] and the RNAi activity in these species would inhibit the gene expression after the formation of double-strand RNA.

Most of the studies about drug resistance in *Leishmania* were performed in the promastigote form, while a limited number of studies were done in amastigotes, the stage responsible for the disease in man. Accordingly, it was observed the ABCC3 overexpression in *L. (L.) infantum* axenic amastigotes resistant to antimony, as already observed in promastigote form of several species of *Leishmania* resistant to metals. Additionally, the transfection of the *ABCC3* gene in *L. (V.) panamensis* conferred resistance to Sb^{V} in intracellular amastigotes; the resistant phenotype could be abrogated by buthionine sulfoximine, a glutathione biosynthesis-specific inhibitor [17]. These data confirmed the first association of ABC transporters with drug resistance in *Leishmania* amastigotes. In a later study, antimony-sensitive and antimony-resistant *L. (L.) donovani* strains isolated from patients with visceral leishmaniasis (Kala-azar), a comparative proteomic analysis was performed; in this case, no ABC protein was differently expressed between the strains, indicating that the mechanism is not associated with the amplification of the *ABCC3* gene or any other *ABC* gene [98]. Besides, customized DNA microarrays have been used in the analysis of ABC transporter genes in antimonial resistance to determine whether other ABC genes are involved in drug resistance. The *ABCC3* gene was also overexpressed in a *L. (L.) infantum* strain resistant to Sb^{III}, compared with Sb^{III}-sensitive parasites, a phenomenon already described in other strains and even in the amastigote stage of this *L. (L.) infantum* mutant [17, 46, 47]. Similar findings were observed in an Sb^{III}-resistant mutant of *L. (L.) infantum* evaluated by proteomic and genomic analyses that contained the ABCC3 overexpressed [99] and in the main species of the parasite that are endemic in Brazil after selection for antimony resistance in vitro [100, 101].

In *T. brucei*, a common phenotype associated with the overexpression of TbMRPA, orthologous to ABCC1 in *Leishmania*, may be found in arsenical-resistant mutants [102]. The resistant phenotype was not significantly increased by the additional overexpression of ODC and/or GSH in this parasite, probably due to the higher cytosolic tripanothione levels when compared to *Leishmania* [102]. Besides TbMRPA, *T. brucei* has another ABCC transporter (TbMRPE, orthologous to ABCC6) involved in the resistance to suramin, a well-established drug used to treat African sleeping sickness [102].

In *Leishmania* six additional ABCC members have already been described; they all belong to the MRP group. Within this cluster ABCC2 (alias PgpB), ABCC1 (PgpC), ABCC5 (PgpD), ABCC4 (PgpE), and ABCC7 (PRP1) are present. Essays of drug susceptibility after gene transfection demonstrated that these ABC proteins have no role in Sb^{III} resistance [12]; only *ABCC7* gene codifies for a protein that confers pentamidine resistance (a second-line drug used in the chemotherapy of the disease) when overexpressed in wild-type promastigotes and in amastigotes of *L. (L.) major* and *L. (L.) amazonensis* [16, 50] (Table 11.2). Moreover, the overexpression of ABCC7 in *L. (L.) infantum* also conferred pentamidine resistance in promastigotes and axenic amastigotes [50].

The PRP1 (pentamidine resistance protein 1) is intracellularly located and is associated with the tubulovesicular element [50, 51] that is linked to the exo- and endocytosis pathways [103–105] (Table 11.2). The role of ABCC7-PRP1 was also investigated in *L. (L.) amazonensis* mutants resistant to pentamidine selected in vitro. Despite its role in pentamidine resistance in a transgenic line overexpressing this gene, no amplification and/or overexpression of *ABCC7* gene was found in the resistant line [16, 50, 67]. In a similar study with *L. (L.) major* mutants resistant to pentamidine, no amplification was either observed, indicating that the mechanism of resistance must not be due to *ABCC7* gene amplification [106]. A search for repetitive elements of DNA in the genomic region of the *ABCC7* gene in *L. (L.) major* demonstrated the absence of repetitive sequences. This fact, combined with the stability of the resistant phenotype in the absence of pentamidine, suggests that stable mutations in one or more genes must be responsible for the pentamidine resistance phenotype [67]. Considering this hypothesis, a mutation in *ABCC7* gene, for example, would increase the transport activity and consequently would confer resistance to the drug. Nevertheless, this hypothesis has not been still proven. The use of whole genome sequencing technology could be useful to reveal potential mutations in this and other resistant parasites [107].

The overexpression of ABCC7 also confers Sb^{III} resistance [12, 16]. However, with the exception of ABCC3, no other ABCC protein was identified overexpressed in antimony-resistant mutants [11, 49].

The cellular localization of all members of *Leishmania* ABCC subfamily was already determined. They are all located intracellularly, although their localization may vary inside the parasite. While ABCC7, ABCC4, and ABCC5 are located in a tubular compartment oriented along the longitudinal axis of the parasite, ABCC1, ABCC2, and ABCC6 are placed in a network of intracellular membranes, while ABCC8 is restricted to the posterior end of the parasite [12]. Among the ABCC proteins, only the ABCC9 protein has not been investigated due to the absence of essential residues in the protein the C motif, according to its sequence and by the fact that this member is restricted to *L. (L.) infantum*. Interestingly, unlike in *Leishmania*, the ABCC proteins in humans or even *T. brucei* are located in the plasma membrane, demonstrating that these proteins can have different cellular localization in other organisms [73, 102].

In the plasma membrane of the parasite was also reported an Sb^{III} efflux system that is not mediated by an ABCC transporter, once all proteins of this subfamily in *Leishmania* are intracellular [12]. So far, the transporter responsible for this antimony efflux system is unknown [108].

Concerning the role of other ABCC transporters in drug resistance, it was demonstrated that ABCC4 and ABCC5 are also associated with antimony resistance, when transfected into a partially revertant cell line of *L. (S.) tarentolae,* initially selected for Sb^{III} resistance, but growing without the drug by several passages [12]. These cells remain resistant to Sb^{III} when grown in the absence of the drug, and their resistance phenotype is increased when ABCC4 or ABCC5 proteins are overexpressed [12]. On the other hand, ABCC3 and ABCC7 transfectants may confer resistance not only to Sb^{III} in the revertant cell line but also in the parental wild-type strain [12]. These observations suggest that ABCC4 and ABCC5 proteins transport Sb^{III} conjugated to thiols, since the level of thiols in this *L. (S.) tarentolae* revertant mutant remains at least threefold higher than in the wild-type parasite.

In *T. cruzi*, two ABC transporters were already characterized: tcpgp1 and tcpgp2 [109, 110], orthologues of ABCC6 and ABCC2 in *Leishmania* respectively [11]. TcABCC2 (tcpgp2) is not related to nifurtimox and benznidazole resistance, the main drugs used in the chemotherapy of Chagas' disease, nor to the several MDR1 substrates, Sb^{V}, Sb^{III} and other metals [109]. Analysis of expression of these *ABC* genes in several *T. cruzi* strains and clones that were either susceptible or naturally resistant to nifurtimox and benznidazole indicated no amplification or overexpression of these ABC transporters [111].

The study of ABCC proteins in drug resistance among field isolates is scarce. It was demonstrated that *ABCC3* gene is amplified in some clinical isolates of *L. (L.) donovani* resistant to Sb^{V} [90, 112]. Contrarily, this was not the case of *ABCC3* and *ABCC7* genes, whose expression was not altered in field isolates of *L. (L.) donovani* resistant to Sb^{V} [113, 114]. Whole genome sequencing of several *L. (L.) donovani* clinical lines isolated from visceral leishmaniasis patients from the same endemic region, which differ in Sb^{III} susceptibility in vitro, was also evaluated [19]. It was observed a low genetic diversity among these isolates and an extensive variation in chromosome copy number, with no specific change in gene copy number or single-nucleotide polymorphism of *ABCC* genes [19]. These findings indicate that ABCC proteins are not necessarily involved in Sb^{III} resistance in *Leishmania* and that these observations may differ among *Leishmania* species, strains and clinical isolates from the field and from resistant lines selected in the laboratory.

11.2.5 ABCG Subfamily's Proteins

The ABCG subfamily, also known as the White family, is another subfamily of ABC transporters with members involved in drug resistance in *Leishmania*. The ABCG proteins are half-transporters that display a reverse topological disposition (NBD-TMD). The main function of these transporters relates to cellular lipids homeostasis and drug resistance [115]. In humans, for example, the ABCG2 protein (also called BCRP and MXR) is involved in a MDR phenotype in some cancer cells, conferring resistance to anthracyclines, mitoxantrone, bisantrene, and camptothecins topotecan [116, 117].

The ABCG subfamily in *Leishmania* is composed of six members, while in the trypanosomes *T. brucei* and *T. cruzi*, there are only four members [11]. Among the six members in *Leishmania*, three members have already been described as able to confer resistance to drugs. Recently, it was demonstrated in *L. (L.) infantum* that the overexpression of *ABCG4* gene is related to resistance to sitamaquine, miltefosine and its analogs edelfosine and perifosine [15] (Table 11.2). Transfectants overexpressing this ABC gene showed a reduction in cytoplasmic concentration of fluorescent phospholipids and labeled miltefosine, suggesting that ABCG4 enhances the outward transport of miltefosine across the plasma membrane. This hypothesis was confirmed by the subcellular localization of ABCG4, restricted to the flagellar pocket and plasma membrane [15] (Table 11.2). On the contrary, *L. (L.) donovani* mutants resistant to sitamaquine are not cross-resistant to miltefosine, indicating that the ABCG4 is not involved in sitamaquine resistance and another mechanism of resistance to this drug may be acting here for this species [118].

ABCG6 is the second protein from the ABCG subfamily characterized in *Leishmania*, and similarly as ABCG4, it is also related to drug resistance. It was demonstrated that *L. (L.) donovani* parasites overexpressing ABCG6 are resistant to camptothecin, a drug that inhibits the topoisomerase I [53] (Table 11.2). The ABCG6 is restricted to the plasma membrane and flagellar pocket, and the camptothecin resistance phenotype is mediated by drug efflux in an ATP-dependent process [53]. These same authors have shown that mutants resistant to this drug had a threefold increase in the mRNA levels of *ABCG6* gene, a result that was not observed for the other ABCG subfamily member, ABCG4, or even for others ABC proteins associated to drug resistance [53]. Moreover, promastigotes as well as axenic amastigotes resistant to camptothecin were neither cross-resistant to general antileishmanial drugs, such as sodium stibogluconate, sodium arsenite, and miltefosine, nor to the topoisomerase inhibitors, dihydrobetulinic acid and baicalein [53]. ABCG6 was also characterized in *L. (L.) infantum*, and unlike *L. (L.) donovani*, the overexpression of ABCG6 in this species confers resistance to camptothecin and cross-resistance to miltefosine, sitamaquine, and chloroquine [54]. On the other hand, these ABCG6 overexpressing transfectants can also transport phospholipids as observed in *L. (L.) donovani* [53, 54].

More recently, the role of ABCG2 in drug resistance in *Leishmania* was demonstrated. Promastigotes overexpressing ABCG2 are resistant to Sb^{III} and As^{III}, but they are not resistant to substrates of the other ABCGs of *Leishmania* (ABCG4 and ABCG6): miltefosine, perifosine, sitamaquine, and chloroquine [52]. ABCG2 is located in intracellular vesicles that fuse with the plasma membrane during the exocytosis. The mechanism of antimony resistance is by sequestering of the drug conjugated to thiols inside vesicles that are eliminated through exocytosis by the flagellar pocket.

The role of the other ABCG proteins (ABCG1, ABCG3, and ABCG5) in drug resistance in *Leishmania* is still unknown. In *T. cruzi*, it was demonstrated that ABCG1 of *T. cruzi* is involved in benznidazole resistance. In strains naturally resistant to this drug, the *ABCG1* gene was overexpressed and several single-nucleotide polymorphisms as compared to the gene of susceptible strain were found [119].

11.2.6 Other ABCs Subfamilies' Proteins

Members of other ABC subfamilies ABCD, ABCE, ABCF, and ABCH proteins have not yet been characterized in *Leishmania*, although it is possible to infer their functions in the parasite based on the study of these transporters in other organisms. In addition, there are other four unclassified ABC proteins in the genome of *Leishmania*. One member of this group of ABC proteins, with no homology with other eukaryotic ABC proteins, was recently related to drug resistance in *Leishmania* [55].

According to in silico analysis, the ABCD subfamily in *Leishmania* contains three members, the same number found in other trypanosomes [11] (Table 11.1). In general, ABCD transporters have a structure of half-transporters (TM-NBD), located in the peroxisome and involved in the import of fatty acids and/or fatty acyl-CoAs into this organelle for their subsequent processing by the β-oxidation pathway [120, 121]. In yeast, two members of this subfamily, PXA1 and PXA2, dimerize to form a functional transporter involved in a very long chain of fatty acid oxidation in the peroxisome [122]. It has also been observed in *C. elegans* that the post-transcriptional silencing of three ABCD transporters disrupted offspring production, suggesting developmental roles of peroxisomal ABC transporters [123]. Nevertheless, *Leishmania* and trypanosomes do not have peroxisomes, but rather glycosomes, organelles which contain enzymes for several processes such as glycolysis, the pentose-phosphate pathway, beta-oxidation of fatty acids, purine salvage, and biosynthetic pathways for pyrimidines, ether-lipids, and squalenes [124]. Up to now, none of the three ABCD members has been characterized, and it is not even known if they play a role in drug resistance. It would be interesting to verify the function/localization of these proteins in *Leishmania*, as well as the strength of their association with glycosomes. These proteins could be prospective targets for new antiparasite drugs, since these organelles are particular to the trypanosomes.

The ABCE and ABCF subfamilies have a typical structure of two fused NBD, and unlike the other members described in this chapter, they are not associated to transport-related processes, since they do not have TM domains [125]. The proteins of these two subfamilies are highly conserved across evolutionary diverse taxa, suggesting a role in fundamental cell biological processes.

The *ABCE1* gene is the most conserved member of the ABC gene family and is one of the most conserved genes in vertebrate and archaeal genomes [126], which suggests that this gene plays an essential role in the biology of the organisms. These proteins are identified by two potential iron/sulfur metal-binding domains in addition to two NBDs [125]. Human ABCE1 was initially identified as an inhibitor of RNase L [127], but recent data indicate that human and yeast ABCE proteins have also a central role in translation initiation [128]. The *T. brucei*'s orthologous was described in its involvement in protein synthesis [129]. Its depletion had an inhibitory effect on parasite growth, confirming that this gene is essential for the parasite growth and survival [129]. Although no study has been conducted to understand the function of ABCE1 in *Leishmania,* it is noteworthy that it may have similar function as observed in trypanosomes.

The ABCF subfamily includes proteins composed of two NBDs, and they are involved in the control of mRNA translation. In *Leishmania*, three ABCF protein-coding genes have been identified in their genomes, the same number observed in the trypanosomes *T. brucei* and *T. cruzi* (Table 11.1) [11]. The human homolog ABCF1 is associated with the ribosome, and it is responsible for the activation of eukaryotic initiation factor 2 (eIF2), a key protein in the process and control of the translation initiation [130]. Its homolog in *Saccharomyces cerevisiae* has a similar role in the activation of the eIF2 [131]. These data indicate that these members have functions that are distinct from those of other non-membrane ABC proteins.

The ABCF proteins of prokaryotes are implicated in resistance to macrolides, a group of antibiotics that inhibit protein synthesis [125], while in yeast, the elevation of the protein levels of the translational elongation factor 3 (EF-3) increases sensitivity to the aminoglycosides paromomycin and hygromycin [132]. Interestingly, the EF-3 factor has similarity to members of the ABCF subfamily and the EF-3 protein from *S. cerevisiae* has higher similarity to the ABCF1 protein of *Leishmania* spp.

Finally, the last subfamily to be considered in this chapter is the subfamily ABCH. Members of this subfamily are present in protozoa such as *Toxoplasma*, *Leishmania*, and *Trypanosoma* [4, 11] (Table 11.1) but are absent in the genomes of plants, worms, yeast, or mammalian [32, 126]. Originally discovered in *Drosophila*, genes of ABCH subfamily encode half-transporters with a NBD-TM organization. This family is considered the most enigmatic subfamily of ABC genes due to their peculiarities [32]. Unlike observed in insects, the members of this subfamily in Trypanosomatidae do not contain transmembrane domains; they have only one NBD domain [11]. Three members are present in the *Leishmania* genome and their functions remain to be described (Table 11.1). A possible involvement of a member of this subfamily in drug resistance has been described in antimony-resistant *L. (L.)*

infantum strain. In these mutants, the *ABCH1* gene was overexpressed when compared with the expression levels found in wild-type parasites [11]. However, it is unclear whether this gene actually has a role in the Sb^{III} resistance, since co-expression of this gene with *ABCC3* did not confer higher resistance to antimony than the *ABCC3* alone.

Besides the eight subfamilies present in *Leishmania*, there are four other ABC transporter members that are not classified in any of those subfamilies [11] (Table 11.1). Their role in the biology of the parasite is unknown, although they are quite conserved among other members of the Trypanosomatidae family, suggesting a common role among these members. Recently, one member of this group of ABC proteins, named as ABCI4, was correlated to drug resistance in *L. (L.) major* [55]. The gene *ABCI4* codes for a half-transporter with a TMD-NBD topology that requires homodimerization to be functional [55]. The ABCI4 protein is located at the plasma membrane and in the mitochondria conferring Sb^{V} and Sb^{III} resistance by the active efflux of metal-conjugated thiols [55] (Table 11.2). This protein also affects the accumulation of porphyrins in the mitochondria probably due to the efflux of these compounds to the cytosol [55].

11.3 Concluding Remarks

Understanding the role of ABC transporters in *Leishmania* is an important strategy for identifying one of the mechanisms of resistance to drugs, since several of these proteins are involved in the protection of the parasite against cytotoxic attack of xenobiotics (Table 11.2). The genome sequencing of the main important species associated with CL, DCL, MCL, and VL (*L. (L.) major*, *L. (L.) amazonensis*, *L. (V.) braziliensis*, and *L. (L.) infantum*, respectively) have contributed to the identification and for a better understanding of the biological functions of ABC proteins, as discussed throughout this chapter [21, 30, 133]. Some practical results can already be extracted from studies demonstrating the potential use of inhibitors of ABC proteins for leishmaniasis chemotherapy, as well as, combined treatment with some inhibitors of ABC transporters to increase the effectiveness of the current drugs [134–136]. One example is glibenclamide, a general blocker of ABC proteins that can produce functional modifications relevant for differentiation, infectivity and survival of the parasite with potential for the therapy of the disease [137–139].

An intrinsic difference in the sensitivity to antimonials, amphotericin-B and miltefosine in *Leishmania* species, strains, and clinical isolates in vitro has been reported [140–148]. These variations in drug susceptibility and resistance may be due to changes in gene copy number and/or single-nucleotide polymorphisms. Different tools available to study gene functions in trypanosomatids have already been described and can be useful for the study of these proteins in drug resistance mechanisms [96, 149]. Furthermore, studies of DNA/RNA microarrays and proteomics have also been demonstrated the role of these proteins in the biology of the parasite and in drug resistance [26, 49, 98, 150].

In the last years, next-generation sequencing has contributed for the identification of drug-resistant markers in the parasite of the genus *Leishmania*. Previous results of resistant mutants selected in vitro or in clinically resistant mutants to the main drugs used in the chemotherapy of the disease have showed gene dosage alterations (amplifications and deletions), changes in chromosome copy number and single-nucleotide polymorphisms [19, 107]. Whole genome sequencing of drug-resistant mutants of *Leishmania* selected in the laboratory for antimony, amphotericin-B and miltefosine did not reveal any mutation or gene dosage alteration in *ABC* genes [84, 99, 151–153], with exception of the *ABCC3* gene that was found amplified in Sb^{III}-resistant mutants of *L. (L.) infantum* and *L. (V.) guyanensis* [99, 100]. Proteomics analysis also revealed changes in the expression of proteins in resistant and susceptible parasites to miltefosine, amphotericin-B and antimony, with no evidence of change in the expression of ABC proteins [154–156].

One must consider that there are multiple mechanisms of drug resistance involved and this fact becomes even more complex if we consider that parasites of genus *Leishmania* have a sexual cycle in the sand fly vector, with the generation of hybrids contributing to phenotypic diversity of resistant parasites in the field [1, 157].

Extensive research over the last several years has allowed the characterization of several ABC transporters involved in metabolism and drug resistance in *Leishmania*. In this chapter, we described the importance of the main pathways responsible for the drug resistance in these parasitic protozoa that is medically important and for which there are very few drugs available. Finally, more studies are necessary to understand the cellular function of these proteins to hasten the development of new drugs against leishmaniasis.

Acknowledgments We thank Glaucia Paranhos for critical reading of the chapter.

References

1. Akopyants NS, Kimblin N, Secundino N, Patrick R, et al. Demonstration of genetic exchange during cyclical development of *Leishmania* in the sand fly vector. Science. 2009;324:265–8.
2. Higgins CF. ABC transporters: from microorganisms to man. Annu Rev Cell Biol. 1992;8: 67–113.
3. Saurin W, Hofnung M, Dassa E. Getting in or out: early segregation between importers and exporters in the evolution of ATP-binding cassette (ABC) transporters. J Mol Evol. 1999;48:22–41.
4. Sauvage V, Aubert D, Escotte-Binet S, Villena I. The role of ATP-binding cassette (ABC) proteins in protozoan parasites. Mol Biochem Parasitol. 2009;167:81–94.
5. Ouellette M, Fase-Fowler F, Borst P. The amplified H circle of methotrexate-resistant *Leishmania tarentolae* contains a novel P-glycoprotein gene. EMBO J. 1990;9:1027–33.
6. Klokouzas A, Shahi S, Hladky SB, Barrand MA, et al. ABC transporters and drug resistance in parasitic protozoa. Int J Antimicrob Agents. 2003;22:301–17.
7. Gottesman MM, Fojo T, Bates SE. Multidrug resistance in cancer: role of ATP-dependent transporters. Nat Rev Cancer. 2002;2:48–58.
8. Walker JE, Saraste M, Runswick MJ, Gay NJ. Distantly related sequences in the alpha- and beta-subunits of ATP synthase, myosin, kinases and other ATP-requiring enzymes and a common nucleotide binding fold. EMBO J. 1982;1:945–51.

9. Endicott JA, Ling V. The biochemistry of P-glycoprotein-mediated multidrug resistance. Annu Rev Biochem. 1989;58:137–71.
10. Gottesman MM, Pastan I. Biochemistry of multidrug resistance mediated by the multidrug transporter. Annu Rev Biochem. 1993;62:385–427.
11. Leprohon P, Legare D, Girard I, Papadopoulou B, et al. Modulation of *Leishmania* ABC protein gene expression through life stages and among drug-resistant parasites. Eukaryot Cell. 2006;5:1713–25.
12. Leprohon P, Legare D, Ouellette M. Intracellular localization of the ABCC proteins of *Leishmania* and their role in resistance to antimonials. Antimicrob Agents Chemother. 2009;53: 2646–9.
13. Allikmets R, Gerrard B, Hutchinson A, Dean M. Characterization of the human ABC superfamily: isolation and mapping of 21 new genes using the expressed sequence tags database. Hum Mol Genet. 1996;5:1649–55.
14. Araujo-Santos JM, Parodi-Talice A, Castanys S, Gamarro F. The overexpression of an intracellular ABCA-like transporter alters phospholipid trafficking in *Leishmania*. Biochem Biophys Res Commun. 2005;330:349–55.
15. Castanys-Munoz E, Alder-Baerens N, Pomorski T, Gamarro F, et al. A novel ATP-binding cassette transporter from *Leishmania* is involved in transport of phosphatidylcholine analogues and resistance to alkyl-phospholipids. Mol Microbiol. 2007;64:1141–53.
16. Coelho AC, Beverley SM, Cotrim PC. Functional genetic identification of PRP1, an ABC transporter superfamily member conferring pentamidine resistance in *Leishmania major*. Mol Biochem Parasitol. 2003;130:83–90.
17. El Fadili K, Messier N, Leprohon P, Roy G, et al. Role of the ABC transporter MRPA (PGPA) in antimony resistance in *Leishmania infantum* axenic and intracellular amastigotes. Antimicrob Agents Chemother. 2005;49:1988–93.
18. Parodi-Talice A, Araujo JM, Torres C, Perez-Victoria JM, et al. The overexpression of a new ABC transporter in *Leishmania* is related to phospholipid trafficking and reduced infectivity. Biochim Biophys Acta. 2003;1612:195–207.
19. Downing T, Imamura H, Decuypere S, Clark TG, et al. Whole genome sequencing of multiple *Leishmania donovani* clinical isolates provides insights into population structure and mechanisms of drug resistance. Genome Res. 2011;21:2143–56.
20. Real F, Vidal RO, Carazzolle MF, Mondego JM, et al. The genome sequence of *Leishmania (Leishmania) amazonensis*: functional annotation and extended analysis of gene models. DNA Res. 2013;20:567–81.
21. Rogers MB, Hilley JD, Dickens NJ, Wilkes J, et al. Chromosome and gene copy number variation allow major structural change between species and strains of *Leishmania*. Genome Res. 2011;21:2129–42.
22. Ivens AC, Lewis SM, Bagherzadeh A, Zhang L, et al. A physical map of the *Leishmania major* Friedlin genome. Genome Res. 1998;8:135–45.
23. Legare D, Hettema E, Ouellette M. The P-glycoprotein-related gene family in *Leishmania*. Mol Biochem Parasitol. 1994;68:81–91.
24. Momen H. Some current problems in the systematics of Trypanosomatids. Int J Parasitol. 2001;31:640–2.
25. Stevens JR, Noyes HA, Schofield CJ, Gibson W. The molecular evolution of Trypanosomatidae. Adv Parasitol. 2001;48:1–56.
26. Brotherton MC, Racine G, Ouameur AA, Leprohon P, et al. Analysis of membrane-enriched and high molecular weight proteins in *Leishmania infantum* promastigotes and axenic amastigotes. J Proteome Res. 2012;11:3974–85.
27. Clayton C, Shapira M. Post-transcriptional regulation of gene expression in trypanosomes and leishmanias. Mol Biochem Parasitol. 2007;156:93–101.
28. Croft SL, Sundar S, Fairlamb AH. Drug resistance in leishmaniasis. Clin Microbiol Rev. 2006; 19:111–26.

29. Albrecht C, Viturro E. The ABCA subfamily—gene and protein structures, functions and associated hereditary diseases. Pflugers Arch. 2007;453:581–9.
30. Peacock CS, Seeger K, Harris D, Murphy L, et al. Comparative genomic analysis of three *Leishmania* species that cause diverse human disease. Nat Genet. 2007;39:839–47.
31. Borst P, Evers R, Kool M, Wijnholds J. A family of drug transporters: the multidrug resistance-associated proteins. J Natl Cancer Inst. 2000;92:1295–302.
32. Dean M, Rzhetsky A, Allikmets R. The human ATP-binding cassette (ABC) transporter superfamily. Genome Res. 2001;11:1156–66.
33. Torres C, Perez-Victoria FJ, Parodi-Talice A, Castanys S, et al. Characterization of an ABCA-like transporter involved in vesicular trafficking in the protozoan parasite *Trypanosoma cruzi*. Mol Microbiol. 2004;54:632–46.
34. Henderson DM, Sifri CD, Rodgers M, Wirth DF, et al. Multidrug resistance in *Leishmania donovani* is conferred by amplification of a gene homologous to the mammalian mdr1 gene. Mol Cell Biol. 1992;12:2855–65.
35. Herget M, Tampe R. Intracellular peptide transporters in human--compartmentalization of the "peptidome". Pflugers Arch. 2007;453:591–600.
36. Katakura K, Iwanami M, Ohtomo H, Fujise H, et al. Structural and functional analysis of the LaMDR1 multidrug resistance gene in *Leishmania amazonensis*. Biochem Biophys Res Commun. 1999;255:289–94.
37. Mitsuhashi N, Miki T, Senbongi H, Yokoi N, et al. MTABC3, a novel mitochondrial ATP-binding cassette protein involved in iron homeostasis. J Biol Chem. 2000;275:17536–40.
38. Sundaram P, Echalier B, Han W, Hull D, et al. ATP-binding cassette transporters are required for efficient RNA interference in Caenorhabditis elegans. Mol Biol Cell. 2006;17:3678–88.
39. Chow LM, Wong AK, Ullman B, Wirth DF. Cloning and functional analysis of an extra-chromosomally amplified multidrug resistance-like gene in *Leishmania enriettii*. Mol Biochem Parasitol. 1993;60:195–208.
40. Hendrickson N, Sifri CD, Henderson DM, Allen T, et al. Molecular characterization of the ldmdr1 multidrug resistance gene from *Leishmania donovani*. Mol Biochem Parasitol. 1993;60:53–64.
41. Katakura K, Fujise H, Takeda K, Kaneko O, et al. Overexpression of LaMDR2, a novel multidrug resistance ATP-binding cassette transporter, causes 5-fluorouracil resistance in *Leishmania amazonensis*. FEBS Lett. 2004;561:207–12.
42. Perez-Victoria JM, Parodi-Talice A, Torres C, Gamarro F, et al. ABC transporters in the protozoan parasite *Leishmania*. Int Microbiol. 2001;4:159–66.
43. Gueiros-Filho FJ, Viola JP, Gomes FC, Farina M, et al. *Leishmania amazonensis*: multidrug resistance in vinblastine-resistant promastigotes is associated with rhodamine 123 efflux, DNA amplification, and RNA overexpression of a *Leishmania* mdr1 gene. Exp Parasitol. 1995;81:480–90.
44. Chiquero MJ, Perez-Victoria JM, O'valle F, Gonzalez-Ros JM, et al. Altered drug membrane permeability in a multidrug-resistant *Leishmania tropica* line. Biochem Pharmacol. 1998;55:131–9.
45. Dodge MA, Waller RF, Chow LM, Zaman MM, et al. Localization and activity of multidrug resistance protein 1 in the secretory pathway of *Leishmania* parasites. Mol Microbiol. 2004;51:1563–75.
46. Callahan HL, Beverley SM. Heavy metal resistance: a new role for P-glycoproteins in *Leishmania*. J Biol Chem. 1991;266:18427–30.
47. Legare D, Richard D, Mukhopadhyay R, Stierhof YD, et al. The *Leishmania* ATP-binding cassette protein PGPA is an intracellular metal-thiol transporter ATPase. J Biol Chem. 2001;276:26301–7.
48. Papadopoulou B, Roy G, Dey S, Rosen BP, et al. Contribution of the *Leishmania* P-glycoprotein-related gene ltpgpA to oxyanion resistance. J Biol Chem. 1994;269:11980–6.

49. Leprohon P, Legare D, Raymond F, Madore E, et al. Gene expression modulation is associated with gene amplification, supernumerary chromosomes and chromosome loss in antimony-resistant *Leishmania infantum*. Nucleic Acids Res. 2009;37:1387–99.
50. Coelho AC, Messier N, Ouellette M, Cotrim PC. Role of the ABC transporter PRP1 (ABCC7) in pentamidine resistance in *Leishmania* amastigotes. Antimicrob Agents Chemother. 2007; 51:3030–2.
51. Coelho AC, Yamashiro-Kanashiro EH, Bastos SF, Mortara RA, et al. Intracellular location of the ABC transporter PRP1 related to pentamidine resistance in *Leishmania major*. Mol Biochem Parasitol. 2006;150:378–83.
52. Perea A, Manzano JI, Castanys S, Gamarro F. The LABCG2 transporter from the protozoan parasite *Leishmania* is involved in antimony resistance. Antimicrob Agents Chemother. 2016; 60:3489–96.
53. Bosedasgupta S, Ganguly A, Roy A, Mukherjee T, et al. A novel ATP-binding cassette transporter, ABCG6 is involved in chemoresistance of *Leishmania*. Mol Biochem Parasitol. 2008;158:176–88.
54. Castanys-Munoz E, Perez-Victoria JM, Gamarro F, Castanys S. Characterization of an ABCG-like transporter from the protozoan parasite *Leishmania* with a role in drug resistance and transbilayer lipid movement. Antimicrob Agents Chemother. 2008;52:3573–9.
55. Manzano JI, Garcia-Hernandez R, Castanys S, Gamarro F. A new ABC half-transporter in *Leishmania major* is involved in resistance to antimony. Antimicrob Agents Chemother. 2013; 57:3719–30.
56. Lee JS, Paull K, Alvarez M, Hose C, et al. Rhodamine efflux patterns predict P-glycoprotein substrates in the National Cancer Institute drug screen. Mol Pharmacol. 1994;46:627–38.
57. Perez-Victoria JM, Perez-Victoria FJ, Parodi-Talice A, Jimenez IA, et al. Alkyl-lysophospholipid resistance in multidrug-resistant *Leishmania tropica* and chemosensitization by a novel P-glycoprotein-like transporter modulator. Antimicrob Agents Chemother. 2001; 45:2468–74.
58. Seifert K, Matu S, Javier Perez-Victoria F, Castanys S, et al. Characterisation of *Leishmania donovani* promastigotes resistant to hexadecylphosphocholine (miltefosine). Int J Antimicrob Agents. 2003;22:380–7.
59. Jha TK, Sundar S, Thakur CP, Bachmann P, et al. Miltefosine, an oral agent, for the treatment of Indian visceral leishmaniasis. N Engl J Med. 1999;341:1795–800.
60. Soto J, Toledo J, Gutierrez P, Nicholls RS, et al. Treatment of American cutaneous leishmaniasis with miltefosine, an oral agent. Clin Infect Dis. 2001;33:E57–61.
61. Lux H, Heise N, Klenner T, Hart D, et al. Ether–lipid (alkyl-phospholipid) metabolism and the mechanism of action of ether–lipid analogues in *Leishmania*. Mol Biochem Parasitol. 2000; 111:1–14.
62. Van Helvoort A, Smith AJ, Sprong H, Fritzsche I, et al. MDR1 P-glycoprotein is a lipid translocase of broad specificity, while MDR3 P-glycoprotein specifically translocates phosphatidylcholine. Cell. 1996;87:507–17.
63. Perez-Victoria FJ, Gamarro F, Ouellette M, Castanys S. Functional cloning of the miltefosine transporter. A novel P-type phospholipid translocase from *Leishmania* involved in drug resistance. J Biol Chem. 2003;278:49965–71.
64. Perez-Victoria FJ, Sanchez-Canete MP, Castanys S, Gamarro F. Phospholipid translocation and miltefosine potency require both *L. donovani* miltefosine transporter and the new protein LdRos3 in *Leishmania* parasites. J Biol Chem. 2006;281:23766–75.
65. Sanchez-Canete MP, Carvalho L, Perez-Victoria FJ, Gamarro F, et al. Low plasma membrane expression of the miltefosine transport complex renders *Leishmania braziliensis* refractory to the drug. Antimicrob Agents Chemother. 2009;53:1305–13.
66. Basselin M, Denise H, Coombs GH, Barrett, MP. Resistance to pentamidine in *Leishmania mexicana* involves exclusion of the drug from the mitochondrion. Antimicrob Agents Chemother. 2002;46:3731–8.
67. Coelho AC, Gentil LG, Da Silveira JF, Cotrim, PC. Characterization of *Leishmania (Leish mania) amazonensis* promastigotes resistant to pentamidine. Exp Parasitol. 2008;120:98–102.

68. Mukherjee A, Padmanabhan PK, Sahani MH, Barrett MP, et al. Roles for mitochondria in pentamidine susceptibility and resistance in *Leishmania donovani*. Mol Biochem Parasitol. 2006;145:1–10.
69. Bitonti AJ, Sjoerdsma A, Mccann PP, Kyle DE, et al. Reversal of chloroquine resistance in malaria parasite *Plasmodium falciparum* by desipramine. Science. 1988;242:1301–3.
70. Wong IL, Chow LM. The role of *Leishmania enriettii* multidrug resistance protein 1 (LeMDR1) in mediating drug resistance is iron-dependent. Mol Biochem Parasitol. 2006;150:278–87.
71. Kispal G, Csere P, Prohl C, Lill R. The mitochondrial proteins Atm1p and Nfs1p are essential for biogenesis of cytosolic Fe/S proteins. EMBO J. 1999;18:3981–9.
72. Martinez-Garcia M, Campos-Salinas J, Cabello-Donayre M, Pineda-Molina E, et al. LmABCB3, an atypical mitochondrial ABC transporter essential for *Leishmania major* virulence, acts in heme and cytosolic iron/sulfur clusters biogenesis. Parasit Vectors. 2016;9:7.
73. Deeley RG, Westlake C, Cole SP. Transmembrane transport of endo- and xenobiotics by mammalian ATP-binding cassette multidrug resistance proteins. Physiol Rev. 2006;86: 849–99.
74. Zhou SF, Wang LL, Di YM, Xue CC, et al. Substrates and inhibitors of human multidrug resistance associated proteins and the implications in drug development. Curr Med Chem. 2008;15:1981–2039.
75. Pompella A, Visvikis A, Paolicchi A, De Tata V, et al. The changing faces of glutathione, a cellular protagonist. Biochem Pharmacol. 2003;66:1499–503.
76. Ouellette M, Legare D, Haimeur A, Grondin K, et al. ABC transporters in *Leishmania* and their role in drug resistance. Drug Resist Updat. 1998;1:43–8.
77. Callahan HL, Beverley SM. A member of the aldoketo reductase family confers methotrexate resistance in *Leishmania*. J Biol Chem. 1992;267:24165–8.
78. Papadopoulou B, Roy G, Ouellette M. A novel antifolate resistance gene on the amplified H circle of *Leishmania*. EMBO J. 1992;11:3601–8.
79. Grondin K, Papadopoulou B, Ouellette M. Homologous recombination between direct repeat sequences yields P-glycoprotein containing amplicons in arsenite resistant *Leishmania*. Nucleic Acids Res. 1993;21:1895–901.
80. Ouellette M, Hettema E, Wust D, Fase-Fowler F, et al. Direct and inverted DNA repeats associated with P-glycoprotein gene amplification in drug resistant *Leishmania*. EMBO J. 1991;10:1009–16.
81. Beverley SM. Gene amplification in *Leishmania*. Annu Rev Microbiol. 1991;45:417–44.
82. Borst P, Ouellette M. New mechanisms of drug resistance in parasitic protozoa. Annu Rev Microbiol. 1995;49:427–60.
83. Mary C, Faraut F, Deniau M, Dereure J, et al. Frequency of drug resistance gene amplification in clinical *Leishmania* strains. Int J Microbiol. 2010;2010:1.
84. Mukherjee A, Boisvert S, Monte-Neto RL, Coelho AC, et al. Telomeric gene deletion and intrachromosomal amplification in antimony-resistant *Leishmania*. Mol Microbiol. 2013;88: 189–202.
85. Papadopoulou B, Roy G, Dey S, Rosen BP, et al. Gene disruption of the P-glycoprotein related gene pgpa of *Leishmania tarentolae*. Biochem Biophys Res Commun. 1996;224:772–8.
86. Haimeur A, Brochu C, Genest P, Papadopoulou B, et al. Amplification of the ABC transporter gene PGPA and increased trypanothione levels in potassium antimonyl tartrate (SbIII) resistant *Leishmania tarentolae*. Mol Biochem Parasitol. 2000;108:131–5.
87. Fairlamb AH, Cerami A. Metabolism and functions of trypanothione in the Kinetoplastida. Annu Rev Microbiol. 1992;46:695–729.
88. Grondin K, Haimeur A, Mukhopadhyay R, Rosen BP, et al. Co-amplification of the gamma-glutamylcysteine synthetase gene gsh1 and of the ABC transporter gene pgpA in arsenite-resistant *Leishmania tarentolae*. EMBO J. 1997;16:3057–65.

89. Haimeur A, Guimond C, Pilote S, Mukhopadhyay R, et al. Elevated levels of polyamines and trypanothione resulting from overexpression of the ornithine decarboxylase gene in arsenite-resistant *Leishmania*. Mol Microbiol. 1999;34:726–35.
90. Mukherjee A, Padmanabhan PK, Singh S, Roy G, et al. Role of ABC transporter MRPA, gamma-glutamylcysteine synthetase and ornithine decarboxylase in natural antimony-resistant isolates of *Leishmania donovani*. J Antimicrob Chemother. 2007;59:204–11.
91. Mukhopadhyay R, Dey S, Xu N, Gage D, et al. Trypanothione overproduction and resistance to antimonials and arsenicals in *Leishmania*. Proc Natl Acad Sci U S A. 1996;93:10383–7.
92. Arana FE, Perez-Victoria JM, Repetto Y, Morello A, et al. Involvement of thiol metabolism in resistance to glucantime in *Leishmania tropica*. Biochem Pharmacol. 1998;56:1201–8.
93. Legare D, Papadopoulou B, Roy G, Mukhopadhyay R, et al. Efflux systems and increased trypanothione levels in arsenite-resistant *Leishmania*. Exp Parasitol. 1997;87:275–82.
94. Dias FC, Ruiz JC, Lopes WC, Squina FM, et al. Organization of H locus conserved repeats in *Leishmania (Viannia) braziliensis* correlates with lack of gene amplification and drug resistance. Parasitol Res. 2007;101:667–76.
95. Lye LF, Owens K, Shi H, Murta SM, et al. Retention and loss of RNA interference pathways in trypanosomatid protozoans. PLoS Pathog. 2010;6:e1001161.
96. Beverley SM. Protozomics: trypanosomatid parasite genetics comes of age. Nat Rev Genet. 2003;4:11–9.
97. Kapler GM, Beverley SM. Transcriptional mapping of the amplified region encoding the dihydrofolate reductase-thymidylate synthase of *Leishmania major* reveals a high density of transcripts, including overlapping and antisense RNAs. Mol Cell Biol. 1989;9:3959–72.
98. Vergnes B, Gourbal B, Girard I, Sundar S, et al. A proteomics screen implicates HSP83 and a small kinetoplastid calpain-related protein in drug resistance in *Leishmania donovani* clinical field isolates by modulating drug-induced programmed cell death. Mol Cell Proteomics. 2007;6:88–101.
99. Brotherton MC, Bourassa S, Leprohon P, Legare D, et al. Proteomic and genomic analyses of antimony resistant *Leishmania infantum* mutant. PLoS One. 2013;8:e81899.
100. Monte-Neto R, Laffitte MC, Leprohon P, Reis P, et al. Intrachromosomal amplification, locus deletion and point mutation in the aquaglyceroporin AQP1 gene in antimony resistant *Leishmania (Viannia) guyanensis*. PLoS Negl Trop Dis. 2015;9:e0003476.
101. Moreira DS, Monte Neto RL, Andrade JM, Santi AM, et al. Molecular characterization of the MRPA transporter and antimony uptake in four new world *Leishmania* spp. susceptible and resistant to antimony. Int J Parasitol Drugs Drug Resist. 2014;3:143–53.
102. Shahi SK, Krauth-Siegel RL, Clayton CE. Overexpression of the putative thiol conjugate transporter TbMRPA causes melarsoprol resistance in *Trypanosoma brucei*. Mol Microbiol. 2002;43:1129–38.
103. Ghedin E, Debrabant A, Engel JC, Dwyer DM. Secretory and endocytic pathways converge in a dynamic endosomal system in a primitive protozoan. Traffic. 2001;2:175–88.
104. Mcconville MJ, Mullin KA, Ilgoutz SC, Teasdale RD. Secretory pathway of trypanosomatid parasites. Microbiol Mol Biol Rev. 2002;66:122–54. table of contents
105. Mullin KA, Foth BJ, Ilgoutz SC, Callaghan JM, et al. Regulated degradation of an endoplasmic reticulum membrane protein in a tubular lysosome in *Leishmania mexicana*. Mol Biol Cell. 2001;12:2364–77.
106. Ellenberger TE, Beverley SM. Multiple drug resistance and conservative amplification of the H region in *Leishmania major*. J Biol Chem. 1989;264:15094–103.
107. Leprohon P, Fernandez-Prada C, Gazanion E, Monte-Neto R, et al. Drug resistance analysis by next generation sequencing in *Leishmania*. Int J Parasitol Drugs Drug Resist. 2015;5:26–35.
108. Dey S, Ouellette M, Lightbody J, Papadopoulou B, et al. An ATP-dependent as(III)-glutathione transport system in membrane vesicles of *Leishmania tarentolae*. Proc Natl Acad Sci U S A. 1996;93:2192–7.
109. Dallagiovanna B, Gamarro F, Castanys S. Molecular characterization of a P-glycoprotein related tcpgp2 gene in *Trypanosoma cruzi*. Mol Biochem Parasitol. 1996;75:145–57.

110. Torres C, Barreiro L, Dallagiovanna B, Gamarro F, et al. Characterization of a new ATP-binding cassette transporter in *Trypanosoma cruzi* associated to a L1Tc retrotransposon. Biochim Biophys Acta. 1999;1489:428–32.
111. Murta SM, Dos Santos WG, Anacleto C, Nirde P, et al. Drug resistance in *Trypanosoma cruzi* is not associated with amplification or overexpression of P-glycoprotein (PGP) genes. Mol Biochem Parasitol. 2001;117:223–8.
112. Mittal MK, Rai S, Gupta S, Ravinder, et al. Characterization of natural antimony resistance in *Leishmania donovani* isolates. Am J Trop Med Hyg. 2007;76:681–8.
113. Singh N, Singh RT, Sundar S. Novel mechanism of drug resistance in kala azar field isolates. J Infect Dis. 2003;188:600–7.
114. Decuypere S, Rijal S, Yardley V, De Doncker S, et al. Gene expression analysis of the mechanism of natural Sb(V) resistance in *Leishmania donovani* isolates from Nepal. Antimicrob Agents Chemother. 2005;49:4616–21.
115. Velamakanni S, Wei SL, Janvilisri T, Van Veen HW. ABCG transporters: structure, substrate specificities and physiological roles : a brief overview. J Bioenerg Biomembr. 2007;39: 465–71.
116. Ejendal KF, Hrycyna CA. Multidrug resistance and cancer: the role of the human ABC transporter ABCG2. Curr Protein Pept Sci. 2002;3:503–11.
117. Leslie EM, Deeley RG, Cole SP. Multidrug resistance proteins: role of P-glycoprotein, MRP1, MRP2, and BCRP (ABCG2) in tissue defense. Toxicol Appl Pharmacol. 2005;204:216–37.
118. Bories C, Cojean S, Huteau F, Loiseau PM. Selection and phenotype characterisation of sitamaquine-resistant promastigotes of *Leishmania donovani*. Biomed Pharmacother. 2008; 62:164–7.
119. Zingales B, Araujo RG, Moreno M, Franco J, et al. A novel ABCG-like transporter of *Trypanosoma cruzi* is involved in natural resistance to benznidazole. Mem Inst Oswaldo Cruz. 2015;110:433–44.
120. Hettema EH, Van Roermund CW, Distel B, Van Den Berg M, et al. The ABC transporter proteins Pat1 and Pat2 are required for import of long-chain fatty acids into peroxisomes of *Saccharomyces cerevisiae*. EMBO J. 1996;15:3813–22.
121. Theodoulou FL, Holdsworth M, Baker A. Peroxisomal ABC transporters. FEBS Lett. 2006; 580:1139–55.
122. Shani N, Valle D. Peroxisomal ABC transporters. Methods Enzymol. 1998;292:753–76.
123. Petriv OI, Pilgrim DB, Rachubinski RA, Titorenko VI. RNA interference of peroxisome-related genes in *C. elegans*: a new model for human peroxisomal disorders. Physiol Genomics. 2002;10:79–91.
124. Michels PA, Bringaud F, Herman M, Hannaert V. Metabolic functions of glycosomes in trypanosomatids. Biochim Biophys Acta. 2006;1763:1463–77.
125. Kerr ID. Sequence analysis of twin ATP binding cassette proteins involved in translational control, antibiotic resistance, and ribonuclease L inhibition. Biochem Biophys Res Commun. 2004; 315:166–73.
126. Dean M, Annilo T. Evolution of the ATP-binding cassette (ABC) transporter superfamily in vertebrates. Annu Rev Genomics Hum Genet. 2005;6:123–42.
127. Zhou A, Hassel BA, Silverman RH. Expression cloning of 2-5A-dependent RNAase: a uniquely regulated mediator of interferon action. Cell. 1993;72:753–65.
128. Chen ZQ, Dong J, Ishimura A, Daar I, et al. The essential vertebrate ABCE1 protein interacts with eukaryotic initiation factors. J Biol Chem. 2006;281:7452–7.
129. Estevez AM, Haile S, Steinbuchel M, Quijada L, et al. Effects of depletion and overexpression of the *Trypanosoma brucei* ribonuclease L inhibitor homologue. Mol Biochem Parasitol. 2004;133:137–41.
130. Tyzack JK, Wang X, Belsham GJ, Proud CG. ABC50 interacts with eukaryotic initiation factor 2 and associates with the ribosome in an ATP-dependent manner. J Biol Chem. 2000; 275: 34131–9.

131. Marton MJ, Vazquez De Aldana CR, Qiu H, Chakraburtty K, et al. Evidence that GCN1 and GCN20, translational regulators of GCN4, function on elongating ribosomes in activation of eIF2alpha kinase GCN2. Mol Cell Biol. 1997;17:4474–89.
132. Sandbaken MG, Lupisella JA, Didomenico B, Chakraburtty K. Protein synthesis in yeast. Structural and functional analysis of the gene encoding elongation factor 3. J Biol Chem. 1990; 265:15838–44.
133. Ivens AC, Peacock CS, Worthey EA, Murphy L, et al. The genome of the kinetoplastid parasite, *Leishmania major*. Science. 2005;309:436–42.
134. Manzano JI, Lecerf-Schmidt F, Lespinasse MA, Di Pietro A, et al. Identification of specific reversal agents for *Leishmania* ABCI4-mediated antimony resistance by flavonoid and trolox derivative screening. J Antimicrob Chemother. 2014;69:664–72.
135. Mookerjee Basu J, Mookerjee A, Banerjee R, Saha M, et al. Inhibition of ABC transporters abolishes antimony resistance in *Leishmania* infection. Antimicrob Agents Chemother. 2008; 52:1080–93.
136. Ponte-Sucre A. Availability and applications of ATP-binding cassette (ABC) transporter blockers. Appl Microbiol Biotechnol. 2007;76:279–86.
137. Padron-Nieves M, Diaz E, Machuca C, Romero A, et al. Glibenclamide modulates glucantime activity and disposition in *Leishmania major*. Exp Parasitol. 2009;121:331–7.
138. Serrano-Martin X, Payares G, Mendoza-Leon A. Glibenclamide, a blocker of K+(ATP) channels, shows antileishmanial activity in experimental murine cutaneous leishmaniasis. Antimicrob Agents Chemother. 2006;50:4214–6.
139. Silva N, Camacho N, Figarella K, Ponte-Sucre A. Cell differentiation and infectivity of *Leishmania mexicana* are inhibited in a strain resistant to an ABC-transporter blocker. Parasitology. 2004;128:629–34.
140. Coelho AC, Trinconi CT, Costa CH, Uliana SR. In vitro and in vivo miltefosine susceptibility of a *Leishmania amazonensis* isolate from a patient with diffuse cutaneous leishmaniasis. PLoS Negl Trop Dis. 2014;8:e2999.
141. Croft SL, Engel J. Miltefosine--discovery of the antileishmanial activity of phospholipid derivatives. Trans R Soc Trop Med Hyg. 2006;100(Suppl 1):S4–8.
142. Espada CR, Ribeiro-Dias F, Dorta ML, Pereira LIA, et al. Susceptibility to Miltefosine in Brazilian clinical isolates of *Leishmania (Viannia) braziliensis*. Am J Trop Med Hyg. 2017; 96(3):656–9.
143. Kumar D, Kulshrestha A, Singh R, Salotra P. In vitro susceptibility of field isolates of *Leishmania donovani* to Miltefosine and amphotericin B: correlation with sodium antimony gluconate susceptibility and implications for treatment in areas of endemicity. Antimicrob Agents Chemother. 2009;53:835–8.
144. Obonaga R, Fernandez OL, Valderrama L, Rubiano LC, et al. Treatment failure and miltefosine susceptibility in dermal leishmaniasis caused by *Leishmania* subgenus Viannia species. Antimicrob Agents Chemother. 2014;58:144–52.
145. Prajapati VK, Mehrotra S, Gautam S, Rai M, et al. In vitro antileishmanial drug susceptibility of clinical isolates from patients with Indian visceral leishmaniasis--status of newly introduced drugs. Am J Trop Med Hyg. 2012;87:655–7.
146. Prajapati VK, Sharma S, Rai M, Ostyn B, et al. In vitro susceptibility of *Leishmania donovani* to miltefosine in Indian visceral leishmaniasis. Am J Trop Med Hyg. 2013;89:750–4.
147. Utaile M, Kassahun A, Abebe T, Hailu A. Susceptibility of clinical isolates of *Leishmania aethiopica* to miltefosine, paromomycin, amphotericin B and sodium stibogluconate using amastigote-macrophage in vitro model. Exp Parasitol. 2013;134:68–75.
148. Zauli-Nascimento RC, Miguel DC, Yokoyama-Yasunaka JK, Pereira LI, et al. *In vitro* sensitivity of *Leishmania (Viannia) braziliensis* and *Leishmania (Leishmania) amazonensis* Brazilian isolates to meglumine antimoniate and amphotericin B. Trop Med Int Health. 2010;15: 68–76.
149. Clayton CE. Genetic manipulation of kinetoplastida. Parasitol Today. 1999;15(9):372–8.

150. Ubeda JM, Legare D, Raymond F, Ouameur AA, et al. Modulation of gene expression in drug resistant *Leishmania* is associated with gene amplification, gene deletion and chromosome aneuploidy. Genome Biol. 2008;9:R115.
151. Coelho AC, Boisvert S, Mukherjee A, Leprohon P, et al. Multiple mutations in heterogeneous miltefosine-resistant *Leishmania major* population as determined by whole genome sequencing. PLoS Negl Trop Dis. 2012;6:e1512.
152. Fernandez-Prada C, Vincent IM, Brotherton MC, Roberts M, et al. Different mutations in a P-type ATPase transporter in *Leishmania* parasites are associated with cross-resistance to two leading drugs by distinct mechanisms. PLoS Negl Trop Dis. 2016;10:e0005171.
153. Shaw CD, Lonchamp J, Downing T, Imamura H, et al. In vitro selection of miltefosine resistance in promastigotes of *Leishmania donovani* from Nepal: genomic and metabolomic characterization. Mol Microbiol. 2016;99:1134–48.
154. Carnielli JB, De Andrade HM, Pires SF, Chapeaurouge AD, et al. Proteomic analysis of the soluble proteomes of miltefosine-sensitive and -resistant *Leishmania infantum chagasi* isolates obtained from Brazilian patients with different treatment outcomes. J Proteomics. 2014;108: 198–208.
155. Matrangolo FS, Liarte DB, Andrade LC, de Melo MF, et al. Comparative proteomic analysis of antimony-resistant and -susceptible *Leishmania braziliensis* and *Leishmania infantum chagasi* lines. Mol Biochem Parasitol. 2013;190:63–75.
156. Walker J, Gongora R, Vasquez JJ, Drummelsmith J, et al. Discovery of factors linked to antimony resistance in *Leishmania panamensis* through differential proteome analysis. Mol Biochem Parasitol. 2012;183:166–76.
157. Rougeron V, De Meeus T, Hide M, Waleckx E, et al. Extreme inbreeding in *Leishmania braziliensis*. Proc Natl Acad Sci USA. 2009;106:10224–9.

Functional Analysis of *Leishmania* Membrane (Non-ABC) Transporters Involved in Drug Resistance

12

Scott M. Landfear

Abstract

Leishmania parasites rely heavily upon membrane transport proteins to deliver essential nutrients from their hosts to the interior of the parasite. Some of these transporters also serve as routes for uptake of drugs used for treatment of leishmaniasis or experimental drugs with potential for development of novel anti-leishmanial therapies. Hence, mutations within the coding regions of such permeases or alterations in the expression of the carrier proteins can confer drug resistance upon the parasites. This chapter reviews the current level of knowledge regarding several classes of membrane transporters known to play roles in uptake or sensitivity to drugs. The increasing knowledge of the "permeome," provided by complete genome sequences of several *Leishmania* species, has advanced considerably our knowledge of how nutrients and drugs or other cytotoxic compounds enter these pathogenic protozoa. Recent genome-wide approaches to functional analysis promise to further our understanding of transporters as determinants of drug sensitivity and resistance.

12.1 Introduction

The phospholipid bilayer of the plasma membrane is poorly permeable to most compounds that are relatively hydrophilic [1]. For this reason, all cells express a panoply of membrane transport proteins (referred to as transporters, permeases, or carriers) and channels that mediate the selective passage of specific compounds or ions across the membrane. Great advances have been made in the past decade or so

S. M. Landfear (✉)
Department of Molecular Microbiology and Immunology, Oregon Health & Science University, Portland, OR, USA
e-mail: landfear@ohsu.edu

A. Ponte-Sucre, M. Padrón-Nieves (eds.), *Drug Resistance in Leishmania Parasites*,
https://doi.org/10.1007/978-3-319-74186-4_12

regarding the structure, function, and biological relevance of transporters and channels. While such advances have been based upon a variety of experimental methods, two notable contributions have been the completion of genome sequences [2] that delimit the large numbers of proteins that may play roles in transport (the permeome) and the increasing number of three-dimensional structures for transporters and channels [3, 4] that have been delivered by both x-ray crystallography and electron diffraction.

Many drugs reach their intracellular targets by being transported across the plasma membrane of the relevant cell. Often, the drugs serve as surrogate substrates that may be structurally related to the natural ligands for some permease. Hence, transporters play a central role in drug delivery in mammals and microorganisms. Since carriers provide a critical route for internalization of such drugs, alterations in the function or level of expression of the relevant transporter can result in resistance to the transported drug. Hence transporters play dual roles regarding drug efficacy, both delivering drugs to their targets and serving as determinants of drug resistance when they do not function properly.

The objective of this chapter is to review the roles of three families of transporters or channels in drug delivery and resistance in *Leishmania* parasites. These three families, the aquaporins, folate permeases, and purine transporters, have been chosen because they play central roles in delivery of drugs and because mechanisms of resistance related to transporter expression or mutation have been investigated in some detail. Aquaporins mediate the uptake of antimonials, still the first-line drugs for treatment of leishmaniasis. Antifolates have not yet been employed effectively against leishmanial infections in vivo, but they do kill these parasites in vitro, and folate transporters have offered remarkable insights into molecular mechanisms of drug resistance. Purine nucleoside and nucleobase transporters import a variety of purine analog drugs or experimental drugs and constitute an important component of the purine salvage pathway that has been of long standing interest for the pharmacology of leishmaniasis.

I have attempted to provide an updated account of these permeases and their roles in drug sensitivity and resistance. It has not, however, been possible to cite every scientific contribution to each field, and I apologize to any authors whose work could not be cited here due to space limitations

12.2 Aquaporins: Sensitivity and Resistance to Antimonials

12.2.1 Delivery of Antimonials to Intracellular Amastigotes

The first-line treatment for *Leishmania* infections in most parts of the world is pentavalent antimony (Sb^V)-containing drugs such as Pentostam (sodium stibogluconate) and Glucantime (meglumine). The mechanism of action of these drugs is not clear, but it has been thought for some time [5, 6] that Sb^V is a prodrug that must be reduced to Sb^{III} to be effective against the parasite. Since *Leishmania* amastigotes are intracellular parasites that reside within phagolysosomal or

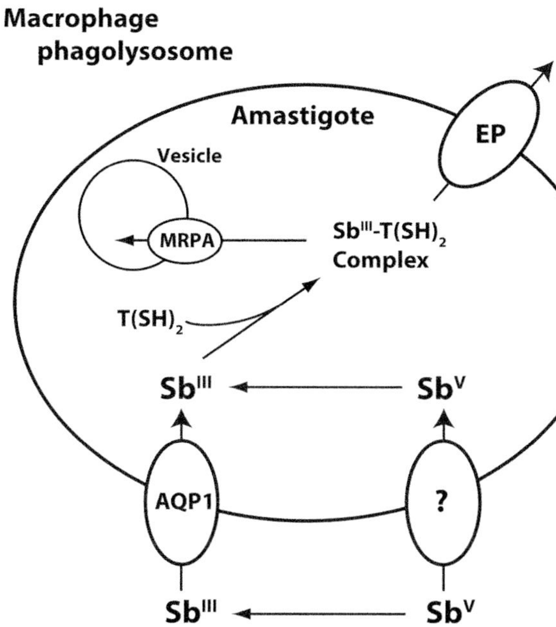

Fig. 12.1 Models for uptake, conjugation, and excretion of Sb^{III} and Sb^{V} by *Leishmania* amastigotes. The large oval represents a *Leishmania* amastigote inside the phagolysosome of a host macrophage. According to the first model for activity of antimonial drugs, Sb^{V} is reduced to Sb^{III} primarily in the macrophage, and Sb^{III} is delivered to the parasite by AQP1. According to the second model, Sb^{V} is imported across the parasite plasma membrane by an unknown protein (?) and then reduced to leishmanicidal Sb^{III} within the amastigote. It is possible that both mechanisms operate in nature. Intraparasitic reduction of Sb^{V} to Sb^{III} may be mediated by thiols. Conjugation of Sb^{III} to the major *Leishmania* thiol trypanothione (T[SH]$_2$) provides a substrate for the MRPA ABC transporter that extrudes the conjugate into an intracellular compartment and another unknown extrusion pump (EP) that exports the conjugate across the plasma membrane, thus providing mechanisms for resistance. This figure is modified from [7]

parasitophorous vesicles within mammalian host macrophages, reduction of Sb^{V} to Sb^{III} could in principle occur within the macrophage, resulting in uptake of Sb^{III} by the parasite, or Sb^{V} could be imported into the parasite and then reduced to Sb^{III} (Fig. 12.1). Despite the ongoing uncertainty regarding the roles of Sb^{III} and Sb^{V}, it is likely that uptake of Sb^{III} by the parasite is a significant route for delivery of antimonial drugs to the intracellular parasite [7]. This conclusion is also consistent with the observation (see below) that the levels of the principal transport protein for Sb^{III} are often, though not universally, reduced in antimony-resistant parasites.

12.2.2 The Aquaporin LmAQP1 Identified as the Major Route for Uptake of Sb^{III} in *Leishmania (L.) major*

Aquaporins are a family of water-permeant channels that have been identified in organisms as diverse as the bacterium *Escherichia coli* and humans [8]. These proteins function as channels, rather than as transporters. While classical aquaporins such as human AQP1, the first functionally expressed member of the aquaporin family [9], flux water as the principal permeant, other members of the family from bacteria to humans mediate the transport of various small solutes such as glycerol and urea in preference to water and are designated "aquaglyceroporins" [10]. The three-dimensional structures of several aquaporin family members have been determined by either electron or X-ray diffraction [11–14]. These proteins consist of six transmembrane helices and two half helices that are located within the membrane (these proteins exhibit a pseudo-inverted duplication between the N-terminal and C-terminal halves). These helices fold into a constricted channel that mediates the permeation of water or of small molecular weight solutes. The observation that aquaglyceroporins from bacteria [15], yeast [16], and humans [17] can flux trivalent metalloids such as As^{III} and Sb^{III} suggested that similar channels in *Leishmania* might serve as routes for uptake of pharmacologically relevant Sb^{III}. As^{III} appears to exist primarily as As(OH)3 in aqueous solution, and this species is thought to function as a molecular mimic of glycerol, explaining the propensity of aquaglyceroporins to transport these metalloids [18]. Indeed, the then emerging sequence of the *L. (L.) major* genome uncovered a gene that encoded a protein homologous to the human aquaglyceroporin AQP9 [19]; the amino acid sequence also shared predicted topology and conserved signature sequences with other aquaglyceroporins (Fig. 12.2). To functionally characterize this new protein, designated LmAQP1, the LmAQP1 gene was overexpressed in promastigotes of several species of *Leishmania* by transfection with an episomal expression vector encompassing the LmAQP1 open reading frame (ORF). These transfectants

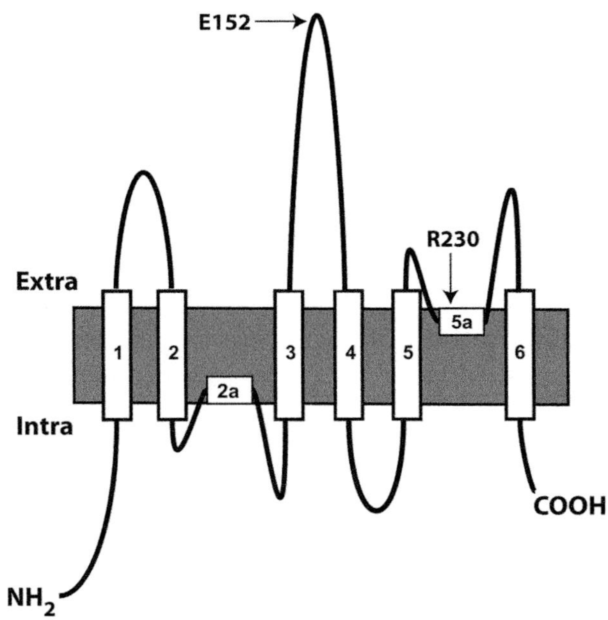

Fig. 12.2 Topology of LmAQP1. The gray rectangle represents the lipid bilayer and the extracellular (extra) and intracellular (intra) surfaces of the membrane. The model, consistent with the experimentally determined three-dimensional structure of several aquaporins, consists of six transmembrane α-helices (numbered rectangles) and two half helices (2a, 5a) that also enter the membrane. Curved black lines represent hydrophilic loops that connect transmembrane domains. E152 and R230 refer to two critical amino acids discussed in the text

exhibited increased sensitivity to As^{III} and Sb^{III} compared to promastigotes transfected with the empty expression vector, and they exhibited greatly increased rates and levels of uptake for these two metalloids. Furthermore, overexpression of LmAQP1 in both wild-type and antimonial-resistant field isolates of *Leishmania (L.) donovani* sensitized these strains to killing by Sb^{V} when the parasites were cultured within macrophages. Notably, when one copy of the LmAQP1 gene was deleted in *L. (L.) major* promastigotes by targeted gene replacement, the "single-knockout" strain exhibited an EC50 for Sb^{III} that was ~15-fold higher than that for wild-type promastigotes. Subsequently, a LmAQP1 null mutant was generated [20] that exhibited greatly reduced uptake of As^{III} and Sb^{III} and a ~30–50-fold increased EC_{50} for each metalloid. All these data support the notion that LmAQP1 is a major route for uptake of Sb^{III} in *Leishmania* parasites and that decreased expression of the LmAQP1 gene can lead to drug resistance. The effect of LmAQP1 expression on sensitivity of intracellular amastigotes to Sb^{V} also supports a principal role for this channel in sensitivity and resistance to clinically relevant antimonials.

12.2.3 Other Properties of LmAQP1

Further biochemical and genetics studies of LmAQP1 have uncovered a number of intriguing biological properties for this aquaglyceroporin. Expression of this channel in *Xenopus* oocytes established that it mediates the flux of water, glycerol, methylglyoxal, dihydroxyacetone, and sugar alcohols, exhibiting a broad permeant specificity [21]. Remarkably, immunofluorescence and immunoelectron microscopy using an anti-LmAQP1 antibody established that the protein is expressed in the flagellar membrane of promastigotes and the flagellar pocket membrane and spongiosum membranes of amastigotes. Furthermore, both promastigotes and axenic amastigotes of *L. (L.) donovani* that were expressing LmAQP1 from an episomal vector showed increased ability to regulate cellular volume in response to hypoosmotic shock, and the LmAQP1 null mutant was impaired in this response [20], suggesting that this channel plays a role in protection of parasites during osmotic stress. In addition, promastigotes overexpressing LmAQP1 migrated more rapidly toward an osmotic gradient than parasites transfected with either the empty expression vector or with LmAQP1 ORFs encompassing mutations at the crucial R230 residue. Hence LmAQP1 appears to play a sensory role in osmotaxis of promastigotes [22], and the flagellar localization may be central to this sensory function.

12.2.4 Role of AQP1 in Mediating Sensitivity and Resistance to Antimonials in Laboratory and Field Isolates

The role of LmAQP1 in mediating uptake of Sb^{III} suggests that the AQP1 proteins encoded by orthologous genes in various *Leishmania* species might be important determinants of sensitivity to this metalloid- and to Sb^{V}-containing drugs. One study

[23] addressed this issue by transfecting an As^{III}-resistant strain of *L. (S.) tarentolae* with a cosmid library of genomic DNA from *L. (L.) mexicana*. One transformant had restored sensitivity to Sb^{III}, and the cosmid present in this clonal line encompassed the LmxAQP1 gene, the ortholog of LmAQP1. As^{III}-resistant mutants transfected with this cosmid exhibited dramatically increased uptake of both As^{III} and Sb^{III} compared to the non-transfected mutant. Furthermore, laboratory-derived mutants of *L. (S.) tarentolae*, *L. (L.) major*, and *L. (L.) infantum* were investigated for altered expression of the AQP1 gene. Reduction in the levels of AQP1 mRNA was observed for many of the mutants, and the reduction in AQP1 mRNA levels also correlated with reduced uptake of Sb^{III}. Several studies also examined the expression of AQP1 mRNA in field isolates of *L. (L.) donovani* from India and Nepal [24–26]. Overall, levels of AQP1 mRNA were often but not always reduced in the drug-resistant compared to drug-sensitive strains. Thus, AQP1 expression appears to be an important but not exclusive determinant for sensitivity to Sb^{V}-containing drugs. The failure of AQP1 expression to correlate completely with antimonial sensitivity is not surprising, as other biochemical pathways are also known to affect sensitivity to antimonials. Thus, the ABC pump MRPA (also called PgpA) sequesters thiol conjugates of As^{III} and Sb^{III} in an intracellular compartment [27] (Fig. 11.1, for detailed description, please see Chap. 11), and upregulation of the pump and of the thiol biosynthetic enzymes [28] can be associated with antimonial resistance. Another metal-thiol extrusion pump (EP) that has not been identified at the molecular level is able to export Sb^{III} across the parasite plasma membrane [29].

Despite this complexity, a recent study has demonstrated that species-specific sensitivity to Sb^{III} correlates with the relative level of expression of AQP1 [30]. Higher innate expression of AQP1 in cutaneous species such as *L. (L.) major* results in greater accumulation of intracellular Sb^{III} and higher metalloid sensitivity compared to visceral species such as *L. (L.) infantum* that express lower levels of AQP1. In a separate study, phosphorylation of LmAQP1 on Thr-197 by MAP kinase 2 was shown to result in stabilization of the channel and an increase in sensitivity of the parasite to Sb^{III} [31]. Remarkably, phosphorylation of Thr-197 also caused relocalization of LmAQP1 from the flagellar membrane to the entire parasite surface.

12.3 Folate Transporters: Mediators of Uptake and Resistance to Methotrexate

12.3.1 Methotrexate: A Model for Development of Drug Resistance in *Leishmania* Parasites

Leishmania do not synthesize folates (conjugated pteridines that consist of a heterocyclic pterin ring linked to *para*-amino benzoic acid and glutamate) and must acquire these nutrients from their hosts [32]. Tetrahydrofolate (THF) is an essential cofactor for the synthesis of thymidylate (Fig. 12.3). Imported folate is reduced to dihydrofolate (DHF), and DHF is reduced to THF by dihydrofolate reductase (DHFR), an enzyme that is joined in a single polypeptide chain to thymidylate

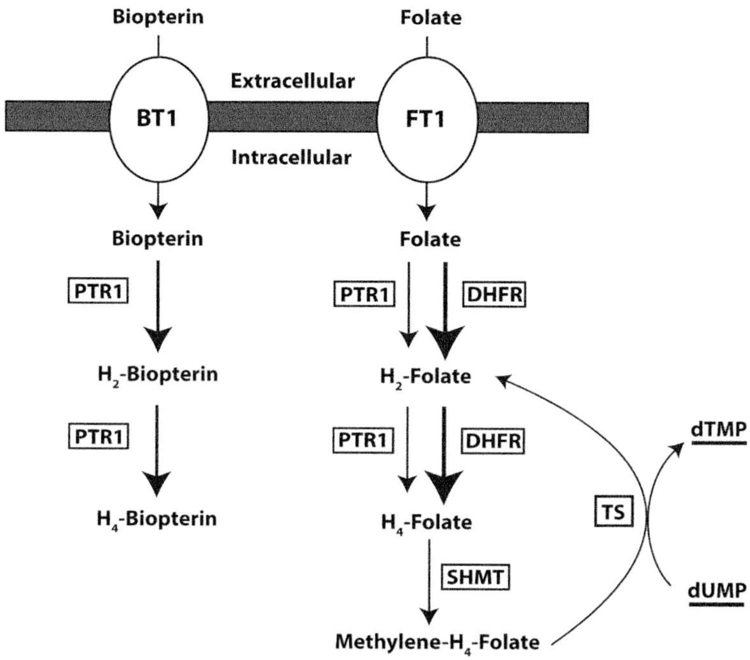

Fig. 12.3 Uptake and metabolism of biopterin and folate in *Leishmania*. Biopterin enters *Leishmania* parasites through the BT1 permease, which can also transport folate but not methotrexate (MTX) with low affinity. The FT1 and FT5 transporters, and possibly other permeases, mediate the high-affinity uptake of folate and MTX. BT1, FT1, and FT5 are members of the folate-biopterin transporter family whose members share significant sequence identity. Folate can be reduced to dihydrofolate (H_2-Folate) and tetrahydro-folate (H_4-Folate) by dihydrofolate reductase (DHFR), and biopterin can be reduced to H_2-biopterin and H_4-biopterin by pterin reductase 1 (PTR1). PTR1 is also able to reduce folates with lower affinity than DHFR. Methylene-H_4-folate, generated from H_4-folate by serine hydroxymethyltransferase (SHMT), is required for conversion by thymidylate synthase (TS) of dUMP to dTMP that is essential for DNA synthesis. MTX is a competitive inhibitor of DHFR and thus prevents synthesis of dTMP. Resistance to MTX can be mediated by alterations in expression of FT1, BT1, DHFR, and PTR1, as discussed in the text. This figure was modified from [32]

synthase (TS) to form the bifunctional DHFR-TS protein. A knockout of the gene encoding DHFR-TS is lethal unless the mutants are supplemented with thymidine, demonstrating the essential role of this enzyme in the synthesis of thymidylate, a crucial precursor for biosynthesis of DNA. In addition, *Leishmania* do not synthesize unconjugated pterins, such as biopterin, and the parasites are also dependent upon salvage of this compound. However, the precise biochemical function of biopterin in *Leishmania* and the reason why it is required for parasite viability is still obscure. Both promastigotes and amastigotes take up high levels of biopterin and folate [32], and they express specific permeases that mediate the transport of these essential nutrients (Fig. 12.3).

The antifolate drug methotrexate (MTX) is a close structural analog of DHF and is a high-affinity competitive inhibitor of DHFR in many organisms. MTX has been employed extensively in anticancer therapy, as it has a pronounced cytotoxic effect against rapidly dividing cells. Antifolates such as trimethoprim and pyrimethamine have also been employed in treatment of bacterial and protozoal (e.g., malaria) infections due to their selective ability to inhibit microbial DHFRs. While MTX is not an effective anti-leishmanial drug due to its higher toxicity against mammalian cells than parasites, it does nonetheless inhibit *Leishmania* DHFR and is toxic to the parasite in vitro. Consequently, MTX has been used extensively as a model for development of drug resistance in these parasites. Of additional importance, the folate pathways are essential for viability of *Leishmania* parasites [33] and are thus targets for development of novel anti-leishmanial drugs.

Early studies on MTX resistance in *L. (L.) major* demonstrated that parasites resistant to 1 mM MTX had a 40-fold increase in DHFR activity and also amplified DNA elements, designated R regions, which were presumed to contain the *DHFR* gene [34]. Indeed, subsequent work confirmed that the amplified R region contained the *DHFR* gene, and overexpression of the enzyme was a common mechanism for development of MTX resistance [35]. Subsequent studies demonstrated that MTX resistance could entail a number of alternate genetic modifications. Thus a line of the kinetoplastid parasite *Crithidia fasciculata* resistant to MTX was shown to be impaired in uptake of [^{3}H]MTX [36], and the MTXA5 line of *L. (L.) donovani*, isolated in a single step by selection for resistance to 1 mM MTX, was deficient in uptake of both folate and MTX [37]. These results provided genetic evidence for a specific folate permease that also mediated the uptake of MTX and whose genetic alteration could confer resistance to this drug. Studies on uptake of labeled folate and MTX in *L. (L.) major* [38], and inhibition by various competitive inhibitors, also suggested that the same transporters mediated the uptake of both compounds, and MTX-resistant mutants in this species were also isolated that had impaired uptake of both folate and MTX. Continued studies on MTX-resistant *L. (L.) major* revealed a third mechanism of resistance, amplification of the H region of DNA that contained a gene encoding an aldo-keto reductase [39–41]. This enzyme, subsequently named pterin reductase 1 (PTR1), was identified as a biopterin reductase that normally reduces biopterin to dihydrobiopterin and dihydrobiopterin to tetrahydrobiopterin [32]. The ability of this enzyme to reduce unconjugated pterins was also accompanied by a more limited ability to reduce folates, yet the enzyme is not susceptible to inhibition by MTX. Thus, amplification of the *PTR1* gene can provide an alternate route for reduction of DHF to THF that is not sensitive to drug inhibition and can thus induce MTX resistance (Fig. 12.3).

12.3.2 Molecular Identification of *Leishmania* Biopterin and Folate Transporters as Members of a Novel Family of Permeases, the Folate-Biopterin Transporter (FBT) Family

The first member of the folate transporter family in *Leishmania* that was identified at the molecular level was the biopterin transporter 1 (BT1), which was isolated from

L. (S.) tarentolae [42] by functional cloning. Wild-type parasites were transfected with a cosmid genomic library, and transfectants that grew on plates containing MTX were analyzed. Mapping of the cosmids indicated that the resistance phenotype corresponded to a single gene, *orfG* [43, 44], and a null mutant of *orfG* was deficient in biopterin uptake. Hence, the ORFG protein was renamed BT1. BT1 also has low-affinity transport capacity for folate, but MTX was not a substrate for BT1. Furthermore, *L. (S.) tarentolae* mutants that were MTX-resistant due to loss of high-affinity folate transport were found to have rearranged the BT1 gene locus leading to higher expression of *BT1* mRNA. This increased expression of BT1 due to such rearrangements was proposed to promote viability of the MTX-resistant parasites by providing an alternate route for salvage of folate that does not import MTX.

Independent studies on *L. (L.) donovani* 1 (LD1) DNA elements, that are spontaneously amplified as extrachromosomal circles in ~15% of *L. (L.) donovani* isolates, also identified *orfG* as a gene encoding a biopterin transporter [45]. Experiments employing *Xenopus* oocytes confirmed that injection of *orfG* cRNA induced biopterin uptake activity, establishing that *orfG* encoded a biopterin transporter that was also renamed biotin transporter 1 (BT1) by this group. Subsequently, a *BT1* null mutant was generated in *L. (L.) donovani* by targeted gene replacement; this mutant was deficient in biopterin uptake, but uptake was restored by complementation with the *BT1* gene on an episomal expression vector.

The ability of BT1 to import folate at high concentrations suggested that other related genes might encode bona fide folate permeases. Employing the *BT1* gene, Ouellette and co-workers [46] demonstrated the existence of a family of cross-hybridizing fragments of genomic DNA in *L. (S.) tarentolae*, suggesting the existence of a family of *BT1*-related genes. Sequencing of an 8.6 kb fragment of genomic DNA demonstrated that it contained two ORFs that were ~40% identical to BT1 encoding proteins designated folate transporters 3 and 5 (FT3 and FT5). Notably, the *FT5* gene corresponded to a fragment of DNA that was absent in a laboratory-generated MTX-resistant line. Ultimately, 14 members of this gene family were detected in the completed *L. (L.) major* genome (www.genedb.org). Expression of FT3 and FT5 in folate transport-deficient MTX-resistant mutants established that FT5 transported radiolabeled folate with high affinity (K_m of 84 nM), but no folate transport activity was detected for FT3. Furthermore, an *FT5* null mutant exhibited reduced uptake of MTX and folate at 50 nM concentrations but did not exhibit reduced uptake if 1 μM substrates were employed. This result suggested that FT5 is a high-affinity folate/MTX transporter but that other lower-affinity transporters exist among this family.

Further studies by the Ouellette group identified a transporter FT1, encoded by another member of the repeated gene family, as the major folate transporter in *L. (L.) infantum* [47]. When the cloned *FT1* gene was transfected into a MTX transport-deficient mutant, it restored uptake of MTX and folic acid to wild-type levels and exhibited a K_m for uptake of folate of ~400 nM, similar to that observed for wild-type parasites. An *FT1* null mutant exhibited a loss of ~75% of folate uptake activity and was highly resistant to MTX. Loss of most but not all folate transport activity in this mutant is consistent with the conclusion that it is the major but not sole folate/MTX transporter.

Structure-function analysis by site-directed mutagenesis has been carried out on FT1 [48] and also on the folate transporter Slr0642 from *Synechocystis* [49], both of which are members of the folate-biopterin transporter (FBT) family of permeases (www.tcdb.org). Slr0642 was expressed in an *Escherichia coli* mutant deficient in synthesis and uptake of folate. A battery of 47 amino acid residues of Slr0642 were chosen for mutagenesis to C or A based upon location in predicted transmembrane domains, charge, or polarity, conservation in other known folate transporters of the FBT family, and predicted location within an aqueous cavity. Mutations in 22 of these amino acids abolished folate uptake without preventing expression of the permease in membranes. A model of the three-dimensional structure of Slr0642 was generated using the crystal structure of the *E. coli* lactose permease, a 12-transmembrane domain protein to which Slr0642 appears to bear structural similarity. Most of these functionally important residues lined a predicted central cavity and were concentrated on the core α-helices H1, H4, H7, and H10. The mutagenesis data were consistent with binding of folate within this cavity at a position roughly equidistant from the extracellular and intracellular surfaces. Furthermore, of the six residues common to the mutagenesis studies on FT1 and Slr0642, five were found to be important or essential for transport in both permeases, suggesting that both folate permeases likely share structural and functional similarities and that the Slr0642 model is likely to be relevant to FT permeases in *Leishmania* species.

12.3.3 Another Member of the FBT Family Is a Transporter for *S*-Adenosylmethionine and for Analogs of this Compound with Potential Antimicrobial Activity

Unexpectedly, another member of the *Leishmania* FBT family, designated AdoMetT1, has been shown [50] to encode a transporter for *S*-adenosylmethionine (AdoMet), a universal methyl donor for methylation of lipids, proteins, nucleic acids, and xenobiotics, and this compound is also utilized in a variety of additional biosynthetic reactions. Sinefungin (SNF) is an analog of AdoMet that is not taken up by mammalian cells but is imported by *Leishmania* parasites and has antimicrobial activity. In *Leishmania* [51] and *T. brucei* [52], SNF competes for uptake with AdoMet, suggesting that these two related compounds share a single transporter.

An unexpected observation demonstrated that a mutant of *L. (S.) tarentolae*, called MTX1000.6, that is resistant to MTX by virtue of decreased accumulation of MTX, is also cross-resistant to SNF. Expression of several *FBT* orthologs from *L. (L.) infantum* in this MTX1000.6 mutant revealed that one of them, *LinJ10_V3.0370*, restored high-level sensitivity to SNF implying that it transported this antimicrobial compound, and this transporter was subsequently named AdoMetT1 to indicate its natural substrate. A null mutant in the *AdoMetT1* gene was generated in *L. (L.) infantum* and shown to be deficient in uptake of labeled

AdoMet, but AdoMet uptake was restored in the null mutant that was complemented with the *AdoMetT1* gene. Furthermore, the *AdoMetT1* null mutant was highly resistant to SNF, and SNF sensitivity was restored by complementation with the *AdoMetT1* gene. Hence, a member of the FBT family transports AdoMet, rather than folate or biopterin, and genetic impairment of this permease confers resistance to AdoMet analogs that are selectively toxic to the parasite but not to mammalian cells.

12.4 Purine Nucleoside and Nucleobase Transporters: Uptake of Cytotoxic Purine Analogs

12.4.1 Purine Salvage in *Leishmania* and Cytotoxic Purine Analogs

One of the distinguishing features of parasitic protozoa such as *Leishmania* is that they do not synthesize purines de novo and rely upon salvage of preformed purines from their hosts [53, 54]. In contrast, most cells of the vertebrate hosts are capable of both de novo synthesis and salvage. Furthermore, purines are essential nutrients for growth of *Leishmania* parasites in defined medium [55, 56]. The first step in purine salvage is the uptake of purine nucleosides or nucleobases across the plasma membrane of the parasite. Subsequently, a battery of purine salvage enzymes [53, 54, 57] interconverts purines and generates purine monophosphates that serve as precursors for synthesis of RNA and DNA and other purine-containing metabolites.

The reliance of parasitic protozoa upon purine salvage has generated considerable interest in targeting the purine salvage pathway for development of novel antiparasitic chemotherapies. In principle, targeting could rely upon two distinct strategies: (1) inhibition of enzymes or transporters that may be critical for purine salvage and (2) incorporation of "subversive substrates" that are taken up by the parasite and selectively utilized by the parasite salvage enzymes to generate a cytotoxic product. The latter approach is particularly relevant to transporters and their roles in drug sensitivity and resistance.

12.4.2 Pyrazolopyrimidines: Purine Analogs That Are Toxic to *Leishmania* or Related Parasites

Pyrazolopyrimidines (PPs) are analogs of naturally occurring purines, either nucleosides or nucleobases, in which the nitrogen in either position 7 or 9 of the purine ring has been translocated to position 8 [58, 59]. Relevant PPs are formycin B (an inosine analog), tubercidin (an adenosine analog), thiopurinol (TPP) and thiopurinol riboside (TPPR), aminopurinol (APP), allopurinol (HPP, a hypoxanthine analog), and allopurinol riboside (HPPR, an inosine analog). These PPs can be taken up by purine transporters, and they are utilized as substrates by the purine salvage enzymes and metabolized to triphosphates. Although the mechanisms of toxicity of these compounds have not been proven, it has been postulated that their incorporation into RNA leads to their cytotoxic effects in parasites. Thus, the specificity that many of these purine analogs exhibit for parasites is ascribed by their ability to be metabolized by the parasite salvage enzymes, whereas they are not efficient substrates for mammalian purine salvage. Tubercidin is one exception, as this compound is toxic to both parasites and mammalian cells. Several of the other PPs have elicited considerable interest in their therapeutic potential as selective antiparasitic agents. Indeed, allopurinol is employed, usually in combination with other drugs, for treatment of leishmaniasis in Central America [60, 61], and formycin B is effective in killing intracellular amastigotes in vitro [62].

12.4.3 Genetic Identification of Purine Nucleoside Transporters in *Leishmania*

Early biochemical studies employing uptake of radiolabeled nucleosides [63] established that two distinct purine nucleoside transport systems existed in *Leishmania* promastigotes, one for adenosine and pyrimidine nucleosides (designated NT1 in subsequent publications [64]) and another for inosine and guanosine (designated NT2 in subsequent publications [65]). To further investigate the existence of two distinct and nonoverlapping transport activities, NT1 and NT2, Ullman and colleagues [66] generated transport-deficient mutants in each permease, confirming that there were separate genes that encoded each of the two transporters. These mutants were isolated by mutagenizing parasites and then selecting for clonal lines that survived in either the cytotoxic adenosine analog tubercidin (the TUBA5 mutant, deficient in NT1, and thus unable to import tubercidin) or the toxic inosine analog formycin B (the FBD5 mutant, deficient in NT2, and thus unable to import formycin B). The TUBA5 and FBD5 mutations were in separate transporter genes, because TUBA5 mutants could still transport guanosine, inosine, and xanthosine, while FBD5 mutants could still transport adenosine and pyrimidine nucleosides. These genetic studies confirmed the existence of two distinct nucleoside uptake systems, NT1 (adenosine, pyrimidine, and tubercidin transporter) and NT2 (guanosine, inosine, xanthosine, and formycin B transporter), and of considerable importance also provided mutants that subsequently proved critical for the molecular cloning of the corresponding transporter genes.

12.4.4 Molecular Cloning and Functional Characterization of Purine Nucleoside Transporter Genes *NT1* and *NT2*

The *NT1* and *NT2* genes were identified in a genetic screen. To identify the *NT1* gene, TUBA5 mutants were transfected with a cosmid genomic library, and the rare transformant that had re-acquired sensitivity to tubercidin was identified by screening through hundreds of transformed lines [64]. Similarly, the *NT2* gene was identified in a parallel screen for transformants that restored formycin B sensitivity to the FBD5 mutant [65]. Analysis of positive cosmids identified two linked genes that encoded two closely related isoforms of the NT1 adenosine/pyrimidine/tubercidin nucleoside transporters, which were designated *LdNT1.1* and *LdNT1.2*. Similarly, a single *LdNT2* ORF (35% identical to *LdNT1* ORFs) restored high-affinity uptake of guanosine, inosine, and xanthosine and sensitivity to formycin B to the FBD5 mutant. Both LdNT1 and LdNT2 are homologous to mammalian "equilibrative nucleoside transporters" (ENTs [67]) and are members of a transporter family designated SLC29 (http://www.bioparadigms.org/slc/menu.asp), whose members have 11 predicted TMDs.

Subsequently, site-directed mutagenesis on LdNT1.1 [68] identified specific amino acids whose mutation strongly impaired transport activity. Hence, these residues are critical for the transport function of the permease. In summary mutation of E94 (in transmembrane domain 2, abbreviated TMD2), K153 (TMD5), or D374 (TMD8), all charged residues predicted to lie within TMDs, either strongly impaired transport activity or altered substrate specificity (e.g., the K153R mutant acquired the ability to take up inosine, which is not a substrate for the wild-type transporter). These results implied that specific charged residues within TMDs of LdNT1.1 play critical roles in transport of nucleoside substrates, possibly mediating direct interactions with these substrates. Parallel studies on LdNT2 have also identified critical residues in this inosine/guanosine/xanthosine permease, especially D389 and R393, both located within TMD8 [69, 70].

12.4.5 Computational Models of NT1 and NT2

While mutagenesis was able to identify functionally important amino acids within LdNT1 and LdNT2, information on the three-dimensional structure of these permeases is also central to understanding their function as transporters. However, no crystal structure has been solved for any member of the SLC29 family. As an alternative approach to understanding the tertiary structure of these permeases, a computational model of the LdNT2 protein was generated by homology modeling [70] to the crystal structure of the glycerol phosphate transporter [71] of *E. coli*. Using this computational model for LdNT2, it was possible to rationalize some of the mutagenesis results for this permease, providing experimental evidence that the model is likely to be a reasonable approximation of the actual structure [70]. Notably, this homology model provided the first indication of how SLC29 family members fold in three dimensions and suggested that their structures were similar to those of major facilitator superfamily (MFS) members [72], represented by such well-studied proteins as the *E. coli* lactose permease [73] and the glycerol phosphate transporter.

In contrast, ab initio modeling, a method that does not rely upon a known crystal structure of another protein but predicts the structure de novo from the physical properties of the constituent amino acids [74], was employed to investigate the structure and function of LdNT1.1. Notably, like the homology model for LdNT2, this ab initio model of LdNT1.1 also predicted an 11 TMD fold that was similar to the structures of 12 TMD MFS members without the final TMD. Furthermore, when the locations of close to 50 site-directed mutants were mapped onto the LdNT1.1

model, those that had strong phenotypes, such as pronounced reduction of transport activity or change in substrate specificity, mapped close to the predicted pore of the structure. In contrast, mutants with modest effects on transport tended to map to locations that were peripheral to the pore. These observations provided a further experimental test for the plausibility of the model.

In silico modeling has also been used to identify regions of LdNT1.1 that serve as "gates" [75] to alternately open and close the permeation pore from the outside or the inside, a fundamental feature of transporters that mediate the passage of solutes across membranes by an "alternating access model." The model in Fig. 12.4 suggested that extracellular tips of TMDs 1, 2, and 7 cluster, together to close off the pore in the closed to the outside—open to the inside conformation. F48 in TMD 1 and W75 in TMD2 were identified as residues likely to mediate these interhelix interactions, and results of cysteine cross-linking and site-directed mutagenesis supported this interpretation [76]. An alternate computational model implicated a cluster of hydrophobic residues in the intracellular ends of TMDs 4, 5, 10, and 11 as the intracellular gate [77] that closes off the pore in the open to the outside—closed

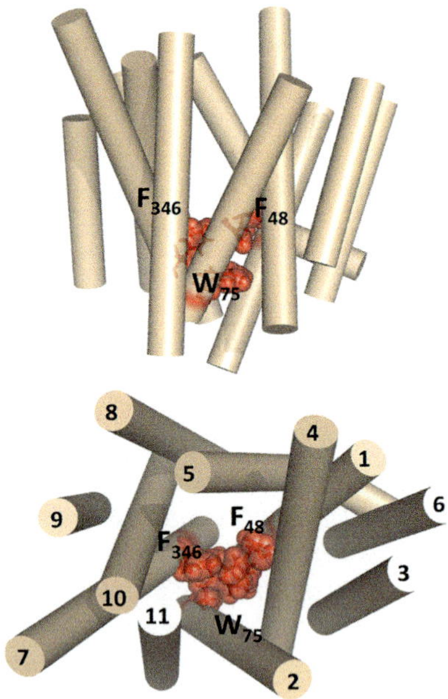

Fig. 12.4 Ab initio model for LdNT1.1. The helix disposition for TMDs 1–11 is shown; loops interconnecting the helices are not indicated. The top image shows a view of the transporter from the side, with the bottom indicating the extracellular surface and the top the cytosolic surface. The bottom image is a view from the cytosol toward the extracellular surface. Hence, the model shows the transporter in an "open to the inside" conformation. The red space-filling densities indicate the predicted positions of F48, W75, and F346. The figure is reproduced from Valdés R1, Arastu-Kapur S, Landfear SM, and Shinde U. An ab initio structural model of a nucleoside permease predicts functionally important residues. J Biol Chem. 2009;284(28):19067–19076. doi: https://doi.org/10.1074/jbc.M109.017947

to the inside confirmation—and this interpretation was also supported by cysteine cross-linking and mutagenesis. The ability of the structural models to predict potential and previously unsuspected functions for specific amino acids, such as F48 and W75, as gating residues further supports the plausibility of the in silico-derived structures.

12.4.6 *Leishmania* Purine Transporters Function as Concentrative Proton Symporters

Mammalian SLC29 nucleoside transporters are "equilibrative" permeases, i.e., transporters that do not concentrate their substrates but simply allow them to flux across the membrane according to the existing concentration gradient [67]. In contrast, SLC29 permeases in parasitic protozoa are high-affinity "concentrative" transporters that are coupled to the proton electrochemical gradient across the plasma membrane [78], which provides the thermodynamic force necessary to concentrate substrates within the cell. This arrangement ensures that parasites are able to capture essential purines from their hosts, even in environments where the concentrations of purines may be low, and it also promotes efficient uptake by the parasite of cytotoxic purine analogs. Two-electrode voltage clamp experiments performed on LdNT1.1, LdNT1.2, and LdNT2 expressed in *Xenopus* oocytes [79] identified inward directed positive currents that were dependent upon purine substrates and that were greatly enhanced at pH values below neutrality. These observations are indicative of cotransport (symport) of protons with purines. The import of positively charged protons into the electronegative environment of the cytosol provides thermodynamic energy that can be coupled to concentrate purines and purine analogs within the parasite. Subsequent electrophysiological experiments on the NT3 and NT4 nucleobase transporters from *L. (L.) major* (see below) confirmed that they are also concentrative proton symporters [80].

12.4.7 Identification of Purine Nucleobase Transporters

While experiments from several groups had identified nucleobase uptake systems in *Leishmania* parasites [81–83], employing uptake of radiolabeled nucleobases in intact promastigotes or amastigotes, the molecular identity of the permeases was initially unknown. The ongoing genome sequencing project for *L. (L.) major* (http://www.genedb.org) uncovered two new ORFs with ~30% identity at the amino acid level to LdNT1 and LdNT2. These genes were subsequently designated *LmaNT3* [84] and *LmaNT4* [85]. Functional expression of LmaNT3 in *Xenopus* oocytes established that it mediated the transport of the purine nucleobases hypoxanthine, xanthine, adenine, and guanine with apparent K_m values of 8–16 μM, similar to the K_m values for purine nucleobase transport in *L. (L.) major* promastigotes. This observation indicated that nucleobase transporters in *Leishmania* were also members of the SLC29 family. Subsequent studies on LmaNT4 indicated that this permease also mediated the uptake of purine nucleobases but with a pH maximum between 5 and 6 [80]. In contrast, LmaNT3 exhibited a pH maximum of ~7. These observations suggest that NT3 is designed to function optimally in the neutral pH conditions experienced by the promastigote stage of the life cycle, whereas NT4 has been optimized to function under the acidic conditions of the macrophage phagolysosome [86] where the disease-causing amastigotes live.

Allopurinol is a hypoxanthine analog that is employed in treatment of leishmaniasis because of its selective cytotoxicity toward the parasites. Nucleobase transporters appear to be the major route for uptake of this drug because (1) the NT3 permease mediates uptake of radiolabeled allopurinol when the transporter is expressed in *Xenopus* oocytes, and (2) a null mutant in the *NT3* gene, Δ*nt3*, exhibits an IC_{50} value for growth inhibition by allopurinol that is ~20-fold higher than that for wild-type parasites, suggesting that the principal mode for import of this drug has been eliminated in this mutant [85].

12.4.8 The TOR Gene and Resistance to Toxic Nucleoside Analogs

An early study by the Detke group [87] examined resistance of *L. (L.) mexicana amazonensis* to toxic nucleosides tubercidin and inosine dialdehyde. Resistant parasites were generated by selection in increasing concentrations of each analog, resulting in the TUB (tubercidin-resistant) and IDA (inosine dialdehyde-resistant) mutants. Both mutants were cross-resistant to tubercidin, inosine dialdehyde, formycin B, and allopurinol riboside. Furthermore, both mutants had greatly reduced transport capacity for guanosine, guanine, and adenine, i.e., for both purine nucleosides and nucleobases. The broad effects of each mutant on both resistance to toxic purine analogs and purine uptake indicated that they did not represent mutations in individual purine transporters but must have affected the uptake of purines and analogs by multiple permeases. Notably, a circular extrachromosomal amplicon of somewhat different size (56 kb in the TUB mutant) was detected in both mutants, and a 9 kb subclone of this circular element was able to confer toxic nucleoside resistance when introduced into the parasites on an expression vector. The authors hypothesized that a single gene, designated *TOR* for toxic nucleoside resistance, was responsible for the resistance phenotype. They furthermore suggested that the *TOR* gene product might interact with multiple purine transporters to modulate their expression or function. In a second paper [88], the genetic element conferring resistance to multiple toxic nucleosides by reducing uptake of various purines was identified within a 2.3 kb fragment of the amplicon, and the sequence of the internal ORF was determined.

In a third paper [89], overexpression of the *TOR* gene from an episomal expression vector was shown to cause retargeting of GFP-NT1 from the plasma membrane to an internal multivesicular tubule lysosome. The level of GFP-NT1 was also greatly reduced. The region of NT1 that interacts with the TOR protein was deduced by overexpressing various segments of NT1 and determining which ones interfered with the ability of TOR to induce resistance to tubercidin. A region (M289-W305) representing part of the large intracellular loop of NT1 between TMDs 6 and 7 was able to "squelch" the effect of overexpressed TOR (i.e., to restore sensitivity to tubercidin), suggesting that this loop interacts with the TOR protein. Furthermore, a deletion of M289-W305 was still functional as an adenosine permease but could no longer be internalized by parasites expressing high levels of TOR, further confirming the importance of this region for interaction with TOR. Studies expressing NT1 in yeast suggested that internalization of NT1 by TOR depends upon ubiquitination of NT1. Hence, the proposed model is that the TOR protein regulates expression of various NTs by binding to the large internal loop and inducing ubiquitination followed by internalization, targeting to the multivesicular lysosome, and degradation. The implication is that this mechanism for regulation of NT levels is likely to be operative under some physiological condition in *Leishmania* parasites; however, the normal biological function of TOR remains to be elucidated. Overall, these studies underscore the importance of expression of purine transporters as determinants of sensitivity and resistance to cytotoxic purine analogs.

12.5 Conclusions and Future Trends

This chapter has focused on a limited number of transporters as mediators of drug sensitivity and resistance in *Leishmania* parasites, those that have clearly implicated roles in uptake of drugs or compounds with cytotoxic activity toward the parasite. A variety of other permeases are likely to play similar roles for other antileishmanials, those that are currently employed therapeutically as well as other drugs that will likely emerge from ongoing drug discovery programs. Prospects for identification of other drug carriers are increased by recently developed genome-wide approaches for identifying determinants of drug sensitivity and resistance. The RIT-Seq method [90], based on genome-wide RNAi libraries, has been employed in *Trypanosoma brucei* as a method for identifying genes involved in conferring sensitivity to drugs. This method has identified both transporters [91] that are known determinants of drug sensitivity and resistance, such as the P2 purine permease that mediates uptake of melarsoprol and pentamidine, and previously unknown determinants, such as *Tb*AQP2 that appears to function as a receptor that can bind pentamidine and internalize the drug via endocytosis [92]. While RNAi is not operative in most species of *Leishmania*, it does occur in the subgenus *Viannia* [93], and RIT-Seq may be feasible for these species. The recently introduced "Cos-Seq" method [94] employs selection with a drug of interest against a population of parasites containing a cosmid genomic library. Cosmids that carry genes whose overexpression confers resistance to drugs are preferentially represented in the population of selected parasites and can be identified by high-throughput sequencing. While permeases involved in drug import would not be selected by this approach, those that mediate export could be identified. Hence it is likely that an increasing number of transporters involved in drug sensitivity and resistance will be identified in *Leishmania* species. In addition, the increasing number of membrane proteins whose structures are being solved and the ability of in silico modeling to extend structural information to permeases with otherwise unknown structures promise to increase our understanding of how drugs interact at the molecular level with transporters that mediate their import or export.

Acknowledgments Preparation of this review was supported by grants AI25920, AI44138, and AI144822 to the author from the National Institutes of Health.

References

1. Van Winkle LJ. Biomembrane Transport. 15th ed. San Diego: Academic Press; 1999.
2. Ivens AC, Peacock CS, Worthey EA, Murphy L, et al. The genome of the kinetoplastid parasite, *Leishmania major*. Science. 2005;309(5733):436–42.
3. Moraes I, Evans G, Sanchez-Weatherby J, Newstead S, et al. Membrane protein structure determination - the next generation. Biochim Biophys Acta. 2014;1838(1 Pt A):78–87.
4. Vinothkumar KR. Membrane protein structures without crystals, by single particle electron cryomicroscopy. Curr Opin Struct Biol. 2015;33:103–14.
5. Goodwin LG, Page JE. A study of the excretion of organic antimonials using a polarographic procedure. Biochem J. 1943;37(2):198–209.
6. Goodwin LG. Pentostam (sodium stibogluconate); a 50-year personal reminiscence. Trans R Soc Trop Med Hyg. 1995;89(3):339–41.
7. Croft SL, Sundar S, Fairlamb AH. Drug resistance in leishmaniasis. Clin Microbiol Rev. 2006;19(1):111–26.
8. King LS, Kozono D, Agre P. From structure to disease: the evolving tale of aquaporin biology. Nat Rev Mol Cell Biol. 2004;5(9):687–98.
9. Preston GM, Agre P. Isolation of the cDNA for erythrocyte integral membrane protein of 28 kilodaltons: member of an ancient channel family. Proc Natl Acad Sci USA. 1991;88 (24):11110–4.
10. Stroud RM, Savage D, Miercke LJ, Lee JK, et al. Selectivity and conductance among the glycerol and water conducting aquaporin family of channels. FEBS Lett. 2003;555(1):79–84.
11. Fu DA, Libson A, Miercke LJ, Weitzman C, et al. Structure of a glycerol-conducting channel and the basis for its selectivity. Science. 2000;290(5491):481–6.
12. Murata K, Mitsuoka K, Hirai T, Walz T, et al. Structural determinants of water permeation through aquaporin-1. Nature. 2000;407(6804):599–605.
13. Sui H, Han BG, Lee LJ, Walian P, et al. Structural basis of water-specific transport through the AQP1 water channel. Nature. 2001;414(6866):872–8.
14. Newby ZE, O'Connell J III, Robles-Colmenares Y, Khademi S, et al. Crystal structure of the aquaglyceroporin PfAQP from the malarial parasite *Plasmodium falciparum*. Nat Struct Mol Biol. 2008;15(6):619–25.
15. Sanders OI, Rensing C, Kuroda M, Mitra B, et al. Antimonite is accumulated by the glycerol facilitator GlpF in *Escherichia coli*. J Bacteriol. 1997;179(10):3365–7.
16. Wysocki R, Chery CC, Wawrzycka D, Van Hulle M, et al. The glycerol channel Fps1p mediates the uptake of arsenite and antimonite in *Saccharomyces cerevisiae*. Mol Microbiol. 2001;40 (6):1391–401.
17. Liu Z, Shen J, Carbrey JM, Mukhopadhyay R, et al. Arsenite transport by mammalian aquaglyceroporins AQP7 and AQP9. Proc Natl Acad Sci USA. 2002;99(9):6053–8.
18. Ramirez-Solis A, Mukopadhyay R, Rosen BP, Stemmler TL. Experimental and theoretical characterization of arsenite in water: insights into the coordination environment of As-O. Inorg Chem. 2004;43(9):2954–9.
19. Gourbal B, Sonuc N, Bhattacharjee H, Legare D, et al. Drug uptake and modulation of drug resistance in *Leishmania* by an aquaglyceroporin. J Biol Chem. 2004;279(30):31010–7.

20. Plourde M, Ubeda JM, Mandal G, Monte-Neto RL, et al. Generation of an aquaglyceroporin AQP1 null mutant in *Leishmania major*. Mol Biochem Parasitol. 2015;201(2):108–11.
21. Figarella K, Uzcategui NL, Zhou Y, LeFurgey A, et al. Biochemical characterization of *Leishmania major* aquaglyceroporin LmAQP1: possible role in volume regulation and osmotaxis. Mol Microbiol. 2007;65(4):1006–17.
22. Leslie G, Barrett M, Burchmore R. *Leishmania mexicana*: promastigotes migrate through osmotic gradients. Exp Parasitol. 2002;102(2):117–20.
23. Marquis N, Gourbal B, Rosen BP, Mukhopadhyay R, et al. Modulation in aquaglyceroporin AQP1 gene transcript levels in drug-resistant *Leishmania*. Mol Microbiol. 2005;57(6):1690–9.
24. Decuypere S, Rijal S, Yardley V, De Doncker S, et al. Gene expression analysis of the mechanism of natural Sb(V) resistance in *Leishmania donovani* isolates from Nepal. Antimicrob Agents Chemother. 2005;49(11):4616–21.
25. Maharjan M, Singh S, Chatterjee M, Madhubala R. Role of aquaglyceroporin (AQP1) gene and drug uptake in antimony-resistant clinical isolates of *Leishmania donovani*. Am J Trop Med Hyg. 2008;79(1):69–75.
26. Mandal S, Maharjan M, Singh S, Chatterjee M, et al. Assessing aquaglyceroporin gene status and expression profile in antimony-susceptible and -resistant clinical isolates of *Leishmania donovani* from India. J Antimicrob Chemother. 2010;65(3):496–507.
27. Legare D, Richard D, Mukhopadhyay R, Stierhof YD, et al. The *Leishmania* ATP-binding cassette protein PGPA is an intracellular metal-thiol transporter ATPase. J Biol Chem. 2001;276 (28):26301–7.
28. Grondin K, Haimeru A, Mukhopadyyay R, Rosen BP, et al. Co-amplification of the γ-glutamylcysteine synthetase gene *gsh1* and of the ABC transporter gene *pgpA* in arsenite-resistant *Leishmania tarentolae*. EMBO J. 1997;16:3057–65.
29. Dey S, Ouellette M, Lightbody J, Papadopoulou B, et al. An ATP-dependent As(III)-glutathione transport system in membrane vesicles of *Leishmania tarentolae*. Proc Natl Acad Sci USA. 1996;93(5):2192–7.
30. Mandal G, Mandal S, Sharma M, Charret KS, et al. Species-specific antimonial sensitivity in *Leishmania* is driven by post-transcriptional regulation of AQP1. PLoS Negl Trop Dis. 2015;9 (2):e0003500.
31. Mandal G, Sharma M, Kruse M, Sander-Juelch C, et al. Modulation of *Leishmania major* aquaglyceroporin activity by a mitogen-activated protein kinase. Mol Microbiol. 2012;85 (6):1204–18.
32. Cunningham ML, Beverley SM. Pteridine salvage throughout the *Leishmania* infectious cycle: implications for antifolate chemotherapy. Mol Biochem Parasitol. 2001;113:199–213.
33. Vickers TJ, Beverley SM. Folate metabolic pathways in *Leishmania*. Essays Biochem. 2011;51:63–80.
34. Coderre JA, Beverley SM, Schimke RT, Santi DV. Overproduction of a bifunctional thymidylate synthetase-dihydrofolate reductase and DNA amplification in methotrexate-resistant *Leishmania tropica*. Proc Natl Acad Sci USA. 1983;80(8):2132–6.
35. Beverley SM, Ellenberger TE, Cordingley JS. Primary structure of the gene encoding the bifunctional dihydrofolate reductase-thymidylate synthase of *Leishmania major*. Proc Natl Acad Sci USA. 1986;83(8):2584–8.
36. Dewes H, Ostergaard HL, Simpson L. Impaired drug uptake in methotrexate resistant *Crithidia fasciculata* without changes in dihydrofolate reductase activity or gene amplification. Mol Biochem Parasitol. 1986;19(2):149–61.
37. Kaur K, Coons T, Emmett K, Ullman B. Methotrexate-resistant *Leishmania donovani* genetically deficient in the folate-methotrexate transporter. J Biol Chem. 1988;263:7020–8.
38. Ellenberger TE, Beverley SM. Biochemistry and regulation of folate and methotrexate transport in *Leishmania major*. J Biol Chem. 1987;262:10053–8.
39. Callahan HL, Beverley SM. A member of the aldoketo reductase family confers methotrexate resistance in *Leishmania*. J Biol Chem. 1992;267:24165–8.

40. Papadopoulou B, Roy G, Ouellette M. Frequent amplification of a short chain dehydrogenase gene as part of circular and linear amplicons in methotrexate resistant *Leishmania*. Nucleic Acids Res. 1993;21(18):4305–12.
41. Papadopoulou B, Roy G, Mourad W, Leblanc E, et al. Changes in folate and pterin metabolism after disruption of the *Leishmania* H locus short chain dehydrogenase gene. J Biol Chem. 1994;269(10):7310–5.
42. Kündig C, Haimeur A, Legare D, Papadopoulou B, et al. Increased transport of pteridines compensates for mutations in the high affinity folate transporter and contributes to methotrexate resistance in the protozoan parasite *Leishmania tarentolae*. EMBO J. 1999;18:2342–51.
43. Myler PJ, Venkataraman GM, Lodes MJ, Stuart KD. A frequently amplified region in *Leishmania* contains a gene conserved in prokaryotes and eukaryotes. Gene1. 1994;48(2):187–93.
44. Segovia M, Ortiz G. LD1 amplifications in *Leishmania*. Parasitol Today. 1997;13(9):342–8.
45. Lemley C, Yan S, Dole VS, Madhubala R, et al. The *Leishmania donovani* LD1 locus gene *ORFG* encodes a biopterin transporter (BT1). Mol Biochem Parasitol. 1999;104:93–105.
46. Richard D, Kundig C, Ouellette M. A new type of high affinity folic acid transporter in the protozoan parasite *Leishmania* and deletion of its gene in methotrexate-resistant cells. J Biol Chem. 2002;277(33):29460–7.
47. Richard D, Leprohon P, Drummelsmith J, Ouellette M. Growth phase regulation of the main folate transporter of *Leishmania infantum* and its role in methotrexate resistance. J Biol Chem. 2004;279(52):54494–501.
48. Dridi L, Haimeur A, Ouellette M. Structure-function analysis of the highly conserved charged residues of the membrane protein FT1, the main folic acid transporter of the protozoan parasite *Leishmania*. Biochem Pharmacol. 2010;79(1):30–8.
49. Eudes A, Kunji ER, Noiriel A, Klaus SM, et al. Identification of transport-critical residues in a folate transporter from the folate-biopterin transporter (FBT) family. J Biol Chem. 2010;285(4):2867–75.
50. Dridi L, Ahmed Ouameur A, Ouellette M. High affinity S-Adenosylmethionine plasma membrane transporter of *Leishmania* is a member of the folate biopterin transporter (FBT) family. J Biol Chem. 2010;285(26):19767–75.
51. Phelouzat M, Basselin M, Lawrence F, Robert-Gero M. Sinefungin shares AdoMet-uptake system to enter *Leishmania donovani* promastigotes. Biochem J. 1995;305:133–7.
52. Goldberg B, Yarlett N, Sufrin J, Lloyd D, et al. A unique transporter of S-adenosylmethionine in African trypanosomes. FASEB J. 1997;11:256–60.
53. Hammond DJ, Gutteridge WE. Purine and pyrimidine metabolism in the Trypanosomatidae. Mol Biochem Parasitol. 1984;13(3):243–61.
54. Carter NS, Yates P, Arendt CS, Boitz JM, et al. Purine and pyrimidine metabolism in *Leishmania*. Adv Exp Med Biol. 2008;625:141–54.
55. Marr JJ, Berens RL, Nelson DJ. Purine metabolism in *Leishmania donovani* and *Leishmania braziliensis*. Biochim Biophys Acta. 1978;544(2):360–71.
56. Iovannisci DM, Ullman B. High efficiency plating method for *Leishmania* promastigotes in semidefined or completely-defined medium. J Parasitol. 1983;69(4):633–6.
57. Boitz JM, Ullman B, Jardim A, Carter NS. Purine salvage in *Leishmania*: complex or simple by design? Trends Parasitol. 2012;28(8):345–52.
58. Ullman B. Pyrazolopyrimidine metabolism in parasitic protozoa. Pharm Res. 1984;1:194–203.
59. Marr JJ. Purine analogs as chemotherapeutic agents in leishmaniasis and American trypanosomiasis. J Lab Clin Med. 1991;118:111–9.
60. Martinez S, Looker DL, Berens RL, Marr JJ. The synergistic action of pyrazolopyrimidines and pentavalent antimony against *Leishmania donovani* and *L. braziliensis*. Am J Trop Med Hyg. 1988;39(3):250–5.
61. Martinez S, Marr JJ. Allopurinol in the treatment of American cutaneous leishmaniasis. N Engl J Med. 1992;326:741–4.
62. Carson DA, Chang KP. Phosphorylation and anti-leishmanial activity of formycin B. Bioch Biophys Res Comm. 1981;100:1377–83.

63. Aronow B, Kaur K, McCartan K, Ullman B. Two high affinity nucleoside transporters in *Leishmania donovani*. Mol Biochem Parasitol. 1987;22:29–37.
64. Vasudevan G, Carter NS, Drew ME, Beverley SM, et al. Cloning of *Leishmania* nucleoside transporter genes by rescue of a transport-deficient mutant. Proc Natl Acad Sci USA. 1998;95:9873–8.
65. Carter NS, Drew ME, Sanchez M, Vasudevan G, et al. Cloning of a novel inosine-guanosine transporter gene from *Leishmania donovani* by functional rescue of a transport-deficient mutant. J Biol Chem. 2000;275:20935–41.
66. Iovannisci DM, Kaur K, Young L, Ullman B. Genetic analysis of nucleoside transport in *Leishmania donovani*. Mol Cell Biol. 1984;4:1013–9.
67. King AE, Ackley MA, Cass CE, Young JD, et al. Nucleoside transporters: from scavengers to novel therapeutic targets. Trends Pharmacol Sci. 2006;27(8):416–25.
68. Valdés R, Liu W, Ullman B, Landfear SM. Comprehensive examination of charged intramembrane residues in a nucleoside transporter. J Biol Chem. 2006;281(32):22647–55.
69. Arastu-Kapur S, Ford E, Ullman B, Carter NS. Functional analysis of an inosine-guanosine transporter from *Leishmania donovani*: the role of conserved residues, aspartate 389 and arginine 393. J Biol Chem. 2003;278(35):33327–33.
70. Arastu-Kapur S, Arendt CS, Purnat T, Carter NS, et al. Second-site suppression of a nonfunctional mutation within the *Leishmania donovani* inosine-guanosine transporter. J Biol Chem. 2005;280(3):2213–9.
71. Huang G, Ulrich PN, Storey M, Johnson D, et al. Proteomic analysis of the acidocalcisome, an organelle conserved from bacteria to human cells. PLoS Pathog. 2014;10(12):e1004555.
72. Saier Jr MH, Beatty T, Goffeau A, Harley KT, et al. The major facilitator superfamily. J Mol Microbiol Biotechnol. 1999;1(2):257–79.
73. Guan L, Kaback HR. Lessons from lactose permease. Annu Rev Biophys Biomol Struct. 2006;35:67–91.
74. Das R, Baker D. Macromolecular modeling with rosetta. Annu Rev Biochem. 2008;77:363–82.
75. Kavanaugh MP. Neurotransmitter transport: models in flux. Proc Natl Acad Sci USA. 1998;95:12737–8.
76. Valdés R, Shinde U, Landfear SM. Cysteine cross-linking defines the extracellular gate for the *Leishmania donovani* nucleoside transporter 1.1 (LdNT1.1). J Biol Chem. 2012;287(53):44036–45.
77. Valdés R, Elferich J, Shinde U, Landfear SM. Identification of the intracellular gate for a member of the equilibrative nucleoside transporter (ENT) family. J Biol Chem. 2014;289:8799–809.
78. Zilberstein D, Philosoph H, Gepstein A. Maintenance of cytoplasmic pH and proton motive force in promastigotes of *Leishmania donovani*. Mol Biochem Parasitol. 1989;36:109–18.
79. Stein A, Vasudevan G, Carter N, Ullman B, et al. Equilibrative nucleoside transporter family members from *Leishmania donovani* are electrogenic proton symporters. J Biol Chem. 2003;278:35127–34.
80. Ortiz D, Sanchez MA, Koch HP, Larsson HP, et al. An acid-activated nucleobase transporter from *Leishmania major*. J Biol Chem. 2009;284:16164–9.
81. Hansen BD, Perez-Arbelo J, Wlakony JF, Hendricks LD. The specificity of purine base and nucleoside uptake in promastigotes of *Leishmania braziliensis panamensis*. Parasitology. 1982;85:271–82.
82. Al-Salabi MI, Wallace LMJ, de Koning HP. A *Leishmania major* nucleobase transporter responsible for allopurinol uptake is a functional homolog of the *Trypanosoma brucei* H2 transporter. Mol Pharmacol. 2003;63:814–20.
83. Al-Salabi MI, de Koning HP. Purine nucleobase transport in amastigotes of *Leishmania mexicana*: involvement in allopurinol uptake. Antimicrob Agents Chemother. 2005;49(9):3682–9.

84. Sanchez M, Tryon R, Vasudevan G, Landfear SM. Functional expression and characterisation of a purine nucleobase transporter gene from *Leishmania major*. Mol Membrane Biol. 2004;21:11–8.
85. Ortiz D, Sanchez MA, Pierce S, Herrmann T, et al. Molecular genetic analysis of purine nucleobase transport in *Leishmania major*. Mol Microbiol. 2007;64:1228–43.
86. Antoine JC, Prina E, Jouanne C, Bongrand P. Parasitophorous vacuoles of *Leishmania amazonensis*-infected macrophages maintain an acidic pH. Infect Immun. 1990;58:779–87.
87. Kerby DR, Detke S. Reduced purine accumulation is encoded on an amplified DNA in *Leishmania mexicana amazonensis* resistant to toxic nucleosides. Mol Biochem Parasitol. 1993;60:171–85.
88. Detke S. Identification of a transcription factor like protein at the TOR locus in *Leishmania mexicana amazonensis*. Mol Biochem Parasitol. 1997;90(2):505–11.
89. Detke S. TOR-induced resistance to toxic adenosine analogs in *Leishmania* brought about by the internalization and degradation of the adenosine permease. Exp Cell Res. 2007;313(9):1963–78.
90. Alsford S, Turner DJ, Obado SO, Sanchez-Flores A, et al. High-throughput phenotyping using parallel sequencing of RNA interference targets in the African tryapnosome. Genome Res. 2011;21:915–24.
91. Alsford S, Eckert S, Baker N, Glover L, et al. High-throughput decoding of antitrypanosomal drug efficacy and resistance. Nature. 2012;482(7384):232–6.
92. Song J, Baker N, Rothert M, Henke B, et al. Pentamidine is not a permeant but a Nanomolar inhibitor of the *Trypanosoma brucei* Aquaglyceroporin-2. PLoS Pathog. 2016;12(2):e1005436.
93. Lye LF, Owens K, Shi H, Murta SM, et al. Retention and loss of RNA interference pathways in trypanosomatid protozoans. PLoS Pathog. 2010;6(10):e1001161.
94. Gazanion E, Fernandez-Prada C, Papadopoulou B, Leprohon P, et al. Cos-Seq for high-throughput identification of drug target and resistance mechanisms in the protozoan parasite *Leishmania*. Proc Natl Acad Sci USA. 2016;113(21):E3012–21.

Part IV

Tools and Strategies to Circumvent Drug Resistance in Leishmania

Bioinformatics in *Leishmania* Drug Design 13

Shishir K. Gupta and Thomas Dandekar

Abstract

Leishmania drug design follows the typical path of the flow of genetic information: By analyzing genome information and considering infection-specific RNA and protein expression, potential targets for drug design and vaccine development are identified. Therefore, to implement successful intervention strategies against *Leishmania* infection, specific features of the process are critical; herein they are described, including specific genome information, good vaccine targets, and classical as well as innovative drug targeting strategies. In addition, a combination of software and web sites has been structured here with references and tools for rapid analysis to rank and examine new target structures in *Leishmania*.

13.1 Introduction

Leishmania is a genus of protozoan parasites that are transmitted by the bite of sand flies and give rise to a range of diseases (collectively known as leishmaniases). Around 350 million people are at risk worldwide (http://www.dndi.org) [1]. Leishmaniasis is a deadly vector-borne infectious disease, a major cause of tropical afflictions, listed as one of the six most important diseases by the World Health Organization regarding Tropical Disease Research (WHO/TDR). The disease is endemic in 98 countries and causes significant morbidity and mortality in Africa, Asia, Latin America, and Mediterranean regions. With 1.3 million new cases

S. K. Gupta
Department of Bioinformatics, Biocenter, University of Würzburg, Würzburg, Germany

T. Dandekar (✉)
Department of Bioinformatics, Biocenter, University of Würzburg, Würzburg, Germany

EMBL Heidelberg, BioComputing Unit, Heidelberg, Germany
e-mail: dandekar@biozentrum.uni-wuerzburg.de

A. Ponte-Sucre, M. Padrón-Nieves (eds.), *Drug Resistance in Leishmania Parasites*,
https://doi.org/10.1007/978-3-319-74186-4_13

reported annually and 20,000–30,000 annual deaths (World Health Organization, September 2016) [2], infection by the insect-transmitted *Leishmania* parasite represents an important global health problem for which there is no vaccine and few partially effective drugs [3]. A total of 21 *Leishmania* spp. have been identified to be pathogenic to human [4]. The pathogen completes its life cycle in insect and human hosts and is transmitted from patient to non-infected individuals by several overlapping species of sand fly vectors; hence, the disease has a complex ecology and epidemiology [5] that is thoroughly described in Chap. 4 from this same volume and will not be discussed in detail herein. However, we would like to stress that although this disease is usually considered as zoonotic, in some geographical areas, infected humans maintain an anthroponotic transmission cycle (human-sand fly-human). In such areas, effective treatment of individual patients can help to control the spread of the parasite (http://www.cdc.gov/parasites/leishmaniasis/epi.html).

The spectrum of diseases caused by *Leishmania* can be categorized broadly into three types: (1) visceral leishmaniasis (VL, Kala-azar), the most lethal form characterized by irregular bouts of fever, substantial weight loss, swelling of the spleen and liver, and anemia; (2) cutaneous leishmaniasis (CL), the most common form of leishmaniasis in which pathogen causes skin lesions, mainly ulcers, on exposed parts of the body, leaving lifelong scars and serious disability; and (3) mucocutaneous leishmaniasis (MCL), with a chronic destruction of mucosal tissue of the nose, mouth, and throat that develops from the cutaneous disease in less than 5% of affected individuals [6]. Leishmaniasis is described on every continent except Australia and Antarctica, although some evidence suggests the presence of midges as potential vectors of *Leishmania* in Australia [7]. Despite their widespread distribution, over 90% of global VL cases occur predominantly in six countries (Bangladesh, Brazil, Ethiopia, Sudan, South Sudan, and India), while most cases (70–75%) of CL mainly occur in ten countries (Afghanistan, Algeria, Brazil, Colombia, Costa Rica, Ethiopia, Iran, North Sudan, Peru, and Syria) [8]. For a detailed description of the pathology of diseases caused by *Leishmania*, please check Chaps. 6, 7, 8, and 9 in this volume.

Regardless lethality of the disease, the progress toward successful prevention and/or treatment is hindered by the challenging and only long-term feasible procedure of vaccine and drug development. Additionally, a strong need for policies to implement educational and health measurements to stop disease transmission is urgently needed. WHO has made significant efforts to improve access to medicines in the poor countries that have the highest burden of cases by reducing the price of two of the five existing medicines for visceral leishmaniasis by 90% for liposomal amphotericin-B (L-AMB) and by 60% for meglumine antimoniate (http://www.who.int/leishmaniasis/en) [9].

There are good standard treatments against leishmaniasis: pentavalent antimony (Sb^V) helps in 90–93% of cases of VL in India, in particular at Uttar Pradesh using one to three intravenous treatment courses. Nevertheless, severe side effects as well as treatment failure and drug resistance (for instance, in India) stress the need for improved and alternative therapies. Currently explored regimes include interferon gamma (IFN-γ) plus intralesional injections antimony (to reduce toxicity). Novel pharmacological options are parenteral drugs such as pentamidine and antifungals such as per os (oral) drugs ketoconazole, itraconazole, and fluconazole. Nevertheless, the drug repertoire to treat leishmaniosis is limited, and all these drugs have severe side effects. Moreover, there is always the risk of emergence of resistance. All this stress the need for novel drug development. To this end, we have highlighted the role of bioinformatics to accelerate the pace of research for prevention and treatment of leishmaniasis in this chapter.

13.2 Genomics, Proteomics, and Transcriptomics

According to phylogenetic analyses, *Leishmania* is divided into three distinct subgenera: the *Leishmania*, the *Viannia*, and the *Sauroleishmania* [10]. For a detailed description of the phylogeny and molecular evolution of this parasite, please check Chap. 2. Most genera of *Leishmania* infect mammal hosts and insect vectors, but *Sauroleishmania* exceptionally infects primarily lizards. The determination of the whole-genome sequences of several *Leishmania* parasites in three distinct subgenera provides the basis for diverse studies of this pathogen. Taking the pathobiology of *Leishmania* into account, the sequenced genomes catalog the full functional repertoire of genes available to the parasite including all enzymes but also regarding virulence and regulatory factors. In addition, this provides the scientific community with an infrastructure for omics level investigations.

The initial efforts for sequencing of *Leishmania* genomes were consolidated in the year 1994 with the establishment of the *Leishmania* Genome Network (LGN) initiative. The sequencing of the first complete genome sequence of *L. major* in the year 2005 [11] was soon followed by genome sequencing of *L. (L.) infantum* and *L. (V.) braziliensis* in the year 2007 [3]. These efforts were successful. In fact, recent years have seen major advances in our understanding of leishmanial biology as the genomes of 15 *Leishmania* genera (several strains) have been sequenced and annotated (Table 13.1) and more are currently being sequenced (www.tritrypdb.org) [13]. From the 15 strains, all are parasitic and lead to *Leishmania* infections in different hosts.

The use of different sequencing technologies (e.g., Roche, Illumina, SOLiD) allowed deep sequencing of *Leishmania* species (genome coverage between 42 and 320 times). This did yield genome assemblies with median contig sizes (N_{50} from 1362 to 302,093). The overall size of these genomes ranges from 23.8 Mb to 35.21 Mb; the genes are organized into 35–37 chromosomes. Table 13.1 does not include the sequenced lizard parasite *L. (S.) tarentolae* [14] and the very recently sequenced genome of *L. (S.) adleri* [15] isolated from the African grass rat (*Arvicanthis niloticus*).

Altogether, these sequenced genomes open exciting new research avenues and opportunities for understanding the genetic basis of *Leishmania* and implied consequences for parasite biology, pathology, and infectivity including interactions with the host. This includes the possibility for large-scale systems biology studies.

Table 13.1 Comparison of sequenced *Leishmania* spp. genomes[a]

	Chromosomes	Genome size (Mb)	Overall G + C content (%)	Assembly level	Assembly accession number	Contig N50
Leishmania (L.) major Friedlin	36	32.8551	59.71	Chromosome	GCA_000002725.2	
Leishmania (L.) major SD 75.1	36	31.2428	59.50	Scaffold	GCA_000250755.2	89,399
Leishmania (L.) major LV39c5	36	32.3275	59.30	Scaffold	GCA_000331345.1	71,814
Leishmania (L.) infantum	36	32.1221	59.5663	Chromosome	GCA_000002875.2	302,093
Leishmania (V.) braziliensis MHOM/BR/75/M2904	35	32.0688	57.73	Chromosome	GCA_000002845.2	63,680
Leishmania (V.) braziliensis MHOM/BR/75/M2903	35	35.2101	57.00	Chromosome	GCA_000340355.2	62,201
Leishmania (L.) Mexicana	34	32.1087	59.7793	Chromosome	GCA_000234665.4	164,930
Leishmania (L.) donovani BPK282A1	36	32.445	59.06	Chromosome	GCA_000227135.2	45,436
Leishmania (L.) donovani MHOM/IN/1983/AG83	36	32.1484	58.21	Chromosome	GCA_001989955.1	19,680
Leishmania (L.) donovani MHOM/IN/1983/AG83	36	32.1964	58.20	Chromosome	GCA_001989975.1	20,549
Leishmania (L.) donovani BHU 1220	36	32.4149	59.04	Chromosome	GCA_000470725.1	41,904
Leishmania (L.) donovani Ld 2001	36	27.4665	55.70	Contig	GCA_000283395.1	3370
Leishmania (L.) donovani Ld 39	36	23.6833	55.80	Contig	GCA_000316305.1	1772
Leishmania (V.) panamensis MHOM/PA/94/PSC-1	35	30.6888	57.39	Chromosome	GCA_000755165.1	97,606
Leishmania (V.) panamensis MHOM/COL/81/L13	35	31.2639	57.50	Scaffold	GCA_000340495.1	22,660
Leishmania sp. MAR LEM2494	36	27.8483	55.80	Contig	GCA_000409445.2	145,034
Leishmania sp. AIIMS/LM/SS/PKDL/LD-974	36	30.814	59.59	Chromosome	GCA_000981925.2	61,709
Leishmania (L.) arabica MPSA/SA/83/JISH220	36	31.2691	59.2021	Chromosome	GCA_000410695.2	51,432

Leishmania (L.) tropica MHOM/IL/1990/ P283	36	32.989	59.3	Scaffold	GCA_000410715.1	32,739
Leishmania (M.) enriettii MCAV/BR/95/ CUR3	36	30.7619	59.104	Chromosome	GCA_000410755.2	101,873
Leishmania (L.) amazonensis		29.0293	59.3	Scaffold	GCA_000438535.1	17,272
Leishmania (L.) gerbilli MRHO/CN/60/ GERBILLI		31.3986	31.3986	Scaffold	GCA_000443025.1	57,008
Leishmania (L.) aethiopica L147	36	31.6308	59.7916	Chromosome	GCA_000444285.2	37,668
Leishmania (V.) peruviana PAB-4377_V1	37	32.9078	56.00	Chromosome	GCA_001403675.1	3917
Leishmania (V.) peruviana LEM-1537_V1	37	33.8902	53.20	Chromosome	GCA_001403695.1	1362

[a]Data source—https://www.ncbi.nlm.nih.gov/genome [12]

Table 13.2 *Leishmania* spp. pathogens and disease manifestations

Pathogen	Main disease manifestation	References
Leishmania (L.) major	Cutaneous *leishmaniasis*	[16]
Leishmania (L.) infantum	Visceral *leishmaniasis*, cutaneous *leishmaniasis*	[17, 18]
Leishmania (V.) braziliensis	Cutaneous *leishmaniasis*, cutaneous *leishmaniasis*, Mucocutaneous *leishmaniasis*	[19–21]
Leishmania (L.) m*exicana*	Cutaneous *leishmaniasis*, diffuse cutaneous *leishmaniasis*	[22, 23]
Leishmania (L.) donovani	Visceral *leishmaniasis*	[24]
Leishmania (V.) panamensis	Cutaneous *leishmaniasis*, Mucocutaneous *leishmaniasis*	[25, 26]
Leishmania Sp. strains AIIMS/LM/SS/PKDL/LD-974 and MAR LEM2494	Post-kala-azar dermal *leishmaniasis*	[27]
Leishmania (L.) turanica	Cutaneous *leishmaniasis*	[28]
Leishmania (L.) arabica	Cutaneous *leishmaniasis*	[29]
Leishmania (L.) tropica	Cutaneous *leishmaniasis*	[30]
Leishmania (M.) enriettii	Cutaneous *leishmaniasis*	[31]
Leishmania (L.) amazonensis	Cutaneous *leishmaniasis*, diffuse cutaneous *leishmaniasis*	[32, 33]
Leishmania (L.) gerbilli	Cutaneous *leishmaniasis*	[34]
Leishmania (L.) aethiopica	Cutaneous *leishmaniasis*, diffuse cutaneous *leishmaniasis*	[35, 36]
Leishmania (V.) peruviana	Cutaneous *leishmaniasis*	[37]

Earlier sequenced *L. (L.) major* primarily causes CL and *L. (L.) donovani* and *L. (L.) infantum* cause VL, whereas recently sequenced *Leishmania* spp. (AIIMS/LM/SS/PKDL/LD-974 and MAR LEM2494) causes post-Kala-azar dermal leishmaniasis, a late cutaneous manifestation of VL (see Chap. 9 of this same volume) (Table 13.2).

The knowledge of the *Leishmania* genomes has furthered important advances in comparative genomic studies, annotating hypothetical genes and finding species-specific genes that could explain the specific pathogenesis and provide potential precise therapeutic targets. Peacock et al. sequenced the genomes of *L. (L.) infantum* and *L. (V.) braziliensis* and reported the first comparative genomics analysis of three *Leishmania* by comparing those of the previously mentioned species with the genome of *L. (L.) major* [3]. The analysis demonstrated a marked conservation of synteny. This means that these different genomes had many genes distributed in the same order (these are called regions of synteny). Genes found to be differentially distributed between the species encoded proteins implicated in host-pathogen interactions, and parasite survival in the macrophage such as GP63 metalloprotease, which interferes with the macrophage signaling during infection, is encoded by a repeated gene cluster that seems to be enlarged fourfold in *L. (V.) braziliensis* as compared with *L. (L.) major* or *L. (L.) infantum* [3, 38, 39].

In 2009, Depledge et al. [40] analyzed the representative proteomes of these three species to reveal conserved genes differentially expressed in the host and suggested that species-specific parasite factors contributing to virulence and pathogenicity in the host may be limited to the products of a small number of differentially distributed genes, or to the differential regulation of conserved genes, either of which are subjected to translational and/or posttranslational control. They concluded that host genetics plays only a minor role in influencing the parasites during macrophage infection which may be significant in determining the clinical outcome of infection. This high degree of synteny in *Leishmania* genomes has been identified by comparative analysis and decomposition of genomes into syntenic blocks.

Increasing the sequencing data of different types of *Leishmania* improves the predictive power to identify conserved drug targets and vaccine candidates which were previously restricted to just three representative *Leishmania* proteomes. Hence, there is a need for the comparative genomics analysis to reevaluate the repository of conserved genes in *Leishmania*. Notably approximately 50% of *Leishmania*-predicted proteomes have no predicted function, and many proteins are annotated as hypothetical proteins. This limits the understanding of the role of these proteins in biological perspective and regarding their value as novel drug or vaccine targets. Once any sequenced eukaryotic genome is demarcated with exons, introns, splicing sites, and other structural annotations, functional annotation of every gene is of fundamental importance [41]. The tremendous amount of data generated by advances in next-generation sequencing projects can be used to generate more reliable annotations together with the annotation of hypothetical proteins. Typical tools and analysis steps for a transcriptome-based reannotation are summarized in [42]. The recently developed database LeishDB consists the updates of *L. (V.) braziliensis* protein-coding genes and noncoding RNAs [43]. The reannotation process implemented here represents an increase of ~26% in protein-coding gene repository of *L. (V.) braziliensis* [43]. Tables 13.1 and 13.3 specify this further for studying *Leishmaniasis*.

The availability of an array of genomes, together with an explosion in microarray and high-throughput transcriptomic sequencing technologies, has facilitated the study of transcriptome responses stimulated by drugs. Several transcriptomic studies have also investigated *Leishmania*-induced regulation of gene expression in infected tissues with the aim to link such responses to disease outcome [68, 69]. Host cell functions are modulated by intracellular pathogens, including *Leishmania*, to actively promote their survival.

Transcriptomic technologies have resulted in rapid expansion of the already substantial plethora of knowledge of the molecular interactions occurring between *Leishmania* and the human host; nevertheless, significant variation in host responses to infection has been described in several studies [70]. Currently, more than 600 gene expression datasets of *Leishmania* are deposited in Gene Expression Omnibus (GEO) database [71] that can be utilized for several studies including the understanding of interaction of *Leishmania* with the host. To analyze this behavior, Beattie et al. used whole-genome array technologies to compare the gene expression profiles of mice macrophages infected with *L. (L.) donovani* to those of uninfected

Table 13.3 Major resources for computational analysis of *Leishmania* spp. pathogens

Resources	Weblinks	Reference
Global distribution maps of the leishmaniases	–	High-resolution evidence-based distribution maps [44]
EuPathDB (the eukaryotic pathogen genomics database)	http://eupathdb.org [45]	Pathogen genomics resource for eukaryotic pathogens [46]
TriTrypDB	http://tritrypdb.org [13]	Integrated genomic and functional genomic resource for *Leishmania* and Trypanosoma [47]
GeneDB (section Kinetoplastid protozoa)	http://www.genedb.org [48]	Curated annotations and sequences of 5 *Leishmania spp.* [49]
trypsNetDB	http://trypsNetDB.org [50]	Experimentally verified as well as predicted protein interactions and annotations for trypanosomatid parasites, includes 7 *Leishmania spp.* [51]
LeishCyc	http://biocyc.org/LEISH/organism-summary?object=LEISH [52]	Biochemical pathways database for *Leishmania (L.) major* [53]
LeishMicrosatDB	http://biomedinformri.com/leishmicrosat [54]	Database of repeat sequences in 6 *Leishmania* spp. [55]
LmSmdB	http://www.nccs.res.in/LmSmdb [56]	Biological networks and regulatory pathways of *Leishmania (L.) major* [57]
Leishmania (L.) amazonensis genome DB	http://bioinfo08.ibi.unicamp.br/leishmania [58]	Sequencing and annotation of the *Leishmania (L.) amazonensis* genome [59]
CPDB[a]	http://cpdbldv.biomedinformri.com [60]	Annotation of cysteine proteases in *Leishmania* [61]
CALP[a]	http://biomedinformri.org/calp [62]	Protein function and families of 4 *Leishmania* spp. [63]
List of putative antileishmanials	https://www.ncbi.nlm.nih.gov/pmc/articles/PMC4247209/ [64]	Drug targets and lead compounds with predicted antileishmanial activity [64]
Leishmania (L.) major metabolic network	http://www.ebi.ac.uk/compneur-srv/biomodels-main/MODEL1507180059 [65]	Genome scale metabolic network of *Leishmania (L.) major* (iAC560) available for flux balance analysis (FBA) [66]
LeishDB	http://www.leishdb.com [67]	Coding gene reannotation and noncoding RNAs in *Leishmania (V.) braziliensis* [43]

[a]Links are broken to the resources. Please contact the authors directly for further information

macrophages, both exposed to inflammatory stimuli such as cytokines [72]. Thus, specific transcriptome datasets for infection studies [73, 74] and drug studies [75, 76] on *Leishmania* are now available in unprecedented detail.

13.3 Interactomics

Protein-protein interaction (PPI) networks are critical determinants for cellular processes. In terms of disease, information about PPIs provide an idea about molecular causes of sicknesses and can offer clues for new therapeutic approaches. The topologically essential proteins of such a system can also be identified by PPI network analysis of pathogen [72]. The PPI network that regulates host and pathogen interaction can highlight the insights of pathogen "attack" and host counter-defense mechanisms.

The in silico methods used for PPI network determination complement the experimental approaches by minimizing the number of false-positive interactions. Several experimental procedures can examine interactions between proteins; however, the methods are usually expensive, labor-intensive, and time consuming. Additionally, experimental methodologies for protein-protein interaction detection yield many results that end up being false negatives and/or positives [77]; these may also be reduced by scrutinizing the data by in silico procedures. The interactomes of several model organisms have been established experimentally. These interactomes can be used as templates to derive PPI networks for other organisms, by means of computational approaches such as interolog and domain interaction methods. Based on and explaining such techniques, we have recently derived the interaction networks of opportunistic pathogens like *Serratia marcescens* [78] and *Aspergillus fumigatus* [79].

Initially host cells suitable for *Leishmania* parasites were widely regarded as highly specific. *Leishmania* were considered as obligate intracellular pathogens of macrophages. Recent studies have confirmed that *Leishmania* has greater degree of promiscuity in host immune cell range such as monocyte, macrophage, dendritic cells, and neutrophils [5, 80]. For a detailed description of the diversity and characteristics of the host cells involved in *Leishmania* infection, please check Chap. 5 in this same volume. In fact, the clinical presentation of *Leishmaniasis* is dependent upon both the parasite species and the host's immune response [5]. The diversity of tropism and disease resulting from infection is one of the hallmarks of the *Leishmania* spp. In the control of infections with all the *Leishmania* spp., host cellular immune mechanisms play a major role. It has become evident from genetic and immunological studies, using a murine model, that the members of the genus *Leishmania* differ in aspects of their "approach" to the host immune system by using number of different virulence factors and the proteins interacting with the host [5].

Interactions of host and pathogen evoke different solutions to the challenges imposed by parasite establishment, survival, and persistence. Understanding the extent of host-pathogen PPIs at system level is increasingly important in ensuring the development of broadly applicable vaccines, drugs, and immunotherapeutic interventions for many diseases including leishmaniasis. The host-pathogen PPI network can be constructed using interolog and domain-based approaches, as previously mentioned. Our group has illustrated and documented the power of this approach by deriving human-fungi PPI networks of *Aspergillus fumigatus* and *Candida albicans* (from the pathogen side) and mouse and man (from the host side) [81]. In our work, we illustrate also the importance of robust filtering and refinement steps in host-pathogen interaction networks to attain biologically relevant relations. The predicted interactions can be ranked further, based on confidence score to validate the top ones experimentally [82, 83]. The confidence score assigned to PPI computationally reflects the likelihood of the correctness of the prediction. In this context, the excellent review article by Kaye and Scott provides a catalog of species-specific *Leishmania* genes important in pathogenesis [5]. Generally, articles validating by experimental means, the host-pathogen interactions at small-scale articles can be retrieved by text-based searches at the PubMed literature repository. For instance, Lieke et al. highlighted the interaction of *Leishmania* surface glycoprotein GP63 and natural killer (NK) cells and expression dynamics of NK cell marker CD56 [84]. Along with the organism-specific literature search, the databases consisting the catalog of host-pathogen PPIs such as HPIDB [85] and PHISTO [86] can be primarily used as a template dataset for interolog predictions. Regarding the host side filtering, genome-wide RNA interference screens are particularly powerful to identify host factors required for pathogenesis. From the pathogen side, virulence factors, effectors, and secretory proteins have key importance to prune the predicted host-pathogen interaction networks.

Exploiting these computational methods further, Rezende et al. [87] established the PPI networks for three *Leishmania* pathogens, *L. (L.) major*, *L. (V.) braziliensis*, and *L. (L.) infantum*. Recently Gazestani et al. experimentally determined the protein complex map of *Trypanosoma brucei* [88] which can be used to refine the computational PPIs of *Leishmania*. In a transcriptome level study, probing the response of liver-resident macrophages (Kupffer cells) in *L. (L.) donovani*-infected mice, Beattie et al. identified a network operating in uninfected Kupffer cells exposed to inflammation that was absent in Kupffer cells coming from the same animal infected with intracellular *Leishmania* [72]. They reported the retinoid X receptor alpha (RXRα) as a key hub in the network that involves in innate resistance of Kupffer cells to *Leishmania* infection. Mining of such hubs is possible by analysis of interactomes. Moreover, the transcriptome or proteome data can be mapped on interactomes and are useful to derive the active networks during the infection.

13.4 Resources

There are several databases and other resources available regarding detailed data on *Leishmania* genome, proteome, metabolome, and interactome. Key resources mainly dedicated to protozoan pathogens including *Leishmania* are listed in Table 13.3.

Different workflows are conceivably exploiting these data, for instance, regarding drug design, vaccine development, lab markers, or patient treatment. This is illustrated in Fig. 13.1. In the following section, we will discuss specific bioinformatics options to improve the challenges that exist in vaccine research, drug design, and treatment of leishmaniasis.

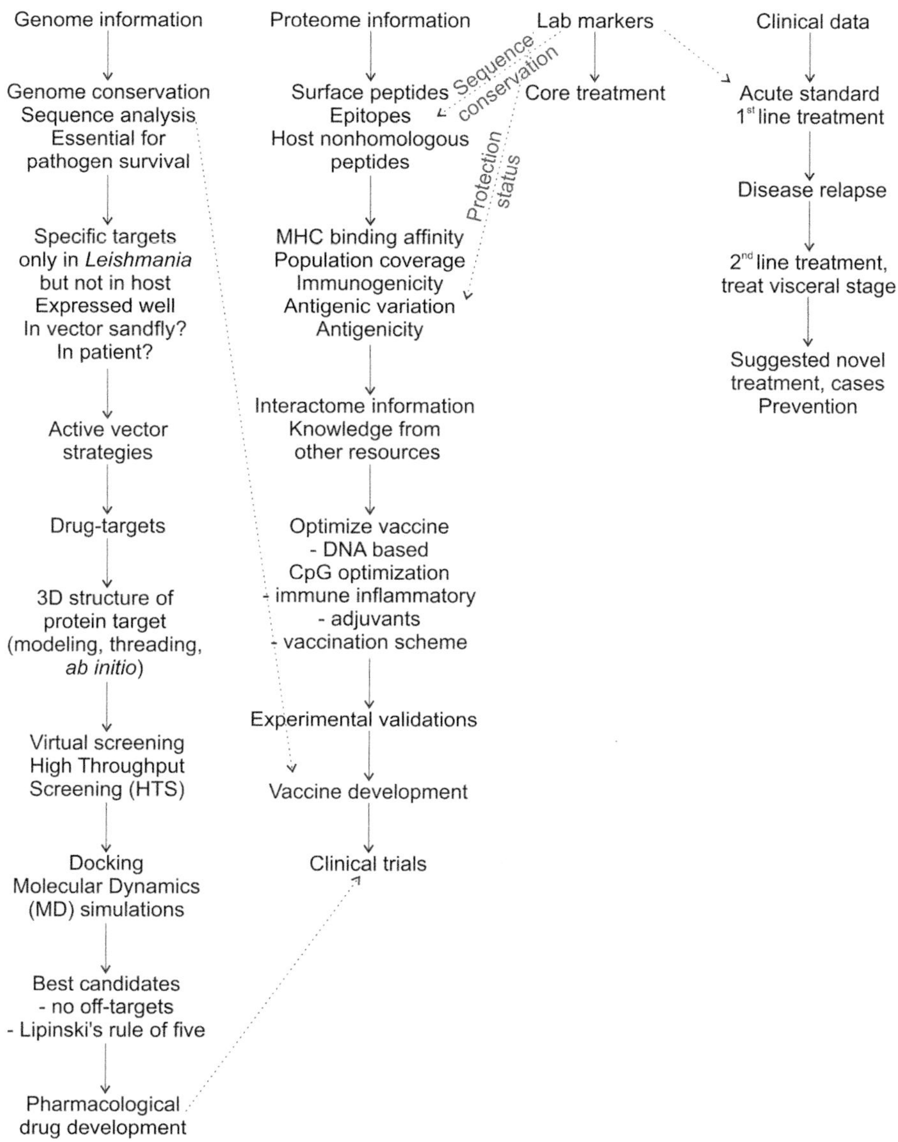

Fig. 13.1 Bioinformatics analysis options regarding Leishmaniosis. The data summarized in Table 13.1 can be used in consecutive steps (arrows) for different workflows exploiting the accumulated data, for instance, regarding drug design and protein target prediction (left), vaccine development (middle), diagnostic lab markers, or patient treatment (right)

13.5 Role of Immune Informatics for Vaccine Development

Vaccines are among the most efficacious and cost-effective tools for reducing morbidity and mortality caused by infectious diseases [89]. Owing to efficient immune evasion mechanisms such as antigenic variation and the intracellular locations of *Leishmania* spp. in the human cells, vaccine design remains challenging in leishmaniasis. The generation of an immunological memory is a prime requirement for effective vaccination. Patients cured from *Leishmania* infections develop lifelong immunity; hence, the prevention of leishmaniasis through prophylactic vaccination is quite feasible as suggested by clinical and experimental evidence. The only proven vaccine agent in humans has been live *L. (L.) major*, and it is discontinued because of the appearance of unacceptable lesions in some recipients [90]. In this regard, Rezvan and Moafi reviewed the strategies for *Leishmania* vaccine development, which can supplement the in silico vaccine designing approach against *Leishmania* spp. before proceeding to in vitro testing [91].

Host immunity against *Leishmania* is mediated via both innate and adaptive immune responses (see Chap. 5). The successful treatment of leishmaniasis depends on efficient elimination of the pathogen by activated macrophages. Internalization of *Leishmania* by macrophages leads to the production of proinflammatory cytokines and pathogen killing [92]. The subversive activity of *Leishmania* parasites in this process is the inhibition of interleukin-12 (IL-12) production, which is necessary for the leishmanicidal activity of macrophages [93], as it leads to upregulation of inducible nitric oxide synthase (iNOS), nitric oxide (NO), and interferon gamma (IFN-γ) [92]. T-lymphocytes play a central role in the generation of protective immune response in many pathogen-mediated infections including *Leishmania*; hence, the identification of peptides that stimulate T-cell responses is a critical requirement for the development of successful epitopic vaccines [94, 95].

The studies have shown that $CD4^+$ T-cells can provide the best protection against *Leishmania* by mediating long-term immunity to *L. (L.) major* infection, even in the absence of persistent parasites [96]. Unfortunately, although several approaches have been taken to develop a vaccine for leishmaniasis, to date none have been successful in humans [92, 95]. In a recent work, Brito et al. used an integrated approach to analyze B- and T-cell epitopes of *Leishmania*, PPI networks, and metabolic pathways and further experimentally validated 20 potential candidates in a murine model [97]. The study suggested the potential of T-cell epitopes over B-cell epitopes as vaccine candidates against leishmaniasis. In this pursuit, to create efficient vaccines for prevention of leishmaniasis, immunoinformatics can play a significant role with the use of computational approaches that aim to identify putative vaccine candidates in the protein-coding genome (proteome) of pathogens like *Leishmania*.

After the successful invention of the first approved multicomponent meningococcal serogroup B (MenB) vaccine (4CMenB, Bexsero®) which started from dry lab to wet lab, immunoinformatics is now poised to deliver more vaccines vindicating its earlier promise [98, 99]: Bexsero is now registered in several countries and was licensed in Europe in January 2013 [100].

Recently, several approaches have been used to improve the potential of computer-aided vaccine designing approaches to identify vaccine candidates: (1) selection of candidates from conserved regions [101], (2) introducing cleavage sites for targeted cleavage of multi-epitopic vaccine [102, 103], (3) using profile methods to analyze biased-ness of predicted epitopes toward profiles of experimentally validated epitopes [101], (4) B-cell epitope prediction by docking [102], (5) - structure-based epitopes [104], (6) CpG optimization [105], (7) homology search against host [94], (8) adding adjuvant such as IL-12 [89], and (9) population coverage analysis [94, 106]; they all enhance the scope of immunoinformatics. Such approaches can be used together to deliver best vaccine candidates against *Leishmania* computationally.

13.6 Computer-Aided Drug Designing (CADD)

Starting from the sequenced genome, one can predict drug targets using a subtractive genomics approach. In brief, first all the proteins in the pathogen are classified based on their importance in a well-characterized corresponding organism (e.g., essential function, even verified by gene knockout as well as less important classes). If there are no experimental data for the gene function and its importance in the pathogen, sequence similarity from proteins of other organisms is used, and the availability of experimental data in these organisms is considered. The "Database of Essential Genes" (DEG database) [107] provides a catalog of functionally important genes in several prokaryotics and eukaryotics ("essential" is not meant in this database in the strict sense used in genetics, but a striking growth effect was observed after modification, choosing specific growth conditions). Based on orthology (i.e., high sequence similarity over most of the protein, indicating similar function), functionally important genes can in this way be correctly annotated in different organisms [78]. Using the DEG database, several important genes have been identified and annotated in *L. (L.) donovani* [108].

Network-based methods can also be applied to determine the topologically central proteins for the organism. Inhibition of such central "hub" proteins can lead to damage of network architecture [109]. It is also feasible to combine both approaches to collect even more potential essential protein targets for pathogen survival [78]. Next the proteins having human homologs are removed from the predicted essential proteins based on sequence similarity, to avoid drug effects on too similar enzymes occurring in the host. One has to stress that there are proven examples where the protein is present also in the human genome but is nevertheless sufficiently different that drugs inhibiting the pathogen molecule will not damage the host protein too much, consequently allowing therapy; this is the case, for instance, regarding methylene blue [110].

The metabolic potential of targets can also be analyzed using the KEGG (Kyoto encyclopedia of genes and genomes) database of metabolic pathways. If the predicted candidates have shown resistance to antibiotics, they can be filtered out. This can now provide a list of putative molecules. Moreover, membrane proteins can be used to find best epitopes for vaccines (predicted high antigenicity of specific membrane protein regions). The collected list contains the presumed drug targets which now can also be searched in the DrugBank to find inhibitors or search for inhibitors known for orthologous (sequence similar) proteins in other organisms, if available. We have recently used such a pipeline to identify the drug targets in *S. marcescens* [78]. In another recent study, we used PPI networks, gene expression data, and metabolism to find new drug targets in *Aspergillus fumigatus* [79].

Such drug-search pipelines can also be extended toward finding inhibitors of potential protein targets. For such an extension, first the three-dimensional (3D) structure of a preferred top-ranked target is determined. If the crystal structure of the protein is not available in the Protein Data Bank (PDB), suitable approaches of computational structure prediction (homology modeling, threading, ab initio or composite modeling) can be applied to determine the 3D structure. Moreover, active sites of enzymes can be determined by modeling, and a library of inhibitors can be screened for finding the best scoring inhibitor using techniques of virtual screening [111]. This is particularly successful if the enzyme or protein to be modeled has a homologous crystal structure. If this is the case, a homology model of the *Leishmania* protein structure can be build based on this known crystal structure. Next, the binding of inhibitors available from public or private drug banks as well as their derivatives may be studied.

We explain this technique analyzing highly toxic ribosome-inactivating proteins [112]. Not only chemical inhibitors but also peptide inhibitors can be modeled and used to inhibit model protein structures [113, 114]. The stability of drug-protein complexes can be further accessed using molecular dynamics simulations [115]. Additionally, the integration of docking techniques with mathematical models can be used to analyze the effectiveness of anti-pathogen therapies [116].

We encourage the reader to consider some of the excellent reviews of the advances in CADD that have been published elsewhere [117–119]. None of the available drugs for the treatment of leishmaniasis (i.e., amphotericin-B, miltefosine, pentavalent antimonials, paromomycin) is satisfactory, and new drugs are required, especially some suitable for rural health systems with limited resources [120]. Sundar et al. assessed the efficacy and safety of three potential short-course combination treatments compared with the standard monotherapy in India [121]. In the comparisons, different groups of patients were treated with (a) amphotericin-B, (b) liposomal amphotericin-B with miltefosine, (c) paromomycin, and (d) paromomycin with miltefosine. They concluded that the two-drug combination treatments for VL were efficacious and safe with fewer adverse events, and with a decreased duration of therapy, thereby encouraging adherence and reducing the emergence of drug-resistant parasites [121].

Such a combination strategy is also powerful regarding malaria treatment [122]. Using computational analysis, Waugh et al. identified a list of prospective compounds which could serve as potential antileishmanial drugs [64]. Broad and general information on drugs and their influence on protein-protein interaction networks including indications, protein targets, and side targets can be accessed from our helpful resource: the drug-minded protein interaction database (DrumPID)

[123] (access link: http://drumpid.bioapps.biozentrum.uni-wuerzburg.de/compounds/index.php).

13.7 Network-Based Drug Target Discovery

The core of network-based drug discovery is a robust prediction of the involved PPI network. In the absence of experimentally derived PPIs, orthology-based methods are mainly used for PPI network predictions. Thus, experimentally available PPIs of related organisms are used to infer the interactions in the orthologous proteins of the *Leishmania*. Another widespread method is the establishment of PPI networks based on domain-domain interactions. In terms of drug target identification, such networks are further analyzed for critical hubs and bottlenecks. Once the target is determined using CADD, potential inhibitors can be predicted. Interestingly, how the proposed inhibitor can affect PPI network structures and how the resulting signals are translated into drug effects can be predicted [124, 125].

There are few reports presenting such drug target identification in *Leishmania* using computational methods. Flórez et al. constructed the first PPI network of the *L. (L.) major* parasite by using a computational approach and proposed potential drug targets for further experimental validation [126]. Later, Rezende et al. constructed PPI networks of *L. (V.) braziliensis*, *L. (L.) infantum*, and *L. (L.) major* which can be used for network-based novel drug target prediction against *Leishmania* [87]. However, considering the resources, genome information, and references reviewed here, we are reasonably optimistic that soon further advances will become possible.

13.8 Conclusion

Leishmania infection is an increasing medical problem in South America but also worldwide [Asia, Africa, Middle East, Southern Europe, few cases even in the USA (Texas, Oklahoma)]. Exploiting the potential of bioinformatics and understanding the extent of the genomic diversity of *Leishmania* genomes will be increasingly important in ensuring the development of broadly applicable and effective vaccines, conserved drug targets, drugs, and other immunotherapeutic interventions. A particular promising approach currently intensively explored is a *Leishmania* vaccine. Currently, drug therapy can cure patients with high success albeit severe medical risks. In particular, the well-known standard drugs against leishmaniasis such as pentavalent antimonials need further alternatives and improvements. Drug resistance observed in certain countries such as India is another motivation for improved drug design and vaccine development against leishmaniasis.

Acknowledgments We thank DFG (TR124/B1) and the land of Bavaria for support.

References

1. http://www.dndi.org
2. World Health Organization, September 2016.
3. Peacock CS, Seeger K, Harris D, Murphy L, et al. Comparative genomic analysis of three *Leishmania* species that cause diverse human disease. Nat Genet. 2007;39(7):839–47.
4. Otranto D, Dantas-Torres F. The prevention of canine leishmaniasis and its impact on public health. Trends Parasitol. 2013;29(7):339–45.
5. Kaye P, Scott P. Leishmaniasis: complexity at the host-pathogen interface. Nat Rev Microbiol. 2011;9(8):604–15.
6. Marsden PD. Mucosal leishmaniasis ("espundia" Escomel, 1911). Trans R Soc Trop Med Hyg. 1986;80(6):859–76.
7. Dougall AM, Alexander B, Holt DC, Harris T, et al. Evidence incriminating midges (Diptera: Ceratopogonidae) as potential vectors of *Leishmania* in Australia. Int J Parasitol. 2011;41 (5):571–9.
8. Alvar J, Velez ID, Bern C, Herrero M, et al. Leishmaniasis worldwide and global estimates of its incidence. PLoS One. 2012;7(5):e35671.
9. http://www.who.int/leishmaniasis/en
10. Bates PA. Transmission of *Leishmania* metacyclic promastigotes by phlebotomine sand flies. Int J Parasitol. 2007;37(10):1097–106.
11. Ivens AC, Peacock CS, Worthey EA, Murphy L, et al. The genome of the kinetoplastid parasite, *Leishmania major*. Science. 2005;309(5733):436–42.
12. https://www.ncbi.nlm.nih.gov/genome
13. www.tritrypdb.org
14. Raymond F, Boisvert S, Roy G, Ritt JF, et al. Genome sequencing of the lizard parasite *Leishmania tarentolae* reveals loss of genes associated to the intracellular stage of human pathogenic species. Nucleic Acids Res. 2012;40(3):1131–47.
15. Coughlan S, Mulhair P, Sanders M, Schonian G, et al. The genome of *Leishmania adleri* from a mammalian host highlights chromosome fission in Sauroleishmania. Sci Rep. 2017;7:43747.
16. Remadi L, Haouas N, Chaara D, Slama D, et al. Clinical presentation of cutaneous leishmaniasis caused by *Leishmania major*. Dermatology. 2016;232(6):752–9.
17. do Rego Lima LV, Santos Ramos PK, Campos MB, dos Santos TV, et al. Preclinical diagnosis of American visceral leishmaniasis during early onset of human *Leishmania (L.) infantum chagasi*-infection. Pathog Glob Health. 2014;108(8):381–4.
18. Castro LS, Franca Ade O, Ferreira Ede C, Hans Filho G, et al. Leishmania infantum as a causative agent of cutaneous leishmaniasis in the state of Mato Grosso Do Sul, Brazil. Rev Inst Med Trop Sao Paulo. 2016;58:23.
19. Alves-Ferreira EV, Toledo JS, De Oliveira AH, Ferreira TR, et al. Differential gene expression and infection profiles of cutaneous and mucosal *Leishmania braziliensis* isolates from the same patient. PLoS Negl Trop Dis. 2015;9(9):e0004018.
20. Farias LH, Rodrigues AP, Silveira FT, Seabra SH, et al. Phosphatidylserine exposure and surface sugars in two *Leishmania (Viannia) braziliensis* strains involved in cutaneous and mucocutaneous leishmaniasis. J Infect Dis. 2013;207(3):537–43.
21. Gomes CM, de Paula NA, Cesetti MV, Roselino AM, Sampaio RN. Mucocutaneous leishmaniasis: accuracy and molecular validation of noninvasive procedures in a L. (V.) braziliensis-endemic area. Diagn Microbiol Infect Dis. 2014;79(4):413–8.
22. Avila-Garcia M, Mancilla-Ramirez J, Segura-Cervantes E, Farfan-Labonne B, et al. Transplacental transmission of cutaneous *Leishmania mexicana* strain in BALB/c mice. Am J Trop Med Hyg. 2013;89(2):354–8.
23. Galindo-Sevilla N, Soto N, Mancilla J, Cerbulo A, et al. Low serum levels of dehydroepiandrosterone and cortisol in human diffuse cutaneous leishmaniasis by *Leishmania mexicana*. Am J Trop Med Hyg. 2007;76(3):566–72.

24. Picado A, Ostyn B, Singh SP, Uranw S, et al. Risk factors for visceral leishmaniasis and asymptomatic *Leishmania donovani* infection in India and Nepal. PLoS One. 2014;9(1): e87641.
25. Morales CA, Palacio J, Rodriguez G, Camargo YC. Zosteriform cutaneous leishmaniasis due to *Leishmania (Viannia) panamensis* and *Leishmania (Viannia) braziliensis*: report of three cases. Biomedica. 2014;34(3):340–4.
26. Ives A, Ronet C, Prevel F, Ruzzante G, et al. *Leishmania* RNA virus controls the severity of mucocutaneous leishmaniasis. Science. 2011;331(6018):775–8.
27. Gupta AK, Srivastava S, Singh A, Singh S. De novo whole-genome sequence and annotation of a *Leishmania* strain isolated from a case of post-kala-azar dermal Leishmaniasis. Genome Announc. 2015;3(4):e00809.
28. Mirzaei A, Schweynoch C, Rouhani S, Parvizi P. Diversity of *Leishmania* species and of strains of *Leishmania major* isolated from desert rodents in different foci of cutaneous leishmaniasis in Iran. Trans R Soc Trop Med Hyg. 2014;108(8):502–12.
29. Peters W, Bryceson A, Evans DA, Neal RA, et al. *Leishmania* infecting man and wild animals in Saudi Arabia. 8. The influence of prior infection with *Leishmania arabica* on challenge with *L. major* in man. Trans R Soc Trop Med Hyg. 1990;84(5):681–9.
30. Eslami G, Hajimohammadi B, Jafari AA, Mirzaei F, et al. Molecular identification of *Leishmania tropica* infections in patients with cutaneous leishmaniasis from an endemic central of Iran. Trop Biomed. 2014;31(4):592–9.
31. Kwakye-Nuako G, Mosore MT, Duplessis C, Bates MD, et al. First isolation of a new species of *Leishmania* responsible for human cutaneous leishmaniasis in Ghana and classification in the *Leishmania enriettii* complex. Int J Parasitol. 2015;45(11):679–84.
32. Yamamoto ES, Campos BL, Jesus JA, Laurenti MD, et al. The effect of ursolic acid on *Leishmania (Leishmania) amazonensis* is related to programed cell death and presents therapeutic potential in experimental cutaneous leishmaniasis. PLoS One. 2015;10(12):e0144946.
33. Coelho AC, Trinconi CT, Costa CH, Uliana SR. In vitro and in vivo miltefosine susceptibility of a *Leishmania amazonensis* isolate from a patient with diffuse cutaneous leishmaniasis. PLoS Negl Trop Dis. 2014;8(7):e2999.
34. Eliseev LN, Strelkova MV, Zherikhina II. The characteristics of the epidemic activation of a natural focus of zoonotic cutaneous leishmaniasis in places with a sympatric dissemination of *Leishmania major, L. turanica* and *L. gerbilli*. Med Parazitol (Mosk). 1991;3:24–9.
35. Negera E, Gadisa E, Hussein J, Engers H, et al. Treatment response of cutaneous leishmaniasis due to *Leishmania aethiopica* to cryotherapy and generic sodium stibogluconate from patients in Silti, Ethiopia. Trans R Soc Trop Med Hyg. 2012;106(8):496–503.
36. Akuffo HO, Fehniger TE, Britton S. Differential recognition of *Leishmania aethiopica* antigens by lymphocytes from patients with local and diffuse cutaneous leishmaniasis. Evidence for antigen-induced immune suppression. J Immunol. 1988;141(7):2461–6.
37. Longoni SS, Marin C, Sanchez-Moreno M. Excreted *Leishmania peruviana* and *Leishmania amazonensis* iron-superoxide dismutase purification: specific antibody detection in Colombian patients with cutaneous leishmaniasis. Free Radic Biol Med. 2014;69:26–34.
38. Isnard A, Shio MT, Olivier M. Impact of *Leishmania* metalloprotease GP63 on macrophage signaling. Front Cell Infect Microbiol. 2012;2:72.
39. Hassani K, Shio MT, Martel C, Faubert D. Absence of metalloprotease GP63 alters the protein content of *Leishmania* exosomes. PLoS One. 2014;9(4):e95007.
40. Depledge DP, Evans KJ, Ivens AC, Aziz N, et al. Comparative expression profiling of *Leishmania*: modulation in gene expression between species and in different host genetic backgrounds. PLoS Negl Trop Dis. 2009;3(7):e476.
41. Gupta SK, Bencurova E, Srivastava M, Pahlavan P. Improving re-annotation of annotated eukaryotic genomes. In: Big data analytics in genomics. Cham: Springer; 2016. p. 171–95.
42. Gupta SK, Kupper M, Ratzka C, Feldhaar H, et al. Scrutinizing the immune defence inventory of *Camponotus floridanus* applying total transcriptome sequencing. BMC Genomics. 2015;16:540.

43. Torres F, Arias-Carrasco R, Caris-Maldonado JC, Barral A, et al. LeishDB: a database of coding gene annotation and non-coding RNAs in *Leishmania braziliensis*. Database. 2017;2017:bax047. https://doi.org/10.1093/database/bax047. [1758-0463 (Electronic)]
44. Pigott DM, Bhatt S, Golding N, Duda KA, et al. Global distribution maps of the leishmaniases. elife. 2014;3
45. http://eupathdb.org
46. Aurrecoechea C, Barreto A, Basenko EY, Brestelli J, et al. EuPathDB: the eukaryotic pathogen genomics database resource. Nucleic Acids Res. 2017;45(D1):D581–d591.
47. Aslett M, Aurrecoechea C, Berriman M, Brestelli J, et al. TriTrypDB: a functional genomic resource for the Trypanosomatidae. Nucleic Acids Res. 2010;38(Database issue):D457–62.
48. http://www.genedb.org
49. Logan-Klumpler FJ, De Silva N, Boehme U, Rogers MB, et al. GeneDB—an annotation database for pathogens. Nucleic Acids Res. 2012;40(Database issue):D98–108.
50. http://trypsNetDB.org
51. Gazestani VH, Yip CW, Nikpour N, Berghuis N, et al. TrypsNetDB: an integrated framework for the functional characterization of trypanosomatid proteins. PLoS Negl Trop Dis. 2017;11 (2):e0005368.
52. http://biocyc.org/LEISH/organism-summary?object=LEISH
53. Saunders EC, MacRae JI, Naderer T, Ng M, et al. LeishCyc: a guide to building a metabolic pathway database and visualization of metabolomic data. Methods Mol Biol. 2012;881:505–29.
54. http://biomedinformri.com/leishmicrosat
55. Dikhit MR, Moharana KC, Sahoo BR, Sahoo GC, et al. LeishMicrosatDB: open source database of repeat sequences detected in six fully sequenced *Leishmania* genomes. Database. 2014;2014:bau078. https://doi.org/10.1093/database/bau078.
56. http://www.nccs.res.in/LmSmdb
57. Patel P, Mandlik V, Singh S. LmSmdB: an integrated database for metabolic and gene regulatory network in *Leishmania* major and *Schistosoma mansoni*. Genom Data. 2016;7:115–8.
58. http://bioinfo08.ibi.unicamp.br/leishmania
59. Real F, Vidal RO, Carazzolle MF, Mondego JM, Costa GG, Herai RH, et al. The genome sequence of *Leishmania (Leishmania) amazonensis*: functional annotation and extended analysis of gene models. DNA Res. 2013;20(6):567–81.
60. http://cpdbldv.biomedinformri.com
61. Rana S, Dikhit MR, Rani M, Moharana KC, Sahoo GC, Das P. CPDB: cysteine protease annotation database in *Leishmania* species. Integr Biol (Camb). 2012;4(11):1351–7.
62. http://biomedinformri.org/calp
63. Dikhit MR, Nathasharma YP, Patel L, Rana SP, et al. A comparative protein function analysis database of different *Leishmania* strains. Bioinformation. 2011;6(1):20–2.
64. Waugh B, Ghosh A, Bhattacharyya D, Ghoshal N, et al. In silico work flow for scaffold hopping in *Leishmania*. BMC Res Notes. 2014;7:802.
65. http://www.ebi.ac.uk/compneur-srv/biomodels-main/MODEL1507180059
66. Chavali AK, Whittemore JD, Eddy JA, Williams KT, et al. Systems analysis of metabolism in the pathogenic trypanosomatid *Leishmania major*. Mol Syst Biol. 2008;4:177.
67. http://www.leishdb.com
68. Hernandez-Santana YE, Ontoria E, Gonzalez-Garcia AC, Quispe-Ricalde MA, et al. The challenge of stability in high-throughput gene expression analysis: comprehensive selection and evaluation of reference genes for BALB/c mice spleen samples in the *Leishmania infantum* infection model. PLoS One. 2016;11(9):e0163219.
69. Patino LH, Ramirez JD. RNA-seq in kinetoplastids: a powerful tool for the understanding of the biology and host-pathogen interactions. Infect Genet Evol. 2017;49:273–82.
70. Kima PE. *Leishmania* molecules that mediate intracellular pathogenesis. Microbes Infect. 2014;16(9):721–6.

71. Clough E, Barrett T. The gene expression omnibus database. Methods Mol Biol. 2016;1418:93–110.
72. Beattie L, d'El-Rei Hermida M, Moore JW, Maroof A, et al. A transcriptomic network identified in uninfected macrophages responding to inflammation controls intracellular pathogen survival. Cell Host Microbe. 2013;14(3):357–68.
73. Fernandes MC, Dillon LA, Belew AT, Bravo HC. Dual transcriptome profiling of *Leishmania*-infected human macrophages reveals distinct reprogramming signatures. MBio. 2016;7(3): e00027.
74. Christensen SM, Dillon LA, Carvalho LP, Passos S, et al. Meta-transcriptome profiling of the human-*Leishmania braziliensis* cutaneous lesion. PLoS Negl Trop Dis. 2016;10(9):e0004992.
75. Kumar D, Singh R, Bhandari V, Kulshrestha A, et al. Biomarkers of antimony resistance: need for expression analysis of multiple genes to distinguish resistance phenotype in clinical isolates of *Leishmania donovani*. Parasitol Res. 2012;111(1):223–30.
76. Schriefer A, Wilson ME, Carvalho EM. Recent developments leading toward a paradigm switch in the diagnostic and therapeutic approach to human leishmaniasis. Curr Opin Infect Dis. 2008;21(5):483–8.
77. Braun P, Tasan M, Dreze M, Barrios-Rodiles M, et al. An experimentally derived confidence score for binary protein-protein interactions. Nat Methods. 2009;6(1):91–7.
78. Gupta SK, Gross R, Dandekar T. An antibiotic target ranking and prioritization pipeline combining sequence, structure and network-based approaches exemplified for *Serratia marcescens*. Gene. 2016;591(1):268–78.
79. Kaltdorf M, Srivastava M, Gupta SK, Liang C, et al. Systematic identification of anti-fungal drug targets by a metabolic network approach. Front Mol Biosci. 2016;3:22.
80. Walker DM, Oghumu S, Gupta G, McGwire BS, et al. Mechanisms of cellular invasion by intracellular parasites. Cell Mol Life Sci. 2014;71(7):1245–63.
81. Remmele CW, Luther CH, Balkenhol J, Dandekar T, et al. Integrated inference and evaluation of host-fungi interaction networks. Front Microbiol. 2015;6:764.
82. Kotlyar M, Pastrello C, Pivetta F, Lo Sardo A, et al. In silico prediction of physical protein interactions and characterization of interactome orphans. Nat Methods. 2015;12(1):79–84.
83. Bader JS, Chaudhuri A, Rothberg JM, Chant J. Gaining confidence in high-throughput protein interaction networks. Nat Biotechnol. 2004;22(1):78–85.
84. Lieke T, Nylen S, Eidsmo L, McMaster WR, et al. *Leishmania* surface protein gp63 binds directly to human natural killer cells and inhibits proliferation. Clin Exp Immunol. 2008;153 (2):221–30.
85. Ammari MG, Gresham CR, McCarthy FM, Nanduri B. HPIDB 2.0: a curated database for host-pathogen interactions. Database. 2016;2016:baw103. https://doi.org/10.1093/database/baw103.
86. Durmus Tekir S, Cakir T, Ardic E, Sayilirbas AS, et al. PHISTO: pathogen-host interaction search tool. Bioinformatics. 2013;29(10):1357–8.
87. Rezende AM, Folador EL, Resende D de M, Ruiz JC. Computational prediction of protein-protein interactions in *Leishmania* predicted proteomes. PLoS One. 2012;7(12):e51304.
88. Gazestani VH, Nikpour N, Mehta V, Najafabadi HS, et al. A protein complex map of *Trypanosoma brucei*. PLoS Negl Trop Dis. 2016;10(3):e0004533.
89. Akhoon BA, Slathia PS, Sharma P, Gupta SK, et al. In silico identification of novel protective VSG antigens expressed by Trypanosoma brucei and an effort for designing a highly immunogenic DNA vaccine using IL-12 as adjuvant. Microb Pathog. 2011;51(1–2):77–87.
90. Murray HW, Berman JD, Davies CR, Saravia NG. Advances in leishmaniasis. Lancet. 2005;366(9496):1561–77.
91. Rezvan H, Moafi M. An overview on *Leishmania* vaccines: a narrative review article. Vet Res Forum. 2015;6(1):1–7.
92. Kedzierski L. Leishmaniasis vaccine: where are we today? J Glob Infect Dis. 2010;2 (2):177–85.

93. Ahuja SS, Reddick RL, Sato N, Montalbo E, et al. Dendritic cell (DC)-based anti-infective strategies: DCs engineered to secrete IL-12 are a potent vaccine in a murine model of an intracellular infection. J Immunol. 1999;163(7):3890–7.
94. Gupta SK, Smita S, Sarangi AN, Srivastava M, et al. In silico CD4+ T-cell epitope prediction and HLA distribution analysis for the potential proteins of *Neisseria meningitidis* serogroup B--a clue for vaccine development. Vaccine. 2010;28(43):7092–7.
95. Costa CH, Peters NC, Maruyama SR, de Brito EC Jr, et al. Vaccines for the leishmaniases: proposals for a research agenda. PLoS Negl Trop Dis. 2011;5(3):e943.
96. Zaph C, Uzonna J, Beverley SM, Scott P. Central memory T cells mediate long-term immunity to *Leishmania major* in the absence of persistent parasites. Nat Med. 2004;10(10):1104–10.
97. Brito RC, Guimaraes FG, Velloso JP, Correa-Oliveira R, et al. Immunoinformatics features linked to *Leishmania* vaccine development: data integration of experimental and in silico studies. Int J Mol Sci. 2017;18(2)
98. Del Tordello E, Serruto D. Functional genomics studies of the human pathogen *Neisseria meningitidis*. Brief Funct Genomics. 2013;12(4):328–40.
99. Gorringe AR, Pajon R. Bexsero: a multicomponent vaccine for prevention of meningococcal disease. Hum Vaccin Immunother. 2012;8(2):174–83.
100. Martin NG, Snape MD. A multicomponent serogroup B meningococcal vaccine is licensed for use in Europe: what do we know, and what are we yet to learn? Expert Rev Vaccines. 2013;12 (8):837–58.
101. Gupta SK, Srivastava M, Akhoon BA, Smita S, et al. Identification of immunogenic consensus T-cell epitopes in globally distributed influenza-A H1N1 neuraminidase. Infect Genet Evol. 2011;11(2):308–19.
102. Gupta SK, Singh A, Srivastava M, Gupta SK, et al. In silico DNA vaccine designing against human papillomavirus (HPV) causing cervical cancer. Vaccine. 2009;28(1):120–31.
103. Ranjbar MM, Gupta SK, Ghorban K, Nabian S, et al. Designing and modeling of complex DNA vaccine based on tropomyosin protein of Boophilus genus tick. Appl Biochem Biotechnol. 2015;175(1):323–39.
104. Gupta SK, Srivastava M, Akhoon BA, Gupta SK, et al. In silico accelerated identification of structurally conserved CD8+ and CD4+ T-cell epitopes in high-risk HPV types. Infect Genet Evol. 2012;12(7):1513–8.
105. Baloria U, Akhoon BA, Gupta SK, Sharma S, et al. In silico proteomic characterization of human epidermal growth factor receptor 2 (HER-2) for the mapping of high affinity antigenic determinants against breast cancer. Amino Acids. 2012;42(4):1349–60.
106. Singh KP, Verma N, Akhoon BA, Bhatt V, et al. Sequence-based approach for rapid identification of cross-clade CD8+ T-cell vaccine candidates from all high-risk HPV strains. 3 Biotech. 2016;6(1):39.
107. Luo H, Lin Y, Gao F, Zhang CT, et al. DEG 10, an update of the database of essential genes that includes both protein-coding genes and noncoding genomic elements. Nucleic Acids Res. 2014;42(Database issue):D574–80.
108. Ravooru N, Ganji S, Sathyanarayanan N, Nagendra HG. In silico analysis of hypothetical proteins unveils putative metabolic pathways and essential genes in *Leishmania donovani*. Front Genet. 2014;5:291.
109. Jeong H, Mason SP, Barabasi AL, Oltvai ZN. Lethality and centrality in protein networks. Nature. 2001;411(6833):41–2.
110. Zirkel J, Cecil A, Schäfer F, Rahlfs S, et al. Analyzing thiol-dependent redox networks in the presence of methylene blue and other antimalarial agents with RT-PCR-supported in silico modeling. Bioinform Biol Insights. 2012;6:287–302.
111. Akhoon BA, Gupta SK, Dhaliwal G, Srivastava M, et al. Virtual screening of specific chemical compounds by exploring E.Coli NAD+−dependent DNA ligase as a target for antibacterial drug discovery. J Mol Model. 2011;17(2):265–73.

112. Srivastava M, Gupta SK, Abhilash PC, Singh N. Structure prediction and binding sites analysis of curcin protein of *Jatropha curcas* using computational approaches. J Mol Model. 2012;18(7):2971–9.
113. Srivastava M, Akhoon BA, Gupta SK, Gupta SK. Development of resistance against blackleg disease in *Brassica oleracea var. botrytis* through in silico methods. Fungal Genet Biol. 2010;47(10):800–8.
114. Akhoon BA, Gupta SK, Verma V, Dhaliwal G, et al. In silico designing and optimization of anti-breast cancer antibody mimetic oligopeptide targeting HER-2 in women. J Mol Graph Model. 2010;28(7):664–9.
115. Akhoon BA, Singh KP, Varshney M, Gupta SK, et al. Understanding the mechanism of atovaquone drug resistance in *Plasmodium falciparum* cytochrome b mutation Y268S using computational methods. PLoS One. 2014;9(10):e110041.
116. Gupta SK, Gupta SK, Smita S, Srivastava M, et al (2011) Computational analysis and modeling the effectiveness of 'Zanamivir' targeting neuraminidase protein in pandemic H1N1 strains. Infect Genet Evol 11 (5):1072–1082.
117. Song CM, Lim SJ, Tong JC. Recent advances in computer-aided drug design. Brief Bioinform. 2009;10(5):579–91.
118. Leelananda SP, Lindert S. Computational methods in drug discovery. Beilstein J Org Chem. 2016;12:2694–718.
119. Sliwoski G, Kothiwale S, Meiler J, Lowe EW Jr. Computational methods in drug discovery. Pharmacol Rev. 2014;66(1):334–95.
120. Field MC, Horn D, Fairlamb AH, Ferguson MA, et al. Anti-trypanosomatid drug discovery: an ongoing challenge and a continuing need. Nat Rev Microbiol. 2017;15(4):217–31.
121. Sundar S, Sinha PK, Rai M, Verma DK, et al. Comparison of short-course multidrug treatment with standard therapy for visceral leishmaniasis in India: an open-label, non-inferiority, randomised controlled trial. Lancet. 2011;377(9764):477–86.
122. Coulibaly B, Pritsch M, Bountogo M, Meissner PE, et al. Efficacy and safety of triple combination therapy with artesunate-amodiaquine-methylene blue for *falciparum* malaria in children: a randomized controlled trial in Burkina Faso. J Infect Dis. 2015;211(5):689–97.
123. Kunz M, Liang C, Nilla S, Cecil A, et al. The drug-minded protein interaction database (DrumPID) for efficient target analysis and drug development. Database. 2016;2016: baw041. https://doi.org/10.1093/database/baw041.
124. Fliri AF, Loging WT, Volkmann RA. Cause-effect relationships in medicine: a protein network perspective. Trends Pharmacol Sci. 2010;31(11):547–55.
125. Iorio F, Saez-Rodriguez J, di Bernardo D. Network based elucidation of drug response: from modulators to targets. BMC Syst Biol. 2013;7:139.

P-Glycoprotein-Like Transporters in *Leishmania*: A Search for Reversal Agents 14

Bruno Pradines

Abstract

Until now, chemotherapy has been the main line of defense against *Leishmania* infections. However, drug use and abuse have resulted in the selection and development of resistance mechanisms which strongly limit the number of antiprotozoal agents that are effective for the treatment of this disease. The emergence and spread of resistance to drugs currently in use and available for leishmaniasis emphasize that new compounds need to be identified and developed and that novel chemotherapeutic targets must be characterized. Mechanisms of drug resistance are often associated with decreased uptake of the drug into the parasite, poor drug activation, physiological alterations in the drug target, and overexpression of drug transporter proteins. One mechanism of resistance to antimony in *Leishmania* involves a decrease in its accumulation by either reduced uptake or increased efflux, mediated by P-glycoprotein (Pgp)-like transporters, which belong to the ATP-binding cassette (ABC) superfamily of proteins. The inhibition of the function of these proteins represents an attractive way to control drug resistance in clinical environments. New natural or synthetic sesquiterpenes, flavonoids, acridonecarboxamide derivative modulators of human Pgp (zosuquidar and elacridar), statins, pyridine analogs, 8-aminoquinolines, or phenothiazines revert in *Leishmania* the resistance phenotype to antimony, pentamidine, sodium stibogluconate, and miltefosine by modulating intracellular drug concentrations. In this chapter, we review some concepts concerning the reversal mechanism of multidrug resistance by the use chemosensitizers which alter the capacity of Pgp.

B. Pradines (✉)
Unité Parasitologie et entomologie, Département des maladies infectieuses, Institut de Recherche Biomédicale des Armées, Marseille, France & Unité de Recherche sur les Maladies Infectieuses et Tropicales Emergentes, Institut Hospitalo-Universitaire (IHU) Méditerranée Infection, UM, Marseille, France

A. Ponte-Sucre, M. Padrón-Nieves (eds.), *Drug Resistance in Leishmania Parasites*,
https://doi.org/10.1007/978-3-319-74186-4_14

14.1 Introduction

Arsenic- and antimony-containing drugs are still the first line of treatment for leishmaniasis. Pentavalent antimonial compounds (Sb^V) remain the choice of treatment for all forms of leishmaniasis, ranging from cutaneous lesions to fatal visceral infections. The emergence and spread of resistance to currently used antileishmanial drugs emphasize the fact that new compounds need to be identified and developed. Resistance to antimonial drugs is everyday more frequently reported [1–3].

A large amount of scientific effort is spent on elucidating the mechanisms underlying this resistance with the hope of restoring/improving the efficacy of existing drugs and of developing new drugs that can bypass resistance mechanisms.

Among the various drug resistance mechanisms identified, those based on drug movement through the membranes appear to play an important role by decreasing the drug concentration at the target sites. The transport proteins of the ATP-binding cassette (ABC) superfamily provide the basis of multidrug resistance in mammalian cancer cells and in pathogenic yeasts, fungi, parasites, and bacteria [4–8]. ABC proteins were also identified in resistance to antileishmanial drugs (see Table 14.1). The ABC transporters are described in Chap. 11.

But all of the ABC families are not associated with antileishmanial drug resistance, such as the ABCA family [9].

The ABCB family includes the multidrug-resistant protein 1 (MDR1) or ABCB4 protein and the multidrug-resistant protein 2 (MDR2) or ABCB2 protein, whose overexpression confers resistance to vinblastine and structurally non-related hydrophobic compounds such as puromycin, adriamycin, doxorubicin, and daunomycin [10–16]. LeMDR1 (LeABCB4) can also affect pentamidine resistance [17]. Additionally, LgMDR1 and LaMDR1 are increased in antimony-resistant strains of *L. (V.) guyanensis* or *L. (L.) amazonensis* [18]. The subcellular location of LeABCB4 and LaABCB2 (LaMDR2) in the tubular structure, a compartment that may correspond to a multivesicular tubule lysosome, suggests that mechanisms of resistance in *Leishmania* are different from those acting in the conventional mammalian efflux pump Pgp MDR1.

The ABCC family includes the multidrug-resistant protein A (MRPA) or P-glycoprotein A (PGPA) or ABCC3; the P-glycoprotein E (PGPE) or ABCC4, associated with resistance to arsenite and antimonial drugs; and the pentamidine resistance protein 1 (PRP1) or ABCC7. ABCC3 and ABCC4 are involved in the resistance of *Leishmania* toward arsenic and antimony compounds [19–22]. Overexpression of ABCC4 and ABCC5 can also confer resistance to antimonial drugs in *L. (S.) tarentolae* [23]. Additionally, field-resistant isolates to antimony exhibit upregulation in ABCC3 (MRPA or PGPA) transcript levels in *L. (L.) donovani*, *L. (V.) braziliensis*, *L. (V.) guyanensis*, *L. (L.) amazonensis*, or *L. (L.) major* (>1.5) [18, 24, 25]. ABCC7 is shown to confer pentamidine resistance in the promastigote and amastigote form of *L. (L.) major* and is cross-resistant to trivalent antimonial drugs when overexpressed [26–28].

The ABCG family includes the ABCG4 and ABCG6 proteins. ABCG4, localized mainly to the parasite plasma membrane, reduced the accumulation of

Table 14.1 ATP-binding cassette (ABC) transporters in *Leishmania* spp.

ABC subfamily	*Leishmania* spp.	Protein	Involvement in drug resistance
ABCA	*L. (L.) infantum*	LiABCA4	No
		LiABCA8	No
	L. (L.) major	LmABCA3	No
		LmABCA4	No
		LmABCA8	No
	L. (L.) tropica	LtrABCA4 or LtrABCA2	No
		LtrABCA8 or LtrABC1.1	No
ABCB PgP cluster	*L. (L.) amanozensis*	LaABCB4 or LaMDR1	Yes
		LaABCB2 or LaMDR2	Yes
	L. (L.) donovani	LdABCB4 or LdMDR1	Yes
	L. (M.) enriettii	LeABCB4 or LeMDR1	Yes
	L. (V.) guyanensis	LgABCB4 or LgMDR1	Yes
	L. (L.) tropica	LtrABCB4 or LtrMDR1	Yes
ABCC MRP cluster	*L. (L.) amazonensis*	LaABCC3 or LaMRPA	Yes
		LaABCC7 or LaPRP1	Yes
	L. (V.) braziliensis	LbABCC3 or LbMRPA	Yes
	L. (L.) donovani	LdABCC3 or LdPGPA or LdMRPA	Yes
	L. (V.) guyanensis	LgABCC3 or LgMRPA	Yes
	L. (L.) infantum	LiABCC3 or LiPGPA or LiMRPA	Yes
		LiABCC4 or LiPGPE	Yes
		LiABCC5	Yes
		LiABCC7 or LiPRP1	Yes
		LiABCC9	?
	L. (L.) major	LmABCC3 or LmPGPA or LmMRPA	Yes
		LmABCC7 or LmPRP1	Yes
	L. (L.) mexicana	LmeABCC3 or LmePGPA or LmeMRPA	Yes
	L. (S.) torentolae	LtABCC2 or LtPGPB	Yes
		LtABCC3 or LtPGPA or LtMRPA	Yes
		LtABCC4 or LtPGPE	Yes
		LtABCC5	Yes
	L. (L.) tropica	LtrABBC4 or LtrPGPE	Yes

(continued)

Table 14.1 (continued)

ABC subfamily	*Leishmania* spp.	Protein	Involvement in drug resistance
ABCG	*L. (L.) donovani*	LdABCG6	Yes
	L. (L.) infantum	LiABCG4	Yes
		LiABCG6	Yes
	L. (L.) major	LmABCG2	Yes

? Not determined

phosphatidylcholine analogs and conferred resistance to alkyl-phospholipids (miltefosine (MIL), edelfosine, and perifosine) when overexpressed. The second ABCG reported, ABCG6, also localized mainly to the parasite plasma membrane, confers resistance to MIL and sitamaquine when overexpressed in *L. (L.) infantum* [29]. ABCG6 confers also resistance to camptothecin and arsenite [30].

The inhibition of the activity of ABC proteins represents an interesting way to control drug resistance. This concept of inhibiting ABC transporters is well studied for malaria [31–33]. *Leishmania* parasites overexpressing ABCG2 are resistant to antimony, as they demonstrate a reduced accumulation of Sb^{III} due to an increase in drug efflux [34].

14.2 Transporter Inhibitors and Modulators of Multidrug Resistance

A number of compounds, e.g., calcium channel blockers, calmodulin antagonists, hydrophobic peptides, protein kinase inhibitors, antibiotics, hormone derivatives, and flavonoids, have been previously described to reverse in vitro multidrug resistance in mammalian cells [35]. They are called modulators or chemosensitizers; those that reverse the multidrug-resistant phenotype in *Leishmania* spp. are listed in Table 14.2.

14.2.1 Calcium Channel Blockers: Verapamil

Some of these compounds, like the L-type voltage-gated channel blocker verapamil, are known to efficiently overcome multidrug-resistant phenotype in vitro, not only in mammalian cells [54–56] but also in some bacteria such as *Mycobacterium* spp. [57, 58] or *Enterococcus* spp. [59] and in parasites such as nematodes like *Haemonchus contortus* [60–62] and protozoa like *Entamoeba histolytica* [63–65] or *Plasmodium falciparum* [66–68]. Verapamil is an inhibitor of the human Pgp (ABCB1) [69].

Previous studies have demonstrated that verapamil increases the in vitro antimony activity on *L. (L.) donovani* [36]. Verapamil shows efficacy in reversing several P-glycoprotein and MRP overexpression-mediated arsenite resistance

Table 14.2 Major multidrug resistance reversal drugs investigated in *Leishmania spp.*

Class of compound and specific modulators	Resistance to	Strains	References
Calcium channel blockers			
Verapamil	Antimonials	*L.(L.) donovani*	[36]
	Pentamidine	*L. (L.) mexicana*	[37]
	Arsenites	*L. (L.) donovani*	[30]
		L. (S.) tarentolae	[38]
	Pirarubicin	*L. (V.) braziliensis*	[39]
		L. (V.) guyanensis	[39]
		L. (L.) mexicana	[39]
		L. (V.) peruviana	[39]
		L. (V.) panamensis	[39]
	Vinblastine	*L. (L.) amazonensis*	[13]
Calmodulin inhibitors: Phenothiazine derivatives			
Chlorpromazine	Antimonials	*L. (L.) donovani*	[40]
		L. (L.) major	[40]
		L. (V.) braziliensis	[39]
		L. (V.) guyanensis	[39]
		L. (L.) mexicana	[39]
	Pentamidine	*L. (L.) mexicana*	[37]
Trifluoperazine, prochlorperazine	Pirarubicin	*L. (V.) braziliensis*	[39]
		L. (V.) guyanensis	[39]
		L. (L.) mexicana	[39]
Thioridazine, trifluoropromazine	Pirarubicin	*L. (V.) braziliensis*	[39]
		L. (V) guyanensis	[39]
		L. (L.) mexicana	[39]
Flavonoids			
Silymarin and silybin derivatives	Daunomycin	*L. (L.) tropica*	[41]
Quercetin	Arsenites	*L. (L.) donovani*	[30]
Synthetic flavonoids	Pentamidine	*L. (L.) donovani*	[42]
		L. (M.) enriettii	[42]
	Sodium stiboglucanate	*L. (L.) donovan*	[42]
		L. (M.) enriettii	[42]
Synthetic flavonoid derivatives	Antimonials	*L. (L.) major*	[43]
Trolox and derivatives	Antimonials	*L. (L.) major*	[43]
Sesquiterpenes			
Dihydro-β-agarofuran sesquiterpenes	Miltefosine	*L. (L.) tropica*	[41]
Sesquiterpene C-3 (agarofuran derivative)	Edelfosine	*L. (L.) tropica*	[41]
	Daunomycin	*L. (L.) tropica*	[41]
Nortriterpene	Daunomycin	*L. (L.) tropica*	[44]
Glycyrrhizic acid	Sodium stiboglucanate	*L. (L.) donovani*	[45]

(continued)

Table 14.2 (continued)

Statins			
Lovastatin	Antimonials	*L. (L.) donovani*	[46]
Pyridine analogs			
PAK104P	Pirarubicin	*L. (V.) braziliensis*	[39]
		L. (V.) guyanensis	[39]
		L. (L.) mexicana	[39]
Oxazolo[3,2-α]pyridine	Daunomycin	*L. (L.) tropica*	[47]
	Miltefosine	*L. (L.) tropica*	[47]
Sulfonylurea			
Glibenclamide	Glucantime	*L. (L.) mexicana*	[48]
		L. (L.) major	[49]
Benzoquinones			
Bis-pyranobenzoquinones	Daunomycin	*L. (L.) tropica*	[50]
Acridine derivatives			
Quinacrine	Pentamidine	*L. (L.) donovani*	[42]
		L. (V.) enriettii	[51]
8-aminoquinolines			
Sitamaquine	Miltefosine	*L. (L.) tropica*	[52]
	Antimonials	*L. (L.) tropica*	[52]
Acridonecarboxamide derivatives			
Elacridar, zosuquidar	Miltefosine	*L. (L.) tropica*	[53]

phenotype in *L. (S.) tarentolae* or *L. (L.) donovani* [30, 38]. The reversion of in vitro drug resistance by verapamil is confirmed in *L. (L.) donovani* clinical isolates resistant to sodium stibogluconate [70]. This drug partially reverses the resistance in vinblastine-resistant *L. (L.) amazonensis*, which show cross-resistance to adriamycin [13]. The energy-dependent efflux of pirarubicin, an anthracycline derivative, is inhibited by verapamil in *L. (V.) braziliensis*, *L. (V.) guyanensis*, *L. (L.) mexicana*, *L. (V.) peruviana*, and *L. (V.) panamensis* [39]. However, verapamil cannot revert the resistance to camptothecin, a cytotoxic quinoline alkaloid which inhibits the DNA enzyme topoisomerase-I [30]. Various studies in cancer cell lines reveal that development of resistance to topoisomerase inhibitors is a multifactorial event including altered transport, modified drug metabolism and detoxification, and change in drug-target interaction. Amino acid substitutions in topoisomerase-I confer camptothecin resistance in *L. (L.) donovani* [71]. The apparent wide substrate specificity of the *Leishmania* transport system suggests that it could be responsible for the intrinsic resistance of parasite promastigotes to drugs. Its physiological relevance is supported by the fact that it was described in at least five different *Leishmania* species. It seems that verapamil regulates drug susceptibility by downregulating Pgp expression in arsenical-resistant *Leishmania* spp. [72]. In tumor cells, the ability of verapamil to modulate multidrug resistance protein 1 (MRP1 or ABCC1)-mediated resistance seems to be link to its effect on the reduced glutathione

(GSH) status [73]. In addition to stimulate MRP1-mediated GSH transport, verapamil modulates MRP1-mediated leukotriene C_4 transport [74].

Verapamil also enhances pentamidine uptake into resistant *L. (L.) mexicana* and also partially reverses the drug resistance phenotype in promastigotes [37], but not in axenic amastigotes [75]. In addition, using nontoxic concentrations of verapamil, a dose-dependent reversion of pentamidine is observed in resistant parasites when compared with those not treated with verapamil in *L. (L.) amazonensis* [27]. However, verapamil has any impact either in drug uptake or drug resistance in *L. (L.) donovani* [76]. This suggests that Pgp-mediated efflux of pentamidine is not operative in *L. (L.) donovani* as it is in *L. (L.) mexicana* or *L. (L.) amazonensis*. PRP1 (ABCC7) is shown to confer pentamidine resistance in the promastigote and amastigote form of *L. (L.) major* and in *L. (L.) infantum* when overexpressed [26, 28], but not in *L. (L.) amazonensis* [27]. No difference in *PRP1* transcript levels is observed between susceptible and resistant *L. (L.) donovani* parasites to Sb^V [77].

The specific Pgp inhibitor cyclosporin-A does not interfere with calcein cell retention (efflux measurement) in *L. (L.) amazonensis*, while verapamil does [78]. These results demonstrate that the drug transport systems expressed in *Leishmania* are susceptible to MRP (ABCC) inhibitors like verapamil, but not to the Pgp (ABCB) inhibitor like cyclosporin-A.

In addition, it seems that verapamil is ineffective in reverting ABCG6 overexpression-mediated resistance in *Leishmania* [30].

14.2.2 Calmodulin Inhibitors: Phenothiazine Derivatives

Phenothiazines and reserpine can also reverse drug resistance in mammalian cells, bacteria, and parasites [79–82]. Phenothiazine drugs, of which chlorpromazine is the leading molecule, are widely used for their antipsychotic, antianxiety, and antiemetic effects. In addition, they also possess protozoacidal activity against amastigotes and promastigotes of *L. (L.) donovani* and *L. (L.) chagasi* in vitro as well as in vivo [83–85]. Chlorpromazine is also an inhibitor of the human Pgp (ABCB1) [69].

Chlorpromazine, trifluoropromazine, thioridazine, trifluoperazine, and prochlorperazine are reported to inhibit the energy-dependent efflux of pirarubicin, an anthracycline derivative, in *L. (V.) braziliensis*, *L. (V.) guyanensis*, and *L. (L.) mexicana* [39]. A synergistic effect between chlorpromazine and N-meglumine antimoniate is observed in multidrug-resistant *L. (L.) donovani* and *L. (L.) major* cells in vitro [40]. The effect of phenothiazine derivatives on *Leishmania* drug transport may be explained by their ability to inhibit the activity of trypanothione reductase [86, 87]. Indeed, if we consider that the reduced form of trypanothione is an important co-factor for the function of the *Leishmania* drug transporter, in the same way as reduced glutathione is required for the MRP1 function [74, 88], phenothiazines may inhibit transport activity by decreasing the intracellular level of reduced trypanothione [39]. However, no significant effect is observed in vivo against amastigotes of *L. (L.) major* and *L. (L.) mexicana*, in cutaneous lesions in mice [40]. The toxic effects reported with the most frequently studied phenothiazine, which is chlorpromazine, have impaired the investigation of other phenothiazines as potential clinical agents.

Prochlorperazine and trifluoperazine enhance pentamidine uptake into resistant *L. (L.) mexicana* and also partially reverse the drug resistance phenotype [37]. However, these drugs have any impact either in drug uptake or drug resistance in *L. (L.) donovani* [76]. This indicates that Pgp-mediated efflux of pentamidine is not operative in *L. (L.) donovani* as it is in *L. (L.) mexicana*, like for verapamil.

14.2.3 Flavonoids

The flavonoid class is constituted by flavones, flavonols, isoflavones, flavanones, and chacones [89]. More than 6500 different flavonoids have been identified from plant sources.

Flavonoids have shown promise to reverse multidrug-resistant phenotypes in *L. (L.) tropica* [41, 42, 90, 91]. Flavonoids constitute a well-known class of natural inhibitors of different proteins [92] with contradictory results concerning their modulation effects on different multidrug-resistant cells [93–95]. They bind to the two cytosolic NBSs of the ABC transporters. The flavanolignan silybin and its hemisynthetic derivatives exhibit good affinity to NBD2 [96]. The flavonoid interactions with the ATP-binding site and a vicinal hydrophobic region [41, 91, 97] cause the inhibition of drug efflux and reverse the resistance to daunomycin in *L. (L.) tropica*. Only flavonoids which bind with high affinity to the cytosolic domain NBD2 are able to both increase daunomycin accumulation in a *L. (L.) tropica* line overexpressing MDR1 (LtrABCB4) and inhibit the parasite growth in the presence of the drug [41]. In addition, flavonoids, such as quercetin a flavone, may modulate the multidrug transporter by decreasing Pgp synthesis and inhibiting the transcriptional activation of the *mdr* gene involved in the susceptibility to daunomycin [53, 98]. Quercetin is a human Pgp (ABCB1), MRP2 (ABCC2), and BCRP (ABCG2) transporter inhibitor [69, 99]. Quercetin reverts the resistance to camptothecin in *L. (L.) donovani* that overexpresses LdABCG6 involved in resistance to camptothecin and arsenite [30] and is associated with reduction of accumulation of alkyl-phospholipid drugs such as MIL in *Leishmania* [29]. Synthetic flavonoid dimmers exhibit a significant reversing activity on pentamidine and sodium stibogluconate resistance in *L. (S.) enriettii* and *L. (L.) donovani* [42]. This modulatory effect is dose-dependent and due to the bivalent nature of the flavonoid compounds. Compared to other MDR inhibitors such as verapamil, reserpine, quinine, quinacrine, and quinidine, these compounds are the only agents that can reverse sodium stibogluconate resistance in *L. (S.) enriettii*. These modulators exhibit reversal activity on pentamidine resistance, comparable to that of reserpine and quinacrine but whatever the level of overexpression of *Lemdr1* gene suggesting that these modulators are not specific to LeABCB4 (LeMDR1). Recently, new compounds derived from aurone, flavones, isoflavones, xanthone, chalcones, and trolox were evaluated against antimony-resistant strains of *L. (L.) major* [43]. Two trolox carboxamides induce reversion of antimony resistance in the promastigote form of *L. (L.) major*. These two compounds are specific reversal agents targeting the *Leishmania* ABCI4 transporter. This transporter belongs to an unclassified group of proteins in the ABC family with no known homology with other eukaryotic ABC proteins but with orthologues in *Trypanosoma brucei* and *Trypanosoma cruzi* [100]. ABCI4 is a protein located in the plasma membrane and mitochondria of the parasite and efflux antimony. Overexpression of ABCI4 confers resistance to antimony.

14.2.4 Sesquiterpenes

Agarofuran sesquiterpenes, e.g., natural compounds isolated from *Maytenus cuzcoina* [101, 102], *M. chubutensis* [91], *M. macroparta* [103], *M. magellanica* [91], *M. apurimacensis*, [104] and *Crossopetalum tonduzii* [105], are new promising reversal agents that overcome the multidrug-resistant phenotype in *Leishmania*, including the resistance to anthracyclines (daunomycin) and alkyl-lysophospholipids (MIL and edelfosine). In *L. (L.) tropica*, dihydro-β-agarofuran sesquiterpenes enhance accumulation of calcein, a Pgp substrate, probably due to Pgp-like transporter inhibition [91]. These compounds bind to the NBD_2 C-terminal of *L. (L.) tropica* Pgp-like transporter, LtrMDR1 (LtrABCB4) [105]. A series of dihydro-β-agarofuran sesquiterpenes isolated from the leaves of *Maytenus cuzcoina* or semisynthetic derivatives have been tested on *L. (L.) tropica* parasites overexpressing Pgp [101]. Three-dimensional quantitative structure-activity relationship using the comparative molecular similarity indices analysis (3D-QSAR/CoMSIA) is employed to characterize the steric, electrostatic, lipophilic, and hydrogen-bond-donor and hydrogen-bond-acceptor requirements of these sesquiterpenes as modulators at Pgp-like transporter. The most salient features of requirements are the H-bond interaction between the substituents at the C-2 and C-6 positions with the receptor. The structure-activity relationship (SAR) suggests that a substituent at the C-2 position seems to be essential for reversal activity in the MDR *Leishmania* line by acting as a H-bond acceptor. The furan ring at the C-6 position seems to form a hydrogen bond with the receptor. The introduction of a carbonyl group, capable of acting as a H-bond acceptor in the H-bond with the receptor, produces a tenfold higher chemosensitization. This suggests a direct interaction with the receptor. These results would be used to design and synthesize more effective and specific new Pgp inhibitors.

Sesquiterpene C-3 remarkably sensitizes multidrug-resistant parasites to MIL and edelfosine by increasing alkyl-lysophospholipid accumulation [53]. Moreover, *mdr1* gene transfections can alter membrane fluidity in mammalian cells and change alkyl-lysophospholipid effects [106, 107].

Nortriterpene, extracted from *Maytenus chubutensis* and *M. magellanica* (Celastraceae family), shows only moderate MDR1 reversal activity in a *L. (L.) tropica* strain overexpressing LtrMDR1, involved in daunomycin resistance [64].

Glycyrrhizic acid, a triterpenoid saponin isolated from the root of the liquorice plant, limits infection with sodium antimony gluconate (SAG)-resistant *L. (L.)*

donovani in combination with SAG treatment [45]. Glycyrrhizic acid enhances antimony retention by inhibition of MRP1 and Pgp expression levels in splenic macrophages from infected mice. Glycyrrhizic acid acts by modulation of host ABC transporters. Glycyrrhizic acid suppresses cell surface expression of MRP1 and Pgp in host macrophages.

14.2.5 Statins: Lovastatin

Statins, 3-hydroxy-3-methylglutaryl-coenzyme A (HMG-CoA) reductase inhibitors, belong to a family of lipid-lowering drugs that are currently used for the control of hyperlipidemia and are considered useful for protection from cardiovascular events. Apart from the cholesterol-lowering activity of statins, the immunomodulatory and pleiotropic effects of statins may significantly impact infection-related survival [108, 109]. Statins interfered with the growth of protozoan parasites in the Trypanosomatidae family, such as *Trypanosoma cruzi* and various *Leishmania* species [110–112].

Statins are also inhibitors of Pgp in cancer cells [66, 113, 114]**.** Additionally, in Plasmodium falciparum, atorvastatin has synergistic effects in combination with antimalarial drugs such as dihydroartemisinin, quinine, or mefloquine [115–117]. atorvastatin acts probably by inhibition of MDR-like proteins, which are involved in malaria resistance.. In *Leishmania*, the combination of the antifungal drug miconazole and lovastatin is synergic in terms of inhibition of promastigote proliferation, macrophage infection, and amastigote number [118]. In promastigote cultures, the effect is more marked in *L. (L.) amazonensis* parasites than *L. (L.) donovani*. But it seems that this effect is due to inhibition of sterol biosynthesis by both lovastatin and miconazole. More recently, lovastatin, which can inhibit both Pgp and MRP1 (ABCC1), allows the accumulation of sodium antimony gluconate in resistant *L. (L.) donovani* and reversion of antimony resistance [46]. Lovastatin can induce not only the retention of antimony compounds but also that of an unrelated chemotherapeutic agent such as doxorubicin in cancer cells.

14.2.6 Pyridine Analog: PAK-104P

A pyridine analog, PAK-104P, was demonstrated in vitro as well as in vivo to inhibit Pgp-mediated multidrug resistance to vincristine, adriamycin, doxorubicin, paclitaxel, and antimonial and arsenical drugs [119–124]. PAK-104P partially reverses the resistance and increases the arsenite accumulation in cancer cells that overexpress MRP1 (ABCC1) [125]. PAK-104P can inhibit both Pgp and MRP [123]. PAK-104P also blocks the energy-dependent efflux of pirarubicin in *L. (V.) braziliensis*, *L. (V.) guyanensis*, and *L. (L.) mexicana* [39]. This compound probably alters the activity of trypanothione reductase and the transport activity by decreasing the intracellular level of reduced trypanothione.

Oxazolo[3,2-α]pyridine derivatives produce a significant reversion of resistance to both MIL and daunomycin in a MDR1 overexpressing *L. (L.) tropica* strain [47].

14.2.7 Sulfonylurea: Glibenclamide

Glibenclamide is a sulfonylurea that inhibits ABC proteins such as Pgp (ABCB1) [69, 126] and MRP1 (ABCC1) of cancer cells [127].

Glibenclamide increases calcein accumulation in *L. (L.) amazonensis*-resistant line, like verapamil [78]. Cyclosporin-A, which is a specific inhibitor of Pgp, doesn't increase calcein accumulation. These results demonstrate that the drug transport systems expressed in *L. (L.) amazonensis* are susceptible to MRP (ABCC) inhibitors like glibenclamide or verapamil, but not to the Pgp (ABCB) inhibitor like cyclosporin-A. The increased expression of MRP1 (ABCC1) at the plasma membrane of the protoplast of *Arabidopsis thaliana* is associated with an increase in the resistance of *Arabidopsis* to Sb^{III} and a decrease of Sb^{III} accumulation in protoplast [128]. The simultaneous administration in vitro of glibenclamide, a human MRP1 (ABCC1) inhibitor, increases the efficacy of Glucantime and decreases the infection rate of infected macrophages by *L. (L.) major* [49]. A fixed concentration of 50 μM glibenclamide in combination with various concentration of Glucantime caused an inhibition of 80–90% in cell growth. The administration of glibenclamide in experimental in vivo settings increases the potency of Glucantime when administered simultaneously and reduces the size of lesions in mice infected with drug-susceptible and drug-resistant *Leishmania* [48]. The Glucantime-glibenclamide combination could represent a novel strategy to fight against *Leishmania* infection.

14.2.8 Acridonecarboxamide Derivatives: Elacridar and Zosuquidar

Acridonecarboxamide derivatives, elacridar (LY335979) and zosuquidar (GF120918), modulators of human P-glycoprotein [129, 130], can overcome Pgp (LtrMDR1 or LtrABCB4)-mediated *Leishmania* MIL resistance by increasing intracellular MIL accumulation [131]. Overexpression of LtrABCB4 is involved in MIL resistance [59]. In addition, ABCG4, localized mainly to the parasite plasma membrane, reduced the accumulation of phosphatidylcholine analogs and conferred resistance to alkyl-phospholipids (MIL, edelfosine, and perifosine) when overexpressed [132]. The second ABCG reported, ABCG6, also localized mainly to the parasite plasma membrane, conferred resistance to MIL and sitamaquine when overexpressed in *L. (L.) infantum* [29]. Overexpression of ABCG6 is associated with reduction of accumulation of alkyl-phospholipid drugs into *Leishmania*.

14.2.9 Dithiocarbamate: Disulfiram

Disulfiram (Antabuse) is used as an adjunct in the treatment of chronic alcoholism. Disulfiram is able to potentiate the antimalarial action of subcurative doses of chloroquine and amodiaquine in *Plasmodium berghei-* and *P. vinckei petteri-*infected mice [133]. Disulfiram inhibits P-glycoproteins by covalently modifying one or more endogenous cysteine residues (Cys1074) in NBD2 [134]. Modification of only one of the Walker A cysteines is sufficient to inactive Pgp [135]. This drug could be effective in combination with Glucantime [136].

14.2.10 Benzoquinones

Bis-pyranobenzoquinones inhibit the activity of Pgp of mammalian cells but not MRP1 (ABCC1) [50]. In addition, these compounds increase the activity of daunorubicin in resistant *L. tropica* line. Bis-pyrano-1,4-benzoquinones are the best modulators in MDR human cancer cells, while bis-pyrano-1,2-benzoquinones exhibit the higher toxicity in combination with daunorubicin in MDR *L. (L.) tropica* line.

14.2.11 Quinacrine

Quinacrine is an acridine derivative with antimalarial, antileishmanial, and antitrypanosomal activities [137–139].

Quinacrine can have a synergistic effect in combination with pentamidine in *L. (M.) enriettii* and in *L. (L.) donovani* [42, 51]. Moreover, quinacrine is only effective in the pentamidine-resistant *Leishmania*, not in the sodium stibogluconate-resistant or vinblastine-resistant parasites [42]. Surprisingly, quinacrine not only restores the susceptibility of resistant parasites to pentamidine but also increases the susceptibility of susceptible parasites. This result suggests that the quinacrine target remains unaltered in susceptible and resistant parasites to pentamidine. Whatever the quinacrine target might be, it cannot be an ABC transporter in *Leishmania*.

14.2.12 8-Aminoquinolines: Sitamaquine

Sitamaquine (WR6026), an 8-aminoquinoline analog, overcomes the MDR1-mediated resistance to MIL by increasing intracellular MIL accumulation in a *L. (L.) tropica* strain overexpressing MDR1 and resistant to MIL [52]. Additionally, sitamaquine also modulates the activity of MRPA, involved in antimony resistance, in resistant *L. (L.) tropica* strain. Sitamaquine reverses MRPA-mediated resistance to antimony.

14.3 Conclusion and Future Trends

Efflux transporters play a key role in the emergence and dissemination of resistant parasites and in the acquisition of additional mechanisms of drug resistance caused by a decrease in intracellular drug concentration. Despite their noticeable divergence in structure and membrane topology, the major efflux systems share a dependence on specific key parameters including (1) the functional assembly of a membrane transporter, (2) the energy required (e.g., ATP, ion antiport, or membrane potential) for active transport, and (3) the presence of affinity sites inside the transporter that are involved in substrate recognition and transport.

The identification of functional domains and the characterization of various interactions with the transported drug may elucidate key parameters that govern efflux activity. At present, some 3D structures have been solved for bacterial drug transporters, and these have allowed the proposal of dynamic and mechanical models for drug transport [140]. The same approach must be used for *Leishmania* infection. Drug-transporter interactions have recently been shown to be an important part of multidrug resistance. In silico modeling is a powerful tool often employed to predict drug properties prior to in vitro and in vivo studies. Modeling efforts are currently being undertaken using both ligand- and transporter-based methods such as structure-activity relationship (SAR) studies, quantitative-SAR (QSAR) studies, hologram QSAR (HQSAR), comparative molecular field analysis (CoMFA) and comparative molecular similarity index analysis (CoMSIA) studies, pharmacophore modeling, homology modeling, and molecular dynamics studies. The most common approaches to discover human ABC substrates and inhibitors are development of QSAR models and SAR. This approach has been carried out in the case of human ATP-transporter multidrug resistance-associated protein 2 (MRP2 or ABCC2) [141]. The goal of QSAR modeling is to construct a mathematical relationship between descriptors and pharmacological activities of compounds. The model can then be used to predict the activity for an untested compound. The goal of SAR is usually to discern the structural features or side groups that directly lead to the desired activity under investigation. In order to use these in silico modeling techniques, compounds need to be screened to find the degree of substrate binding to inhibition. Until now, there are no or very few inhibitors or substrate datasets available for ABC transporters in *Leishmania* in literature. Some compounds with inhibitory effects toward human ABCB1 (Pgp) and ABCC1 (MRP1) transporters were studied by pharmacophore modeling, docking, and 3D QSAR to described the binding preferences of these proteins [142]. Docking of selective inhibitors into the Pgp binding cavity by the use of a structural model based on the recently resolved Pgp structure confirms the Pgp pharmacophore features identified and reveals the interactions of some functional groups and atoms in the structures with particular protein residues. However, due to the complex nature of the applied methods, useful interpretation of the models that can be directly translated into chemical structures by the medicinal chemist is rather difficult.

The aim of these efforts is to decipher the molecular basis of drug transport, to explain how differences in chemical structures modify interactions with the transporter, or to elucidate how the transporter functions in general. In addition, original molecules have been demonstrated to restore the antileishmanial activity of drugs that are pump substrates, and these studies make it possible to identify pharmacophoric groups that are involved in efflux inhibition.

These data are crucial for the design of (1) new antileishmanial molecules that are devoid of efflux-substrate characteristics and can reach a normal intracellular accumulation level and (2) new compounds that have strong efflux pump affinity associated with a high inhibitor capability and block the pump, restoring the intracellular concentration of antileishmanial drugs.

The most prevalent mechanisms of resistance in *Leishmania* are mutations of proteins involved in the drug transport (uptake or efflux) and amplification of transporter genes. The role of ABC transporters in drug resistance in *Leishmania* is well established. Several modulators have been described to reverse multidrug resistance in vitro in *Leishmania*. Most of these drugs remain to be evaluated in vivo. Hence, clinical evaluation of therapeutic regimens is now required to validate the efficacy of these promising compounds or combinations for the treatment of leishmaniasis.

Another perspective is to modulate proteins which participate to the regulation of the expression of the level of MDR1 in *Leishmania*. Silent information regulator 2 (Sir2) is involved in *Leishmania* survival by preventing programmed cell death [143]. Sir2 plays a role in regulating the expression of MDR1 and thereby amphotericin-B (AMB) efflux from the resistant *L. (L.) donovani* [144]. Inhibition or deletion of Sir2 allele shows decreased expression levels of MDR1 and lower efflux of AMB in resistant parasites. In contrast, Sir2 overexpression in susceptible parasites leads to resistant phenotype associated with reduced activity of AMB, increased drug efflux, and increased mRNA level of MDR1. Sir2 will be used as a potent drug target for *Leishmania* treatment.

References

1. Faraut-Gambarelli F, Piarroux R, Deniau M, Giusiano B, et al. *In vitro* and *in vivo* resistance of *Leishmania infantum* to meglumine antimoniate: a study of 37 strains collected from patients with visceral Leishmaniasis. Antimicrob Agents Chemother. 1997;41:827–30.
2. Jackson JE, Tally JD, Ellis WY, Mebrahtu YB, et al. Quantitative *in vitro* drug potency and drug susceptibility evaluation of *Leishmania spp.* from patients unresponsive to pentavalent antimony therapy. Am J Trop Med Hyg. 1990;43:464–80.
3. Sundar S, More DK, Singh MK, Singh VP, et al. Failure of pentavalent antimony in visceral Leishmaniasis in India: report from the center of the Indian epidemic. Clin Infect Dis. 2000;31:1104–7.
4. Ambudkar SV, Rosen BP, Gottesman MM. Workshop on ABC transporters and human diseases. Drug Resist Update. 2000;3:51–4.
5. Lage H. ABC-transporters: implications on drug resistance from microorganisms to human cancers. Int J Antimicrob Ag. 2003;22:188–99.
6. Rubio JP, Cowman AF. The ATP-binding cassette (ABC) gene family of *Plasmodium falciparum*. Parasitol Today. 1996;12:135–40.

7. Schuetzer-Muehlbauer M, Willinger B, Egner R, Ecker G, et al. Reversal of antifungal resistance mediated by ABC efflux pumps from *Candida albicans* functionally expressed in yeast. Int J Antimicrob Ag. 2003;22:291–300.
8. Sparreboom A, Danesi R, Ando Y, Chan J, et al. Pharmacogenomics of ABC transporters and its role in cancer chemotherapy. Drug Resist Update. 2003;6:71–84.
9. Araujo-Santos JM, Parodi-Talice A, Castanys S, Gamarro F. The overexpression of an intracellular ABCA-like transporter alters phospholipid trafficking in *Leishmania*. Biochem Biophys Res Commun. 2005;330:349–55.
10. Chiquero MJ, Perez-Victoria JM, O'Valle F, Gonzales-Ros JM, et al. Altered drug membrane permeability in a multidrug-resistant *Leishmania tropica* line. Biochem Pharmacol. 1998;55:131–9.
11. Chow LM, Wong AK, Ullman B, Wirth DF. Cloning and functional analysis of an extrachromosomally amplified multidrug resistance-like gene in *Leishmania enriettii*. Mol Biochem Parasitol. 1993;60:195–208.
12. Gamarro F, Chiquero MJ, Amador MV, Lagare D, et al. P-glycoprotein overexpression in methotrexate-resistant *Leishmania tropica*. Biochem Pharmacol. 1994;47:1939–47.
13. Gueiros-Filho FJ, Viola JPB, Gomes FCA, Farina M, et al. *Leishmania amazonensis*: multidrug resistance in vinblastine-resistant promastigotes is associated with rhodamine 123 efflux, DNA amplification, and RNA overexpression of a *Leishmania mdr1* gene. Exp Parasitol. 1995;81:480–90.
14. Henderson DM, Sifri CD, Rodgers M, Wirth DF, et al. Multidrug resistance in *Leishmania donovani* is conferred by amplification of a gene homologous to the mammalian *mdr1* gene. Mol Cell Biol. 1992;12:2855–65.
15. Katakura K, Iwanami M, Ohtomo H, Fujise H, et al. Structural and functional analysis of the *LaMDR1* multidrug resistance gene in *Leishmania amazonensis*. Biochem Biophys Res Commun. 1999;255:289–94.
16. Mary C, Faraut F, Deniau M, Dereure J, et al. Frequency of drug resistance gene amplification in clinical *Leishmania* strains. Int J Microbiol. 2010;2010:819060.
17. Wong ILK, Chow LMC. The role of *Leishmania enriettii* multidrug resistance protein 1 (LeMDR1) in mediating drug resistance is iron-dependent. Mol Biochem Parasitol. 2006;150:278–87.
18. Moreira DS, Monte Neto RL, Andrade JM, Santi AMM, et al. Molecular characterization of the MRPA transporter and antimony uptake in four new world *Leishmania spp.* susceptible and resistant to antimony. Int J Parasitol Drugs Drugs Resist. 2013;3:143–53.
19. Borst P, Ouellette M. New mechanisms of drug resistance in parasitic protozoa. Annu Rev Microbiol. 1995;49:427–60.
20. Chow LMC, Volkman K. *Plasmodium* and *Leishmania*: the role of mdr genes in mediating drug resistance. Exp Parasitol. 1998;90:135–41.
21. El Fadili K, Messier N, Leprohon P, Roy G, et al. Role of the ABC transporter MRPA (PGPA) in antimony resistance in *Leishmania infantum* axenic and intracellular amastigotes. Antimicrob Agents Chemother. 2005;49:1988–93.
22. Ouellette M, Legare D, Papadopoulou B. Multidrug resistance and ABC transporters in parasitic protozoan. J Mol Microbiol Biotechnol. 2001;3:201–6.
23. Leprohon P, Legare D, Ouellette M. Intracellular localization of the ABCC proteins of *Leishmania* and their role in resistance to antimonials. Antimicrob Agents Chemother. 2009;53:2646–9.
24. Rai S, Goel SK, Dwivedi UN, Sundar S, et al. Role of efflux pumps and intracellular thiols in natural antimony resistant isolates of *Leishmania donovani*. PLoS One. 2013;8:74862.
25. Soleimanifard S, Arjmand R, Saberi S, Khamesipour A, et al. P-glycoprotein a gene expression in glucantime-resistant and sensitive *Leishmania major* (MRHO/IR/75/ER). Iranian. J Parasitol. 2014;9:423–8.

26. Coelho AC, Beverley SM, Cotrim PC. Functional genetic identification of PRP1, an ABC transporter superfamily member conferring pentamidine resistance in *Leishmania major*. Mol Biochem Parasitol. 2003;130:83–90.
27. Coelho AC, Gentil LG, Franco da Silveira J, Cotrim PC. Characterization of *Leishmania (Leishmania) amazonensis* promastigotes resistant to pentamidine. Exp Parasitol. 2008;120:98–102.
28. Coelho AC, Messier N, Ouellette M, Cotrim PC. Role of the ABC transporter PRP1 (ABCC7) in pentamidine resistance in *Leishmania* amastigotes. Antimicrob Agents Chemother. 2007;51:3030–2.
29. Castanys-Munoz E, Perez-Victoria JM, Gamarro F, Castanys S. Characterization of an ABCG-like transporter from the protozoan parasite *Leishmania* with a role in drug resistance and transbilayer lipid movement. Antimicrob Agents Chemother. 2008;52:3573–9.
30. Bose Dasgupta S, Ganguly A, Roy A, Mukherjee T, et al. A novel ATP-binding cassette transporter, ABCG6 is involved in chemoresistance in *Leishmania*. Mol Biochem Parasitol. 2008;158:176–88.
31. Alibert-Franco S, Pradines B, Mahamoud A, Davin-Regli A, et al. Efflux mechanism, an attractive target to combat *Plasmodium falciparum* and *Pseudomonas aeruginosa*. Curr Med Chem. 2009;16:301–17.
32. Henry M, Alibert S, Orlandi-Pradines E, Bogreau H, et al. Chloroquine resistance reversal agents as promising antimalarial drugs. Curr Drug Targets. 2006;7:935–48.
33. Henry M, Alibert S, Rogier C, Barbe J, et al. Inhibition of efflux of quinolines as new therapic strategy in malaria. Curr Top Med Chem. 2008;8:563–78.
34. Perea A, Manzano JI, Castanys S, Gamarro F. The LABCG2 transporter from the protozoan parasite *Leishmania* is involved in antimony resistance. Antimicrob Agents Chemother. 2016;60:3489–96.
35. Stein WD. Reversers of the multidrug resistance transporter P-glycoprotein. Curr Opin Investig Drugs. 2002;3:812–7.
36. Neal RA, van Bueren J, McCoy NG, Iwobi M. Reversal of drug resistance in *Trypanosoma cruzi* and *Leishmania donovani* by verapamil. Trans R Soc Trop Med Hyg. 1989;83:197–8.
37. Basselin M, Denise H, Coombs GH, Barrett MP. Resistance to pentamidine in *Leishmania mexicana* involves exclusion of the drug from the mitochondrion. Antimicrob Agents Chemother. 2002;46:3731–8.
38. Dey S, Papadopoulou B, Haimeur A, Roy G, et al. High level arsenite resistance in *Leishmania tarentolae* is mediated by an active extrusion system. Mol Biochem Parasitol. 1994;67:49–57.
39. Essodaïgui M, Frézard F, Moreira ESA, Dagger F, et al. Energy-dependent efflux from *Leishmania* promastigotes of substrates of the mammalian multidrug resistance pumps. Mol Biochem Parasitol. 1999;100:73–84.
40. el-On J, Rubinstein N, Kernbaum S, Schnur LF. *In vitro* and *in vivo* anti-leishmanial activity of chlorpromazine alone and combined with N-meglumine antimonate. Ann Trop Med Parasitol. 1986;80:509–17.
41. Perez-Victoria JM, Chiquero MJ, Conseil G, Dayan G, et al. Correlation between the affinity of flavonoids binding to the cytosolic site of *Leishmania tropica* multidrug transporter and their efficiency to revert parasite resistance to daunomycin. Biochemistry. 1999;38:1736–43.
42. Wong ILK, Chan KF, Burkett BA, Zhao Y, et al. Flavonoid dimmers as bivalent modulators for pentamidine and sodium stibogluconate resistance in *Leishmania*. Antimicrob Agent Chemother. 2007;51:930–40.
43. Manzano JI, Lecerf-Schmidt F, Lespinasse MA, Di Pietro A, et al. Identification of specific reversal agents for *Leishmania* ABCI4-mediated antimony resistance by flavonoid and trolox derivatives screening. J Antimicrob Agents. 2014;69:664–72.
44. Kennedy ML, Lianos GG, Castanys S, Gamarro F, et al. Terpenoids from *Maytenus species* and assessment of their reversal activity against a multidrug-resistant *Leishmania tropica* line. Chem Biodivers. 2011;8:2291–8.

45. Bhattacharjee A, Majumder S, Majumdar SB, Choudhuri SK, et al. Co-administration of glycyrrhizic acid with the antileishmanial drug sodium antimony gluconate (SAG) cures SAG-resistant visceral leishmaniasis. Int J Antimicrob Agents. 2015;45:268–77.
46. Mookerjee Basu J, Mookerjee A, Banerjee R, Saha M, et al. Inhibition of ABC transporters abolished antimony resistance in *Leishmania* infection. Antimicrob Agents Chemother. 2008;52:1080–93.
47. Caballero E, Manzano JI, Puebla P, Castanys S, et al. Oxazolo[3,2-]pyridine. A new structural scaffold for the reversal of multi-drug resistance in *Leishmania*. Bioorg Med Chem Lett. 2012;22:6272–5.
48. Serrano-Martin X, Payares G, Mendoza-Leon A. Glibenclamide, a blocker of $K^{+}{}_{ATP}$ channels, shows antilesihmanial activity in experimental murine cutaneous leishmaniasis. Antimicrob Agents Chemother. 2006;50:4214–6.
49. Padron-Nieves M, Diaz E, Machuca C, Romero A, et al. Glibenclamide modulates glucantime activity and disposition in *Leishmania major*. Exp Parasitol. 2009;121:331–7.
50. Jimenez-Alonso S, Perez-Lomas AL, Estevez-Braun A, Munoz Martinez F, et al. Bis-pyranobenzoquinones as a new family of reversal agents of the multidrug resistance phenotype mediated by P-glycoprotein in mammalian cells and the protozoan parasite *Leishmania*. J Med Chem. 2008;51:7132–43.
51. Wong ILK, Chan KF, Zhao Y, Hang Chan T, et al. Quinacrine and a novel apigenin dimmer can synergistically increase the pentamidine susceptibility of the protozoan parasite *Leishmania*. J Antimicrob Chemother. 2009;63:1179–90.
52. Perez-Victoria JM, Bavchvarov BI, Torrecillas IR, Martinez-Garcia M, et al. Sitamaquine overcomes ABC-mediated resistance to miltefosine and antimony in *Leishmania*. Antimicrob Agents Chemother. 2011;55:3838–44.
53. Perez-Victoria JM, Perez-Victoria FJ, Parodi-Talice A, Jimenez IA, et al. Alkyl-lysophospholipid resistance in multidrug-resistant *Leishmania tropica* and chemosensitization by a novel P-glycoprotein-like transporter modulator. Antimicrob Agents Chemother. 2001;45:2468–74.
54. Tsuruo T, Iida H, Tsukagoshi S, Sakurai Y. Overcoming of vincristine resistance in P388 leukemia *in vivo* and *in vitro* through enhanced cytotoxicity of vincristine and vinblastine by verapamil. Cancer Res. 1981;41:1967–72.
55. Tsuruo T, Iida H, Tsukagoshi S, Sakurai Y. Potentiation of vincristine and adriamycin effects in human hemopoietic tumor cell lines by calcium antagonists and calmodulin inhibitors. Cancer Res. 1983;43:2267–72.
56. Twentyman PR, Fox NE, Bleehen NM. Drug resistance in human lung cancer cell lines: cross-resistance studies and effects of the calcium transport blocker, verapamil. J Radiat Oncol Biol Phys. 1986;12:1355–8.
57. Banerjee SK, Bhatt K, Rana S, Misra P, et al. Involvement of an efflux system in mediating high level of fluoroquinolone resistance in *Mycobacterium smegmatis*. Biochem Biophys Res Commun. 1996;226:362–8.
58. Choudhuri BS, Bahkta S, Barik R, Basu J, et al. Overexpression and functional characterization of an ABC (ATP-binding cassette) transporter encoded by the genes *drrA* and *drrB* of *Mycobacterium tuberculosis*. Biochem J. 2002;367:279–85.
59. Jonas BM, Murray BE, Weinstock GM. Characterization of emeA, a NorA homolog and multidrug resistance efflux pump, in *Enterococcus faecalis*. Antimicrob Agents Chemother. 2001;45:3574–9.
60. Beugnet F, Gauthey M, Kerboeuf D. Partial *in vitro* reversal of benzimidazole resistance by the free-living stages of *Haemonchus contortus* with verapamil. Vet Rec. 1997;141:575–6.
61. Kerboeuf D, Blackhall W, Kaminsky R, von Samson-Himmelstjerna G. P-glycoprotein in helminths: function and perspectives for antihelminthic treatment and reversal of resistance. Int J Antimicrob Agents. 2003;22:332–46.

62. Kerboeuf D, Chambrier P, Le Vern Y, Aycardi J. Flow cytometry analysis of drug transport mechanisms in *Haemonchus contortus* susceptible or resistant to antihelminthics. Parasitol Res. 1999;85:118–23.
63. Ayala P, Samuelson J, Wirth D, Orozco E. *Entamoeba histolytica*: physiology of multidrug resistance. Exp Parasitol. 1990;71:169–75.
64. Banuelos C, Orozco E, Gomez C, Gonzales A, et al. Cellular location and function of the P-glycoprotein (EhPgp) in *Entamoeba histolytica* multidrug-resistant trophozoites. Microb Drug Resist. 2002;8:291–300.
65. Orozco E, Lopez C, Gomez C, Perez DG, et al. Multidrug resistance in the protozoan parasite *Entamoeba histolytica*. Parasitol Int. 2002;51:353–9.
66. Holtzman CW, Wiggings BS, Spinler SA. Role of P-glycoprotein in statin drug interaction. Pharmacotherapy. 2006;26:1601–7.
67. Millet J, Torrentino-Madamet M, Alibert S, Rogier C, et al. Dihydroethanoanthracene derivatives as *in vitro* malarial chloroquine resistance reversal agents. Antimicrob Agents Chemother. 2004;48:2753–6.
68. Pradines B, Alibert-Franco S, Houdoin C, Mosnier J, et al. *In vitro* reversal of chloroquine resistance in *Plasmodium falciparum* with dihydroethanoanthracene derivatives. Am J Trop Med Hyg. 2002;66:661–6.
69. Matsson P, Pedersen JM, Norinder U, Bergstrom CAS, et al. Identification of novel specific and general inhibitors of the three major human ATP-binding cassette transporters, P-gp, BCRP and MRP2 among registered drugs. Pharm Res. 2009;26:1816–31.
70. Valiathan R, Dubey ML, Mahajan RC, Malla N. *Leishmania donovani*: effect of verapamil on *in vitro* susceptibility of promastigote and amastigote stages of Indian clinical isolates to sodium stibogluconate. Exp Parasitol. 2006;114:103–8.
71. Marquis JF, Hardy I, Olivier M. Topoisomerase I amino acid substitutions, Gly185Arg and Asp325Glu, confer camptothecin resistance in *Leishmania donovani*. Antimicrob Agents Chemother. 2005;49:1441–6.
72. Kaur J, Dey CS. Putative P-glycoprotein expression in arsenite-resistant *Leishmania donovani* down-regulated by verapamil. Biochem Biophys Res Commun. 2000;271:615–9.
73. Loe DW, Deeley RG, Cole SPC. Verapamil stimulates glutathione transport by the 190-kDa multidrug resistance protein 1 (MRP1). J Pharmacol Exp Ther. 2000;293:530–8.
74. Loe DW, Oleschuk CJ, Deeley RG, Cole SPC. Structure-activity studies of verapamil analogs that modulate transport of Leukotriene C4 and reduced glutathione by multidrug resistance protein MRP1. Biochem Biophys Res Commun. 2000;275:795–803.
75. Sereno D, Lemesre JL. *In vitro* life cycle of pentamidine-resistant amastigotes: stability of the chemoresistant phenotypes is dependent on the level of resistance induced. Antimicrob Agents Chemother. 1997;41:1898–903.
76. Mukherjee A, Padmanabhan PK, Sahani MH, Barrett MP, et al. Roles for mitochondria in pentamidine susceptibility and resistance in *Leishmania donovani*. Mol Biochem Parasitol. 2006;145:1–10.
77. Decuypere S, Rijal S, Yardley V, De Doncker S, et al. Gene expression analysis of the mechanism of natural Sb(V) resistance in *Leishmania donovani* isolates from Nepal. Antimicrob Agents Chemother. 2005;49:4616–21.
78. Machuca C, Rodriguez A, Herrera M, Silva S, et al. *Leishmania amazonensis*: metabolic adaptations induced by resistance to an ABC transporter blocker. Exp Parasitol. 2006;114:1–9.
79. Ford JM, Prozialeck WC, Hait WN. Structural features determining activity of phenothiazines and related drugs for inhibition of cell growth and reversal of multidrug resistance. Mol Pharmacol. 1989;35:105–15.
80. Henry M, Alibert S, Baragatti M, Mosnier J, et al. Dihydroethanoanthracene derivatives reverse *in vitro* quinoline resistance in *Plasmodium falciparum* malaria. Med Chem. 2008;4:426–37.

81. Molnar J, Hever A, Falka I, Ocsovski I, et al. Inhibition of the transport function of membrane proteins by some substituted phenothiazines in *E. coli* and multidrug resistant tumor cells. Anticancer Res. 1997;17:481–6.
82. Pearce HL, Safa AR, Bac NJ, Winter MA, et al. Essential features of the P-glycoprotein pharmacophore as defined by a series of reserpine analogs that modulate multidrug resistance. Proc Natl Acad Sci U S A. 1989;86:5128–32.
83. Pearson RD, Manian AA, Hall D, Harcus JL, et al. Antileishmanial activity of chlorpromazine. Antimicrob Agents Chemother. 1984;25:571–4.
84. Pearson RD, Manian AA, Harcus JL, Hall D, et al. Lethal effect of phenothiazine neuroleptics on the pathogenic protozoan *Leishmania donovani*. Science. 1982;217:369–71.
85. Werbovetz KA, Lehnert EK, MacDonald TL, Pearson RD. Cytotoxicity of acridine compounds for *Leishmania* promastigotes *in vitro*. Antimicrob Agents Chemother. 1992;36:495–7.
86. Chan C, Yin H, Garforth J, McKie JH, et al. Phenothiazines inhibitors of trypanothione reductase as potential antitrypanosomal and antileishmanial drugs. J Med Chem. 1998;41:148–56.
87. Khan MO, Austin SE, Chan C, Yin H, et al. Use of an additional hydrophobic binding site, the Z site, in the rational design of a new class of stronger trypanothione reductase inhibitor, quaternary alkylammonium phenothiazines. J Med Chem. 2000;43:3148–56.
88. Loe DW, Stewart RK, Massey TE, Deeley RG, et al. ATP-dependent transport of aflatoxin B1, and its glutathione conjugates by the product of the multidrug resistance protein (*MRP*) gene. Mol Pharmacol. 1997;51:1034–41.
89. di Pietro A, Conseil G, Perez-Victoria JM, Dayan G, et al. Modulation by flavonoids of cell multidrug resistance mediated by P-glycoprotein and related ABC transporters. Cell Mol Life Sci. 2002;59:307–22.
90. Kennedy ML, Cortes F, Pinero JE, Castanys S, et al. Leishmanicidal and reversal multidrug resistance constituents from *Aeonium lindleyi*. Planta Med. 2011;77:77–80.
91. Kennedy ML, Cortés-Selva F, Perez-Victoria JM, Jimenez IA, et al. Chemosensitization of a multidrug-resistant *Leishmania tropica* line by new sesquiterpenes from *Maytenus magellanica* and *Maytenus chubutensis*. J Med Chem. 2001;44:4668–76.
92. Shapiro AB, Ling V. Effect of quercetin on Hoechst 33342 transport by purified and reconstituted P-glycoprotein. Biochem Pharmacol. 1997;53:587–96.
93. Critchfield JW, Welsh CJ, Phang JM, Yeh GC. Modulation of adriamycin accumulation and efflux by flavonoids in HCT-15 colon cells. Activation of P-glycoprotein as a putative mechanism. Biochem Pharmacol. 1994;48:1437–45.
94. Ferté J, Kühnel JM, Chapuis G, Rolland Y, et al. Flavonoid-related modulators of multidrug resistance: synthesis, pharmacological activity, and structure-activity relationships. J Med Chem. 1999;42:478–89.
95. Shapiro AB, Ling V. Positively cooperative sites for drug transport by P-glycoprotein with distinct drug specificities. Eur J Biochem. 1997;250:130–7.
96. Maitrejean M, Comte G, Barron D, El Kirat K, et al. The flavanolignan silybin and its hemisynthetic derivatives, a novel series of potential modulators of P-glycoprotein. Bioorg Med Chem Lett. 2000;10:157–60.
97. Conseil G, Baubichon-Cortay H, Dayan G, Jault JM, et al. Flavonoids: a class of modulators with bifunctional interactions at vicinal ATP- and steroid-binding sites on mouse P-glycoprotein. Proc Natl Acad Sci U S A. 1998;95:9831–6.
98. Kioka N, Hosokawa N, Komano T, Hirayoshi K, et al. Quercetin, a bioflavonoid, inhibits the increase of human multidrug resistance gene (MDR1) expression caused by arsenite. FEBS Lett. 1992;301:307–9.
99. Yoshikawa M, Ikegami Y, Sano K. Transport of SN-38 by the wild type of human ABC transporter ABCG2 and its inhibition by quercetin, a natural flavonoid. J Exp Ther Oncol. 2004;4:25–35.

100. Manzano JI, Garcia-Hernandez R, Castanys S, Gamarro F. A new ABC half-transporter in *Leishmania* is involved in resistance to antimony. Antimicrob Agents Chemother. 2013;57:3719–30.
101. Cortes-Selva F, Campillo M, Reyes CP, Jimenez IA, et al. SAR studies of dihydro-beta-agarofuran sesquiterpenes as inhibitors of the multidrug-resistance phenotype in a *Leishmania tropica* line overexpressing a P-glycoprotein-like transporter. J Med Chem. 2004;47:576–87.
102. Cortes-Selva F, Munoz-Martinez F, Ilias A, Jimenez IA, et al. Functional expression of a multidrug P-glycoprotein transporter of *Leishmania*. Biochem Biophys Res Commun. 2005;329:502–7.
103. Cortes-Selva F, Jimenez IA, Munoz-Martinez F, Campillo M, et al. Dihydro-beta-agarofuran sesquiterpenes: a new class of reversal agents of the multidrug resistance phenotype mediated by P-glycoprotein in the protozoan parasite *Leishmania*. Curr Pharm Des. 2005;11:3125–39.
104. Delgado-Mendez P, Herrera N, Chavez H, Estevez-Braun A, et al. New terpenoids from Maytenus apurimacensis as MDR reversal agents in the parasite *Leishmania*. Bioorg Med Chem. 2008;16:1425–30.
105. Perez-Victoria JM, Tincusi BM, Jimanez IA, Bazzocchi IL, et al. New natural sesquiterpenes as modulators of daunomycin resistance in a multidrug-resistant *Leishmania tropica* line. J. Med Chem. 1999;42:4388–93.
106. Callaghan R, van Gorkom LC, Epand RM. A comparison of membrane properties and composition between cell lines selected and transfected for multi-drug resistance. Br J Cancer. 1992;66:781–6.
107. Storme GA, Berdel WE, van Blitterswijk WJ, Bruyneel EA, et al. Antiinvasive effect of racemic 1-O-octadecyl-2-O-methylglycero-phosphocholine on MO4 mouse fibrosarcoma cells *in vitro*. Cancer Res. 1985;45:351–7.
108. Greenwood J, Mason JC. Statins and the vascular endothelial inflammatory response. Trends Immunol. 2007;28:88–98.
109. Terblanche M, Almog Y, Rosenson RS, Smith TS, et al. Statins: panacea for sepsis? Lancet Infect Dis. 2006;6:242–8.
110. Montalvetti A, Pena-Diaz J, Hurtado R, Ruiz-Perez LM, et al. Characterization and regulation of *Leishmania major* 3-hydroxy-methyl-glutaryl—CoA reductase. Biochem J. 2000;349:27–34.
111. Urbina JA, Lazardi K, Marchan E, Visbal G, et al. Mevinolin (lovastatin) potentiates the antiproliferative effects of ketoconazole and terbinafine against *Trypanosoma (Schizotrypanum) cruzi*: *in vitro* and *in vivo* studies. Antimicrob Agents Chemother. 1993;37:580–91.
112. Yokoyama K, Trobridge P, Buckner FS, Scholten J, et al. The effects of protein farnesyltransferase inhibitors on trypanosomatids: inhibition of protein farnesylation and cell growth. Mol Biochem Parasitol. 1998;94:87–97.
113. Wang E, Casciano CN, Clement RP, Johnson WW. HMG-CoA reductase inhibitors (statins) characterized as direct inhibitors of P-glycoprotein. Pharm Res. 2001;18:800–6.
114. Wu X, Whitfield I, Stewart BH. Atorvastatin transport in the Caco-2 cell model: contributions of P-glycoprotein and the proton-monocarboxylic acid co-transporter. Pharm Res. 2000;17:209–15.
115. Parquet V, Henry M, Wurtz N, Dormoi J, et al. Atorvastatin as a potential anti-malarial drug: *in vitro* synergy in combinational therapy with quinine against *Plasmodium falciparum*. Malar J. 2010;9:139.
116. Savini H, Souraud JB, Briolant S, Baret E, et al. Atorvastatin as a potential antimalarial drug: *in vitro* synergy in combinational therapy with dihydroartemisinin. Antimicrob Agents Chemother. 2010;54:966–7.
117. Wurtz N, Briolant S, Gil M, Parquet V, et al. Synergy of mefloquine activity with atorvastatin, but not chloroquine and monodesethylamodiaquine, and association with the *pfmdr1* gene. J Antimicrob Chemother. 2010;65:1387–94.

118. Haughan PA, Chance ML, Goad LJ. Synergism *in vitro* of lovastatin and miconazole as anti-leishmanial agents. Biochem Pharmacol. 1992;44:2199–206.
119. Chen ZS, Mutoh M, Sumizawa T, Furukawa T, et al. Reversal of heavy metal resistance in multidrug-resistant human KB carcinoma cells. Biochem Biophys Res Commun. 1997;236:586–90.
120. Chuman Y, Chen ZS, Seto K, Sumizawa T, et al. Reversal of MRP-mediated vincristine resistance in KB cells by buthionine sulfoximine in combination with PAK-104P. Cancer Lett. 1998;129:69–76.
121. Kitasono M, Okumura H, Ikeda R, Sumizawa T, et al. Reversal of LRP-associated drug resistance in colon carcinoma SW-620 cells. Int J Cancer. 2001;91:126–31.
122. Shudo N, Mizoguchi T, Kiyosue T, Arita M, et al. Two pyridine analogues with more effective ability to reverse multidrug resistance and with lower calcium blocking activity than their dihydropyridine counterparts. Cancer Res. 1990;50:3055–61.
123. Sumizawa T, Chen ZS, Chuman Y, Seto K, et al. Reversal of multidrug resistance-associated protein-mediated drug resistance by the pyridine analog PAK-104P. Mol Pharmacol. 1997;51:399–405.
124. Vanhoefer U, Cao S, Minderman H, Toth K, et al. PAK-104P, a pyridine analogue, reverses paclitaxel and doxorubicin resistance in cell lines and nude mice bearing xenografts that overexpress the multidrug resistance protein. Clin Cancer Res. 1996;2:369–77.
125. Tachiwada T, Chen ZS, Che XF, Matsumoto M, et al. Isolation and characterization of arsenite-resistant human epidermoid carcinoma KB cells. Oncol Rep. 2007;18:721–7.
126. Golstein PE, Boom A, van geffel J, Jacobs P, et al. P-glycoprotein inhibition by glibenclamide and related compounds. Pflugers Arch. 1999;437:652–60.
127. Conseil G, Deeley RG, Cole SPC. Role of two adjancent cytoplasmic tyrosine residues in MRP1 (ABCC1) transport activity and sensitivity to sulfonylureas. Biochem Pharmacol. 2005;69:451–61.
128. Gayet L, Picault N, Cazalé AC, Beyly A, et al. Transport of antimony salts by *Arabidopsis thaliana* protoplasts over-expressing the human multidrug resistance-associated protein 1 (MRP1/ABCC1). FEBS Lett. 2006;580:6891–7.
129. Dantzig AH, Law KL, Cao J, Starling JJ. Reversal of multidrug resistance by the P-glycoprotein modulator, LY335979, from the bench to the clinic. Curr Med Chem. 2001;8:39–50.
130. Hyafil F, Vergely C, Du Vignaud P, Grand-Perret T. *In vitro* and *in vivo* reversal of multidrug resistance by GF120918, an acridonecarboxamide derivative. Cancer Res. 1993;53:4595–602.
131. Perez-Victoria JM, Cortes-Selva F, Parodi-Talice A, Bavchvarov BI, et al. Combination of suboptimal doses of inhibitors targeting different domains of LtrMDR1 efficiently overcomes resistance of *Leishmania spp.* to miltefosine by inhibiting drug efflux. Antimicrob Agent Chemother. 2006;50:3102–10.
132. Castanys-Munoz E, Alder-Baerens N, Pomorski T, Gamarro F, et al. A novel ATP-binding cassette transporter from *Leishmania* is involved in transport of phosphatidylcholine analogues and resistance to alkyl-phospholipids. Mol Microbiol. 2007;64:1141–53.
133. Deharo E, Barkan D, Krugliak M, Golenser J, et al. Potentialization of the antimalarial action of chloroquine in rodent malaria by drugs known to reduce cellular glutathione levels. Biochem Pharmacol. 2003;66:809–17.
134. Loo TW, Bartlett MC, Clarke DM. Disulfiram metabolites permanently inactivate the human multidrug resistance P-glycoprotein. Mol Pharm. 2004;1:426–33.
135. Loo TW, Clarke DM. Blockage of drug resistance *in vitro* by disulfiram, a drug used to treat alcoholism. J Natl Cancer Inst. 2000;92:898–902.
136. Namazi MR. Potential utility of disulfiram against leishmaniasis. Indian J Med Res. 2008;127:193–4.
137. Gamage SA, Figgitt DP, Wojcik SJ, Ralph RK, et al. Structure-activity relationships for the antileishmanial and trypanosomal activities of 1′-substituted 9-anilinoacridines. J Med Chem. 1997;40:2634–42.

138. Gamage SA, Tepsiri N, Wilairat P, Wojcik SJ, et al. Synthesis and *in vitro* evaluation of 9-anilino-3,6-diaminoacridines active against a multidrug-resistant strain of the malaria parasite *Plasmodium falciparum*. J Med Chem. 1994;37:1486–94.
139. Girault S, Grellier P, Berecibar A, Maes L, et al. Antimalarial, antitrypanosomal, and antileishmanial activities and cytotoxicity of bis(9-amino-6-chloro-2-methoxyacridines): influence of the linker. J Med Chem. 2000;43:2646–54.
140. Seeger MA, Schiefner A, Eicher T, Verrey F. Structural asymmetry of AcrB trimer suggests a peristaltic pump mechanism. Science. 2006;313:1295–8.
141. Pedersen JM, Matsson P, Bergstrom CA, Norinder U, et al. Prediction and identification of drug interactions with the human ATP-binding cassette transporter multidrug-resistance associated protein 2 (MRP2; ABCC2). J Med Chem. 2008;51:3275–87.
142. Pajeva IK, Globisch C, Wiese M. Combined pharmacophore modeling, docking, and 3D QSAR studies of ABCB1 and ABCC1 transporter inhibitors. Chem Med Chem. 2009;4:1883–96.
143. Vergnes B, Sereno D, Madjidian-Sereno N, Lesmesre JL, Ouaissi A. Cytoplasmic SIR2 homologue overexpression promotes survival of *Leishmania* parasites by preventing programmed cell death. Gene. 2002;296:139–50.
144. Purkait B, Singh R, Wasnik K, Das S, et al. Up-regulation of silent information regulator 2 (Sir2) is associated with amphotericin B resistance in clinical isolates of *Leishmania donovani*. J Antimicrob Agents. 2015;70:1343–56.

The Concept of Fitness in *Leishmania* 15

Manu Vanaerschot, Franck Dumetz, Marlene Jara,
Jean-Claude Dujardin, and Alicia Ponte-Sucre

Abstract

A pathogen's fitness relates to all biological processes that ensure its survival, reproduction, and transmission in specific conditions. These often include the presence of drugs, forcing pathogens to adapt and develop drug resistance in order to survive. The acquisition of a drug-resistant trait usually comes at a cost, making drug-resistant parasites less fit than their wild-type counterparts. This has important implications on the development of drug resistance and on the frequency of treatment failure cases in endemic regions. Treatment failure in patients suffering from leishmaniasis has been observed for most antileishmanials, but could not always be correlated to drug resistance of the infecting parasite. One similitude of both pentavalent antimonial and miltefosine treatment failure, however, relates to changes in parasite fitness. In the specific case of *Leishmania donovani*, for example, this may contrast with the usual fitness cost observed in natural drug-resistant organisms and highlights parasite fitness as an important

M. Vanaerschot (✉)
Department of Microbiology and Immunology, Columbia University College of Physicians and Surgeons, New York, NY, USA

F. Dumetz
Department of Biomedical Sciences, Institute of Tropical Medicine, Antwerp, Belgium

M. Jara
Instituto de Medicina Tropical Alexander von Humboldt, Universidad Peruana Cayetano Heredia, Lima, Peru

J.-C. Dujardin
Department of Biomedical Sciences, Institute of Tropical Medicine, Antwerp, Belgium

Department of Biomedical Sciences, Faculty of Pharmaceutical, Biomedical and Veterinary Sciences, University of Antwerp, Antwerpen, Belgium

A. Ponte-Sucre
Laboratory of Molecular Physiology, Institute of Experimental Medicine, Luis Razetti School of Medicine, Faculty of Medicine, Universidad Central de Venezuela Caracas, Caracas, Venezuela

A. Ponte-Sucre, M. Padrón-Nieves (eds.), *Drug Resistance in Leishmania Parasites*,
https://doi.org/10.1007/978-3-319-74186-4_15

contributor to treatment failure in visceral leishmaniasis in the Indian subcontinent. In this final chapter, we will canvass the knowns and the unknowns of *Leishmania* fitness at different parasite life stages and for different *Leishmania* species and discuss its relevance for the development and spread of drug resistance and/or treatment failure in the field. We will also propose new research avenues for leishmaniasis drug development and control in the context of current elimination efforts.

15.1 Introduction

Viruses were pioneers as target for studies of the concept of fitness. In these organisms, fitness was initially defined as their ability to successfully survive, reproduce, and infect in a defined environment [1–4]. For *Leishmania*, the concept was initially related to proficiency; i.e., the complex integrated skills that allow *Leishmania* to successfully replicate and cause the disease [5]. As the life cycle of *Leishmania* oscillates between two life stages that occur in a specific host—promastigotes develop in the insect vector and amastigotes develop in mammalian hosts—*Leishmania* adapted to these environments by undergoing several developmental stages; each bears specific traits to guarantee survival, reproduction, and ultimately, transmission to a new host. The fitness of *Leishmania* is thus the amalgamation of its success in all these processes combined (reviewed in [6]). Although many of the determinants involved in these processes are becoming more and more appreciated, only few are well understood. These include determinants specific to parasite life stages such as promastigote metacyclogenesis and amastigote survival in host cells (Sect. 15.2.1) and molecular traits that contribute to the parasite's adaptive skills during its whole life cycle (Sect. 15.2.2).

Importantly, the fitness of an organism is not only dependent on that organism itself but also on the environment in which it lives (Fig. 15.1). In the case of *Leishmania*, this includes host factors such as immunity and nutritional status, whether or not the parasite can hide in certain tissues (Sect. 15.2.1.1), to even dynamic global trends that may enhance the chance for emerging infectious diseases to occur and expand swiftly [7], discussed in Sect. 15.2.3. The interaction of all these fitness determinants is complex and eventually results in the capacity of the parasite to be transmitted and to infect the next host, where it may cause disease, a process originally defined as virulence. Virulence has been used as one of the foremost markers for fitness in *Leishmania* since its expression constitutes the mechanism per excellence that permits the "survival of the best," guaranteeing successful transmission to the next host [8]. Virulence is important at both the promastigote and the amastigote stage. Its function is evident at the dynamic interface that allows integrity but at the same time guarantees communication between the organism and its host.

More recent contributor to this environment are drugs. Drugs can dramatically alter the fitness landscape for *Leishmania* parasites, selecting *Leishmania* sub-populations

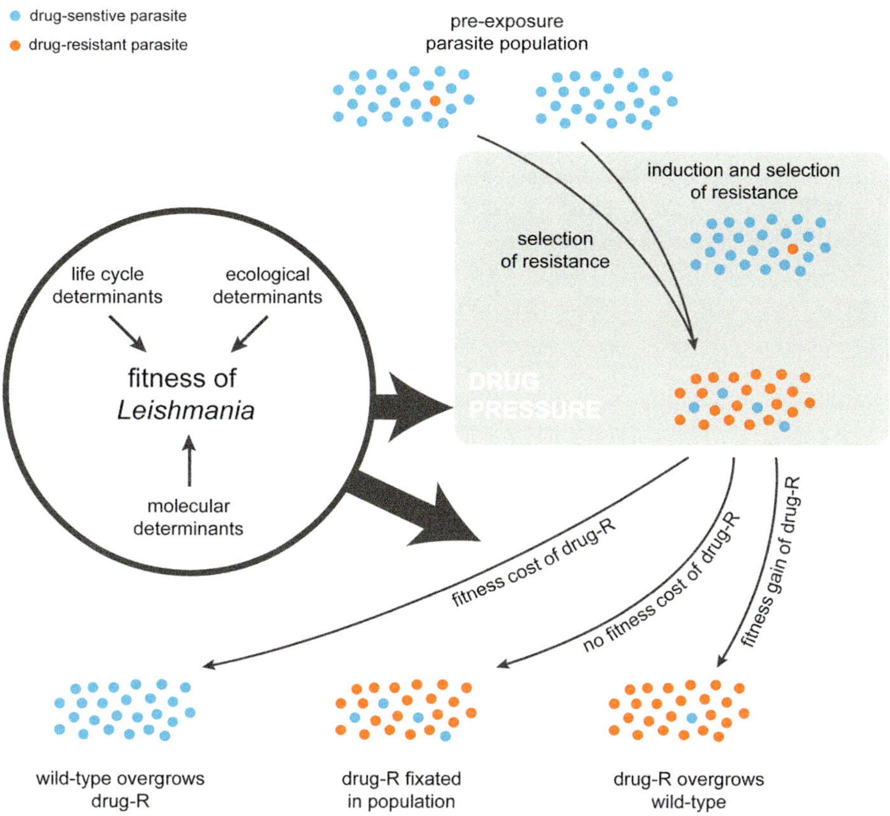

Fig. 15.1 The impact of parasite fitness on the evolution of parasite populations in the presence of drugs. Drug pressure selects for drug-resistant parasites, but sensitive parasites may potentially overcome drug treatment through mechanisms related to quiescence or hiding in niches where drug levels are low—thus without developing a classic drug-resistant phenotype. Once drug pressure is relieved again (due to, e.g., changes in treatment policies), their fitness compared to wild-type drug-sensitive parasites will decide on their future success in the population

that are able to survive drug pressure thanks to specific physiological traits—this will be discussed in Sect. 15.3 of this chapter.

15.2 The Knowns and Unknowns of *Leishmania* Fitness

15.2.1 Life Cycle Determinants

15.2.1.1 Amastigotes

Once an infected sand fly bites a mammalian host, parasites and sand fly saliva components are inoculated into the skin and invade mononuclear phagocytes in which they will develop into amastigotes. This may lead to two different outcomes:

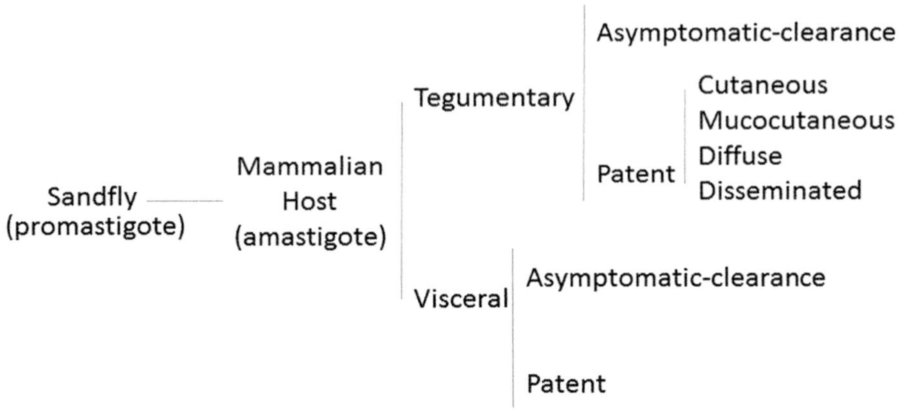

Fig. 15.2 Spectrum of clinical manifestations that may result from *Leishmania* infection in New and Old World *leishmaniasis*

either the host immune system successfully controls the infection, resulting in an asymptomatic infection, or the infection becomes patent, resulting in mild or severe disease (Fig. 15.2).

Classically, amastigotes are defined as the non-motile, parasitic forms with an ovoid or spherical body, a rod-shaped kinetoplast and a rudimentary, retracted flagellum arising from a basal body. This developmental form paradoxically lives in the immune cells that constitute the primary defense against invasion by foreign organisms, suggesting that through evolution, *Leishmania* has successfully learned to adapt to the stressful environment constituted by the intracellular milieu. *Leishmania* amastigotes are experts at exploiting host cell processes to establish infection and persist in several tissues. Although infected cells favor the immediate control of intracellular pathogens, the intracellular milieu constitutes a pathogenic protective space that drives the adaptive response of the parasite and allows it to display its florid pathogenic potential [9] and divert host mechanisms that would otherwise lead to parasite killing.

1. Immune System

   Once the primary parasite-host interaction occurs, the immune system initiates its activity with the aim to control the infection. The final end of its function could represent control of the disease, with or without sterilization, eventually leading to the asymptomatic character of the infection, or to a patent infection, either tegumentary or visceral. *Leishmania* is a versatile organism with diverse host defense evading mechanisms [9]. These host manipulation skills of the parasite are key to its survival and replication inside host cells. While being phagocytosed, *Leishmania* ensures that it is not recognized as a foreign organism by the host cell by interacting with specific surface receptors expressed by host neutrophils, dendritic cells, macrophages, and monocytes (reviewed in [10]). This is exemplified by *L. (L.) amazonensis*, which causes diffuse cutaneous leishmaniasis (CL), a true anergic form of tegumentary leishmaniasis. As described by Zerpa et al. (Chap. 8), an initial local lesion may be the origin of the spread of parasites by lymphatic and hematic means, with the subsequent inhibition of specific cellular immunity. *L. (L.) amazonensis* expose phosphatidylserine on their surface, a signal to host immune cells to phagocytose harmless agents. This "apoptotic mimicry" of *L. (L.) amazonensis* allows it to silently enter the mononuclear phagocytes in which it multiplies, without activating the immune system, and is thought to have evolved from a few parasites with altruistic behavior for the greater good of the overall parasite population in a host—a trait that was fixated throughout the parasite's evolution [11]. Once inside the host cell, the biggest threat to the parasite is the production of reactive oxygen and nitrogen species by this host cell. However, specific molecular features of *Leishmania* will protect it from these immune effector molecules (see Sect. 15.2.2). Additionally, *Leishmania* actively inhibits the host cell from producing these toxic molecules. One intriguing example of specific molecular features triggered by *Leishmania* is the parasite-mediated activation of the host cell phosphatase SHP-1 that will inhibit host cell pathways that would normally lead to mounting an adequate anti-parasite immune response, including the production of ROS and RNS (reviewed in [6]).

The parasite also affects the immune system at a more systemic level: infected macrophages can produce high levels of activating cytokines like tumor necrosis factor α, interleukin-1, or the down-modulatory interleukin-10 and transforming growth factor-β [9]. Additionally, the parasite contributes to confuse host cells from their functions, by expressing, for example, decoy molecules on their surface or excreting molecules into the host cell that disturb cell signaling pathways [6, 12]. By affecting physiological functions of the host cell, the parasite ultimately determines its own fate and that of the host (possibly causing disease). For example, the amastigote form of the parasite can influence the phosphorylation state of host molecules, as well as the activity of mitogen-activated protein kinases [13, 14]; additionally, it can inhibit the production of superoxide and nitric oxide by infected macrophages [15], as well as macrophage activation by interferon-γ [16, 17]. Last but not least, their presence inside macrophages is effective to prevent the action of interleukin-12 [10, 18]. All these events occur upon internalization of the parasite into the parasitophorous vacuole in the newly infected host cell. However, the signaling mechanisms and pathways that are essential to prevent amastigotes disappearance and to guarantee their survival and replication inside the parasitophorous vacuole are not yet fully elucidated [10]. Since chemotherapy, especially with drugs like antimonials that need a competent immune system to exert their action mechanism, decreases the parasite load in the patient, the host immune system might be able to retake control and mount an effective response [6]. An interesting example of how determinant the immune system is on the outcome of the disease is exemplified in visceral leishmaniasis (VL) and post-Kala-azar dermal leishmaniasis (PKDL). In this case, the continuous presence of T-regulatory cells and their selective recruitment to the infected sites play a critical role in the persistence of a residual parasite burden [19]. This continuous presence can result in visceral disease relapse after apparent cure or the development of post-Kala-azar dermal leishmaniasis [20]. On the other hand, MIL does not require a potent immune system to fully exert its action but has been reported to positively affect the immune status of VL patients [21]. Immunomodulation may thus also depend on the parasite load in the patient: a higher parasite load likely further boosts the immunomodulatory effects that are already intrinsic to any *Leishmania*.

2. Niches and Quiescence

   Leishmania parasites are ancient eukaryotic organisms that have evolved into a species that has a higher diversity and adaptive capacity than its hosts. This is especially important since intracellular parasitism (rare, obligatory) associates with challenges that if not conquered mean the senescence of an organism and at the end, of a species. Thus, parasites must invade host cells successfully and be able to escape or divert intracellular mechanisms that would otherwise clear intracellular invaders. The used mechanisms include programmed cell death either by apoptosis or autophagy and machineries related to the activation of immunity like production of reactive oxygen-nitrogen intermediates and lysosomal degradation [22]. Moreover, host surveillance such as Toll-like receptors and intracellular sensor systems impose an additional challenge that intracellular parasites must overcome [22].

   This means that a determinant factor that modulates the outcome of the invasion produced by *Leishmania* depends on its ability to infect alternative tissue niches within the vertebrate host, less accessible not only to the surveillance systems but also to drugs. In fact, amastigotes either remain in the original site of infection (as in the case of CL) or disseminate to other teguments (as in mucocutaneous leishmaniasis (MCL) or disseminated leishmaniasis) or to the viscera (as in VL) [20]. Interestingly, parasites are capable of invading sites other than those expected to be affected, albeit at lower levels and hereby remaining unnoticed. These places may function as hidden niches that can be (re)activated at a later moment. As such, *Leishmania* DNA has been described to be present in the bloodstream [23], in urine [24], and in apparently healthy mucosa [25] of patients suffering from cutaneous and MCL. More interestingly, as the *Leishmania* kinetoplast DNA degrades rapidly [26], this observed DNA should originate from living or recently dead parasites. In VL patients, parasites have been found in the blood [27] and skin as evidenced by the emergence of post-Kala-azar dermal leishmaniasis [28]. Interestingly, both MCL and PKDL are examples of leishmanial disease that appears many years after apparent cure. Yet the tissues and organs that are targeted are either very well perfused in the case of MCL (the mucosa) or not so perfused in the case of post-Kala-azar dermal leishmaniasis (the skin). This imposes a controversial discussion since hiding in a well-perfused tissue might result in a higher exposure to the immune system, while hiding in a less perfused organ could imply hiding from the immune system.

   Host cells of *Leishmania* include macrophages, neutrophils, and dendritic cells. Upon initial infection, neutrophils are recruited to the site of sand fly bite and survival within these cells will determine the fate of the parasite. Inside neutrophils, *Leishmania* parasites establish vacuoles that avoid lysosome fusion thus providing a protective environment for survival, if not replication. The parasite might also invade tissue cells like fibroblasts or Langerhans cells that support growth but are less able to clear parasites, perhaps due to the restricted microbicidal capacity of these host cells [22, 29]. Upon time, less neutrophils and more macrophages are infected, resulting in an active infection [22]. Amastigotes are thought to be metabolically less active than promastigotes. This is exemplified by the longer doubling time for axenic amastigotes (4 days) and amastigotes from lesions (12 days) compared to promastigotes (9 h) [30]. There is also experimental evidence showing that *Leishmania* amastigote transcription [31, 32] and translation [30, 33, 34] are significantly decreased in the amastigote stage, coinciding with lower levels of polysomes observed in axenic amastigotes [34]. Amastigotes also have a downregulated metabolism. The uptake and utilization of amino acids and glucose is diminished [35]. At the energetic level, amastigotes have lower levels of ATP than promastigotes, probably due to their attenuated oxidative phosphorylation and lower oxygen consumption [36]. Although such studies should also be performed on intracellular amastigotes, these results imply that amastigotes (or a subset of them) could be in a quiescent state, living on their reserves. This has been shown to be the case in the chronic stage of *L. (L.) major* infection in a murine model after the lesion is self-cured: persistent amastigotes could be divided into a population of amastigotes that grow at the same rate (60% of total) and another population that shows no evidence of active growth (40% of total) [37]. Interestingly, both dividing and non- or slow-dividing cells resided in the same host cells, being macrophages and dendritic cells. Quiescence among amastigotes or/and other niches of infection could be critical factors to hide from the host's immune system and eventually promote the parasite's survival.

15.2.1.2 Promastigotes

When a female sand fly bites an infected host, it will engorge *Leishmania* amastigotes and amastigote-containing cells together with the blood. These amastigotes will then transform to slender flagellated promastigotes in the abdominal midgut of the sand fly, where they need to overcome several bottlenecks in order to continue the parasite's life cycle. Alkalinization, changes in the midgut, and a decrease in the level of proteolytic activity promote the development of promastigotes in the gut of sand flies, meaning that growth and differentiation within the sand fly are linked to changes in pH, sugars, and among others, AA levels that might even modulate migration from preceding gastrointestinal portions into the cardio-esophageal valve [38, 39]. Gut epithelial cells of the sand fly will secrete a chitinous matrix that will form a peritrophic membrane encircling the blood meal and the engorged parasites, but promastigote-secreted chitinases will cause it to break down sooner than normal to allow migration of parasites to the anterior part of the sand fly [40]. To avoid excretion with the rest of the digested blood meal, promastigotes attach themselves to the microvillar lining by their flagellum (reviewed in [41]). Over the course of a few days, they will migrate to the thoracic midgut and the stomodeal valve and will undergo a transformation from dividing non-infective promastigotes into nondividing infective metacyclic promastigotes, a process called metacyclogenesis [42]. This process is of the utmost importance for *Leishmania*, as only these metacyclic parasites will be able to successfully initiate infection of the mammalian host later on. In the anterior midgut of the sand fly, promastigotes will secrete a gel-like substance to create a plug that fills the anterior midgut and extends to the stomodeal valve into the foregut [43]. When the sand fly wants to feed, it will first have to regurgitate to overcome the obstruction by the plug, hereby expelling (metacyclic) promastigotes into the skin of the host [44] and allowing the life cycle to continue. Breaking through the peritrophic membrane, attaching to the midgut to avoid excretion and metacyclogenesis are processes that are initiated by the parasite. However, while undergoing these developmental steps, the parasite needs to continuously defend against the sand fly immune system (reactive oxidative and nitrosative stress) and compete for resources with the normal flora of the sand fly.

This complex play of various *Leishmania* promastigote differentiation stages in the metabolically different locations in the sand fly and the complexity of sand fly studies itself have hampered our understanding of the exact detail of the fitness actors at play in this part of the parasite's life cycle. However, the in vivo transmission model of *Leishmania* development that has been developed using hamsters and *Lutzomyia longipalpis* sand flies [45] opens new avenues for fitness studies, including the promastigote stages in their natural environment.

15.2.2 Molecular Determinants

Leishmania belongs to the trypanosomatid family [46], implying among others two particularities: (1) at the genome expression level, all trypanosomatids transcribe their genes in long transcripts that contain several genes, also called polycistronic expression [47, 48], and (2) at the biochemical level, they use trypanothione (two glutathione molecules linked by spermidine) as the main regulator of their intracellular reducing environment and to detoxify the cell, in contrast to other eukaryotes that only have the less powerful glutathione [49–51]. To adapt to the poor flexibility of polycistronic expression, *Leishmania* developed multiple and unique, genomic adaptations among trypanosomatids. *Leishmania* is constitutively mosaic aneuploid, meaning that a given chromosome may have a different copy number, or somy, within different cells in the total population, going from monosomy (one copy of the chromosome) to pentasomy (five copies of the chromosome) [52–54]. Evidences that mosaic aneuploidy is also present at the amastigote stage were recently described in *L. (L.) donovani* parasites isolated from hamsters [55]. This creates a vast diversity within the population, providing a high adaptive capacity of the parasite population to various kinds of stress, including drugs [56]. This adaptive capacity provided by somy variation and SNP selection was exemplified when selecting for MIL resistance in vitro: first a somy reduction of chromosome 13 carrying the *L. (L.) donovani* MIL transporter (LdMT) appeared, secondly a LdMT deletion on one chromosome, and thirdly a nonlethal mutation on the second LdMT allele that provided good levels of resistance [57]. *Leishmania* also has other features related to genome flexibility. As such, the parasite can generate local gene copy number variations (CNV) through linear or circular extrachromosomal amplification, using direct and inverted DNA repeats [58], as well as intrachromosomal amplification (ICA) [59, 60]. This phenomenon was observed in in vitro laboratory parasites selected against many different drugs such as arsenic [61], antimonials [60, 62, 63], amphotericin-B (AMB) [64], methotrexate [65–67], and other non-antileishmanial drugs [68, 69], highlighting that this mechanism is one of the main adaptive features of *Leishmania*. Interestingly, 94% of the clinical isolates from the lowland of the Indian subcontinent assessed in a genetic diversity study showed two different ICAs, and two epidemic clones that carry these ICAs showed to have propagated successfully in India. Parasites not carrying these ICAs were also present in the Indian subcontinent but were restricted to one restricted area, the Nepalese highlands, and seemed less fit to spread throughout the Indian subcontinent [59]. This highlights once more the importance of ICAs for the parasite's adaptive capacity to survive environmental stress, be it the presence of drugs (Dumetz F. et al., unpublished data) or other selective pressures.

Such ICAs may indeed affect the metabolomic profile of the parasite: the same clinical antimonial-resistant (SSG-R) parasites from Nepal that carried an ICA at the level of argininosuccinate synthase [59], the enzyme catalyzing the transformation of citrulline in argininosuccinate, also displayed a significant increase in their argininosuccinate content as identified by metabolome studies [70]. Notably, argininosuccinate is a metabolite that is part of the urea cycle and is, among others, a basic component of the pathway that eventually leads to putrescine and trypanothione synthesis.

Trypanothione is the main active defense system of *Leishmania* against reactive oxygen and nitrogen stress (ROS/RNS). The parasite will encounter oxidative and nitrosative stress throughout its life cycle as a promastigote and an amastigote, but ROS/RNS can also be induced by drugs such as pentavalent antimonials (Sb^V), for example. *Leishmania*'s redox system consists of a cascade of enzymes with trypanothione as the main reducing agent (Fig. 15.3). When ROS and RNS are detoxified by members of this cascade (either trypanothione itself (H_2O_2 [71],

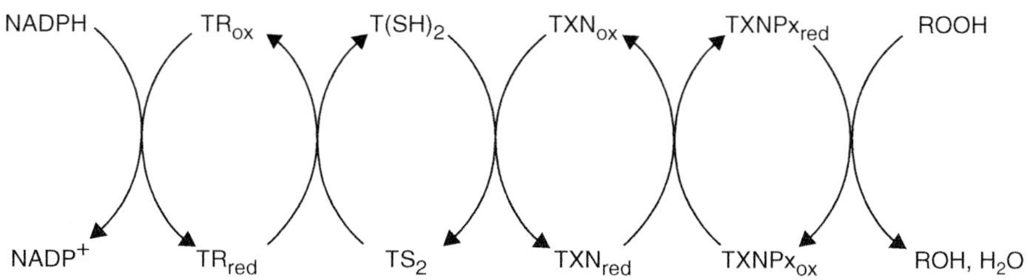

Fig. 15.3 The NADPH-dependent redox cascade with trypanothione (TSH2) as the central reductant. *TR* trypanothione reductase, *TXN* tryparedoxin, *TXNPx*, tryparedoxin peroxidase

NO. [72, 73]), tryparedoxin or tryparedoxin peroxidase (H_2O_2 [74, 75], $ONOO^-$ [74, 76], H_2O_2 + NO. [77]), the flavoenzyme trypanothione reductase (TR) will replenish the pool of reduced trypanothione (T[SH]2) from oxidized trypanothione (T[S]2) using NADPH as an electron donor (Fig. 15.3). TR is therefore thought to be a central and very important enzyme for the intracellular survival of *Leishmania* [78–80].

When promastigotes were put under pressure with Sb^{III}, which is the toxic reduced form of Sb^{V} that is the core component of SSG, many intermediates of the trypanothione pathway were found to be upregulated [70, 81–83], confirming earlier investigations carried out at the protein level in different *L. (L.) donovani* strains from the Indian subcontinent where an upregulation of the enzymes of the thiol pathway was observed in SSG-R *L. (L.) donovani* [84] and *L. (L.) infantum* [85] parasites.

Studies on the metabolomic profile of MIL-resistant *L. (L.) donovani* showed a large modification of the lipid composition, probably due to the mechanism of action of MIL on the membrane, but also an increase of the metabolites implicated in the thiol pathway [57, 86]. The lipid composition is also found to be changed in parasites resistant to drugs without a clear link to lipid metabolism: unsaturated phosphatidylcholine lipids and phosphatidylethanolamine were increased in SSG-R versus SSG-S parasites, suggesting an extensive change in the membrane composition of SSG-R parasites [87].

Interestingly, studies on in vitro selection of resistance against a combination of drugs identified that this requires different adaptations compared to resistance against just either one of the drugs in that combination [82]. One common factor, however, was the pivotal role of pathways regulating protection against oxidative stress and membrane composition [82]. These molecular traits of *Leishmania* are thus considered to be important molecular determinants of the parasite's adaptive capacity and therefore also its fitness.

15.2.3 Epidemiological Determinants

When talking about epidemiological determinants that might affect the fitness of *Leishmania*, we have to be aware of the fact that nowadays, and regarding the spectrum of leishmaniasis, CL and VL have undoubtedly a wider geographical distribution than before; additionally, the higher leishmaniasis incidence is a result of risk factors that can also be determinant for changes in fitness and virulence of the parasite [88]. In fact, changes in environmental conditions (i.e., temperature), human behavior (nutrition, misuse of drugs), immunogenic patient profile (co-infection with HIV), and genetic factors (parasite species) might determine the fate of the parasite-host interaction affecting directly the interplay between these two fundamental actors in the development of the disease.

Regarding climate models, it is well recognized that there will be a global average increase of air temperatures from 1 °C to 4 °C by 2100 [89]—more than ever before [90]. A consequent modification of species occurrence and distribution will occur with up to 37% of all existent species "committed to extinction" due to climate change [91]. For parasites, an increase in organisms' virulence and transmission rates are the most commonly described responses to rising temperatures [92, 93], implying that some parasites might become more successful and increase their fitness compared to earlier times. Examples exist in the bacterial parasite *Pasteuria ramosa* [94] or in the tapeworm *Schistocephalus solidus* with increased castration rates of *Daphnia magna* or growth rates in three-spined sticklebacks at higher temperatures [95]. Besides direct effects on hosts and/or parasites, if the global warming changes parasite virulence and/or host resistance in an asynchronous way, the interactions among both organisms will also be affected [96]. Environmental changes can therefore induce adaptive peaks (different host species) to occur closer in time, easing the transfer to a new host by a proportion of the parasite population, afterward followed by the rest of the same population. These changes in the environment can also facilitate the invasion of more species that then become potential hosts suitable for ecological fitting of the parasite [90].

The importance of the immune system status for disease development can be understood by evaluating the effect of co-infection between HIV and leishmaniasis. In fact, HIV is changing the nature of the human infection, the response to treatment, and the epidemiology of leishmaniasis in different geographical areas including Africa, Europe, and Brazil. HIV patients are immunosuppressed, and treatment of VL in such patients requires a long course of treatment, resulting in an increased risk of relapse and a high chance on the development of drug resistance. Further suppression of the immune system by HIV exacerbates the situation. Both diseases drive each other at least in experimental settings [97], and patients suffering from both diseases simultaneously have higher parasite burdens and weaker or absent immune responses. This causes them to respond slowly to treatment with antimonials (SSG) [98], and their clinical improvement does not correlate with parasite clearance from splenic aspirate smears, resulting in about 60% of the patients showing relapse within 1 year, and with any antileishmanial drug used [99], with secondary resistance being common to all of them [100, 101]. As under experimental settings, the vector *Phlebotomus ariasi*, common in southern Europe, can become infected by feeding on HIV-*Leishmania* co-infected patients [97]. Questions arise about whether or not these patients can provide a human reservoir prone to modulate the epidemiology of the disease in southern Europe. This is a fundamental question since without HIV, VL patients are not infectious to this sand fly. As the courses of drug treatment should be increased concomitantly, an open question is whether this condition can lead to the emergence of primary drug resistance [102].

In the American leishmaniasis context, it is interesting to note that parasites of the *Viannia* subgenus may be infected by a specific virus (*Leishmania* RNA virus-1 or LRV1) that successfully impairs the host immune response to *Leishmania* and promotes parasite persistence [103]. In *L. (V.) braziliensis*, the presence of the RNA virus was shown to be associated to the development of mucosal disease [104] and even treatment failure [105]. The importance of viral or bacterial endosymbiosis and how this may shape the genome and the fitness of the parasite remains to be further studied.

This takes us to the discussion that among the factors that are determinant to the outcome of an initial infection with *Leishmania*, the species constitutes one of the strongest predictors for the development of a given clinical form of disease. This is clearly exemplified in American leishmaniasis. *L. (V.) braziliensis* and *L. (L.) amazonensis* infections lead mostly to tegumentary forms of disease, while *L. (L.) infantum* has the potential to induce visceral disease. Even more, strain differences within the same species might also be associated with a given clinical form of disease [9]. As described elsewhere [20], in Peru, patients infected with *L. (V.) guyanensis* are generally more responsive to SSG than patients infected with *L. (V.) braziliensis* [106], while the opposite result was observed in Brazil [107]. In Venezuela, diffuse CL patients infected with either *L. (L.) amazonensis* or *L. (L.) mexicana* comprise a poor response to SSG [108] (Chap. 8 by Zerpa et al.). These results reveal the important role of the different epidemiological and genetic diversity of New World *Leishmania* on treatment outcome of American tegumentary leishmaniasis.

A final determinant that we will briefly discuss relates to the fact that the response to treatment in the New World differs significantly from that in the Old World, an issue that further reflects the multifactorial character of the disease. As previously mentioned, drug, host, and parasite factors contribute to the final outcome [109]. Old World leishmaniasis has a more homogeneous therapeutic outcome, except when caused by *L. (L.) aethiopica*, compared to New World leishmaniasis, where therapeutic responses are mixed and unpredictable. This implies that treatment guidelines have to be evaluated on a global basis, taking into account the vast differences between Old and New World leishmaniasis [106, 109]. This also implies a different rationale for researchers looking for determinant factors that contribute to treatment outcome, as drug resistance could be partially responsible for treatment failure, but additional factors like the epidemiological complexity of the disease due to the diversity of etiological agents and their (epi-) genetic features may dramatically complicate the panorama, especially for American tegumentary leishmaniasis treatment. However, other issues can determine the response to treatment and we will briefly refer to them herein.

Substandard product levels constitute the inevitable consequence of inadequate local regulation of pharmaceutical companies and the lack of good manufacturing practices in many countries [110]. Drugs with substandard concentrations of the active ingredient determine a poor response to treatment and can increase the risk of spread of drug-resistant (drug-R) pathogens [111]. Similarly, inadequate dosage (even higher dosage than needed) is also a positive factor that could be a selective factor for the selection of resistant parasites occurring in a patient [20, 112].

Additionally, poor hygienic measures and transmission control in clinics and hospitals in the developing world, the natural niche for leishmaniasis, lead to environmentally suboptimal disposition of the medicaments. The threat from these (and other) released medicaments is illustrated by the existence of a large reservoir of resistance genes present in the human microflora. These genes could serve as donors for the transfer of genes to human pathogens by means of horizontal gene transfer.

Little is known about the role of horizontal gene transfer in poor response to drugs in parasites like *Leishmania*. Nevertheless, we cannot exclude the relevance that this mechanism might have in this parasite [113].

15.3 Leishmaniasis Treatment Failure and Fitness

15.3.1 Fitness Cost or Not

The presence of drugs has a dramatic impact on parasite fitness and therefore also on the equilibrium that exists within the parasite populations in a region where drugs are deployed. In fact, although "fitness cost" is the most common feature observed in nature as a result of drug resistance expression, "fitness compensation" is also observed in such circumstances (reviewed in [114]).

Parasite populations under drug pressure can result in either the selection of pre-existing resistant variants that were circulating in the field or in the induction of new variants emerging under drug pressure. The level of drug pressure will play an important role in the emergence and/or spread of drug-R parasites. As mentioned earlier, both substandard drug levels and higher dosages than what is required may result in a high selective pressure for pre-existing drug-R parasites [115].

Drug pressure on a parasite population may result in parasites with a drug-R trait that may have an originally lower relative fitness compared to others in natural no-drug conditions. However, they will become more successful than drug-sensitive (drug-S) parasites in drugged conditions. This capacity to better withstand drugs may be related to genetic factors that prevent the drug from acting on its target or to factors that enable the parasite to more easily adapt to drug pressure compared to its counterparts. As discussed earlier in this chapter, these factors can be species dependent. Assessing the fitness of drug-S and drug-R parasites can therefore shed more light on the life span of a drug, as a rise in drug-R parasites leads to a more frequent appearance of treatment failure, which may eventually lead to the drug being too inefficacious to justify further use. However, the acquisition of a drug-R trait generally comes at a cost [114]. This fitness cost will make the drug-R parasite less fit compared to wild-type parasites when the drug pressure on the parasite population is low or even absent [20]. Since most *Leishmania* parasites hide in reservoirs that are generally untreated, such as asymptomatics or PKDL-patients for *L. (L.) donovani* and animal reservoirs for *L. (V.) braziliensis*, the relative fitness of drug-S and drug-R parasites in no-drug conditions will have a major impact on the speed by which drug-R parasites will spread in a parasite population.

Such a fitness assessment is hard to make and requires adequate in vitro and in vivo tools and, even more important, a set of *Leishmania* strains that are representative for the region of interest. In the context of drug resistance studies, clinical drug-R strains or strains from treatment failure patients may not always be available and therefore require substitution by strains that are made resistant in the lab. Although the resistance mechanisms generated in the lab may differ from those in the field, they do provide insights into how a drug works and what the parasite's options are to become resistant. The fitness effects related to these resistance mechanisms, however, may play out very different when induced in an in vitro context compared to being naturally generated in a patient. This is mainly due to the lack of immune factors, different host cell niches, and other fitness determinants described earlier that are missing in a simplified in vitro context.

In the last few years, there was an appreciable upsurge of fitness studies in the context of both natural and in vitro drug resistance. Comparing a set of clinical *L. (L.) donovani* SSG-S and SSG-R strains, an increase in metacyclogenesis [116] and an increased fitness in infected mice were observed for SSG-R lines compared to SSG-S lines [117, 118]. Since SSG interacts with the immune system to reduce the parasite load in the patient, it was hypothesized that the parasite adapted to the host immune system while adapting to the drug, leading to the traits that are suggestive of a higher fitness compared to wild-type drug-S strains [6, 20]. This was further substantiated by several studies that identified specific host manipulation skills of clinical SSG-R strains that can be directly related to the increased fitness of these strains in vivo (reviewed in [20]). Interestingly, the majority of these clinical SSG-R strains isolated from SSG-treatment failure patients belong to a specific genetic group of parasites (ISC5) that has expanded significantly in the Indian subcontinent, even at times when SSG was no longer the first-line treatment [119]. This observation was confirmed by mathematical modeling studies showing that SSG-R strains must have had an increased fitness compared to SSG-S strains in order to explain their success in the field [120, 121]. Recent reports, however, indicate that the genotype related to SSG-R parasites is decreasing in prevalence since 2013 [122], possibly due to other treatment options (such as MIL) wiping out genetic diversity and reshaping the landscape of *Leishmania* genotypes circulating in the field. Initially, the higher fitness that was described for *L. (L.) donovani* SSG-R versus SSG-S strains was thought to be a unique case due to the combination of a highly adaptive parasite and a drug that interacts closely with the immune system. However, when testing clinical *L. (L.) donovani* strains from patients that failed the more recently introduced MIL treatment, an increased metacyclogenesis that translated into higher in vitro infection levels was again observed—this despite the lack of a clear in vitro miltefosine-resistant (MIL-R) phenotype in these clinical lines [123]. Phenotypes linked to an increased fitness might thus be a common trait of *L. (L.) donovani* parasites that are able to overcome drug treatment. This is further supported by studies on *L. (L.) donovani* lines that were in vitro generated to be resistant to various single and combination treatment regiments, showing a generally higher competitive fitness of resistant lines compared to their wild-type [124]. These studies identified a higher promastigote survival rate in conditions of starvation, a higher tolerance to heat shock and pH stress, and an increased survival rate in in vitro macrophages [124]. Although some of these traits seemed to be absent in some drug-R lines (combinations with amphotericin-B and the MIL-R line), there is a general trend toward a fitness increase of *L. (L.) donovani* drug-R lines, even when generated in vitro.

However, what is true for one *Leishmania* species is not necessarily true for another. In the closely related *L. (L.) infantum* species, for example, studies on in vitro-induced MIL-R lines did not reveal similar trends [125]. These strains did not display the increased metacyclogenesis rate and even showed a lower in vitro survival rate than the wild-type control, contrasting with the *L. (L.) donovani* findings described earlier [123]. The induced *L. (L.) infantum* MIL-R line showed a similar susceptibility to nitrosative stress as the wild-type control but showed a lower capacity to induce IL-10 production in in vitro-infected macrophages [125]. While it is hard to compare fitness results between species and experimental designs due to differences in protocols, some studies have compared the effect of several drugs using the same (model) system. While the in vitro-induced MIL-R line did not show a difference or a lower in vitro infection level, the same study reports that in vitro-induced paromomycin-R lines did show a better in vitro and in vivo growth at the amastigote level and a higher tolerance for nitrosative stress, without a clear influence of metacyclogenesis as defined in their setup. Induced IL-10 levels remained unchanged in paromomycin-R vs wild-type *L. (L.) infantum* lines [125]. Also, in *L. (L.) majo* made resistant in vitro to MIL, the MIL-R strain proliferated at comparable rates as wild-type parasites and exhibited similar responses regarding programmed cell death. Interestingly, metacyclogenesis was increased in MIL-resistant *L. (L.) major*, although they proved to be less virulent both in vitro and in vivo. These results thus suggest that development of experimental resistance to MIL did not lead to an increased competitive fitness in *L. (L.) major* [126].

Assessing the fitness of drug-R or treatment failure parasites reaches an even higher level of complexity in the case of American tegumentary leishmaniasis, which comprises infection of many different *Leishmania* species. Here, treatment outcome is largely affected by the infecting (tolerant) species, although it is not clear if true parasite adaptation to the drug exists [20]. In fact, as previously mentioned, *L. (V.) guyanensis*-infected patients in Peru respond better to SSG than those infected with *L. (V.) braziliensis* [106], but the opposite occurs for Brazilian patients [107]. Venezuelan *L. (L.) amazonensis*- or *L. (L.) mexicana*-infected diffuse CL patients also often show a poor response to SSG [127–130] (Chap. 8 by Zerpa et al.). Another complicating factor for New World leishmaniasis is the existence of hybrids. The analysis of *L. (V.) braziliensis-L. (V.) peruviana* hybrids suggests that they display a growth capacity (growth rate and cell density at stationary phase) similar to that of wild-type *L. (V.) peruviana* parasites but significantly lower than that of *L. (V.) braziliensis*, thus suggesting a lower fitness of the hybrids in comparison to the *L. (V.) braziliensis* wild-type parasites [131]. How these hybrids relate to parasite fitness in the context of drug resistance and treatment failure requires more research. However, it is clear that this vast variety of *Leishmania* species and their different epidemiological and genetic context in New World *Leishmania* has a major impact on treatment outcome and makes an assessment of the fitness effects of drug resistance and tolerance in New World *Leishmania* species even more complicated than for Old World *Leishmania* species.

Another factor that affects parasite fitness and treatment outcome in New World leishmaniasis is superinfection of *L. (V.) braziliensis* by the *Leishmania* RNA virus (LRV) [132]. Taylor et al. in 1998 developed a mathematical model explaining that a lower infectivity of superparasitized parasites might exist in contrast to the potential benefit of being infected by an organism that encodes functions as resistance to antibiotics. This is common in nature as pathogens might be infected either by plasmids, viruses, or parasites [133]. *Leishmania* superinfected with LRV has been associated with failure of SSG-treatment, most likely due to RNA factors that modulate the host's immune system, ensuring survival of *L. (V.) braziliensis* and therefore also the virus it carries. Although superinfection of *Leishmania* by a virus might induce a fitness cost in the absence of drugs, it seems to result into an advantage when the patient in which it resides is being treated.

The previously discussed quiescent-like state among amastigotes could affect their drug tolerance compared to promastigotes: if the drug depends on the action of a metabolic pathway that is downregulated in a quiescent stage or if the drug enters the cell through a transporter and this transporter is downregulated in a quiescent stage, this will result in an increased tolerance to the drug for the population with a quiescent phenotype [134]. For example, *L. (L.) amazonensis* and *L. (L.) mexicana* amastigotes have shown to be more tolerant to treatment with trivalent antimonials (Sb^{III}), which enter the cell through the aquaglyceroporin 1 transporter, compared to their respective promastigotes [135]. *L. (L.) mexicana* amastigotes are also more tolerant to exposure to pentamidine, a drug which interferes with the synthesis of DNA and the morphology of kinetoplast DNA [136, 137]. Larger studies comparing the IC_{50}s of promastigotes and amastigotes should be performed in order to extend these segregated observations. From another perspective, quiescence might explain the survival of a small population of amastigotes inside the tissue that, because of their low metabolic status, are drug tolerant or indifferent even when the majority of the population is susceptible.

Host tissue niche preference may also affect treatment outcome, as drug distribution might differ between different niches, possibly resulting in sublethal or irregular drug exposure of amastigotes and apparent clinical cure of the patient. Such niches might then serve as foci from where infection can spread again and result in PKDL or MCL [20, 138], as described earlier (Chap. 8 by Zerpa et al.). The presence of other niches of infection could explain the survival of *Leishmania* despite treatment of the host, but the fact that in most of the cases the amastigotes remain in the original lesion indicates that quiescence could be an important strategy of *Leishmania* to survive the drug pressure and the immune system.

Oversimplification of the process by which drug-R lines are selected in nature sometimes leads to the difficulty by which in vitro or in vivo experimental resistance can be attained being interpreted as an argument against fitness benefits in natural drug-R lines [125]. It is important to stress that in the field, drug-R phenotypes are selected in the context of immune systems (which are different than those of common in vivo VL models), transmission through sand flies, and additional challenges for the parasite that are not present in in vitro or in vivo selection systems in the lab. When the parasite is developing drug resistance in the field, these natural bottlenecks will also serve as positive filters for those drug-R parasites that have the best combination of traits to survive all bottlenecks. This series of bottlenecks gives the opportunity to rare variants to become successful and may result in different traits emerging in natural drugged populations compared to lab parasite populations.

15.3.2 Drug Discoveries and Control Perspectives

The treatment of leishmaniasis has long relied on drugs based on ancient compounds with known curative but also toxic effects, such as SSG. MIL and AMB, the two most recent additions to the antileishmaniasis drug arsenal, were originally developed as antineoplastic or antifungal compounds, respectively. New compounds are in the pipeline but are not likely to evolve into an actual therapy option in the next few years to come. The search for new drugs against leishmaniasis, being a neglected tropical disease, has been hampered by the lack of public and private interest ever since the parasite was discovered. However, the lack of funding that this entailed was not the only limiting factor for drug discovery projects. The intracellular lifestyle of *Leishmania* amastigotes, the only life stage that reproduces in the host, severely complicated the development of large-scale leishmaniasis drug discovery pipelines as intracellular amastigotes could not be easily cultured in vitro. Methods to grow amastigotes extracellularly (axenic amastigotes), which are to certain extents similar to the naturally occurring intracellular amastigotes, have been developed and further optimization of these culture protocols recently allowed high-throughput screening with a high predictability of leishmanicidal intracellular activity [139]. In addition, recent efforts developed an in vitro model that allowed replication of actual intracellular amastigotes in THP-1 cells [140], providing a model that is much closer to natural infections than the axenic models, allowing *Leishmania* to grow intracellularly, invade new host cells, etc. This intracellular model may therefore also allow assessing these fitness determinants at a higher throughput. Evaluating the fitness of natural wild-type parasites and parasites resistant to experimental compounds can provide a better insight into the effect that introducing a drug in a certain geographical context may have on the local parasite population and the spread of a possible resistant phenotype. Such studies are rarely performed at early stages of leishmaniasis drug discovery but could now be encouraged by the development of such higher-throughput assays.

All monotherapies but one, AMB, have succumbed to a rise in treatment failure rates several years after their introduction. This is due to treatment failure-inducing parasites having a competitive fitness over wild-type parasites when under treatment. This not only entails that they continue to replicate in the host in the presence of the drug, but also that they are able to spread to the vector, undergo the different promastigote development stages, and eventually infect new hosts. A better knowledge on the factors important for parasite fitness in both the mammalian host and the vector might contribute to the development of innovative treatment regimens that disturb this fitness advantage of resistant parasites in a parasite population under treatment. One could consider treating patients with a combination of one or more drugs that aim to cure the patient and one other compound that has the sole purpose of allowing easy emergence of a specific resistance mechanism that induces a fitness defect at the level of promastigote development in parasites that somehow survived exposure to the curative partner drug. This would prevent drug-R parasites to undergo full development in the vector, impeding their transmission to new hosts and preventing the spread of drug-R parasites. This is of course easier said than done, as it requires the identification of factors that are important for parasite development in both the host and the vector and subsequently the identification of a chemical compound able to induce a specific genetic change in the parasite that results in resistance in the mammalian host and impedes promastigote development in the vector. Nevertheless, innovative treatment schemes such as these exploit the parasite's ability to become drug-R but provide the benefit of prolonging the life span of the other drugs that are part of the combination treatment regimen. Designing more treatment schemes that directly affect parasite fitness in wild-type and drug-R parasites may be a way forward in rational drug design pipelines. In the *Leishmania* field, the importance of studying drug-R parasites in drug discovery projects has only recently gained more attention [141].

However, rational drug design and drug use are only two of several important aspects in leishmaniasis control. As such, the Kala-azar Elimination Program in the Indian subcontinent relied on early diagnosis, adequate treatment, and vector control. While early diagnosis and adequate treatment are pivotal to cure patients, mathematical modeling has shown that it has only little effect on eventual control of the disease at the population level, i.e., reducing infection incidence [120]. Building further upon this mathematical model, studies have estimated that 10 years of sustained suboptimal insecticidal residual spraying would be required to reach the VL elimination goal [142]. These studies highlight that transmission is a major contributor to the fitness of *Leishmania* and emphasize the importance of affecting parasite development in the vector, either by killing the vector itself or by preventing parasite development in this vector. A better understanding of the fitness factors related to promastigote development and how they can be affected may thus provide powerful new tools for leishmaniasis control.

15.4 Conclusion

Leishmania is a parasite with remarkable adaptive skills, posing major challenges for its control in endemic areas. Fitness of drug-R versus drug-S parasites plays an important role in shaping future parasite populations, and understanding the processes involved is pivotal to allow the design of new treatment strategies that defy the parasite's capacity to render new drugs useless through the development of drug resistance. It is encouraging that fitness studies are more and more performed in the context of drug resistance. However, it is often difficult to compare results between species or even between studies on the same species due to varying epidemiological and genetic contexts. Current advances in genome editing (CRISPR/Cas9) in combination with a detailed knowledge of resistance mechanisms should now allow to create genetically paired clinical isolates in the lab with and without these resistance determinants. This is well exemplified by a combinatorial genetic modeling study that focused on a quadruple *Plasmodium* mutant resistant to chloroquine [143]. Through the creation of a battery of genetically engineered mutants, fitness studies on each of these and implementation of all data into a mathematical model, the mutational trajectory that led to this successful mutant could be reconstructed. Comparing such genetic mutants or revertants with their wild types will provide more insight into the exact fitness consequences of the phenotype, how it might have evolved and allow a more straightforward comparison of results obtained in different systems.

References

1. Bates M, Wrin T, Huang W, Petropoulos C, et al. Practical applications of viral fitness in clinical practice. Curr Opin Infect Dis. 2003;16:11–8.
2. Geretti AM. The clinical significance of viral fitness. J HIV Ther. 2005;10:6–10.
3. Quiñones-Mateu ME, Arts EJ. Virus fitness: concept, quantification, and application to HIV population dynamics. Curr Top Microbiol Immunol. 2006;299:83–140.
4. Andino R, Domingo E. Viral quasispecies. Virology. 2015;479–480:46–51.
5. Natera S, Machuca C, Padrón-Nieves M, Romero A, et al. *Leishmania* spp.: proficiency of drug-resistant parasites. Int J Antimicrob Agents. 2007;29:637–42.
6. Vanaerschot M, Decuypere S, Berg M, Roy S, et al. Drug-resistant microorganisms with a higher fitness – can medicines boost pathogens? Crit Rev. Microbiol. 2012;39:1–11.
7. Semenza JC, Rocklöv J, Penttinen P, Lindgren E. Observed and projected drivers of emerging infectious diseases in Europe. Ann N Y Acad Sci. 2016;1382:73–83.
8. Debrabant A, Nakhasi H. Programmed cell death in trypanosomatids: is it an altruistic mechanism for survival of the fittest? Kinetoplastid Biol Dis. 2003;2:7.
9. Gollob KJ, Viana AG, Dutra WO. Immunoregulation in human American leishmaniasis: balancing pathology and protection. Parasite Immunol. 2014;36:367–76.
10. Kima PE. The amastigote forms of *Leishmania* are experts at exploiting host cell processes to establish infection and persist. Int J Parasitol. 2007;37:1087–96.
11. El-Hani C, Borges VM, Wanderley JLM, Barcinski MA. Apoptosis and apoptotic mimicry in *Leishmania*: an evolutionary perspective. Front Cell Infect Microbiol. 2012;2:96.
12. Bogdan C, Röllinghoff M. The immune response to *Leishmania*: mechanisms of parasite control and evasion. Int J Parasitol. 1998;28:121–34.
13. Boggiatto PM, Jie F, Ghosh M, Gibson-Corley KN, et al. Altered dendritic cell phenotype in response to *Leishmania amazonensis* amastigote infection is mediated by MAP kinase, ERK. Am J Pathol. 2009;174:1818–26.
14. Xin L, Li K, Soong L. Down-regulation of dendritic cell signaling pathways by *Leishmania amazonensis* amastigotes. Mol Immunol. 2008;45:3371–82.
15. Van Assche T, Deschacht M, da Luz RAI, Maes L, et al. *Leishmania*–macrophage interactions: Insights into the redox biology. Free Radic Biol Med. 2011;51:337–51.
16. Matte C, Descoteaux A. *Leishmania donovani* amastigotes impair gamma interferon-induced STAT1alpha nuclear translocation by blocking the interaction between STAT1alpha and importin-alpha5. Infect Immun. 2010;78:3736–43.
17. Abu-Dayyeh I, Hassani K, Westra ER, Mottram JC, et al. Comparative study of the ability of *Leishmania mexicana* promastigotes and amastigotes to alter macrophage signaling and functions. Infect Immun. 2010;78:2438–45.
18. Ruhland A, Kima PE. Activation of PI3K/Akt signaling has a dominant negative effect on IL-12 production by macrophages infected with *Leishmania amazonensis* promastigotes. Exp Parasitol. 2009;122:28–36.

19. Rai AK, Thakur CP, Singh A, Seth T, et al. Regulatory T cells suppress T cell activation at the pathologic site of human visceral leishmaniasis. PLoS One. 2012;7:e31551.
20. Vanaerschot M, Dumetz F, Roy S, Ponte-Sucre A, et al. Treatment failure in leishmaniasis: drug-resistance or another (epi-) phenotype? Expert Rev Anti Infect Ther. 2014;12:937–46.
21. Ghosh M, Roy K, Roy S. Immunomodulatory effects of antileishmanial drugs. J Antimicrob Chemother. 2013;68:2834–8.
22. David Sibley L. Invasion and intracellular survival by protozoan parasites. Immunol Rev. 2011;240:72–91.
23. Guevara P, Rojas E, Gonzalez N, Scorza JV, et al. Presence of *Leishmania braziliensis* in blood samples from cured patients or at different stages of immunotherapy. Clin Diagn Lab Immunol. 1994;1:385–9.
24. Veland N, Espinosa D, Valencia BM, Ramos AP, et al. Polymerase chain reaction detection of *Leishmania* kDNA from the urine of Peruvian patients with cutaneous and mucocutaneous leishmaniasis. Am J Trop Med Hyg. 2011;84:556–61.
25. Figueroa RA, Lozano LE, Romero IC, Cardona MT, et al. Detection of *Leishmania* in unaffected mucosal tissues of patients with cutaneous leishmaniasis caused by *Leishmania* (*Viannia*) species. J Infect Dis. 2009;200:638–46.
26. Prina E, Roux E, Mattei D, Milon G. *Leishmania* DNA is rapidly degraded following parasite death: an analysis by microscopy and real-time PCR. Microbes Infect. 2007;9:1307–15.
27. Deborggraeve S, Boelaert M, Rijal S, De Doncker S, et al. Diagnostic accuracy of a new *Leishmania* PCR for clinical visceral leishmaniasis in Nepal and its role in diagnosis of disease. Trop Med Int Heal. 2008;13:1378–83.
28. Mukhopadhyay D, Dalton JE, Kaye PM, Chatterjee M. Post kala-azar dermal leishmaniasis: an unresolved mystery. Trends Parasitol. 2014;30:65–74.
29. Bogdan C, Donhauser N, Döring R, Röllinghoff M, et al. Fibroblasts as host cells in latent leishmaniosis. J Exp Med. 2000;191:2121–30.
30. Kloehn J, Saunders EC, O'Callaghan S, Dagley MJ, et al. Characterization of metabolically quiescent *Leishmania* parasites in murine lesions using heavy water labeling. PLOS Pathog. 2015;11:e1004683.
31. Alcolea PJ, Alonso A, Gómez MJ, Moreno I, et al. Transcriptomics throughout the life cycle of *Leishmania infantum*: High down-regulation rate in the amastigote stage. Int J Parasitol. 2010;40:1497–516.
32. Michel G, Ferrua B, Lang T, Maddugoda MP, et al. Luciferase-expressing *Leishmania infantum* allows the monitoring of amastigote population size, in vivo, ex vivo and in vitro. PLoS Negl Trop Dis. 2011;5:e1323.
33. Biyani N, Madhubala R. Quantitative proteomic profiling of the promastigotes and the intracellular amastigotes of *Leishmania donovani* isolates identifies novel proteins having a role in *Leishmania* differentiation and intracellular survival. Biochim Biophys Acta. 2012;1824:1342–50.
34. Cloutier S, Laverdière M, Chou M-N, Boilard N, et al. (2012) Translational control through eIF2alpha phosphorylation during the *Leishmania* differentiation process. PLoS One 7: e35085.
35. Saunders EC, Ng WW, Kloehn J, Chambers JM, et al. Induction of a stringent metabolic response in intracellular stages of *Leishmania mexicana* leads to increased dependence on mitochondrial metabolism. PLoS Pathog. 2014;10:e1003888.
36. Mondal S, Roy JJ, Bera T. Characterization of mitochondrial bioenergetic functions between two forms of *Leishmania donovani* – a comparative analysis. J Bioenerg Biomembr. 2014;46:395–402.
37. Mandell MA, Beverley SM. Continual renewal and replication of persistent *Leishmania major* parasites in concomitantly immune hosts. Proc Natl Acad Sci USA. 2017;114:E801–10.
38. Dillon RJ, Ivens AC, Churcher C, Holroyd N, et al. Analysis of ESTs from *Lutzomyia longipalpis* sand flies and their contribution toward understanding the insect-parasite relationship. Genomics. 2006;88:831–40.

39. Diaz E, Zacarias AK, Pérez S, Vanegas O, et al. Effect of aliphatic, monocarboxylic, dicarboxylic, heterocyclic and sulphur-containing amino acids on *Leishmania* spp. chemotaxis. Parasitology. 2015;142:1621–30.
40. Schlein Y, Jacobson RL, Shlomai J. Chitinase secreted by *Leishmania* functions in the sandfly vector. Proceedings Biol Sci. 1991;245:121–6.
41. Sacks DL, Melby PC. Animal models for the analysis of immune responses to leishmaniasis. Curr Protoc Immunol. 2001. https://doi.org/10.1002/0471142735.im1902s108.
42. da Silva R, Sacks DL. Metacyclogenesis is a major determinant of *Leishmania* promastigote virulence and attenuation. Infect Immun. 1987;55:2802–6.
43. Rogers ME, Chance ML, Bates PA. The role of promastigote secretory gel in the origin and transmission of the infective stage of *Leishmania mexicana* by the sandfly *Lutzomyia longipalpis*. Parasitology. 2002;124:495–507.
44. Bates PA. Transmission of *Leishmania* metacyclic promastigotes by phlebotomine sand flies. Int J Parasitol. 2007;37:1097–106.
45. Aslan H, Dey R, Meneses C, Castrovinci P, et al. A new model of progressive visceral leishmaniasis in hamsters by natural transmission via bites of vector sand flies. J Infect Dis. 2013;207:1328–38.
46. Barrett MP, Burchmore RJ, Stich A, Lazzari JO, et al. The trypanosomiases. Lancet. 2003;362:1469–80.
47. Sunkin SM, Kiser P, Myler PJ, Stuart K. The size difference between *Leishmania major* friedlin chromosome one homologues is localized to sub-telomeric repeats at one chromosomal end. Mol Biochem Parasitol. 2000;109:1–15.
48. Johnson PJ, Kooter JM, Borst P. Inactivation of transcription by UV irradiation of *T. brucei* provides evidence for a multicistronic transcription unit including a VSG gene. Cell. 1987;51:273–81.
49. Fairlamb AH, Blackburn P, Ulrich P, Chait BT, et al. Trypanothione: a novel bis(glutathionyl) spermidine cofactor for glutathione reductase in trypanosomatids. Science. 1985;227:1485–7.
50. Mukhopadhyay R, Dey S, Xu N, Gage D, et al. Trypanothione overproduction and resistance to antimonials and arsenicals in *Leishmania*. Proc Natl Acad Sci U S A. 1996;93:10383–7.
51. Krauth-Siegel RL, Meiering SK, Schmidt H. The Parasite-Specific Trypanothione Metabolism of *Trypanosoma* and *Leishmania*. Biol Chem. 2003;384:539–49.
52. Sterkers Y, Lachaud L, Crobu L, Bastien P, et al. FISH analysis reveals aneuploidy and continual generation of chromosomal mosaicism in *Leishmania major*. Cell Microbiol. 2011;13:274–83.
53. Sterkers Y, Lachaud L, Bourgeois N, Crobu L, et al. Novel insights into genome plasticity in Eukaryotes: mosaic aneuploidy in *Leishmania*. Mol Microbiol. 2012;86:15–23.
54. Lachaud L, Bourgeois N, Kuk N, Morelle C, et al. Constitutive mosaic aneuploidy is a unique genetic feature widespread in the *Leishmania* genus. Microbes Infect. 2013:2–7.
55. Prieto-Barja P, Pesher P, Bussotti G, Dumetz F, et al. Haplotype selection as an adaptive mechanism in the protozoan pathogen *Leishmania donovani*. Nat Ecol Evol. 2017;1:1961–9.
56. Leprohon P, Légaré D, Raymond F, Madore E, et al. Gene expression modulation is associated with gene amplification, supernumerary chromosomes and chromosome loss in antimony-resistant *Leishmania infantum*. Nucleic Acids Res. 2009;37:1387–99.
57. Shaw CD, Lonchamp J, Downing T, Imamura H, et al. In vitro selection of miltefosine resistance in promastigotes of *Leishmania donovani* from Nepal: genomic and metabolomic characterization. Mol Microbiol. 2016;99:1134–48.
58. Ubeda J-M, Raymond F, Mukherjee A, Plourde M, et al. Genome-wide stochastic adaptive DNA amplification at direct and inverted DNA repeats in the parasite *Leishmania*. PLoS Biol. 2014;12:e1001868.
59. Imamura H, Downing T, Van den Broeck F, Sanders MJ, et al. Evolutionary genomics of epidemic visceral leishmaniasis in the Indian subcontinent. Elife. 2016;5:1–39.

60. Monte-Neto R, Laffitte M-CN, Leprohon P, Reis P, et al. Intrachromosomal amplification, locus deletion and point mutation in the aquaglyceroporin AQP1 gene in antimony resistant *Leishmania (Viannia) guyanensis*. PLoS Negl Trop Dis. 2015;9:e0003476.
61. Grondin K, Papadopoulou B, Ouellette M. Homologous recombination between direct repeat sequences yields P-glycoprotein containing amplicons in arsenite resistant *Leishmania*. Nucleic Acids Res. 1993;21:1895–901.
62. Moreira DS, Monte Neto RL, Andrade JM, Santi AMM, et al. Molecular characterization of the MRPA transporter and antimony uptake in four New World *Leishmania* spp. susceptible and resistant to antimony. Int J Parasitol Drugs Drug Resist. 2013;3:143–53.
63. Brotherton M-C, Bourassa S, Leprohon P, Légaré D, et al. Proteomic and genomic analyses of antimony resistant *Leishmania infantum* mutant. PLoS One. 2013;8:e81899.
64. Singh A, Papadopoulou B, Ouellette M. Gene amplification in amphotericin B-resistant *Leishmania tarentolae*. Exp Parasitol. 2001;99:141–7.
65. Papadopoulou B, Roy G, Ouellette M. Frequent amplification of a short chain dehydrogenase gene as part of circular and linear amplicons in methotrexate resistant *Leishmania*. Nucleic Acids Res. 1993;21:4305–12.
66. Grondin K, Roy G, Ouellette M. Formation of extrachromosomal circular amplicons with direct or inverted duplications in drug-resistant *Leishmania tarentolae*. Mol Cell Biol. 1996;16:3587–95.
67. Ubeda J-M, Légaré D, Raymond F, Ouameur AA, et al. Modulation of gene expression in drug resistant *Leishmania* is associated with gene amplification, gene deletion and chromosome aneuploidy. Genome Biol. 2008;9:R115.
68. Ritt J-F, Raymond F, Leprohon P, Légaré D, et al. Gene amplification and point mutations in pyrimidine metabolic genes in 5-fluorouracil resistant *Leishmania infantum*. PLoS Negl Trop Dis. 2013;7:e2564.
69. Kumar P, Lodge R, Raymond F, Ritt J-F, et al. Gene expression modulation and the molecular mechanisms involved in Nelfinavir resistance in *Leishmania donovani* axenic amastigotes. Mol Microbiol. 2013:1–18.
70. Berg M, Vanaerschot M, Jankevics A, Cuypers B, et al. Metabolic adaptations of *Leishmania donovani* in relation to differentiation, drug resistance, and drug pressure. Mol Microbiol. 2013;90:428–42.
71. Ariyanayagam MR, Fairlamb AH. Ovothiol and trypanothione as antioxidants in trypanosomatids. Mol Biochem Parasitol. 2001;115:189–98.
72. Bocedi A, Dawood KF, Fabrini R, Federici G, et al. (2010) Trypanothione efficiently intercepts nitric oxide as a harmless iron complex in trypanosomatid parasites. FASEB J 24:1035–1042.
73. Romão PRT, Tovar J, Fonseca SG, Moraes RH, et al. Glutathione and the redox control system trypanothione/trypanothione reductase are involved in the protection of *Leishmania* spp. against nitrosothiol-induced cytotoxicity. Brazilian J Med Biol Res Rev. 2006;39:355–63.
74. Piñeyro MD, Arcari T, Robello C, Radi R, et al. Tryparedoxin peroxidases from Trypanosoma cruzi: High efficiency in the catalytic elimination of hydrogen peroxide and peroxynitrite. Arch Biochem Biophys. 2011;507:287–95.
75. Flohé L, Budde H, Bruns K, Castro H, et al. Tryparedoxin peroxidase of *Leishmania donovani*: molecular cloning, heterologous expression, specificity, and catalytic mechanism. Arch Biochem Biophys. 2002;397:324–35.
76. Alvarez MN, Peluffo G, Piacenza L, Radi R. Intraphagosomal peroxynitrite as a macrophage-derived cytotoxin against internalized Trypanosoma cruzi: consequences for oxidative killing and role of microbial peroxiredoxins in infectivity. J Biol Chem. 2011;286:6627–40.
77. Iyer JP, Kaprakkaden A, Choudhary ML, Shaha C. Crucial role of cytosolic tryparedoxin peroxidase in *Leishmania donovani* survival, drug response and virulence. Mol Microbiol. 2008;68:372–91.

78. Dumas C, Ouellette M, Tovar J, Cunningham ML, et al. Disruption of the trypanothione reductase gene of *Leishmania* decreases its ability to survive oxidative stress in macrophages. EMBO J. 1997;16:2590–8.
79. Tovar J, Wilkinson S, Mottram JC, Fairlamb AH. Evidence that trypanothione reductase is an essential enzyme in *Leishmania* by targeted replacement of the tryA gene locus. Mol Microbiol. 1998;29:653–60.
80. Tovar J, Cunningham ML, Smith AC, Croft SL, et al. Down-regulation of *Leishmania donovani* trypanothione reductase by heterologous expression of a trans-dominant mutant homologue: effect on parasite intracellular survival. Proc Natl Acad Sci USA. 1998;95:5311–6.
81. Rojo D, Canuto GAB, Castilho-Martins EA, Tavares MFM, et al. A multiplatform metabolomic approach to the basis of antimonial action and resistance in *Leishmania infantum*. PLoS One. 2015;10:1–20.
82. Berg M, Garcia-Hernandez R, Cuypers B, Vanaerschot M, et al. Experimental resistance to drug combinations in *Leishmania donovani*: Metabolic and phenotypic adaptations. Antimicrob Agents Chemother. 2015;59:2242–55.
83. Decuypere S, Rijal S, Yardley V, De Doncker S, et al. Gene expression analysis of the mechanism of natural Sb(V) resistance in *Leishmania donovani* isolates from Nepal. Antimicrob Agents Chemother. 2005;49:4616–21.
84. Decuypere S, Vanaerschot M, Brunker K, Imamura H, et al. Molecular mechanisms of drug resistance in natural *Leishmania* populations vary with genetic background. PLoS Negl Trop Dis. 2012;6:e1514.
85. Gómez Pérez V, García-Hernandez R, Corpas-López V, Tomás AM, et al. Decreased antimony uptake and overexpression of genes of thiol metabolism are associated with drug resistance in a canine isolate of *Leishmania infantum*. Int J Parasitol Drugs Drug Resist. 2016;6:133–9.
86. Canuto GAB, Castilho-Martins EA, Tavares MFM, Rivas L, et al. Multi-analytical platform metabolomic approach to study miltefosine mechanism of action and resistance in *Leishmania*. Anal Bioanal Chem. 2014;406:3459–76.
87. t'Kindt R, R a S, Jankevics A, Brunker K, et al. Metabolomics to unveil and understand phenotypic diversity between pathogen populations. PLoS Negl Trop Dis. 2010;e904:4.
88. Oryan A, Shirian S, Tabandeh M-R, Hatam G-R, et al. Genetic diversity of *Leishmania major* strains isolated from different clinical forms of cutaneous leishmaniasis in southern Iran based on minicircle kDNA. Infect Genet Evol. 2013;19:226–31.
89. CoreWriting Team, Pachauri RK. Climate Change 2014: Synthesis report contributions of working groups I, II and III to the fifth assessment report of the IPCC. 2014.
90. Brunner FS, Eizaguirre C. Can environmental change affect host/parasite-mediated speciation? Zoology. 2016;119:384–94.
91. Thomas CD, Cameron A, Green RE, Bakkenes M, et al. Extinction risk from climate change. Nature. 2004;427:145–8.
92. Kutz SJ, Hoberg EP, Polley L, Jenkins EJ. Global warming is changing the dynamics of Arctic host-parasite systems. Proc Biol Sci. 2005;272:2571–6.
93. Larsen MH, Mouritsen KN. Temperature–parasitism synergy alters intertidal soft-bottom community structure. J Exp Mar Bio Ecol. 2014;460:109–19.
94. Mitchell SE, Rogers ES, Little TJ, Read AF. Host-parasite and genotype-by-environment interactions: temperature modifies potential for selection by a sterilizing pathogen. Evolution. 2005;59:70–80.
95. Macnab V, Barber I. Some (worms) like it hot: fish parasites grow faster in warmer water, and alter host thermal preferences. Glob Chang Biol. 2012;18:1540–8.
96. Scharsack JP, Schweyen H, Schmidt AM, Dittmar J, et al. Population genetic dynamics of three-spined sticklebacks (*Gasterosteus aculeatus*) in anthropogenic altered habitats. Ecol Evol. 2012;2:1122–43.

97. Jiménez M, Alvar J, Tibayrenc M. *Leishmania infantum* is clonal in AIDS patients too: epidemiological implications. AIDS. 1997;11:569–73.
98. Rosenthal E, Marty P, Poizot-Martin I, Reynes J, et al. Visceral leishmaniasis and HIV-1 co-infection in southern France. Trans R Soc Trop Med Hyg. 1995;89:159–62.
99. Lopez-Velez R, Perez-Molina JA, Guerrero A, Baquero F, et al. Clinico epidemiologic characteristics, prognostic factors, and survival analysis of patients coinfected with human immunodeficiency virus and *Leishmania* in an area of Madrid, Spain. Am J Trop Med Hyg. 1998;58:436–43.
100. Bryceson AD, Chulay JD, Ho M, Mugambii M, et al. Visceral leishmaniasis unresponsive to antimonial drugs. I. Clinical and immunological studies. Trans R Soc Trop Med Hyg. 1985;79:700–4.
101. Davidson RN, Di Martino L, Gradoni L, Giacchino R, et al. Liposomal amphotericin B (AmBisome) in Mediterranean visceral leishmaniasis: a multi-centre trial. Q J Med. 1994;87:75–81.
102. Bryceson A. Current issues in the treatment of visceral leishmaniasis. Med Microbiol Immunol. 2001;190:81–4.
103. Ives A, Ronet C, Prevel F, Ruzzante G, Fuertes-Marraco S, et al. *Leishmania* RNA virus controls the severity of mucocutaneous leishmaniasis. Science. 2011;331(6018):775–8.
104. Cantanhêde LM, da Silva Júnior CF, Ito MM, Felipin KP, et al. Further evidence of an association between the presence of *Leishmania* RNA virus 1 and the mucosal manifestations in tegumentary leishmaniasis patients. PLoS Negl Trop Dis. 2015;9:e0004079.
105. Adaui V, Lye L-F, Akopyants NS, Zimic M, et al. Association of the endobiont double-stranded RNA virus LRV1 with treatment failure for human leishmaniasis caused by *Leishmania braziliensis* in Peru and Bolivia. J Infect Dis. 2016;213:112–21.
106. Arevalo J, Ramirez L, Adaui V, Zimic M, et al. Influence of *Leishmania* (*Viannia*) species on the response to antimonial treatment in patients with American tegumentary leishmaniasis. J Infect Dis. 2007;195:1846–51.
107. Romero GA, Guerra MV, Paes MG, Macêdo VO. Comparison of cutaneous leishmaniasis due to *Leishmania* (*Viannia*) *braziliensis* and *L.* (*V.*) *guyanensis* in Brazil: therapeutic response to meglumine antimoniate. Am J Trop Med Hyg. 2001;65:456–65.
108. Zerpa O, Convit J. Diffuse cutaneous leishmaniasis in Venezuela. Gaz méd Bahia. 2009;79:30–4.
109. Goto H, Lindoso JAL. Current diagnosis and treatment of cutaneous and mucocutaneous leishmaniasis. Expert Rev Anti Infect Ther. 2010;8:419–33.
110. Ponte-Sucre A, Diaz E, Padrón-Nieves M. The concept of fitness and drug resistance in *Leishmania*. In: Ponte-Sucre A, Diaz E, Padrón-Nieves M, editors. Drug Resist. *Leishmania* parasites, Consequences, molecular mechanisms and possible treatments. Vienna: Springer; 2013. p. 431–49.
111. Newton PN, Green MD, Fernández FM. Impact of poor-quality medicines in the "developing" world. Trends Pharmacol Sci. 2010;31:99–101.
112. de Mello CX, de Oliveira Schubach A, de Oliveira RVC, Conceição-Silva F, et al. Comparison of the sensitivity of imprint and scraping techniques in the diagnosis of American tegumentary leishmaniasis in a referral centre in Rio de Janeiro, Brazil. Parasitol Res. 2011;109:927–33.
113. Andersson DI, Hughes D. Antibiotic resistance and its cost: is it possible to reverse resistance? Nat Rev Microbiol. 2010;8:260–71.
114. Melnyk AH, Wong A, Kassen R. The fitness costs of antibiotic resistance mutations. Evol Appl. 2015;8:273–83.
115. Huijben S, Bell AS, Sim DG, Tomasello D, et al. Aggressive chemotherapy and the selection of drug resistant pathogens. PLoS Pathog. 2013;9:e1003578.
116. Ouakad M, Vanaerschot M, Rijal S, Sundar S, et al. Increased metacyclogenesis of antimony-resistant *Leishmania donovani* clinical lines. Parasitology. 2011;138:1392–9.
117. Vanaerschot M, Maes I, Ouakad M, Adaui V, et al. Linking in vitro and in vivo survival of clinical *Leishmania donovani* strains. PLoS One. 2010;5:e12211.

118. Vanaerschot M, de Doncker S, Rijal S, Maes L, et al. Antimonial resistance in *Leishmania donovani* is associated with increased in vivo parasite burden. PLoS One. 2011;6:e23120.
119. Imamura H, Downing T, Van den Broeck F, Sanders MJ, et al. Evolutionary genomics of epidemic visceral leishmaniasis in the Indian subcontinent. Elife. 2016;5:e12613. https://doi.org/10.7554/eLife.12613.
120. Stauch A, Sarkar RR, Picado A, Ostyn B, et al. Visceral leishmaniasis in the Indian subcontinent: modelling epidemiology and control. PLoS Negl Trop Dis. 2011;5:e1405.
121. Stauch A, Duerr HP, Dujardin JC, Vanaerschot M, et al. Treatment of visceral leishmaniasis: model-based analyses on the spread of antimony-resistant *L. donovani* in Bihar, India. PLoS Negl Trop Dis. 2012;6(12):e1973.
122. Rai K, Bhattarai NR, Vanaerschot M, Imamura H, et al. Single locus genotyping to track *Leishmania donovani* in the Indian subcontinent: Application in Nepal. PLoS Negl Trop Dis. 2017;11:e0005420.
123. Rai K, Cuypers B, Bhattarai NR, Uranw S, et al. Relapse after treatment with Miltefosine for visceral Leishmaniasis is associated with increased infectivity of the infecting *Leishmania donovani* strain. MBio. 2013;4:e00611-13.
124. García-Hernández R, Gómez-Pérez V, Castanys S, Gamarro F. Fitness of *Leishmania donovani* parasites resistant to drug combinations. PLoS Negl Trop Dis. 2015;9:e0003704.
125. Hendrickx S, Beyers J, Mondelaers A, Eberhardt E, et al. Evidence of a drug-specific impact of experimentally selected paromomycin and miltefosine resistance on parasite fitness in *Leishmania infantum*. J Antimicrob Chemother. 2016;71:1914–21.
126. Turner KG, Vacchina P, Robles-Murguia M, Wadsworth M, et al. Fitness and phenotypic characterization of Miltefosine-resistant *Leishmania major*. PLoS Negl Trop Dis. 2015;9: e0003948.
127. Padrón-Nieves M, Machuca C, Díaz E, Cotrim P, et al. Correlation between glucose uptake and membrane potential in *Leishmania* parasites isolated from DCL patients with therapeutic failure: a proof of concept. Parasitol Res. 2014;113:2121–8.
128. Ponte-Sucre A. Leishmaniasis, the biology of a parasite. In: Ponte-Sucre A, Padron Nieves M, editors. Drug Resist. *Leishmania* parasites, Consequences, molecular mechanisms and possible treatments. Vienna: Springer; 2013. p. 1–12.
129. Padron-Nieves M, Ponte-Sucre A. Marcadores de resistencia en *Leishmania*: susceptibilidad in vitro a drogas leishmanicidas vs. retencion de calceina en aislados de pacientes venezolanos con leishmaniasis cutanea difusa. Arch Venez Farmacol y Ter. 2015;32:29–33.
130. Zerpa O, Ulrich M, Blanco B, Polegre M, et al. Diffuse cutaneous leishmaniasis responds to miltefosine but then relapses. Br J Dermatol. 2007;156:1328–35.
131. Torrico MC, De Doncker S, Arevalo J, Le Ray D, et al. In vitro promastigote fitness of putative *Leishmania (Viannia) braziliensis/Leishmania* (Viannia) *peruviana* hybrids. Acta Trop. 1999;72:99–110.
132. Hartley M-A, Ronet C, Zangger H, Beverley SM, et al. *Leishmania* RNA virus: when the host pays the toll. Front Cell Infect Microbiol. 2012;2:99.
133. Taylor DR, Jarosz AM, Fulbright DW, Lenski RE. The acquisition of hypovirulence in host-pathogen systems with three trophic levels. Am Nat. 1998;151:343–55.
134. Maisonneuve E, Gerdes K. Molecular mechanisms underlying bacterial persisters. Cell. 2014;157:539–48.
135. Sereno D, Lemesre JL. Axenically cultured amastigote forms as an in vitro model for investigation of antileishmanial agents. Antimicrob Agents Chemother. 1997;41:972–6.
136. Callahan HL, Portal AC, Devereaux R, Grogl M. An axenic amastigote system for drug screening. Antimicrob Agents Chemother. 1997;41:818–22.
137. Kaur G, Rajput B. Comparative analysis of the omics technologies used to study antimonial, amphotericin B, and pentamidine resistance in *Leishmania*. J Parasitol Res. 2014;2014:1–11.
138. Saravia NG, Weigle K, Segura I, Giannini SH, et al. Recurrent lesions in human *Leishmania braziliensis* infection--reactivation or reinfection? Lancet (London, England). 1990;336:398–402.

139. Nühs A, De Rycker M, Manthri S, Comer E, et al. Development and validation of a novel *Leishmania donovani* screening cascade for high-throughput screening using a novel axenic assay with high predictivity of leishmanicidal intracellular activity. PLoS Negl Trop Dis. 2015;9:e0004094.
140. Tegazzini D, Díaz R, Aguilar F, Peña I, et al. A replicative in vitro assay for drug discovery against *Leishmania donovani*. Antimicrob Agents Chemother. 2016;60:3524–32.
141. Hefnawy A, Berg M, Dujardin J-C, De Muylder G. Exploiting knowledge on *Leishmania* drug resistance to support the quest for new drugs. Trends Parasitol. 2017;33:162–74.
142. Le Rutte EA, Coffeng LE, Bontje DM, Hasker EC, et al. Feasibility of eliminating visceral leishmaniasis from the Indian subcontinent: explorations with a set of deterministic age-structured transmission models. Parasit Vectors. 2016;9:24.
143. Gabryszewski SJ, Modchang C, Musset L, Chookajorn T, et al. Combinatorial genetic modeling of pfcrt -mediated drug resistance evolution in *Plasmodium falciparum*. Mol Biol Evol. 2016;33:1554–70.

Index

A. Ponte-Sucre, M. Padrón-Nieves (eds.), *Drug Resistance in Leishmania Parasites*,
https://doi.org/10.1007/978-3-319-74186-4

D

E

Printed by Printforce, the Netherlands